VARIORUM COLLECTED STUDIES SERIES

Essays on Early Medieval Mathematics

Menso Folkerts

Essays on Early Medieval Mathematics

The Latin Tradition

LONDON AND NEW YORK

First published 2003 by Ashgate Publishing

2 Park Square, Milton Park, Abingdon, Oxfordshire OX14 4RN
711 Third Avenue, New York, NY 10017

Routledge is an imprint of the Taylor & Francis Group, an informa business

First issued in paperback 2018

British Library Cataloguing-in-Publication Data
Folkerts, Menso
    Essays on early medieval mathematics : the Latin tradition.
    – (Variorum collected studies series)
    1. Mathematics, Medieval 2. Mathematics – Europe – History –
    To 1500
    I. Title
    510. 9'4'09021

US Library of Congress Cataloging-in-Publication Data
Folkerts, Menso.
    Essays on early medieval mathematics : the Latin tradition / Menso
    Folkerts.
       p. cm. -- (Variorum collected studies series ; 751)
    Includes bibliographical references and index (alk. paper).
    1. Mathematics, Medieval. I. Title. II. Collected studies ; CS751.
    QA23.F65 2003
    510'.94'0902--dc21   2002043951

ISBN 978-0-86078-895-9 (hbk)
ISBN 978-1-138-37544-4 (pbk)

VARIORUM COLLECTED STUDIES SERIES CS751

Contents

Contents

This volume contains xiv + 366 pages

Publisher's Note

The articles in this volume, as in all others in the Variorum Collected Studies Series, have not been given a new, continuous pagination. In order to avoid confusion, and to facilitate their use where these same studies have been referred to elsewhere, the original pagination has been maintained wherever possible.

Each article has been given a Roman number in order of appearance, as listed in the Contents. This number is repeated on each page and is quoted in the index entries.

Preface

Contrary to what is sometimes asserted, the sciences were by no means dead in the western Middle Ages before the translations from Arabic in the twelfth century. The eleven articles included in this volume show the truth of this assertion for mathematics (in the sense of arithmetic, algebra and geometry) of the eighth to the 11th century. Some of the articles contain critical editions of mathematical texts written in this time. Five of the eleven articles are simply reprints; and of the remainder some have been altered to accommodate new results, some are translated from German, usually with small changes, and two are new contributions, albeit depending on previously published works.

The volume begins with a survey of the development of mathematics in the western Middle Ages (item I). This article is not restricted to the time before the translations from Arabic, but ranges over the most important topics in the period from 500 to 1500.

Item II is on the mathematical content of the *Corpus agrimensorum*, the writings of the Roman land surveyors. These writings were collected in late Antiquity and were principally about the practical geometry of measuring land, though other matters were also considered, e.g. formulae for determining the areas of regular polygons. The *agrimensores* drew their knowledge of geometry from the practical geometrical writings of the Greeks, among others Heron (1st c. A.D.). The *Corpus agrimensorum* was widely known in the Middle Ages and was an important source of geometrical knowledge, at least before the 12th century.

Probably the oldest surviving mathematical text from the western Middle Ages is the *De arithmeticis propositionibus*, written not later than the 9th century. This dating stands even if the ascription to the Venerable Bede (672/3 to 735), which has only slight justification, is false. Three of the four parts of this treatise concern mathematical games between two people ("think of a number" etc.) and give procedures to find the numbers involved. The fourth section, which gives rules for the addition of positive and negative numbers, is unique in western Europe before the 15th century. Item III is concerned with the transmission of the text, the affiliation of the manuscripts, the question of authorship, the date of composition and the importance of this treatise in the history of mathematics. It includes a critical edition of the text based on all known manuscripts. Since the appearance of the article in "Sudhoffs Archiv" (1972) five new manuscripts have been discovered and the article presented here is greatly enlarged.

Of somewhat later origin than the *De arithmeticis propositionibus* is a text with the title *Propositiones ad acuendos iuvenes* ("Problems to sharpen the minds of youths") which was very probably written by Alcuin (ca. 735 to 804), Charlemagne's educational adviser. This text, too, stands in the long tradition of recreational mathematics. It contains 56 problems, some of them similar to those found in late Greek and Byzantine collections. There are also connections to the Roman tradition. Some problems are similar to problems in the Arabic-Islamic culture. On the other hand, some items, such as the wolf-goat-cabbage puzzle, seem to have first appeared in the *Propositiones*. The *Propositiones* had great influence: items from the collection reappeared in collections of problems in the later Middle Ages and in early modern times. Item IV is a reprint of the critical edition first published in 1978. Its introduction treats the transmission, the content, the authorship and the sources of the text. A summary of this introduction, which is in German, appears in English, in an updated and shorter form, as item V. *Inter alia*, it contains information about two newly discovered manuscripts.

Item VI is about the names and forms of the numerals on the abacus in the tradition of Gerbert of Aurillac. Gerbert, later Pope Sylvester II (d. 1003), was largely responsible for introducing a type of abacus (calculation board) at the end of the 10th century. In this type of abacus numbers were represented not by a multitude of counters each representing a unit – such as were used by the Greeks and Romans –, but by counters on which the numerical values were marked (i. e. nine different characters). It should be noted that these characters representing the numbers from 1 to 9 were not used either by merchants or by astronomers, but were restricted to the monasteries. In item VI it is shown that it is very probable that the value on the counter was indicated by a form of the Hindu-Arabic numerals which Gerbert came to know on a journey to the Spanish March. This form of the abacus had its own tradition, which ended at the beginning of the twelfth century. The number characters are to be found in some writings on this type of abacus as well as in special illustrations of the device; in both types of source there were special names for the symbols and some of these names indicate an Arabic origin.

Items VII to IX concern the geometrical treatise written by Boethius about 500 A.D. and its fate in the Middle Ages. Boethius' *Geometry*, which was apparently a translation of Euclid's *Elements*, is lost in its original form. What we have are extracts transmitted in four different texts: in the so-called "Geometry I" and "Geometry II" ascribed to Boethius; in manuscripts of the Roman land surveyors; and in a special redaction of Cassiodorus' *Institutiones*. These compendia – with the exception of "Geometry II" – were composed in the eighth century in Corbie. Item VII is a survey of the history of Boethius' *Geometry* and its importance in the Middle Ages. Item VIII concerns the "Geometry I" and presents, for the first time, an edition of the so-called "Altercatio", a dialogue about geometrical questions which comprises part of the "Geometry I" and which is largely based on writings by St. Augustine. In item IX the contents and sources of the "Geometry II" are discussed. This text was put together from three sources in the second quarter of the 11th century in Lorraine: the translation of Euclid by Boethius; parts of the *Corpus agrimensorum*; and finally from a redaction of Gerbert's treatise on the abacus, in which Gerbert's Hindu-Arabic numerals and their names appear.

The "Geometry II" attributed to Boethius is closely connected with a treatise on squaring the circle which Franco of Liège wrote about 1050 A.D. Item X is a reprint of the edition of Franco's text, originally published by A. J. E. M. Smeur and myself in 1976. In this article we also collected all available information about Franco's life and work and summarized the mathematical contents of the work and its sources. Franco's treatise is the only extant work of the western Middle Ages on the problem of squaring the circle before Archimedean works on this topic were made available in translations from the Arabic. Readers may perhaps prefer to read Part II, which discusses the text and its author, before the edition itself, in Part I.

Item XI is on "Rithmomachia", a medieval mathematical game. Its elements go back to the Pythagorean theory of numbers which was known in the Latin Middle Ages mostly through Boethius' *Arithmetic*. The game was played on a double chess-board with pieces which were inscribed on their upper surfaces with special numbers. These numbers were generated from even and odd digits by using Pythagorean numerical ratios. "Rithmomachia" was invented in the first half of the 11th century in Southern Germany and was well known from that time up to the beginning of the 18th century. The article describes the elements of the game, its rules and its history. The known 32 texts, with some indication of their contents and a list of the manuscripts and printed editions, are mentioned.

It is a pleasure to thank Richard Lorch (IGN Munich) for helping translate many German articles into English, Gerhard Brey for technical help with computer programmes used in typesetting this book, Paul Kunitzsch for his constant help and encouragement, and finally Dr. John Smedley for including the book in his series and for his kindly advice.

MENSO FOLKERTS

Munich
August 2002

Acknowledgements

Grateful acknowledgement is made to the following for kindly permitting the reproduction of articles originally published by them: Vandenhoeck & Ruprecht Verlagsbuchhandlung, Göttingen (II); Österreichische Akademie der Wissenschaften, Vienna (V); Prof. Flavio G. Nuvolone, Fribourg (VI); Peter Lang AG, Europäischer Verlag der Wissenschaften, Berne (VII); Erich Schmidt Verlag, Berlin (VIII); Prof. Robert Halleux, editor of *Archives Internationales d'Histoire des Sciences*, Liège (X). The photographs in item VI are reprinted with permission of the Deutsche Staatsbibliothek zu Berlin, Preußischer Kulturbesitz (Ms. lat. oct. 162, f. 6r, 73v, 74r); Burgerbibliothek, Berne (Hs. 250, f. 1r); Leiden, University Library (Ms. Voss. lat. oct. 95, f. 6v); Oxford, Bodleian Library (Ms. Hatton 108, f. 4v–5r, 5v–6r); Oxford, St. John's College (Ms. 17, f. 48v–49r); Bibliothèque nationale de France, Paris (Ms. lat. 8663, f. 49v); Bibliothèque municipale, Rouen (Ms. A. 489, f. 68v–69r); Biblioteca Apostolica Vaticana, Vatican City (Ms. lat. 644, f. 77v–78r).

I

The Importance of the Latin Middle Ages for the Development of Mathematics*

This paper does not claim to give a detailed survey of medieval mathematics. Besides a very summary account of the state of research this paper will point to the principal areas in which advances have been made in the last decades. In the main I shall restrict myself to geometry and arithmetic, though I shall also mention algebra – of trigonometry I shall say nothing. This paper concerns, *inter alia*, the distinction, often made in the Middle Ages, between theoretical and practical mathematics.

Many modern histories of mathematics present the medieval Latin period much as it was done in the 1920s. True, by the endeavours of Paul Tannery, Johan Ludvig Heiberg, Maximilian Curtze, Axel Anthon Björnbo and others most of the principal texts had become available and had been evaluated. About 1900 one of the main focuses of the history of mathematics was the Latin Middle Ages, but between the wars other periods took prominence, e. g. mathematics before the Greeks. A new flowering of the history of medieval mathematics began in the 1950s with the researches of Marshall Clagett and his followers: a series of hitherto unedited text were made available and it was shown how close the connections were between mathematics, physics and philosophy. Unfortunately, Clagett's followers have themselves had few followers and it appears that a new impulse in the field is not to be expected from the United States. However, in Europe a large number of texts have been edited by Hubert L. L. Busard, among them all the important translations and redactions of Euclid's *Elements*.

First let us consider the geometry before the translations from Arabic in the 12th century. It is generally known that few writings of mathematical content circulated in the early Middle Ages. This may be explained by the slight interest of the Romans in theoretical mathematics and by

*This is the English translation, with some minor changes, of the article "Die Bedeutung des lateinischen Mittelalters für die Entwicklung der Mathematik – Forschungsstand und Probleme", in: C. Hünemörder (ed.), *Wissenschaftsgeschichte heute. Ansprachen und wissenschaftliche Vorträge zum 25jährigen Bestehen des Instituts für Geschichte der Naturwissenschaften, Mathematik und Technik der Universität Hamburg*, Stuttgart 1987, pp. 87–114.

the sufficiency of a very little arithmetical and geometrical knowledge for daily life. Relatively well known were the writings of the Roman land surveyors (*agrimensores*)[1]. These treatises were mostly on legal and religious matters, but some of the writers, above all Balbus and also Epaphroditus and Vitruvius Rufus, also give geometrical procedures. Thus Balbus' treatise informs the reader on units of length, area and volume, on the forms of boundary lines, and on various plane figures, especially triangles, quadrilaterals and circles. Some definitions were given, obviously in order to explain the concepts which a surveyor needs. Here the connection with practical matters is unmistakable. The treatise of Epaphroditus and Vitruvius Rufus shows, by numerical examples, how the various triangles, quadrilaterals, polygons and the circle should be measured – these formulas are not restricted to the areas of the figures, but apply also to the perimeters and in the case of triangles to the heights and the sections of the base made by the perpendicular from the opposite point. As in Babylonian texts, the solutions were given as a series of instructions, like recipes. It was, however, possible to generalize the solutions to accommodate other numerical values. Only some of the problems were those which would be encountered in daily life. Much space is taken by the regular polygons. In modern terms, the area $F_n(a)$ of the n-sided polygon (here n-gon) with the side a is given (incorrectly) as the ath polygonal number of order n (here p_n^a):

$$F_n(a) = p_n^a = \frac{(n-2)a^2 - (n-4)a}{2} \quad (n \geq 3,\ a \geq 1).$$

One example[2] is the determination of the side a of the 11-gon from its area $F_{11}(a) = 415$. The "recipe" gives:

$$a = \frac{\sqrt{72 \times 415 + 49} + 7}{18}.$$

This is correct, because a is the solution of the quadratic equation $F_{11} = 415 = (9a^2 - 7a)/2$. Of course, this and similar problems have nothing to do with practical life. They reflect, at a low level, the Greek solution of quadratic equations and the Pythagorean theory of figured numbers – i. e. theoretical mathematics. Thus the writings of the *agrimensores* not

---

[1] Most of these texts, including that of Balbus, are still available only in the edition of Blume / Lachmann / Rudorff, 'Feldmesser'. The treatise of Epaphroditus and Vitruvius Rufus is available in Bubnov, 'Opera mathematica', pp. 518–551.

[2] Bubnov, 'Opera mathematica', p. 543, no. 27c.

only treat practical matters needed by surveyors, but also are concerned with knowledge of theoretical mathematics.

Other witnesses to theoretical mathematics in the early Middle Ages were excerpts from Euclid's *Elements*. Some of these were taken from Boethius' translation[3] and some were taken from encyclopedias of late antiquity or the early Middle Ages (for example, Martianus Capella, Cassiodorus, Isidore of Seville). None of these excerpts went further than book 5 of the *Elements* or give any hint of the logical structure of the Greek proofs; there are proofs only for the first three propositions of book 1 and these are found only in one group of excerpts.

The history of the transmission of the *Corpus agrimensorum* in the early Middle Ages is now in large measure known. Berthold L. Ullman showed in an article published in 1964[4] that practically all geometrical texts were collected in Corbie in the 7th to 8th centuries and in Corbie new compilations were made: in this monastery, the geometrical centre of the Western world, there were *inter alia* manuscripts of the Roman land surveyors, the geometrical excerpts from Columella and copies of Martianus Capella's encyclopedia. In Corbie the so-called *Geometry I* attributed to Boethius was compiled in the 8th/9th century – a compilation from various sources (Boethius' *Arithmetic*, his translation of Euclid, the *Corpus agrimensorum* etc.). This text was one of the most influential geometrical treatises before the translations from the Arabic.

For what purpose did the monks of Corbie transcribe or compile such texts? Ullman has convincingly shown that the mathematical writings of the *Corpus agrimensorum* did not survive primarily because they were needed for the instruction of future surveyors, but because they were used for teaching geometry in the quadrivium. In this matter they were different from other technical writings, e.g. on medicine or agriculture, which were principally used for practical ends.

The early Middle Ages drew its knowledge of geometry mostly from the *Corpus agrimensorum*. On such works was based the so-called *Geometria incerti auctoris* (Geometry of an unknown author), which was widely known in this period[5]. At one time it was thought to be part of Gerbert's *Geometry*. Nowadays we know that it is older and was written in the 9th or 10th century. In the second part it contains numerous problems that are in the tradition of the *agrimensores*: above all, calcu-

---

[3] Editions of all existing excerpts and a reconstruction of the original texts can be found in Folkerts, 'Geometrie II', pp. 176–217.

[4] Ullman, 'Geometry'.

[5] Edited in Bubnov, 'Opera mathematica', pp. 317–365.

lations about triangles, quadrilaterals and polygons, about the circle and about simple solids. Still more interesting is the first part, which treats matters such as the determination of the breadth of rivers, the height of mountains or of towers and the depth of wells, with or without practical aids. Various instruments are named[6] and it appears that the procedures described, which are largely based on similar triangles, were really used in practice.

One of the most important geometrical texts written before the 12th century is Gerbert's *Geometry*. This text, as we have it, is incomplete and the length of the original is not known. Gerbert used a manuscript of the *Corpus agrimensorum* (which he found in 983 when he was in Bobbio), Calcidius' commentary on Plato's *Timaeus*, Augustine's *De quantitate animae*, Boethius' *Arithmetic* and his commentary on Aristotle's *Categories*, Macrobius' commentary on the *Somnium Scipionis* and Martianus Capella's encyclopedia. An exact analysis of this work has yet to be made, though it has been edited for over 100 years[7]. It contains explanations of the basic terms of geometry, goes into the different weights and measures and how to convert one into another, treats the various sorts of angle – here he reworks material from Euclid –, and finally gives procedures to solve triangles and quadrilaterals. In this last part material from the *agrimensores* is used. Gerbert tries to present the connections among the topics he treats; his didactic endeavours are everywhere evident. It seems likely that this book, too, was meant less for practical application than as a textbook of geometry in monastery schools. It was apparently used by Gerbert himself in the cathedral school at Rheims, of which he was at one time director.

In some of the manuscripts of the Geometry attributed to Gerbert there are hints about geometrical discussions that took place among scholars in Lorraine at the beginning of the 11th century[8]. The main questions were about the sum of the angles of a triangle and about the definitions of interior and exterior angle. Sometimes the interior angle was identified with the acute angle and the exterior angle with the obtuse. This throws some light on the poor condition of the transmitted text of the *Elements*. This so-called "angle dispute" appears in the correspondence, of about 1025, between Radulph, a monk of Liège, and Ragimbold, a monk of Cologne. The question of the exterior and interior

---

[6]*horoscopus* (= astrolabe?), *virgula, arundo, orthogonium, hasta, speculum, arcus cum sagitta et filo*.

[7]In Bubnov, 'Opera mathematica', pp. 48–97.

[8]The existing letters are edited in Tannery / Clerval, 'Correspondance'.

angle of a triangle was decided in a text which is transmitted in only one manuscript[9]. In this text the most important theorems about equality and the sum of angles were experimentally demonstrated by cutting the angles out and laying them one over the other or one beside the other. This anonymous text and Gerbert's *Geometry* were both known to a monk of Liège named Franco, who wrote about 1045 a treatise on squaring the circle[10]. Apart from a powerful iteration procedure for finding square roots, Franco's work contains nothing new mathematically and, of course, cannot be compared with Archimedes' treatment on squaring the circle. It is interesting mostly because one can recognize from it which geometrical problems were treated in the monasteries of this time.

Thus in the first half of the eleventh century there was a group of scholars in Lorraine who had a special interest in geometrical problems. This has been recognized only in the last few decades. The initiator of this intellectual activity was Gerbert. About this time – after 1025 and before the middle of the century – the so-called *Geometry II* attributed to Boethius was written. For this compilation its unknown author used Gerbert's treatise on the abacus, a codex of the *agrimensores* and one of the collections of Euclid's excerpts of the Boethius tradition; his method of compilation is now to a large extent clear[11]. The style of the treatise is similar to Franco's; it is possible, but not certain, that Franco himself put the *Geometry II* together.

It seems remarkable that calculation with the abacus should be described in a geometrical treatise. In fact, the abacus was counted as part of geometry in the early Middle Ages. We note that the abacus was at this time called *mensa Pythagorica* and that this table could also be used for geometrical diagrams[12]. We will come later to the history of the abacus when considering arithmetic.

The above-mentioned texts are the most important geometrical works available in the early Middle Ages. As we have seen, they served, despite some items that could have been used for practical purposes, as textbooks for teaching the quadrivium. This is scarcely to be wondered at, for they were written or copied in monasteries. They contained, however, knowledge necessary for practicioners, i. e. surveyors – for example the conversion of units of measure and elementary determination of areas.

---

[9] Edited in Hofmann, 'Winkelstreit'.

[10] See Folkerts / Smeur, 'Franco'.

[11] See the edition and analysis in Folkerts, 'Geometrie II', and this volume, item IX.

[12] For the early history of the abacus see Bergmann, 'Innovationen'.

What additional material was available to the practical geometer, cannot be guessed from the existing geometrical texts.

For medieval mathematics in the West the 12th century was a turning point: this was the century in which a mass of new knowledge from the Arabic became suddenly available. It was not really new knowledge, but for the most part the works of the Greeks and in smaller measure Indian and genuine Arabic works based on Greek and Indian sources. Nothing will be said here about the methods of translation. In this field the fundamental work of Charles Homer Haskins[13] has still not been superseded; a good survey is given by David D. Lindberg[14]. Suffice it to say here that after the reconquest of Toledo (1085) a series of translators and compilers began their work in this city and in other parts of Spain. Scholars from all over Europe worked together; they were often supported in their endeavours by Jews, Mozarabs or converted Muslims. Names like Gerard of Cremona, Adelard of Bath, Robert of Chester, Hermann of Carinthia and Plato of Tivoli point up the international character of this movement.

Through these translations the most important geometrical texts of the Greeks became again accessible. Above all we should mention Euclid's *Elements*, of which only parts of the first five books (without proofs) were known until about 1120. In the 12th century the whole of the *Elements* was translated three times from the Arabic and several times reworked. Some commentaries on Euclid were also translated into Latin[15]. Of Archimedes' writings the *Measurement of the Circle* and parts of the *Sphere and Cylinder* – as well as some Archimedean material in Arabic writings[16] – were made available. An almost complete translation of Archimedes, direct from the Greek, was made by William of Moerbeke in 1269, but was not widely known. In the last few decades Marshall Clagett has published his monumental *Archimedes in the Middle Ages*. This includes practically all witnesses to the Archimedes tradition from the Middle Ages to the 16th century; the texts are edited and translated with full commentary[17]. It is clear from these texts that the medieval copyists and the reworkers of Archimedean material had considerable problems in understanding the material. They tried to understand the logical structure

---

[13] Haskins, 'Studies'.

[14] "The Transmission of Greek and Arabic Learning to the West", in Lindberg, 'Science', pp. 52–90.

[15] al-Nayrīzī; Muḥammad ibn 'Abd al-Bāqī.

[16] Banū Mūsā; *De curvis superficiebus*.

[17] Clagett, 'Archimedes'.

of the proofs and where Archimedes silently assumed that a statement was obvious, they often filled out the proof by using material taken from Euclid. For instance, when Archimedes applied the so-called method of exhaustion, he silently assumed properties of continuity, the convergence of certain series and the existence of fourth proportionals. The Western commentators justified these assumptions, often by reference to Euclid's *Elements*, books 5 and 10. Few scholars understood Archimedes in the later Middle Ages and those who did, did not develop his results to find something new.

Of the work of the third great Greek mathematician, Apollonius, only fragments were known in the medieval West[18]. In general, his theory of conic sections was seldom mentioned at this time. When theorems about the parabola and hyperbola were applied, it was almost always in connection with texts on optics[19]. Pappus' *Collection* was not translated into Latin and only fragments of its contents were known through secondary transmission. In contrast, the *Spherics* of Theodosius and that of Menelaus were known in several versions, some of them annotated[20].

To come back to Euclid, it was mentioned above that besides complete translations there were also reworkings of the text. One may ask what the aim of these reworkings was[21]. The most obvious characteristic of the Latin redactions consists not in the omission of parts or failures to understand certain passages, but rather in the numerous additions. These often take the form of new theorems or discussions of the assumptions, but also insertion of reference to earlier theorems. The motif behind many of them was the attempt to bring Euclid's work into a more convenient didactic form. There was a clear tendency towards a "textbook" of the *Elements*, which can be seen already in Arabic redactions and even as far back as Theon of Alexandria. The growing interest in didacticism can be seen for instance in the discussions about the construction of a proof. The characteristic parts of a proof are indicated, hints are given on how to draw the constructions or how three-dimensional figures should be drawn, and similar propositions are referred to. Also we can see how the strict Greek separation of number and magnitude was gradually weakened.

---

[18] Edited in Heiberg, 'Apollonii ...', pp. LXXV–LXXX.

[19] The most important were those by Ibn al-Haytham (Alhazen) and by Witelo. The relevant texts were edited in Clagett, 'Archimedes', vol. 4 (1980).

[20] The medieval Latin translations of Theodosius and Menelaus have not been edited.

[21] A good survey on the main characteristics of the medieval Latin translations and reworkings of Euclid's *Elements* is Murdoch, 'Medieval Euclid'.

For the principle that theorems in book 5 should not be used to prove propositions in the arithmetic books 7–9 disappeared, and inadequate numerical "proofs" are to be found in theorems dealing with general quantities.

The general preoccupation with assumptions is remarkable: everywhere, even in the middle of a proof, new axioms would be included, so that all possible gaps in the proof would be closed, and on logical correctness a great value is placed. This, too, agrees with the didactical aims. But this concern with the assumptions not only served to make the *Elements* easier for all those who had to study them in the medieval faculty of arts. They were also supposed to lead to results that could be applied to external, above all philosophical, problems. In this respect the most important questions were those of incommensurability, the horn-angle (or "angle of contingence") between circumference and tangent of a circle and the divisibility of quantities. All these problems are in the end concerned with properties of the infinite and continuity which so much bothered the medieval philosophers. Thus the Arabic-Latin translations and redactions delivered to the West a textbook Euclid which agreed well with the current scholastic interests and not only in the mathematical circle.

With the above mentioned problems, which arose from the study of Euclid's *Elements*, the most important mathematical areas with which the scholastics occupied themselves are delineated. In the 14th and 15th centuries they created new results in mathematics which in some cases exceeded the Greek or Arabic achievement. They had no intention of forming algorithms to solve problems of geometry or physics. Rather they wanted, with the help of mathematics, to explain some general quantitative properties of space, time and motion. These considerations must always be considered in connection with Aristotelian physics.

Perhaps we may note the following details: in the first half of the 14th century Thomas Bradwardine in his *De continuo* asked the question whether the continuum may be divided in infinitum or whether there existed a smallest part or indivisible and, if the latter, whether there was a finite or infinite number of these "atoms". Through one-to-one correspondences between the elements of sets (finite or infinite), for instance between the points of a diameter of a circle and the corresponding points of the semi-circumference (by dropping perpendiculars), he proved that the continuum does not consist of atoms. His use of set-theoretical paradoxes foreshadowed future discussions.

The passage in Euclid about the "angle of contingence"[22] was made by Campanus, about 1260, into a subject for a long excursus. He proved that the angle of contingence is smaller than any angle between straight lines, but claimed that it was nonetheless not equal to zero. On this assumption Campanus asserted that the so-called "mean value principle" was false. According to this principle, which was used for squaring the circle since the time of the Greeks, a continuous quantity that goes from a smaller to a larger value also goes through all the values in between. The angle of contingence was hotly disputed until the 17th century.

In connection with the theory of proportions of book 5 and 7 of the *Elements* Bradwardine presented in his *Tractatus de proportionibus*[23] an elaborate theory of compound proportions. It was his intention to find a rule, according to Aristotle's physics, for velocity v in terms of force F and resistance R. The law he found involved an exponential: $\frac{F}{R} = n^v$. His work gave Nicole Oresme the impulse to develop the theory further. He formulated numerous rules for operating with fractional ratios and finally developed an algorithm for fractional ratios – in modern terms, he extended the concept of powers from integers to positive fractions as exponents[24].

Finally, I should like to mention the theory of the "latitude of forms" (*latitudo formarum*)[25], which was also treated in the 14th century, especially at Merton College, Oxford (Heytesbury, Swineshead etc.), and the university of Paris (Nicole Oresme etc.). This theory treated the change of the qualities in the Aristotelian sense (e. g. heat, density, colours, velocity of a motion). Such change was symbolically represented by Oresme in diagrams in which the length represents the "extension" (*extensio*; usually time) and the breadth the "intensity" (*intensio*; the degree of the quality at a specified part of the extension, e. g. of the time). The most important example is the variation of speed (the *intensio*) with time (the *extensio*). The area of the figure is a measure for the total quantity of the quality (in the example of speed varying with time the quantity is the distance traversed). In the case of uniform acceleration starting from zero the diagram takes the form of a triangle. In Merton College it was recognized and proved that the quantity of this motion, measured by the area of the triangle, is the same as though there were constant motion at half

---

[22] *Elements*, book 3, prop. 16.

[23] Edited in Crosby, 'Bradwardine'.

[24] See Grant, 'Oresme'.

[25] Other names: variation of intensities, configuration of qualities. The basic treatises by Oresme on this questions are edited in Clagett, 'Oresme'.

the final velocity (which would be measured by the area of a rectangle); and thus the distance traversed would be the same in the two cases[26]. From this so-called "Merton Theorem" we may recognize the modern formula $s = \frac{g}{2}t^2$ for uniformly accelerated motion. Other calculations of the variation of qualities led, in the 14th century, to results in the summation of infinite series. For example, Richard Swineshead treated a variation of quality in which the time interval of 1 unit is divided into $\frac{1}{2}$, $\frac{1}{4}$, $\dots$, $\frac{1}{2^n}$, $\dots$ and the intensities increase from interval to interval according to the arithmetical series 1, 2, 3, $\dots$ He proved that the motion was equivalent to that which would have occurred had the motion in the second interval operated in the entire interval, i. e.

$$\frac{1}{2} \cdot 1 + \frac{1}{2^2} \cdot 2 + \frac{1}{2^3} \cdot 3 + \dots = 2 \,.$$

In a similar way the convergence of other series, even those that led to logarithms, and the divergence of the harmonic series were proved.

These mathematical achievements are among the greatest in mathematics of the medieval West. They were induced by the study of Euclid's *Elements* and the physics of Aristotle and his commentators. Of course, these were not problems that interested the man in the street; they were part and parcel of university teaching in the faculty of arts. Treatises by Oresme, Bradwardine and others were not seldom mentioned as the subjects of lectures held in the 14th and (even more) in the 15th centuries: in the 14th century in Paris and Oxford; in the 15th century at many universities, for instance Vienna, Cracow and (after 1480) also in Leipzig. Again, the number of manuscripts extant today shows that these works were more often copied than one would have expected.

Well before the reception of Arabic science, medieval scholars had begun to distinguish between the theory of a science and its applications. In the *Didascalicon*, written at the beginning of the 12th century[27], Hugh of St. Victor separated, according to ancient precepts, the theoretical from the practical sciences. He applied this distinction to geometry: there was a theoretical, speculative part, which only functions with rational thought (*sola rationis speculatione*), and a practical, active, part, which uses instruments. The geometrical treatises just mentioned would fall into the first category. Hugh himself wrote a *Practica geometriae*, which in both title and content inspired texts in the following generations[28]. Hugh

---

[26]For the different formulations and proofs of the Merton theorem see Clagett, 'Mechanics', chapter 5.

[27]Edited in Buttimer, 'Didascalicon'.

[28]Edited in Baron, 'Practica', and Baron, 'Opera'.

treated the measure of heights (*altimetria*) and the determination of areas (*planimetria*) and of volumes (*cosmimetria*). The geometry which we find in his *Practica* is very similar to that of the *agrimensores*: it is in the tradition of the Romans. It is therefore remarkable that the measurement of angles is performed with the astrolabe, which comes from the Arabs.

It may surprise us that the tradition of these treatises was unbroken and almost without Arabic influence until the 15th or 16th centuries. Tracts on some forms of the quadrant and the *Practica geometriae* of Dominicus de Clavasio belong to this tradition. Robertus Anglicus of Montpellier describes, about 1270, the application of another instrument which did come from the Arabs, the so-called "Old quadrant" (*Quadrans vetus*), to practical ends[29]. In presentation this work is of the same scientific standing as Hugh's. It is surprising that Euclid's *Elements*, which at that time was well known, is not used, though the elementary parts of books 1 and 3 are used implicitly, for instance in the construction of an isosceles triangle. That this writing is really directed to practicioners, is to be seen in the way that the quadrant is divided into degrees: it is first divided into halves, then each half into thirds, then each third again into thirds, and finally each part into fifths. On how this subdivision is to be made, nothing is said; the problems of the division into three or five parts are not addressed.

Only in the first half of the 14th century did Euclid's *Elements* influence the style and content of the treatises on practical geometry. A good example is the writing of Dominicus de Clavasio (1346), which was particularly widely known[30]. Since all instruments and procedures of practical geometry depend in the last analysis on the similarity of triangles, and so on proportions, Dominicus begins with four *suppositiones* on ratios and on the determination of unknown elements in a simple proportion. In the problems of measurements that he handles Dominicus takes care to couch it in a "scientific" way, in that he expresses a proof in the manner of Euclid and using the logical structures usual in Euclid. This outer form appears, however, rather artificial and the content is not essentially new. But we should mention that Dominicus understood the problem of squaring the circle when he said that the ratio of the circumference of a circle to the diameter is about $3\frac{1}{7}$ to 1, but that there is no exact ratio[31]:

---

[29] Edited in Hahn, 'Quadrans vetus'.

[30] Edited in Busard, 'Practica Geometriae'.

[31] Busard, 'Practica Geometriae', p. 556.

> When I say that a circle is measured by a square, I do not
> intend to speak demonstratively (*non intendo demonstrative
> loqui*), but I only wish to show how the area may be found
> so that no noticeable error remains.

Dominicus' work influenced later tracts, e. g. the *Geometria Culmensis*, which was written at the end of the 14th century and soon translated into German[32]. As mentioned above, these and similar works used only the similarity of triangles in their procedures. Progress was only made when trigonometry was applied – and in the West this did not happen until the 15th century.

A second invigorating element came from Arabic algebra. Algebra used in geometrical matters is to be seen in translations of the 12th century. For example, Abū Bakr in his *Liber mensurationum*[33] applied algebraic solutions to problems about areas which lead to equations of the first and second degrees. Problem 55, for instance, runs:

> When somebody says to you, "the side (of a rhombus) is
> 10 and the shorter diameter is 12"; how long is the longer
> diameter?

This problem leads, by Pythagoras' theorem, to a simple quadratic equation. Other problems are essentially algebraic, but are presented as geometrical, e. g.:

> When somebody says to you, "I have taken away the side
> from an area and there remains 90"; how big is the side?

It is here only a matter of solving the quadratic equation $x^2 - x = 90$ and has no direct connection with a geometrical problem, because quantities of different dimensions are subtracted. The formulation and method of solution are strongly reminiscent of Babylonian problems.

Many of the problems of the *Liber mensurationum* are also to be found in the treatise on practical geometry written by the Jewish mathematician Savasorda and translated, under the title *Liber embadorum*, into Latin by Plato of Tivoli in 1145[34]. In turn, this work influenced the *Practica geometriae* of Leonardo of Pisa (Fibonacci) written in 1220[35]. Also, parts of the *De arte mensurandi*, written by John de Muris in the

---

[32] Edited in Mendthal, 'Geometria Culmensis'.
[33] Edited in Busard, 'Algèbre'.
[34] Edited in Curtze, 'Savasorda'.
[35] Edited in Boncompagni, 'Scritti', pp. 1–224.

14th century[36], are in this tradition. Leonardo says he wrote his *Practica geometriae*[37]

> so that those who wish to work according to geometrical demonstrations and those who wish to work according to common use, in the lay approach, with (various) measurements (*in dimensionibus*), will find a perfect document in the eight distinctions of this art, which are explained below.

He thus addresses both laymen looking only for geometrical recipes and those interested in theoretical mathematics. For the latter the following topics were especially intended: finding square roots, calculation with expressions involving roots (section 2); finding cube roots (section 5); *subtilitates*, *inter alia* the inscribing of a regular pentagon in an isosceles triangle and in a circle (section 8). Topics directed more to the practicioners include: calculation of the area of rectangles with sides expressed in integers, fractions and various units of length (section 1); measurement of various-shaped fields, i. e. triangles, quadrilaterals, polygons, circles (section 3); division of fields in a given ratio (section 4); volumes of solids (section 6); problems of surveying with a quadrant (section 7). Thus the *Practica geometriae* is indeed a *perfectum documentum* of practical geometry: only the higher geometry of Euclid, Archimedes and Apollonius is – naturally – missing.

Let us make a summary: apart from the theoretical geometrical writings, which were clearly principally used in university teaching and which could only exert influence after the translations of the 12th century, there was a series of practical texts on geometry. They are in the tradition of the Roman *agrimensores* and take little from the Arabs. Methods that were essentially the procedures of the *agrimensores* are still to be found in manuscripts of the 15th century and in books on geometry of the 16th and 17th centuries. In some, admittedly not very numerous, geometrical writings there is clear influence from Arabic treatises, e. g. that of Abū Bakr. These so-called *misāḥa* texts treat the science of the comparison of geometrical quantities and how such comparisons are to be made. Only exceptionally – e. g. in Leonardo's work – is the *misāḥa* tradition mixed with that of the practicioners. But further investigations are necessary to establish the connections between the various texts: a complete history of practical geometry in the Middle Ages has yet to be written.

---

[36] Edited in Busard, 'De arte mensurandi'.

[37] Boncompagni, 'Scritti', p. 1.

To estimate the importance of practical geometry in the Middle Ages one should look not only at those texts that call themselves "Practica geometriae", but also at the remarks by contemporary authors who concerned themselves with the classification of the sciences (e. g. Hugh of St. Victor, Dominicus Gundissalinus). I can here only point to the excellent monograph by Stephen K. Victor[38]. He considers the question of whether these "practical geometries" were really applied to practical ends[39]. Of course, this question is hard to answer, because among skilled manual workers and craftsmen knowledge was in the main transmitted orally. Clearly the "practical" geometries did indeed contain parts that could be applied in practice, e. g. on metrology, measurement of distances, determination of areas, making and using instruments. That they were in fact applied is likely. For if reports of how measurements were made or how buildings were erected are compared with the methods that are indicated in the practical geometries, the similarities are clear enough. For instance the surveyors active in Belgium in the 12th and 13th centuries gave the sizes of estates and smallholdings in the same units that are found in the practical geometries. In the planning and construction of towns in the High Middle Ages a right-angled grid was used for the streets and buildings. Sometimes towns were built with a circular city wall and it was important to know how many houses with equal rectangular ground plan would fit in. For both problems there was information in the practical geometries. Architects could take information from practical geometries, since there is a remark in the sketch-book of Villard de Honnecourt (ca. 1225–1250) about three diagrams that they were "taken from geometry" (*sunt estraites de geometrie*). The diagrams concerned the determination of the height of a tower, of the breadth of a river and of a window – typical problems in the treatises on practical geometry. Thus it is possible that even architects knew these texts. Certainly they were written for practically oriented readers. Since so many copies were made, even when better geometries were available, it is clear that these works fulfilled a pedagogic aim.

The relation between theory and practice in medieval geometry is shown in particularly strong light by the problem of measuring the volume of a barrel[40]. In practical geometries of the 10th to the 14th century this problem is solved in various ways: the barrel is approximated by one

---

[38] "The Place of Practical Geometry in the Middle Ages", in Victor, 'Practical Geometry', pp. 1–73.

[39] Victor, 'Practical Geometry', pp. 53–73.

[40] See Folkerts, 'Visierkunst'.

or two frusta of cones or by a cylinder – in the latter case the area of
the circular base of the cylinder is either the arithmetical mean of two
or three of the principal circles (top, base and middle) or simply the
area of the base. There was no consistent procedure. Since the problem
was only one of many in solid geometry, perhaps it was not considered
particularly interesting at the time. This was changed in the 14th and
even more in the 15th century when, because of urbanization and increased trade, more barrels and their contents (particularly wine) were
transported from one place to another. At this time the towns and prin-
cipalities, which gained much by taxing the wine-trade, and also the
merchants and consumers needed to be able to determine the contents
of the barrels. In the larger cities there were official gaugers (in German:
"Visierer"). Their task was principally to measure the volume of barrels
and secondarily to oversee weights and measures. Their social and legal
standing will not be discussed here. Suffice it to say that many of them
were trained manual workers who possessed no theoretical knowledge of
the art of measurement. To teach them the art of gauging, special tracts
on gauging (in German: "Visiertraktate") were written; they are mentioned from the middle of the 14th century, but they are extant only from
about the middle of the 15th century. About a hundred manuscripts of
the 15th and 16th centuries containing such tracts, to say nothing of
numerous printed texts, are known to me. Most of them were written in
South Germany and Austria, and in the 16th century also in Belgium.
The measurement of barrels was occasionally also treated in some universities (Erfurt, Vienna, Cracow). But in general the tracts on gauging
were written by practitioners, mostly by *Rechenmeister*, who had no connection with a university. From the procedures the practical origin may
be recognized immediately: the measurement was mechanized by means
of a gauging rod ("Visierrute"). There were two different types of rod: the
"square" and the "cubic". On the "square rod" there were two scales. On
the scale of depth the square roots of the unit of measures were written;
on the scale of length were the multiples of the unit. The units of length
and of depth were determined from a standard cylinder of unit volume.
The number taken as the depth is found by measuring the depth of the
two bases and the depth at the bung-hole and taking the average; the
number taken as the length is found by direct measurement. The volume
is then equal to the product of the numbers taken as length and depth. In
this procedure the fact was used that the volume of a cylinder is proportional to the length and to the square of the diameter of the base, though
this was not explicitly mentioned: only "recipes" were given for making

the two scales of the square-rod and how to use them to find the volume of the barrel. It seems that the cubic-rod was developed in Austria. With this rod only one measurement was necessary: the rod was put through the bung-hole obliquely so that the end was at the furthermost point of the barrel and the number on the scale at the bung-hole was read off. The scale was constructed with the help of a table of cube roots. The calibration was made with a standard barrel of unit volume which had proportional diameter of the base, depth at the bung-hole and length. In contrast to the square-rod, the cubic-rod was only appropriate for barrels of the same shape. The question of the validity of this procedure induced Kepler to investigate the accuracy of the measurement. Thus he was led to write his *Stereometria doliorum* (1615), the first attempt to treat scientifically the measurement of barrels and at the same time an important first step in the development of the integral calculus.

The art of gauging is a good example of practice preceding theory. The needs of trade made it necessary to be able to find quickly and simply the volumes of barrels. To this end it was not necessary to have any geometrical knowledge; in particular it was not necessary to have understood the "difficult" squaring of a circle. By using the rods a mathematical calculation was replaced by a purely mechanical procedure: it was only necessary to read off a value from a scale or at the worst to multiply two numbers. The problem was transformed from calculation to the construction of the scales of the rods; and this, too, was reduced by specialists to "recipes". This phenomenon is typical of mathematics and other branches of knowledge of the 15th to 17th century: analogous are the sector compass and the astrolabe[41]. At first there were no doubts about the correctness of the procedure. Some 300 years later, after the first mention of the tracts on gauging, Kepler was the first to try to found the theory of the measurement of barrels. With him begins a new, scientific, era of "doliometry".

Now some remarks on theoretical and practical aspects of medieval arithmetical texts. The most important arithmetic writing available in the West in the early Middle Ages was Boethius' *Arithmetic*[42]. It is essentially a derivative of Nicomachus' *Arithmetic* and belongs to Boethius' translation program of about 500 A.D., in which he tried to preserve Greek knowledge for the West – in a world in which knowledge of Greek was rapidly vanishing. This work was in fact read and exerted a great

---

[41]For the history of the sector compass see Schneider, 'Proportionalzirkel'.

[42]Edited in Friedlein, 'De institutione arithmetica', pp. 3–173, and Guillaumin, 'Boèce'.

influence up to the Renaissance. It was studied in all universities; and over two hundred extant manuscripts testify to the importance of this mathematical text. Boethius' *Arithmetic* contains essential elements of Pythagorean theory of numbers, e. g. the division of the (natural) numbers into even and odd, prime numbers, perfect numbers; classification of numerical ratios, which were also used in the theory of music; in book 2 he treats the polygonal and polyhedral numbers (numbers represented by regular polygonal or polyhedral arrays of points), and finally the three means (arithmetic, geometric, harmonic). The work is accordingly completely irrelevant for practical calculation. Why was it none the less so often copied? One reason is perhaps to be found at the beginning of book 1, chapter 2, where Boethius says[43]:

> Everything which is constructed by the primeval nature of things is seen as formed by means of numbers. For this was the main principle in the mind of the Creator.

This thought goes well with the contemporary Christian understanding of the purpose of knowledge. It corresponds well with the much-quoted biblical passage that God has "ordered all things in measure and number and weight"[44]. In later centuries there was developed from Boethius' number theory a number symbolism (which appears extraordinary to us) by which an allegorical sense was attached to the natural numbers. Terms like *mysteria numerorum* or *numeri sacrati* betoken the mysterious hiddenness of the meaning of the number and its holiness. A number is taken as an allegorical sign and the holiness hidden in the number is made visible: thus the secret is revealed[45]. A series of works *De numeris* were written which belong to this type; they begin with a tract ascribed to Isidore and continue into the 13th and 14th centuries[46]. In these texts Boethius-like mathematical contents are connected with the meaning of the numbers in the Bible. For example, the number 6 is characterized as perfect because it is equal to the sum of its proper parts; other numbers for which this is not true are, however, also perfect because their use in the Bible suggests this. Such totally unconnected properties of numbers

---

[43] Friedlein, 'De institutione arithmetica', p. 12, lines 14–17.

[44] Wisdom of Solomon 11, 20.

[45] See Hellgardt, 'Zahlensymbolik', who mentions further books and articles on this subject.

[46] Editions of the most important texts have been published by Hanne Lange, in: *Université de Copenhague. Cahiers de l'Institut du Moyen-Age Grec et Latin*, vols. 29, 32, 40, Copenhague 1978, 1979, 1981.

– mathematical demonstration and biblical symbolism – is hard for us to understand, but was evidently acceptable for the whole of the Middle Ages. The great number of texts and manuscripts of this kind – which incidentally have not all received sufficient scholarly attention – show how widespread these mathematically poor treatises were. Thus number and arithmetic serve to interprete the Bible. Occasionally, geometry took over the same function. For instance, Hugh of St. Victor applied the determination of volumes of cuboids to the determination of the size of the Ark and to prove that all animals would fit into it[47].

As mentioned above, these texts had nothing to do with practical calculation. If one wanted to calculate, one had to use the clumsy Roman numerals. For multiplication the reckoning table (*calculus*) of Victorius of about 450 A.D. was used, which contained the multiples of one to fifty and of the Roman fractions[48]. These fractions were expressed with the impractical and elaborate Roman fractions to base 12. The so-called "finger numbers", which are known above all through the description by the Venerable Bede[49], did not, apparently, serve the purpose of calculation, rather, the various positions of the fingers represented the various numbers and the system was simply a way of recording numbers. In the monasteries there was scarcely any necessity for practical calculation. The most important application was, without doubt, the determination of Easter (*computus*). Since this depends on finding the first full moon after the vernal equinox and the Sunday that follows this, it was necessary to take into consideration the cycle of the Moon (19 solar years or 235 lunations) and to make use of the Julian calendar, in which the days of a week on a specific date vary, but repeat themselves every $4 \times 7 = 28$ years. To make the correspondences between the year and solar and lunar cycles, it was necessary to divide the number of the year by 19 and 28 and find the remainders. The problems associated with *computus* were solved at the latest in Bede's tract *De temporum ratione* (725)[50]. Later "recipes" were formulated for the various stages of the calculation of Easter; mnemonic aids were devised, which were copied up until the time of the printed calendar (about 1500).

---

[47] *De arca Noe morali*, edited in Migne, *Patrologia Latina* 176, cols. 628f.

[48] Edited in Friedlein, 'Der Calculus' and 'Victorii Calculus'.

[49] They form the first chapter of his *De temporum ratione*, but are also transmitted in many manuscripts as a separate text.

[50] A good survey on the history of the *computus* is in the introduction of Jones, 'Bedae opera'.

For computistic calculation Roman numerals were entirely satisfactory. Since the 10th century another means of calculation was available: the abacus. This was a reckoning board with colums on which counters were placed; these were marked with the Hindu-Arabic numerals. Gerbert used such an abacus and wrote a treatise on calculation with it. Recent investigations, however, have shown that the calculation board with inscribed counters was already in use in Western Europe before Gerbert's time[51]. Calculation with this board was taught in monasteries until the first half of the 12th century. There are texts of at least 10 authors, albeit mostly in a few manuscripts, on this subject and there are several anonymous texts. This type of abacus was apparently known mostly in ecclesiastical circles and was seldom used for practical purposes.

At the beginning of the 11th century a numerical game, which is based on Boethius' *Arithmetic*, was invented: the so-called "Rithmomachia". The origins of this game have only recently been discovered: Arno Borst has shown[52] that, starting with a competition in the ecclesiastical world, this game was invented by one Asilo of Würzburg about 1030; further he has traced the history of the game in the following centuries. In the next ten decades six different reworkings of Asilo's texts appeared[53]. A definitive treatise was written about 1130 by Fortolf, who made the first detailed description of the rules of the game[54]. Even in the later Middle Ages the game enjoyed a great popularity, as may be seen from the numerous still unpublished manuscript texts from the 12th to the 16th centuries and from a series of printed works from the 15th to the 17th centuries[55]. Without going into details, we may note that from the 14th century on there was a series of commentaries and even critical remarks on the value of the various rules. In the later time the game was popular mostly in court circles. That the game was also treated in the universities is shown by a manuscript in Jena of about 1520, which contains notes from lectures and other sources by a Cracow student, and by the famous manuscript C 80 of the Sächsische Landesbibliothek in Dresden, which contains *inter alia* lectures given at the Leipzig university in the 1480s.

---

[51] See Bergmann, 'Innovationen', pp. 176–185.

[52] See his fundamental monograph (Borst, 'Zahlenkampfspiel').

[53] Hermannus Contractus (about 1040); the so-called "Lütticher Anonymus" (about 1070); Odo of Tournai (about 1090); the so-called "Regensburger Anonymus" (about 1090); the "Fränkischer Kompilator" (about 1100); and the "Bayerischer Kompilator" (about 1105).

[54] All texts mentioned above have been edited in Borst, 'Zahlenkampfspiel'.

[55] See Folkerts, 'Rithmomachia', in this volume (item no. XI).

The translations from Arabic had no influence whatever on *rithmo-machia*. But, of course, they did influence the arithmetic sciences in general. Number theory and ratio theory have already been mentioned in connection with the translations of Euclid from Arabic. Here we will leave algebra aside, although there were interesting developments in the 15th and 16th centuries – for instance the development of symbolism in about 1450 to 1550 and the solution of the cubic equation in Italy in the first half of the 16th century. More important for the practitioners was the introduction of the Hindu-Arabic numerals and calculation with them, beginning with the translation of al-Khwārizmī's *Arithmetic* in the first half of the 12th century. The new arithmetic made slow progress among merchants and it was some centuries before it was generally used by them. The main reasons for the slowness of acceptance were the abstract nature of the place value system and the logical difficulty that "zero", which in fact means "nothing", can increase the value of a digit ten or hundred times. In late medieval manuscripts there are often tables that show the values of the various digits according to their decimal place; and there are often mistakes in these tables. Also, the forms of the numerals had to be learned. In some manuscripts written in Northern Germany there are explanations of how to write them. One of them[56] begins: "Unum dat vinger, duo treppe, sed tria sustert"[57] and ends, referring to the number 10: "... Si vingerken desit, ringelken nichil significabit"[58]. The vernacular was thus used to illustrate the forms of the numerals.

Such aids were fairly common in the 15th century, when it was clearly a question of bringing the Arabic numerals to the common people. Already in the 13th and 14th centuries numerous elementary treatises were written, evidently with the same intention, but at this time for the use of students: the so-called "algorismus" treatises, i. e. derivatives of al-Khwārizmī's *Arithmetic*. The most ubiquitous texts of this sort were by Johannes de Sacrobosco and Alexander de Villa Dei: there are several hundred copies of the Sacrobosco text, which was the most popular mathematical writing of the entire Middle Ages[59]. A typical *algorismus* treatise contains the following: how numbers were written, starting with digits; the six basic operations (addition, subtraction, multiplication, division;

---

[56] Wolfenbüttel, Herzog August Bibliothek, Cod. Guelf. 1189 Helmst., f. 189v (from 1462).

[57] "One looks like a finger, two like stairs, but three like a pig's tail."

[58] "If the finger is missing, the little ring will mean 'nothing'."

[59] The text has been edited in Pedersen, 'Petri Philomenae ...', pp. 174–201. Pedersen also gives an edition of the commentary by Petrus de Dacia.

and also doubling and halving); extraction of square and cube roots; and the summation of arithmetic and geometric progressions. Thus they were not practical instruction for merchants, but intended for students in the faculty of arts in universities.

If a merchant wished to learn book-keeping and calculation, he went not to the universities but to the "maestri d'abbaco", from the 13th century in Italy, or "Rechenmeister", from about 1400 in Germany and other parts of Europe. They wrote almost exclusively in the vernacular, not Latin, and did not teach at universities, but in private schools, most of which were supported by the towns[60]. So much has been written about the "Rechenmeister" that it is not necessary to go into detail here.

The textbooks of the "Rechenmeister" contained, besides practical instruction, items of purely theoretical interest. In historical times there have always been collections of mathematical problems, some of which arose from practical circumstances, but many were for intellectual recreation. The term "recreational mathematics" has been applied to such problems. Its aim was to make teaching mathematics more pleasant. The oldest medieval Western collection of this kind is the *Propositiones ad acuendos iuvenes* attributed to Alcuin, which was written in the 9th century[61]. This collection is in the Roman tradition, though Byzantine and even some Arabic influence may be traced. Some of the problems, such as the "transport problems" (like the known problem of the wolf, goat and cabbage), are to be seen for the first time in the *Propositiones.* This text was the first of several collections which are transmitted in medieval manuscripts. In these collections the problems are often of a limited number of standard types, though usually differing in detail. Evidently, these types were somehow transmitted, but it is not yet clear how. We do know that these anonymous collections were transmitted in ecclesiastical circles and accordingly tended to omit problems derived from practical considerations. Many of these problems found their way into the textbooks of the "Rechenmeister", where they were mixed with problems of mercantile practice – such as the constitution of alloys in coins, calculation of interest or currency exchange[62]. Even the textbooks of today contain problems of this sort.

---

[60] A catalogue of all known Italian texts can be found in Van Egmond, 'Practical Mathematics'.

[61] Edited in Folkerts, 'Aufgabensammlung' (reprinted in this book, item V).

[62] The types of problems which occur in mathematical collections of the Middle Ages and later have been categorized by Kurt Vogel, who also gave a list of their occurence; see Tropfke, 'Elementarmathematik', pp. 513–660.

It is to be hoped that most of the research of the last decades has been mentioned in the above. What is urgently needed is critical editions of the many texts which are not yet adequately edited. Even elementary quantitative information, such as an account of how many manuscripts there are of the various texts, where they were written and where they were read, would be a beginning.

Bibliography

R. Baron (ed.), "Hvgonis de Sancto Victore *Practica Geometriae*", *Osiris* 12 (1956), 176–224.

R. Baron (ed.), *Hugonis de Sancto Victore opera propaedeutica*, Notre Dame 1966.

W. Bergmann, *Innovationen im Quadrivium des 10. und 11. Jahrhunderts. Studien zur Einführung von Astrolab und Abakus im lateinischen Mittelalter*, Stuttgart 1985.

F. Blume, K. Lachmann, A. Rudorff (eds.), *Die Schriften der römischen Feldmesser. Band 1*, Berlin 1848.

B. Boncompagni (ed.), *Scritti di Leonardo Pisano*, vol. 2, Rome 1862.

A. Borst, *Das mittelalterliche Zahlenkampfspiel*, Heidelberg 1986. (= *Supplemente zu den Sitzungsberichten der Heidelberger Akademie der Wissenschaften, Philosophisch-historische Klasse*, Jg. 1986, Bd. 5.)

N. Bubnov, *Gerberti postea Silvestri II papae opera mathematica*, Berlin 1899.

H. L. L. Busard (ed.), "The Practica Geometriae of Dominicus de Clavasio", *Archive for History of Exact Sciences* 2 (1965), 520–575.

H. L. L. Busard (ed.), "L'Algèbre au Moyen Age. Le 'Liber mensurationum' d'Abū Bekr", *Journal des Savants* (1968), 65–124.

H. L. L. Busard (ed.), *Johannes de Muris, De arte mensurandi. A Geometrical Handbook of the Fourteenth Century*, Stuttgart 1998.

Ch. H. Buttimer (ed.), *Hugonis de Sancto Victore Didascalicon de studio legendi. A Critical Text*, Washington 1939.

M. Clagett, *The Science of Mechanics in the Middle Ages*, Madison (Wisc.) 1959.

M. Clagett, *Archimedes in the Middle Ages*, vol. 1, Madison (Wisc.) 1964; vols. 2–5, Philadelphia 1976–1984.

M. Clagett, *Nicole Oresme and the Medieval Geometry of Qualities and Motions*, Madison / Milwaukee / London 1968.

H. L. Crosby (ed.), *Thomas of Bradwardine. His Tractatus de Proportionibus*, Madison (Wisc.) 1955.

M. Curtze (ed.), "Der 'Liber embadorum' des Savasorda in der Übersetzung des Plato von Tivoli", in *Urkunden zur Geschichte der Mathematik im Mittelalter und der Renaissance, 1. Teil*, Leipzig 1902, pp. 1–183.

M. Folkerts, *"Boethius" Geometrie II, ein mathematisches Lehrbuch des Mittelalters*, Wiesbaden 1970.

M. Folkerts, "Die Entwicklung und Bedeutung der Visierkunst als Beispiel der praktischen Mathematik der frühen Neuzeit", *Humanismus und Technik* 18 (1974), 1–41.

M. Folkerts (ed.), "Die älteste mathematische Aufgabensammlung in lateinischer Sprache. Die Alkuin zugeschriebenen Propositiones ad acuendos iuvenes. Überlieferung, Inhalt, Kritische Edition", *Denkschriften der Österreichischen Akademie der Wissenschaften, mathematisch-naturwissenschaftliche Klasse*, 116. Band, 6. Abhandlung, pp. 13–80, Vienna 1978. (Reprinted in this volume, item V.)

M. Folkerts, A. J. E. M. Smeur, "A treatise on the squaring of the circle by Franco of Liège, of about 1050", *Archives Internationales d'Histoire des Sciences* 26 (1976), 59–105, 225–253. (Reprinted in this volume, item X.)

G. Friedlein (ed.), *Anicii Manlii Torquati Severini Boetii de institutione arithmetica libri duo, de institutione musica libri quinque, accedit geometria quae fertur Boetii*, Leipzig 1867.

G. Friedlein, "Der Calculus des Victorius", *Zeitschrift für Mathematik und Physik* 16 (1871), 42–79, 253f.

G. Friedlein, "Victorii Calculus ex codice Vaticano editus", *Bullettino Boncompagni* 4 (1871), 443–463.

E. Grant (ed.), *Nicole Oresme*, De proportionibus proportionum *and* Ad pauca respicientes. *Edited with Introductions, English Translations, and Critical Notes*, Madison / Milwaukee / London 1966.

J.-Y. Guillaumin (ed.), *Boèce. Institution arithmétique*, Paris 1995.

N. Hahn (ed.), *Medieval Mensuration:* Quadrans vetus *and* Geometrie Due Sunt Partes Principales . . . , Philadelphia 1982.

Ch. H. Haskins, *Studies in the History of Mediaeval Science*, 2nd ed., Cambridge (Mass.) 1927.

J. L. Heiberg (ed.), *Apollonii Pergaei quae Graece exstant cum commentariis antiquis*, vol. 2, Leipzig 1893.

E. Hellgardt, "Zahlensymbolik", in *Reallexikon der deutschen Literaturgeschichte*, 2nd ed., vol. 4, Berlin / New York 1984, pp. 947–957.

J. E. Hofmann, "Zum Winkelstreit der rheinischen Scholastiker in der ersten Hälfte des 11. Jahrhunderts", in *Abhandlungen der Preußischen Akademie der Wissenschaften, Jahrgang 1942, Mathematisch-naturwissenschaftliche Klasse, Nr. 6*, Berlin 1942.

Ch. W. Jones, *Bedae opera de temporibus*, Cambridge (Mass.) 1943.

D. Lindberg (ed.), *Science in the Middle Ages*, Chicago / London 1978.

H. Mendthal (ed.), *Geometria Culmensis. Ein agronomischer Tractat aus der Zeit des Hochmeisters Conrad von Jungingen (1393–1407)*, Leipzig 1886.

J.-P. Migne, *Patrologia Latina*, vol. 176, Paris 1854.

J. E. Murdoch, "The Medieval Euclid: Salient Aspects of the Translations of the *Elements* by Adelard of Bath and Campanus of Novara", in *XIIe Congrès International d'Histoire des Sciences. Colloques*, Paris 1968, pp. 67–94.

F. S. Pedersen (ed.), *Petri Philomenae de Dacia et Petri de S. Audomaro opera quadrivialia. Pars I*, Copenhagen 1983. (= *Corpus philosophorum Danicorum medii aevi*, X.1.)

I. Schneider, *Der Proportionalzirkel, ein universelles Analogrecheninstrument der Vergangenheit*, Munich 1970.

P. Tannery, A. Clerval, "Une correspondance d'écolàtres du XIe siècle", *Notices et extraits des manuscrits de la Bibliothèque Nationale et autres bibliothèques* 36 (1901), 487–543. Reprinted in P. Tannery, *Mémoires scientifiques*, vol. 5, Toulouse / Paris 1922, pp. 229–303.

J. Tropfke, *Geschichte der Elementarmathematik*, 4th ed., vol. 1 (ed. by K. Vogel, K. Reich, H. Gericke), Berlin / New York 1980.

B. L. Ullman, "Geometry in the Mediaeval Quadrivium", in *Studi di bibliografia e di storia in onore di Tammaro de Marinis*, vol. 4, Rome 1964, pp. 263–285.

W. Van Egmond, *Practical Mathematics in the Italian Renaissance. A Catalog of Italian Abbacus Manuscripts and Printed Books to 1600*, Florence 1980.

S. K. Victor, *Practical Geometry in the High Middle Ages*, Philadelphia 1979.

II

Mathematische Probleme im Corpus agrimensorum

Als in der zweiten Hälfte des vorigen Jahrhunderts die Mathematikgeschichte ihre erste Blüte erlebte – man kann sogar sagen, daß sie erst in dieser Zeit zu einer wissenschaftlichen Disziplin wurde –, gehörten die Schriften der römischen Agrimensoren zu denjenigen Texten, mit denen man sich gleich am Anfang intensiver beschäftigte. Die erste kritische Ausgabe von F. BLUME/K. LACHMANN/A. RUDORFF war 1848/1852 erschienen (Bd. 1 im folgenden zitiert als: LACHMANN) und lieferte erstmals für einen großen Teil der Schriften eine brauchbare Textgrundlage. 1864/66 gab Friedrich HULTSCH, der sich in der Wissenschaftsgeschichte vor allem als Herausgeber der Schriften des Pappos und des Heron einen Namen erwarb, die Metrologicorum Scriptorum Reliquiae heraus; 1872 erschien sein Artikel Gromatici in ERSCH/GRUBERS Allgemeiner Encyklopädie der Wissenschaften und Künste[1]. Richtungsweisend wurde Moritz CANTORS Monographie Die römischen Agrimensoren und ihre Stellung in der Geschichte der Feldmeßkunst, Leipzig 1875 (im folgenden zitiert als: CANTOR, Agrimensoren); sie stellt den ersten Versuch dar, die Mathematik der Agrimensoren ausführlich zu behandeln und die Quellen ihres Wissens zu ermitteln, gibt ferner Hinweise auf die Benutzung ähnlicher Methoden in mathematischen Schriften des Mittelalters und bringt außerdem eine – wenn auch nur mäßig gute – Ausgabe des Textes von Epaphroditus und Vitruvius Rufus, der von LACHMANN nicht ediert worden war. CANTORS Buch wurde sofort die maßgebliche Abhandlung über die Mathematik der Agrimensoren und ist es eigentlich bis heute geblieben, auch deshalb, weil er wesentliche Teile daraus in seine vierbändigen Vorlesungen über Geschichte der Mathematik übernahm, die lange Zeit hindurch fast so etwas wie die Bibel für die Mathematikgeschichte waren.

Bei allen Verdiensten um die Erforschung der Mathematik der Agrimensoren, die CANTOR sich erworben hat, muß man doch auf einen empfindlichen Mangel hinweisen: Er stellte keine eigenen Untersuchungen zur Überlieferungsgeschichte des Corpus agrimensorum an, sondern übernahm die – nicht immer zutreffenden – Ergebnisse LACHMANNS. Dadurch sind CANTORS Folgerungen mindestens in Teilen nicht korrekt. Insbesondere ignorierte er die Forschungen Nikolai BUBNOVS, die ihm zugänglich waren und die er für seine Vorlesungen hätte berücksichtigen können: BUBNOV hat bekanntlich in seiner grundlegenden Monographie Gerberti postea Silvestri II papae Opera

---

1 1. Sektion, 92. Teil, S. 97–105.

Mathematica (972–1003), Berlin 1899 (im folgenden zitiert als: BUBNOV), nicht nur eine Edition der Schriften Gerberts vorgelegt, sondern als Appendix VII eine Monographie zum Corpus agrimensorum angefügt, die auf umfangreichen Handschriftenstudien beruht und erstmals im großen und ganzen korrekt die Überlieferungsgeschichte wiedergibt, insbesondere die Stellung der Hauptklassen AB, PG, der Mischklasse EF und der Exzerptenhandschriften im Umkreis der sogenannten „Geometrie I" des Pseudo-Boethius. BUBNOV erstellte ferner kritische Ausgaben einiger mathematischer Texte im Corpus agrimensorum, insbesondere von Epaphroditus und Vitruvius Rufus und vom Liber podismi.

Die sehr sorgfältigen Arbeiten von Carl THULIN, insbesondere seine drei Aufsätze über Handschriften und Überlieferungsgeschichte des Corpus agrimensorum[2], werden hier nur am Rande erwähnt, weil sie zwar für die Abhängigkeit der Handschriften die auch heute noch maßgebliche Grundlage bilden, aber über den Inhalt der Texte keine neuen Informationen liefern. Auch THULINS Neuausgabe des Corpus betrifft nicht die mathematischen Texte.

Es mag vielleicht überraschen, daß nach CANTOR und BUBNOV keine wichtigen mathematikhistorischen Spezialarbeiten zum Corpus agrimensorum mehr erschienen sind. Dies hängt sicher damit zusammen, daß die kontinuierliche Entwicklung der Mathematikgeschichte durch den 1. Weltkrieg unterbrochen wurde. Nach dem Krieg wandte man sich anderen, vermeintlich interessanteren und sicher auch spektakuläreren Themen zu, etwa der Erforschung der vorgriechischen Mathematik. Die Mathematik der Römer und des frühen Mittelalters fristet seitdem ein Schattendasein in mathematikhistorischen Darstellungen. Schlimmer noch: Die wenigen Arbeiten, die sich mit diesem Thema beschäftigen, pflegen sich auf CANTOR zu berufen, ohne BUBNOVS Ergebnisse zu berücksichtigen. Auch hierfür lassen sich leicht Gründe finden: Es ist für heutige Mathematikhistoriker ziemlich mühselig, BUBNOVS Buch zu benutzen, nicht nur, weil es durchgehend auf Latein geschrieben ist, sondern auch, weil wesentliche Informationen oft in d n Anmerkungen versteckt sind und weil viele Aussagen, die BUBNOV im Hauptteil macht, im Anhang 7, der erst nach Drucklegung des Werks abgeschlossen wurde, aufgehoben bzw. relativiert werden, so daß man ständig in den Addenda et corrigenda nachschauen muß, ob die Informationen des 1. Teils noch zutreffen. Eine zweite Ursache dafür, daß sich BUBNOVS Ansichten nicht durchsetzten, liegt in CANTORS Stellung zu BUBNOV begründet: CANTOR konnte sich nicht damit abfinden, daß BUBNOV ihm in bezug auf die Boethius zugeschriebene Geometrie II widersprach, die CANTOR für echt hielt,

---

2 Die Handschriften des Corpus agrimensorum Romanorum, in: Abh. d. K. Preuß. Ak. d. Wiss., phil.-hist. Kl., 1911, Nr. 2 (im folgenden zitiert als: THULIN, Handschriften); Zur Überlieferungsgeschichte des Corpus agrimensorum. Exzerptenhandschriften und Kompendien, in: Göteborgs Kungl. Vetensk. och Vitterh.-Samh. Handlingar, Fjärde följden, XIV. 1, Göteborg 1911 (im folgenden zitiert als: THULIN, Überlieferungsgeschichte); Humanistische Handschriften des Corpus agrimensorum Romanorum, in: Rh. Mus. 66, 1911, 417–451.

während Bubnov überzeugend nachwies, daß es sich um eine Fälschung handelt, die erst nach Gerbert entstanden sein kann. Die Frage nach der Echtheit ist bekanntlich für die Geschichte unseres (indisch-arabischen) Zahlsystems wichtig. Um seine Position zu halten, versuchte Cantor, Bubnovs Arbeiten auf diesem Gebiet geringschätzig darzustellen, und da man trotz der Kritik an Cantors Vorlesungen, die etwa Eneström erhob, dieses Werk als maßgebliche Darstellung der Mathematikgeschichte ansah, wurde Bubnovs Leistung ins Abseits gedrängt.

Ich glaube, es war nötig, auf die mathematikhistorischen Arbeiten zum Corpus agrimensorum relativ ausführlich einzugehen, einerseits, um zu zeigen, daß unser heutiges Wissen immer noch durch Cantor geprägt ist, andrerseits, um zu begründen, daß eine Darstellung der Mathematik der Agrimensoren von den Informationen auszugehen hat, die Bubnov liefert. Insofern wird sich mein Vortrag wesentlich auf Bubnov stützen, daneben aber natürlich auch Cantors Ergebnisse einbeziehen. Ich benutze ferner die kürzlich erschienene Darstellung von H. Gericke: Mathematik im Abendland[3]. Gericke räumt im Abschnitt: Mathematik bei den Römern auch den Agrimensoren ihren verdienten Platz ein und analysiert dabei Texte, die bei Bubnov ediert, danach aber nicht weiter beachtet wurden.

Das Ziel meiner Untersuchung ist es, die wesentlichen mathematischen Inhalte in den Schriften des Corpus agrimensorum darzustellen. Dabei werde ich mich hauptsächlich auf solche Texte beschränken, die in den ältesten Handschriften (AB) existieren, allerdings auch kurz auf Schriften eingehen, die nur in der jüngeren Redaktion (PG), den Mischhandschriften (EFN) und den Exzerptenhandschriften (X^I, X^II) überliefert werden. Im Anschluß daran werde ich etwas über die möglichen Quellen sagen.

Vorweg noch einige Bemerkungen über das, was ich hier unter „Mathematik" verstehe. Zunächst das, was ich nicht behandeln möchte: Es ist nicht meine Absicht, auf technische Details bei den Vermessungen der Agrimensoren einzugehen. So werde ich nicht die Methoden schildern, den *decumanus* zu finden, Methoden, die ja auch im Corpus agrimensorum dargestellt werden[4]. Ich werde auch nicht erklären, wie man die dazu senkrechte Linie, den *cardo*, absteckte, und somit auch nichts über die Verwendung der Groma sagen. Vielmehr beschränke ich mich darauf, die geometrischen Probleme zu charakterisieren, die sich in den Texten der Gromatiker finden. Im einzelnen werde ich folgende Schriften behandeln:

1. „Pyrrhus"
2. Varro-Fragmente
3. Balbus, Expositio et ratio omnium formarum
4. Fragment aus der Geometrie des Frontinus
5. Liber podismi
6. Bruchstück über den Inhalt von Körpern
7. Epaphroditus und Vitruvius Rufus
8. Fluminis varatio

---

3 Berlin usw. (Springer-Verlag), 1990.
4 Hyginus; siehe Cantor, Agrimensoren, S. 67–69.

1. „PYRRHUS".
   Überlieferung: (A)J. Ed. BUBNOV 494.

Zunächst zu „Pyrrhus": Beim Fragment, das Pyrrhus zugeschrieben wird, kann ich mich kurz fassen. Die eine erhaltene Zeile: *Mensurarum sunt genera tria, rectum Planum solidum. Rectum ...* ist geometrisches Allgemeingut, und ob Pyrrhus wirklich der Name des Autors sein soll, ist angesichts des korrupten Textes des Arcerianus und seiner Abschriften unklar[5].

2 a. VARRO (?).
   Überlieferung: (A), E(F)N. Ed. BUBNOV 495,22–503,17.

Jetzt zu Texten, die Varro zugeschrieben werden können: In A steht am Ende eines Quaternio (Sp.184) der Titel der Geometrie Varros: *Incipit liber Marci Barronis de geometria ad Rufum feliciter Silbium*[6]. Der Text der folgenden Blätter ist verloren, doch glaubt BUBNOV, daß in E (28,19–35, 23) Reste von Varros Geometrie vorhanden sind[7]. Er schließt dies daraus, daß in A offenbar Varros Geometrie vorhanden war und daß die genannten Exzerpte in E die einzigen geometrischen Teile dieser Handschrift sind, die sich nicht im heutigen Text der Klasse AB wiederfinden lassen. Marcus Terentius Varro (116–27 v.Chr.) verfaßte bekanntlich eine Enzyklopädie De disciplinis in 9 Büchern, von denen Buch 4 die Geometrie und Buch 5 die Arithmetik betraf. Es ist gut möglich, daß die Bruchstücke, die BUBNOV nach E veröffentlichte (S.495,22–503,17), aus diesen Büchern stammen. Dann wären es die ältesten erhaltenen Texte zur römischen Mathematik.

Die Bruchstücke behandeln folgende Fragen[8]:

a) Benennung und Umrechnung der Längen- und Flächenmaße, bezogen auf die Grundeinheit Fuß[9]. Dieser Abschnitt wurde wahrscheinlich durch einen Text eingeleitet, der sich nur in dem erst kürzlich wiederaufgetauchten Codex Nansianus (N) findet: *Mensurarum agrestium vocabula ferme hec sunt: ...*[10].

b) Kreisaufgaben: Bestimmung des Umfangs aus dem Durchmesser durch Multiplikation mit 3 1/7, also unter Benutzung des archimedischen

---

5 THULIN, Handschriften, S.17; BUBNOV, S.418 (Nr.1), 494.
6 THULIN, Handschriften, S.16.
7 BUBNOV, S.418 f.
8 Nach GERICKE: Mathematik im Abendland, S.38–42.
9 BUBNOV §9–10; S.498,1–499, 22.
10 Siehe M.FOLKERTS: Zur Überlieferung der Agrimensoren: Schrijvers bisher verschollener „Codex Nansianus", in: Rhein. Mus.f. Philol., NF112, 1969, 53–70, hier S.65.

Werts[11]; Berechnung des Durchmessers aus dem Umfang, indem durch 22 dividiert und mit 7 multipliziert wird[12].

c) Bestimmung der Fläche eines *ager cuneatus* (wir würden sagen: rechtwinkligen Trapezes) in *iugera* (1 *iugerum* ist etwa 25 Ar) nach der Formel

$$F = \frac{a+b}{2} \cdot c^{13}$$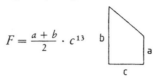

Auch die Aufgaben 7 und 8[14] behandeln Trapeze, und zwar ein gleichschenkliges und ein unregelmäßiges. Die Flächenformel ist dieselbe; allerdings wird beim gleichschenkligen Trapez nicht angegeben, wie die Höhe berechnet wird, und das allgemeine Trapez ist überbestimmt.

d) Bestimmung der Fläche eines gleichseitigen Dreiecks[15]. Der Text ist lückenhaft; auch hier wird nicht angegeben, wie die Höhe berechnet wird.

e) Bestimmung der Fläche eines Dreiecks mit den Seiten 14, 13, 15[16]. Dieses Dreieck hat eine ganzzahlige Höhe und ganzzahlige Abschnitte auf der Grundseite. Es wird aber nicht angegeben, wie die Höhe berechnet wird.

f) In den Aufgaben 11–23[17] werden Äcker behandelt mit bestimmter Länge a, Breite b und Fläche F, wobei a und b in Fuß, F in *iugera* gemessen sind, so daß stets der Umrechnungsfaktor 28800 (1 *iugerum* = 28800 Quadratfuß) auftritt. In den Aufgaben 11–13 ist F gegeben; zwischen a und b bestehen die Beziehungen: $a = b$ (Aufg. 11); $a = 4b$ (Aufg. 12); a ist bekannt, b gesucht (Aufg. 13).

In den Aufgaben 14, 19–23 ist a gegeben, b und F sind gesucht. Dabei wird verlangt:
$b = F^3$ (Aufg. 14), $b = F^2$ (Aufg. 19), $b = F^4$ (Aufg. 20), $b = F^5$ (Aufg. 21), $b = F^6$ (Aufg. 22), $b = (3 F)^2$ (Aufg. 23).
Es müssen also, um F zu bestimmen, Wurzeln gezogen werden, die in den gebrachten Beispielen stets aufgehen, weil man offenbar von F ausgegangen ist, um die Werte von a zu bestimmen. – Auffällig sind die griechischen Potenzbezeichnungen *dynamus* (x^2), *kybus* (x^3), *dynamodynamus* (x^4), *dynamokybus* (x^5) und *kybokybus* (x^6), die man in der griechischen Mathematik erst bei Diophant (3. Jh. n. Chr.) systematisch findet, und auffällig ist auch die Aufgabenstellung, die keinen Praxisbezug erkennen läßt.

Die Aufgaben 15–18 haben mit Polygonalzahlen zu tun. Im Corpus agrimensorum geht man häufig davon aus, daß die Fläche des regelmäßigen n-Ecks der Seite a mit der a-ten n-Eckzahl p_n^a gleichgesetzt wird:

$$F_n(a) = p_n^a = \frac{(n-2)\,a^2 - (n-4)\,a}{2} \qquad (n \geq 3,\ a \geq 2).$$

11 Bubnov § 1–2; S. 495, 22–496, 3.   12 Bubnov § 3; S. 496, 4–6.
13 Bubnov § 4; S. 496, 9–14. Dasselbe Verfahren und derselbe Name auch bei Columella; siehe S. 326.
14 Bubnov, S. 497, 10–20.   15 Bubnov § 5; S. 496, 15–497, 3.
16 Bubnov § 6; S. 497, 4–8.   17 Bubnov, S. 499, 23–503, 17.

Dies führt zu den Formeln

$$F_3(a) = \frac{1}{2}(a^2 + a),\ F_4(a) = a^2,\ F_5(a) = \frac{1}{2}(3\,a^2 - a),\ F_6(a) = 2\,a^2 - a$$

usw. für das gleichseitige Dreieck, Viereck, Fünfeck, Sechseck, also zu Formeln, die nur für das Quadrat stimmen. Die Polygonalzahlen sind bekanntlich ein wesentliches Element der pythagoreischen Zahlentheorie und waren während der ganzen griechischen Antike verbreitet. Vor allem durch Boethius' Arithmetik, eine lateinische Bearbeitung der griechischen Schrift des Nikomachos (ca. 100 n. Chr.), waren diese Zahlen ebenso wie die darauf aufbauenden Pyramidalzahlen auch im westlichen Mittelalter überall bekannt. Im Sinne der Pythagoreer gehörten die figurierten Zahlen der theoretischen Arithmetik an; es wäre für sie undenkbar gewesen, sie für Flächenbestimmungen zu benutzen. Mir ist nicht bekannt, daß die Polygonalzahlen vor den Texten des Corpus agrimensorum für Flächenbestimmungen gebraucht wurden; in den mathematischen Schriften der Gromatiker geschieht dies jedoch sehr oft.

2 b. VARRO (?).
  Überlieferung: B. Ed. Bubnov 503–508.

Mit den Polygonalzahlen hat noch ein zweites geometrisches Fragment zu tun, das BUBNOV Varro zuschreibt, weil es inhaltlich mit dem ersten zusammenhängt. Es wird im Arcerianus B überliefert[18]. Im ersten Kapitel dieses Textes wird angegeben, wie für Dreiecke, Vierecke usw. bis zu den Zehnecken die entsprechenden Polygonalzahlen gemäß obiger Formel gefunden werden können, im zweiten Kapitel umgekehrt die Berechnung der Seite auf der Polygonalzahl nach der auf S. 320 angegebenen Formel[19]. Im Gegensatz zu diesen rohen Vorschriften wird in Kapitel 3[20] die exakte Flächenformel für das Achteck, also mit korrekter Bestimmung der Höhe

$$h = \sqrt{\frac{s^2}{2} + \frac{s}{2}},$$

angegeben. Diese Formel begegnet uns auch in Herons Metrika[21]. – Es folgt ein Abschnitt[22], in dem zunächst der Satz Euklid II, 4 zitiert wird, nach dem ein Quadrat in zwei Teilquadrate und zwei Rechtecke (entsprechend der Formel $(a + b)^2 = a^2 + b^2 + 2\,ab$) zerlegt werden kann, und dann diese Zerlegung am Beispiel eines Quadrats mit der Seitenlänge 17 = 10 + 7 verifiziert wird. Die letzten beiden Abschnitte des Fragments[23] sind so verderbt, daß der Sinn unklar bleibt.

---

18 Spalte 157–164; ed. BUBNOV, S. 503–508.
19 BUBNOV, S. 504, 6 – 505, 20.
20 BUBNOV, S. 505, 22 – 506, 2.
21 Buch I, Kapitel 21.
22 BUBNOV, § 4; S. 506, 3–507, 2.
23 § 5: *Concha instructoria super circinum* …, § 6: *Camara infra circinum* …: BUBNOV, S. 507, 3 – 508, 24.

3. BALBUS: Expositio et ratio omnium formarum.
Überlieferung: (B)JV, PG, EFN, X^I, X^{II}, Y. Ed. LACHMANN 91,1–107,9.

Die Schrift des Balbus: Expositio et ratio omnium formarum umfaßt in LACHMANNS Ausgabe 16 Seiten[24]. Balbus begleitete nach seinen eigenen Worten Trajan auf dessen Feldzug gegen die Daker und wurde nach dem Sieg Trajans mit Vermessungsarbeiten in dem eroberten Gebiet betraut; er muß demnach um 100 n. Chr. gelebt haben. Das Wort *forma* im Titel bedeutet nicht „Grundriß", sondern „Figur", und die Schrift ist keine Anleitung zum Zeichnen, sondern eine Darlegung über die geometrischen Figuren. Sie ist die älteste erhaltene Abhandlung in lateinischer Sprache über die Anfangsgründe der Geometrie. Balbus verfaßte sie, weil es ihm schrecklich (*foedum*) vorkam, wenn er auf die Frage, wie viele Arten von Winkeln es gebe, antworten mußte: „viele"[25], und beabsichtigte, in dieser Schrift all die Dinge zu erklären, mit denen professionelle Feldmesser zu tun hatten[26]. Tatsächlich beginnt das Werk mit der Aufzählung der Maße; danach werden die geometrischen Grundbegriffe: Punkt, Linie, Fläche, Körper, Winkel, Figur erörtert. Man erkennt Anklänge an die Definitionen des Euklid, aber das Ganze ist praxisorientierter. So werden bei den Linien drei *genera* unterschieden: *rectum, circumferens, flexuosum*[27], wobei *circumferens* sich offenbar auf den Kreisumfang bezieht, während *flexuosus* jede irgendwie krumme Linie bezeichnen kann. Ein anderer Unterschied zu Euklid: Bei Euklid ist eine „Grenze" das, worin etwas endet, bei Balbus dagegen das, bis wohin das Eigentumsrecht reicht[28]. Die Erklärungen und Einteilungen erinnern oft an Heron. Die Schrift des Balbus enthält keinen Abschnitt über die rechnende Geometrie, etwa über die Berechnung von Flächeninhalten. Da das Werk aber nur bruchstückhaft überliefert wird, könnte am Ende ein entsprechender Abschnitt verlorengegangen sein.

4. Fragment aus der Geometrie des FRONTINUS.
Überlieferung: (B) JV. Ed. LACHMANN 107,10–108,8.

In LACHMANNS Ausgabe folgt nach dem eigentlichen Balbus-Text noch ein Stück, das nach THULINS Ansicht[29] ein Fragment oder Auszug aus der Geometrie des Frontinus ist[30]. Dieser Abschnitt, der nur in den Humanistenhandschriften JV überliefert wird, beschäftigt sich u. a. mit Figuren, die in Kreise einbeschrieben werden. Der Text befindet sich in einem korrupten

---

24 S.91,1–107,9.
25 LACHMANN, S.93,11–13.
26 *rerum ad professionem nostram pertinentium, in quantum potui occupatus, species qualitates condiciones modos et numeros excussi:* LACHMANN, S.93,13–15.
27 LACHMANN, S.99,3 f.
28 *Extremitas est quousque uni cuique possidendi ius concessum est, aut quousque quisque suum servat:* LACHMANN, S.98,3 f.
29 THULIN, Handschriften, S.23.
30 LACHMANN, S.107,10–108,8.

Zustand; obwohl der Name „Euklid" genannt wird, kann von exakter Mathematik keine Rede sein.

5. Liber podismi.

Überlieferung: A, EFN, X^I, X^{II}. Ed. LACHMANN 295,16 – 301,14; BUBNOV 510-516.

Der Autor des sogenannten Liber podismi[31] ist unbekannt. BUBNOV macht wahrscheinlich, daß die verbreitete Annahme, ein im übrigen unbekannter Marcus Iunius Nipsus sei der Autor, nicht begründet ist[32]. Der kurze Text beginnt mit einer Aufzählung der drei Arten der *mensurae* und der *anguli*. Es folgen sechs Aufgaben der ebenen Geometrie. Bei der ersten[33] geht es darum, bei einem stumpfwinkligen Dreieck die *eiectura* und Höhe zu berechnen, wobei die *eiectura* die Strecke zwischen dem Eckpunkt des Dreiecks und Fußpunkt der außerhalb des Dreiecks liegenden Höhe ist. Das Verfahren stimmt genau, die Zahlenwerte mit einer Ausnahme (18 statt 17) mit Heron überein[34].

$$x = \frac{b^2 - a^2 - c^2}{2\,a\,c} \qquad (a = 10,\ b = 18,\ c = 9).$$

Eine der interessantesten Aufgaben ist die zweiteilige Nr. 2[35]: Es geht zunächst darum, im rechtwinkligen Dreieck die Katheten a, b zu berechnen, wenn die Hypotenuse $c = 25$ und die Fläche $F = 150$ gegeben sind. Die Vorschrift besagt: Bestimme $\sqrt{c^2 + 4F}$ $(= a + b)$ und $\sqrt{c^2 - 4F}$ $(= a - b)$. Dann ist a gleich der halben Summe dieser Terme. Im zweiten Teil sind $a + b$, c und F gegeben. Wie zuvor wird $\sqrt{c^2 - 4F}$ $(= a - b)$ bestimmt und daraus durch Addition mit $a + b$ die gesuchte Strecke a. Das Verfahren benutzt den Satz des Pythagoras $a^2 + b^2 = c^2$, die Flächenformel des rechtwinkligen Dreiecks $4F = 2\,a\,b$ und die binomische Formel $(a \pm b)^2 = a^2 \pm 2\,a\,b + b^2$. Eine Parallele bei Heron oder in anderen frühen Schriften ist mir nicht bekannt. Das Verfahren erinnert sehr stark an babylonische Aufgaben in Verbindung mit der Lösung quadratischer Gleichungen; auch dort werden oft zwei Unbekannte x und y berechnet, indem man zunächst $x + y$ und $x - y$ bestimmt und dann die Summe bzw. Differenz dieser Terme bildet.

Über die letzten vier Aufgaben möchte ich nicht sehr viel sagen, da sie sich alle auch in Herons Geometrie finden. Bei Nr. 3 geht es darum, ein spitzwinkliges Dreieck aus den Seiten 13, 14, 15 zu berechnen. Dies geschieht analog wie in Aufgabe 1 beim stumpfwinkligen Dreieck[36]. – In Nr. 4

---

31 ed. LACHMANN, S. 295, 16–301,14; BUBNOV, S. 510–516.
32 BUBNOV, S. 510, Anm. 1. 33 BUBNOV, S. 510, 12 – 511, 11.
34 Geometria, § 12, 33–36: Heronis Alexandrini opera quae supersunt omnia (im folgenden zitiert: Heron, Opera), Bd. 4, Leipzig 1912, S. 250-252.
35 BUBNOV, S. 511, 12 – 513, 6.
36 BUBNOV, S. 513, 6–514, 3; Heron, Opera 4, S. 238-240, § 12, 9–13.

sucht man, ausgehend von einer geraden bzw. ungeraden Zahl a, pythagoreische Zahlentripel a, b, c. Bei ungeradem a bildet man $1/2\,(a^2{-}1) = b$. Dann ist a, b, c mit $b + 1 = c$ ein pythagoreisches Tripel[37]. Man benutzt hierbei den Satz des Pythagoras und die Identität

$$a^2 + \left(\frac{a^2-1}{2}\right)^2 = \left(\frac{a^2+1}{2}\right)^2.$$

Auch an dieser Stelle gibt es Anklänge an babylonische Methoden; bekanntlich besaßen die Babylonier ein ähnlich strukturiertes Verfahren, um pythagoreische Zahlen zu finden. – Die Aufgabe 5 lehrt die Bestimmung der Fläche eines beliebigen Dreiecks mit Hilfe der sogenannten Heronischen Dreiecksformel. Auch hier gibt es eine Parallele bei Heron, der an anderer Stelle diese Formel beweist[38]. Es ist bemerkenswert, daß im Liber podismi ein Dreieck mit den Seiten 6, 8, 10 als Beispiel genommen wird, also ein rechtwinkliges, bei dem die Fläche natürlich einfacher ohne die Heronische Dreiecksformel bestimmt werden könnte. – Die letzte (6.) Aufgabe lehrt die Berechnung der Höhe im rechtwinkligen Dreieck, wenn Katheten und Hypotenuse bekannt sind. Die Formel $h = ab/c$ benutzt die Ähnlichkeit des Teildreiecks zum ganzen Dreieck (also $h : b = a : c$)[39]. Mit dieser Aufgabe bricht der Liber podismi unvollständig ab.

6. Bruchstück über den Inhalt von Körpern.
   Überlieferung: EF. Ed. LACHMANN 296,4–26.

LACHMANN hat in die Ausgabe des Liber podismi einen Abschnitt aufgenommen, den BUBNOV als fremden Zusatz erkannt hat[40]. Bei diesem Bruchstück, das nur in der EFN-Klasse überliefert wird, geht es darum, den Inhalt von Piscinen, Fässern und (runden) Kalköfen zu bestimmen. Bei den (quaderförmig gedachten) *piscinae vel lacus* wird natürlich das Produkt aus Länge, Breite, Höhe bestimmt. Bei der Berechnung des Faßinhalts wird offenbar (der Text ist lückenhaft) das arithmetische Mittel aus den drei Durchmessern (unten, in der Mitte, oben) gebildet, quadriert und dann mit 11/14 und der Höhe multipliziert[41]. Bei den *calcaria* verfährt man ebenso wie bei den *dolei*.

7. EPAPHRODITUS und VITRUVIUS RUFUS.
   Überlieferung: A, EN, X^I, X^{II}, Y. Ed. BUBNOV 516–551.

Kommen wir jetzt zur Schrift des Epaphroditus und Vitruvius Rufus, dem umfangreichsten mathematischen Traktat im Corpus agrimensorum. Er ge-

---

37 BUBNOV, S.514,4–14; ähnlich Heron, Geometrie, Opera 4, S.218, §8,1 für ungerade Zahlen („Methode des Pythagoras"), und S.220, §9,1 für gerade Zahlen („Methode Platons").
38 BUBNOV, S.515,1–15; vgl. Heron, Opera 4, S.248–250, §12, 30–32. Beweis in den Metrika: Opera 3, Leipzig 1903, S.18–24.
39 BUBNOV, S.515,16–516,5; vgl. Heron, Opera 4, S.218–220, §8,3.
40 LACHMANN, S.296,4–26; BUBNOV, S.422, Nr.10.
41 LACHMANN, S.296,9–15.

hört der ältesten Redaktion an, da er in A vorhanden ist, wird aber auch in EN, X und Y überliefert. Bubnov, der den Text kritisch edierte[42], glaubt, Epaphroditus sei ein griechischer Geometer, kein Feldmesser, gewesen, und es gebe keinen Grund, sein Werk später als das 2. Jahrhundert anzusetzen. Dabei spielt für Bubnov eine entscheidende Rolle, daß das Wort *aera* („gegebene Zahl, von der eine Rechnung ausgeht"), vorkommt, das bei Nonius (3. Jh.) erklärt wird und demnach damals schon nicht mehr gebräuchlich war. Vitruvius Rufus, der in der Überschrift als *architecton* bezeichnet wird, könne, so Bubnov, durchaus mit dem Verfasser von De architectura identisch sein. Nach Bubnov[43] dürfte der überlieferte Text aus zwei Büchern stammen, von denen er das erste dem Epaphroditus, das zweite dem Vitruvius Rufus zuschreibt; die Reihenfolge der Kapitel ist gestört[44]. Beide Bücher hängen nach Bubnovs Meinung von Heron ab. Tatsächlich lassen sich für fast alle Kapitel Parallelen in Herons Schriften, vor allem in seiner Geometrie, finden.

Es ist hier natürlich unnötig, den Inhalt der Schrift in allen Einzelheiten wiederzugeben. Vielmehr genügt es, die Themen kurz zu nennen. Dabei gehe ich nach der vermutlichen ursprünglichen Reihenfolge vor, die Bubnov rekonstruiert hat.

Buch I beginnt mit einer Glosse, in der die Namen der verschiedenen Dreiecke genannt werden (I 1; § 14). Es folgt die Berechnung des gleichschenkligen Dreiecks (I 2; § 15), der Diagonalen im Rechteck (I 3; § 16) und der Höhe im allgemeinen Dreieck (I 4; § 17). Die nächsten Aufgaben behandeln das rechtwinklige Dreieck (I 5; § 1), das rechtwinklige Trapez (I 6; § 2) und die Flächenberechnung des gleichseitigen Dreiecks (I 7; § 3). Es schließt sich das Verfahren an, pythagoreische Zahlentripel zu finden, wenn man von einer geraden Zahl ausgeht (I 8; § 18) – dasselbe Verfahren, das wir schon aus dem Liber podismi kennen. Die Bestimmung der Höhe und Fläche des spitzwinkligen Dreiecks (I 10; § 20) wiederholt die in I 4 geschilderte Methode. All diese Verfahren haben ihre Entsprechung bei Heron. Nicht bei Heron dagegen finden sich die Aufgaben I 9 (§ 19) und I 11–18 (§ 21–28). Sie betreffen das reguläre Dreieck und die Polygone vom Fünf- bis hin zum Zwölfeck. In jedem Fall wird nach der schon bekannten Formel für die *a*-te *n*-Eckzahl p_n^a die „Fläche" $F_n(a)$ des Polygons ausgerechnet:

$$F_n(a) = p_n^a = {}^1\!/_2 [(n-2) a^2 - (n-4) a]$$

und umgekehrt aus der „Fläche" $F_n(a)$ die Seite a des Polygons:

$$a = \frac{\sqrt{8(n-2) F_n(a) + (n-4)^2} + (n-4)}{2(n-2)}.$$

Außerdem wird aus den Polygonalzahlen p_n^a die entsprechende Pyramidalzahl $P_n^a = p_n^1 + p_n^2 + \ldots + p_n^a$ nach der Formel

$$P_n^a = {}^1\!/_6 [(2 p_n^a + a)(a+1)]$$

---

42 Bubnov, S. 516–551.
43 S. 518 f., Anm. 1.
44 Buch I: Kap. 14–17, 1–3, 18–34; Buch II: Kap. 4–13, 39–44, vielleicht 35–38.

bestimmt. Diese Formeln stehen, wie erwähnt, nicht bei Heron, aber in py-
thagoreisch orientierten Traktaten über theoretische Mathematik, etwa bei
Nikomachos. – Die Fragmente aus dem 1. Buch werden abgeschlossen durch
einige Aufgaben zum Kreis, Halbkreis und zur Kugel (I 19; § 29, bis I 24;
§ 34). Sie benutzen für π den Wert 22/7 (I 19,20,22,23). In I 21 wird die
Kreisfläche F aus Umfang U und Durchmesser d nach der Formel $F =$
$^{1}/_{2}\,U \cdot {}^{1}/_{2}\,d$ bestimmt. All dies steht auch bei Heron. Neu ist die letzte Aufgabe, in ein rechtwinkliges Dreieck einen Kreis einzubeschreiben (I 24; § 34).
Sein Durchmesser ist gleich der Summe der Katheten, vermindert um die
Hypotenuse.

Das 2. Buch, das nach Bubnov von Vitruv stammen könnte[45], beginnt mit
drei Aufgaben zum rechtwinkligen Dreieck (II 1, § 4, und II 3, § 6) und Quadrat (II 2, § 5). Diese bringen gegenüber Heron nichts Neues. Es folgen zwei
Aufgaben, bei denen es darum geht, die Zahl der Bäume auf einem rechtek-
kigen Acker zu bestimmen, wenn zwischen den Reihen jeweils 5 Fuß Zwi-
schenraum besteht (II 4,5; § 7,8). Diese Aufgaben haben bei Heron keine
Entsprechung. Nach der Bestimmung des rechtwinkligen Trapezes (II 6; § 9)
und einer simplen Berechnung des Rhombus nach heronischer Manier (II 7;
§ 10) folgen drei Aufgaben, die sich mit der Ermittlung der Oberfläche eines
Berges befassen (II 8–10; § 11–13) und die man bei Heron nicht findet. Der
Berg wird dabei als Kegelstumpf gedacht. Aus den Umfängen am Fuß und
am Gipfel (in II 9 zusätzlich noch in der Mitte) nimmt man das arithmetische Mittel, multipliziert es mit dem *ascensus montis*, also der Seitenlinie des
Kegelstumpfes, und erhält die Fläche. Bei der letzten Aufgabe handelt es
sich um einen schiefen Berg, der links und rechts unterschiedlich ansteigt.
Man bildet das Mittel aus den beiden *ascensus* und rechnet dann wie zuvor.

Es folgen ein paar Paragraphen, die sich auf die Unterteilung und Arten
von Maßen sowie die verschiedenen Winkel beziehen (II 11–14; § 35–38).
Diese Teile erinnern an Balbus. Neu hingegen scheinen zwei Abschnitte zu
sein, die Formeln für die Summe der Quadratzahlen (II 15; § 39) und der
Kubikzahlen (II 18; § 42) angeben:

$$1^2 + 2^2 + \ldots + n^2 = {}^{1}/_{6}\,n\,(n+1)\,(2n+1) \qquad \text{und}$$
$$1^3 + 2^3 + \ldots + n^3 = [{}^{1}/_{2}\,n\,(n+1)]^2 = (1+2+\ldots+n)^2.$$

Diese Formeln waren schon den Pythagoreern bekannt, die sie mit Hilfe von
Steinchen bewiesen. Zumindest die Formel für die Kubikzahlen läßt sich so-
gar schon bei den Babyloniern nachweisen, die ebenfalls mit figurierten
Zahlen gearbeitet haben dürften.

Zwei weitere Abschnitte beschäftigen sich mit der 4. bzw. 3. Potenz einer
Zahl (II 16,17; § 40,41). Interessanter als das Rechnen selbst, bei dem lediglich die Grundzahl mehrfach mit sich selbst multipliziert wird, ist die Terminologie *decem in dynamum* bzw. *in quibo*. Es werden für die Potenzen also

---

45 Tatsächlich gibt es in II 19 = § 43 die Formulierung *decumbe in dentes,* die an *procumbatur
in dentes* erinnert (Vitruv, De architectura, VIII, 1, 1).

dieselben aus dem Griechischen stammenden Ausdrücke benutzt wie im Varro(?)-Fragment.

Von den beiden abschließenden Aufgaben (II 19-20; § 43-44) behandelt die letzte die Umrechnung Quadratfuß – *iugera* und ist daher wenig interessant. Dagegen verdient der vorangehende Paragraph Beachtung. Dort wird die Höhe eines Gegenstandes ohne Benutzung der Schattenlänge bestimmt. Wie CANTOR wahrscheinlich machte[46], benutzte der hier angesprochene Beobachter vermutlich ein gleichschenklig-rechtwinkliges Dreieck als Visierinstrument, bei dem die Hypotenuse als Visierachse diente.

Fassen wir zusammen: Die meisten Aufgaben bei Epaphroditus und Vitruvius Rufus haben Entsprechungen bei Heron. Ausnahmen bilden die arithmetischen Formeln zur Bestimmung der Polygonalzahlen, der Seiten aus den Polygonalzahlen, der Pyramidalzahlen sowie die Summenformeln für Quadrat- und Kubikzahlen. In der Geometrie sind drei Probleme neu: Bestimmung der Zahl der Bäume auf einem rechteckigen Acker, der Oberfläche eines Berges und die Höhenmessung ohne Benutzung der *umbra*. Diese drei Aufgaben werden seit dem 8. Jahrhundert zu Standardproblemen der angewandten Geometrie.

8. Fluminis varatio.
Überlieferung: A, EFN. Ed. LACHMANN 285,5 - 286,10.

Das letzte nach meiner Definition „mathematische" Stück in der ältesten Rezension der Agrimensoren (AB) ist die Fluminis varatio, die in A und EFN überliefert ist[47]. In ihr geht es darum, die Breite eines (unzugänglichen) Flusses mit Hilfe von Standlinien durch Visieren zu bestimmen. Man kann schließlich auf dem Feld vor dem Fluß ein Dreieck abstecken, das mit dem anvisierten Dreieck kongruent ist, und dadurch die Breite des Flusses direkt messen. Dieses Verfahren macht einen sehr altertümlichen Eindruck; es erinnert an die Thales zugeschriebene Methode, die Entfernung der Schiffe auf dem Meer mit Hilfe von Kongruenzsätzen zu bestimmen.

9. Fragment über Felder.
Überlieferung: EFN. Ed. LACHMANN 290,6-16.

In der Mischklasse EFN gibt es ein kurzes geometrisches Fragment, das LACHMANN ohne Grund der Limitis repositio eingefügt hat[48]. Der von LACHMANN fragmentarisch herausgegebene Text und auch die Lesarten der Handschriften F und N ermöglichen keine eindeutige Interpretation[49]. Offenbar geht der Autor von einem gleichseitigen Dreieck mit der Seitenlänge 20 aus. Legt man die übliche Näherung $7/4$ für $\sqrt{3}$ zugrunde, so hat dieses

---

46 CANTOR, Agrimensoren, S. 120.
47 ed. LACHMANN, S. 285,5 - 286,10.
48 LACHMANN, S. 290,6-16.
49 Die Lesarten der drei Handschriften sind im Anhang wiedergegeben.

die Höhe 17$^1/_2$ und die Fläche 175. 52$^1/_2$ · 10 wäre also die Fläche von drei solchen Dreiecken, und die Multiplikation mit 2 würde die Figur zu einem Sechseck (nicht: Fünfeck) ergänzen. Im folgenden ist der Schreiber vermutlich in eine andere Aufgabe hineingeraten, da jetzt von einem Dreieck und einem Trapez die Rede ist. BUBNOV hält es für möglich, daß auch dieses Bruchstück auf Varro zurückgeht[50].

10. Über Sechseck und Achteck.
   Überlieferung: X^I, X$^{II.}$. Ed. BUBNOV 552–553.

Ein weiteres geometrisches Fragment, das – mindestens nach BUBNOVS Meinung – ebenfalls varronisch sein könnte[51], wird nur in den Exzerptenhandschriften X^I und X^{II} überliefert[52]. Es besteht aus drei Teilen. Zunächst wird ein Konstruktionsverfahren für ein Achteck angegeben, bei der der erforderliche Winkel von 135° entsteht, indem man ein gleichschenklig-rechtwinkliges Dreieck konstruiert und die Hypotenuse verlängert. Der zweite Teil lehrt die übliche Konstruktion des Sechsecks durch sechsmaliges Abtragen des Radius auf dem Umfang. Schließlich wird angegeben, wie man aus einem Quadrat ein reguläres Achteck bildet, indem man um die Eckpunkte Kreise mit der halben Diagonalenlänge als Radius schlägt.

11 a. Euklidexzerpte.
   Überlieferung: PG, X, Y.

Es gibt noch ein paar zusätzliche mathematische Texte in der jüngeren Handschriftenklasse PG. Wie ULLMAN gezeigt hat[53], entstand diese Redaktion in Corbie, dem Zentrum der Geometrie in der vorkarolingischen Zeit. In PG gibt es folgende mathematische Texte, die nicht in der AB-Klasse vorhanden sind:
   Zunächst einmal wurden aus Euklids Elementen die Definitionen, Postulate, Axiome und die Propositionen 1–3 des 1. Buches nach der Übersetzung des Boethius aufgenommen[54]. Sie bilden eine der vier Quellen, durch die wir Teile von Boethius' Übersetzung rekonstruieren können; die anderen drei sind Exzerpte aus Cassiodors Institutiones und die beiden dem Boethius zugeschriebenen Geometrien, von denen die frühere, die in den Agrimensorenhandschriften der Klasse X überliefert wird, ebenfalls in Corbie entstand. Insgesamt besitzen wir die meisten Enuntiationen der Bücher I–IV und die Definitionen von Buch V[55].

---

50 BUBNOV, S. 421.
51 BUBNOV, S. 421.
52 Herausgegeben von BUBNOV, S. 552 f.
53 B. L. ULLMAN: Geometry in the mediaeval quadrivium, in: Studi di bibliografia e di storia in onore di Tammaro de Marinis, IV, Roma 1964, S. 263–285.
54 PG, Nr. 13; auch in X^I.
55 Siehe hierzu M. FOLKERTS: „Boethius" Geometrie II. Ein mathematisches Lehrbuch des Mittelalters. Wiesbaden (Franz Steiner Verlag) 1970.

11 b. De iugeribus metiendis.
      Überlieferung: PG, X^I, X^{II}, Y. Ed. LACHMANN 354, 1–356, 20.

Die anonyme Schrift De iugeribus metiendis, die in PG, aber auch in X^I und X^{II} überliefert wird[56], enthält 10 oder 11 Aufgaben, bei denen es darum geht, die Flächen verschieden geformter Äcker zu berechnen[57]. Maßeinheit ist 1 *iugerum*, d. h. ein Flächenstück von 240 × 120 Fuß, also 28 800 Quadratfuß. Die Schrift bringt inhaltlich nichts Neues: In Aufgabe 1 (S. 354, 2–10) geht es um die Berechnung rechteckiger bzw. quadratischer Flächen. Bei der 2. Aufgabe (S. 354, 11–15) wird die Fläche eines Kreises vom gegebenen Umfang bestimmt, indem ein Viertel des Umfangs quadriert wird; offenbar liegt hier die naive Vorstellung zugrunde, die Fläche des Kreises sei gleich groß wie die eines umfanggleichen Quadrats. – Ähnlich ungenau ist die Vorschrift für die Flächenbestimmung eines gleichseitigen Dreiecks (S. 354, 16–19): Man bilde das Produkt aus Seitenlänge und halber Seitenlänge ($F = a \cdot {}^1\!/_2 a$). Vermutlich liegt hier die uralte, schon den Ägyptern und Babyloniern bekannte Formel für die Flächenbestimmung des Vierecks zugrunde, nach der man die arithmetischen Mittel der gegenüberliegenden Seiten multipliziert, also das unregelmäßige Viereck durch ein Rechteck approximiert:

$$(1) \qquad F = {}^1\!/_2 (a + c) \cdot {}^1\!/_2 (b + d) \,.$$

Setzt man $d = 0$, so erhält man die ebenfalls in vorgriechischer Zeit bekannte Formel für das Dreieck

$$(2) \qquad F = {}^1\!/_2 (a + c) \cdot {}^1\!/_2 b \,,$$

und dies führt beim gleichseitigen Dreieck zur hier benutzten Formel. – Dieselbe Formel wird in Aufgabe 4 (S. 354, 20–24) für einen Rhombus (*caput bubulum*, also ochsenkopfförmiges Feld) benutzt, der aus zwei gleichseitigen Dreiecken zusammengesetzt ist, so daß dessen Fläche $F = a^2$ ist, d. h. genauso groß wie die eines entsprechenden Quadrats. – Der Fall des allgemeinen Vierecks (*ager inaequalis*) wird in der nächsten Aufgabe (S. 355, 1–7) behandelt, auch hier nach der Formel (1). – Aufgabe 6 (S. 355, 8–13) beschäftigt sich mit einem mondförmigen Feld (*ager lunatus*), das oben eine bestimmte Breite hat, unten aber punktförmig ist. Auch hier wird die schon bekannte Dreiecksformel (2) benutzt. – Die Kreisfläche wird nach der Formel

$$F = d^2 \cdot 11/14$$

berechnet (Aufgabe 8; S. 355, 21–27) und entsprechend der Halbkreis, wobei bezeichnenderweise der Durchmesser *basis,* der Radius *curvaturae latitudo* heißt (Aufgabe 7; S. 355, 14–20). – Es folgt die Fläche eines Kreisabschnitts

---

56 PG, Nr. 17; siehe THULIN, Handschriften, S. 54.
57 ed. LACHMANN, S. 354, 1–356, 20; siehe CANTOR, Agrimensoren, S. 135–138.

(*ager minor quam semicirculus*: Aufg. 9, S. 356, 1-10), für die eine Näherungs-
formel angegeben wird:

$$F = \frac{1}{2}(a + b)\,b + (\frac{1}{2}a)^2 : 14.$$

Diese Formel ist aus der groben Näherung $F = \frac{1}{2}(a + b)\,b$ entstanden, die
nach Heron schon „die Alten" kannten[58]. Tatsächlich kommt diese Formel
schon in einem ägyptischen demotischen Papyrus vor, der im 3. Jahrhundert
v. Chr. oder früher geschrieben wurde[59]. Später hat man dann – so berichtet
Heron[60] – noch das Korrekturglied $(\frac{1}{2}a)^2 : 14$ hinzugefügt, um eine auch für
den Halbkreis richtige Formel zu erhalten. – Die vorletzte Aufgabe (Nr. 10;
S. 356, 11-20) behandelt die Fläche eines sechseckigen Feldes. Dieses wird
berechnet nach der Formel

$$F = \left(\frac{a^2}{3} + \frac{a^2}{10}\right) \cdot 6 \,.$$

Man nimmt für die Fläche des gleichseitigen Dreiecks also die Formel

$$F = \left(\frac{a^2}{3} + \frac{a^2}{6}\right)\left\langle = \frac{13}{30}\,a^2\right\rangle$$

an, so daß man auf die nicht schlechte Näherung $\frac{26}{15}$ für $\sqrt{3}$ kommt. Dieselbe
Formel

$$F = a^2\left(\frac{1}{3} + \frac{1}{10}\right)$$

für die Fläche des gleichseitigen Dreiecks wird auch in Herons Geometrie
benutzt[61]. – Den Schluß bildet die Inhaltsbestimmung einer quaderförmigen
archa nach der richtigen Formel $V = abc$. Dieser Text dürfte aufgrund der
handschriftlichen Überlieferung[62] wohl zu einer anderen Schrift gehören.
   Insgesamt verrät der Text De iugeribus metiendis die Hand eines in der
wissenschaftlichen Mathematik ungebildeten Verfassers. Er versucht, jede
Figur durch Mittelbildung einem Rechteck anzunähern, auch dort, wo dies
keinen Sinn ergibt, etwa bei kreis- oder mondförmigen Gebilden. Dies ist
aus der Sicht der Mathematik sicher der roheste Text im Corpus agrimensorum.

12. Auszüge aus COLUMELLA, De re rustica.
    Überlieferung: X[II].

   Es fällt auf, daß die meisten Aufgaben bzw. Formeln aus De iugeribus
metiendis sich auch in einem anderen Text finden, der erst relativ spät –
nach ULLMANS überzeugenden Argumenten in Corbie – in das Corpus agri-

---

58 Metrika I, 30; Opera 3, S. 72-74.
59 Richard A. PARKER: Demotic mathematical papyri, Providence, R. I./London 1972, Aufgabe 36.
60 Metrika I, 31; Opera 3, S. 74-76.   61 Heron, Opera 4, S. 222-224.
62 Er steht in PG bzw. X[II] an einer anderen Stelle; siehe BUBNOV, S. 422, Nr. 11. LACHMANN
   hat ihn zweimal ediert: S. 353, 6-9 und S. 356, 21-24.

mensorum aufgenommen wurde: Ich meine die mathematischen Exzerpte aus Columellas De re rustica, die in einigen Exzerptenhandschriften der Klasse X^{II} überliefert werden[63]. Columella, der unter Kaiser Claudius schrieb, behandelt zu Beginn von Buch V auch die Themen der Feldmessung, die er nach seinen eigenen Worten von anderen übernommen hat[64]. Columella geht zunächst auf die verschiedenen Ackermaße und ihre Umrechnungen sowie auf die Einteilung des *iugerum* ein. In Kapitel 2 werden dann neun Aufgaben gestellt und gelöst, die sich mit der Berechnung verschiedenförmiger Äcker befassen. Die Methoden habe ich schon an anderer Stelle erwähnt, insbesondere bei der Betrachtung von De iugeribus metiendis, so daß es hier genügt, kurze Andeutungen zu machen:

Die Aufgaben 1-3 behandeln die Flächenberechnung von Quadrat, Rechteck und Trapez (*ager cuneatus*). Dann folgt das gleichseitige Dreieck mit der Flächenformel $F = \frac{a^2}{3} + \frac{a^2}{10}$. Die Fläche des rechtwinkligen Dreiecks ist das halbe Produkt der Katheten (Aufg. 5). Für Kreis- und Halbkreisfläche (Aufgaben 6 und 7) werden dieselben Formeln wie in De iugeribus metiendis angegeben, im Fall des Halbkreises auch dieselben Namen (*curvaturae latitudo, basis*). Auch die Fläche eines Kreisabschnitts, der kleiner als ein Halbkreis ist (Aufg. 8), und des Sechsecks (Aufg. 9) hat seine Entsprechung in De iugeribus metiendis.

13. Auszüge aus IsIDOR und Exzerpte De mensuris, De ponderibus.

Nicht näher eingehen möchte ich auf Auszüge aus Isidor, Origines, die naturgemäß erst spät in das Corpus agrimensorum aufgenommen worden sein können. Vorhanden sind Exzerpte De geometria (= Origines III 10-14) in Handschriften der X^{I}-Klasse[65] und andere De mensuris (= Origines XV, Kapitel 14, 15, 13) in PG (Nr. 22), X^{I}, X^{II} und Y[66]. Von ähnlicher Qualität sind die Abschnitte De mensuris, De ponderibus und De mensuris in liquidis[67], die in (P)G (nach Nr. 23) und X^{II} vorkommen.

* * *

Ich habe versucht, Ihnen einen Eindruck über die mathematischen Probleme zu geben, die in den Schriften der römischen Feldmesser vorkommen. Wenn man aus den vielen Einzelheiten etwas Allgemeineres herausholen möchte, so kann man zunächst feststellen, daß in fast allen Fällen keine Formeln in unserem Sinn gegeben werden, sondern Musterbeispiele, aus denen man al-

---

63 Siehe THULIN, Überlieferungsgeschichte, S. 16, 17; BUBNOV, S. 423, Nr. 20.
64 V 1, 4: *desque veniam, si quid in eo fuerit erratum, cuius scientiam mihi non vindico.*
65 THULIN, Überlieferungsgeschichte, S. 9; BUBNOV, S. 423, Nr. 18.
66 ed. LACHMANN, S. 366, 10 – 370, 1. Siehe THULIN, Handschriften, S. 55; Überlieferungsgeschichte, S. 10, 11, 15, 16 f., 45; BUBNOV, S. 423, Nr. 19.
67 ed. LACHMANN, S. 371-376. Siehe THULIN, Handschriften, S. 55 unten, und Überlieferungsgeschichte, S. 15.

lerdings das Vorgehen auch für andere Zahlenwerte erkennt. Selbstverständlich enthalten die Schriften Informationen, die der Feldmesser für seine Arbeit braucht, z. B. über die Berechnung der Flächen von Dreiecken, Trapezen und allgemeinen Vierecken, Angaben über die verschiedenen Maße und ihre Umrechnungen, auch Höhenberechnungen und Längenbestimmungen bei unzugänglichen Punkten, wie sie etwa in der Fluminis varatio gelehrt werden. Es fällt aber auf, daß darüber hinaus auch eine Reihe von Fragen behandelt werden, die für den Praktiker uninteressant sind. Ich denke hier vor allem an die Berechnung der Polygone, aber auch an Aufgaben, die mit dem Kreis zusammenhängen, die manchmal praxisfremd sind. Was haben sie in einer Textsammlung für Agrimensoren zu suchen?

Diese Frage ist sicherlich nicht von der nach den Quellen zu trennen. Spätestens seit CANTOR ist es üblich, zu behaupten, das mathematische Wissen, das im Corpus agrimensorum präsentiert wird, stamme ganz oder fast ausschließlich von Heron. Tatsächlich finden sich die meisten Regeln auch in Schriften, die Heron zugeschrieben werden, aber so einfach ist meines Erachtens die Situation nicht: Zunächst einmal wissen wir heute, daß Heron um 60 n. Chr. lebte[68], während man noch vor einigen Jahrzehnten geneigt war, seine Lebenszeit wesentlich früher anzusetzen. Dies bedeutet, daß einige Schriften im Corpus agrimensorum, z. B. das Werk des Balbus, fast zeitgleich mit Heron sind, während die Texte des Varro (falls sie wirklich von ihm stammen) sogar noch vor Heron liegen. Eine einseitige Abhängigkeit von Heron ist also wenig wahrscheinlich. Dazu kommt, daß einige Schriften, die Herons Namen tragen, Überarbeitungen aus byzantinischer Zeit sind: Die Dioptra und die Metrika, vielleicht auch die Definitiones, scheinen in der überlieferten Fassung einigermaßen original zu sein, während die Geometrika, die die größten Anklänge zu den Schriften im Corpus agrimensorum zeigen, in der vorliegenden Form nicht von Heron stammen, sondern wohl eine Bearbeitung aus dem 10. Jahrhundert sind[69].

Im übrigen sind viele Rechenverfahren, die wir im Corpus agrimensorum finden, nicht nur bei Heron nachweisbar, sondern gehen auf eine viel ältere Zeit zurück. Ich erinnere nur an die grobe Formel für die Bestimmung der Fläche des allgemeinen Vierecks und die daraus abgeleitete Dreiecksformel, die uns schon in ägyptischen und babylonischen Texten begegnet, oder an die Berechnung der pythagoreischen Tripel und der Summe der Kubikzahlen, die ebenfalls schon den Babyloniern bekannt war. Hier muß man in größeren Zusammenhängen denken. Das Vermessungswesen ist ja so alt wie die geometrische Wissenschaft selbst. Wir kennen die ägyptischen Harpedonapten (Seilspanner), die beim Bau der Tempel und Pyramiden und beim Vermessen der Felder beteiligt waren. Wir wissen, daß schon die Ägypter die Himmelsrichtungen bestimmen konnten, Fundamente nivellierten und Böschungswinkel einhielten. Das Verfahren, unregelmäßige Flächenstücke in

---

68 Er berichtet über eine Mondfinsternis, die im Jahre 62 n. Chr. stattfand: Dioptra 35;
Opera 3, S. 302.
69 T. L. HEATH: A History of Greek Mathematics, Bd. 2, Oxford 1931, S. 318.

Dreiecke, Rechtecke und Trapeze zu zerlegen, um die Fläche bestimmen zu können, ist durch Pläne bei den Ägyptern und Babyloniern belegt. Griechische Geräte für das Nivellieren und Visieren, z. B. Chorobates und Asterikon, ferner die Dioptra, die Heron beschreibt, sind bekannt. Dies alles bedeutet, daß die Geometrie, die im Corpus agrimensorum gelehrt wird, in einer langen Tradition steht, die nur deshalb nicht in Einzelheiten rekonstruiert werden kann, weil vor Heron keine bedeutenden zusammenhängenden Texte dieser Art erhalten sind. Die Situation ist hier ganz ähnlich wie bei der Logistik, der praktischen Arithmetik der Griechen: Auch dort sind wir gezwungen, aus späteren Zeugnissen, oft erst aus byzantinischer Zeit, Rückschlüsse auf das griechische Wissen zu ziehen. Es war in der griechischen Antike eben nicht üblich, derartige Texte aufzuschreiben bzw. aufzubewahren.

Ich habe betont, daß die mathematischen Texte des Corpus agrimensorum auch Informationen enthalten, die für die praktische Arbeit des Vermessers unwichtig waren. Tatsächlich gibt es eine Reihe von Elementen aus der theoretischen, der wissenschaftlichen Mathematik der Griechen. Dies beginnt bei den Definitionen, die entfernt an Euklid erinnern, setzt sich fort bei den griechischen Potenzbezeichnungen, die hier – falls es sich bei dem Text wirklich um ein Fragment aus Varro handelt – früher als in den erhaltenen griechischen Texten selbst auftauchen, und führt schließlich zu den Abschnitten über figurierte Zahlen bei Epaphroditus und Vitruvius Rufus. Um die Bedeutung dieser theoretischen Teile besser einschätzen zu können, müßte man mehr über das Selbstverständnis der Geometrie in der griechisch-römischen Antike wissen. Klar ist jedenfalls, daß sich die Ansichten über den Zweck der Geometrie seit der griechischen Klassik geändert haben. Während Euklid in den Elementen nur Linie, gerade Linie, Kreis, die verschiedenen Winkel, Dreiecke und Vierecke definierte und alle Eigenschaften von Figuren auf die Eigenschaften von Geraden und Kreisen zurückführte, die in den Postulaten festgelegt waren, hat Heron in seinen Definitiones auch andere Linien erklärt und eine systematische Ordnung erreicht, die nach Proklos[70] schon von Geminos (1. Jh. v. Chr.) angestrebt worden war. GERICKE sagt treffend[71]: „Gegenüber der Deduktion von Sätzen aus Axiomen tritt jetzt die Beschreibung der Mannigfaltigkeit der Formen in den Vordergrund, ähnlich wie bei Nikomachos die Beschreibung der Arten der Zahlen." In diesem Sinne finden wir im Corpus agrimensorum immer wieder eine geordnete Aufzählung und Beschreibung der Begriffe und Figuren. Dies ist jetzt der theoretische Teil der Geometrie, nicht mehr, wie bei Euklid, eine Deduktion aus den Grundbegriffen mit Hilfe mathematischer Methoden. Es ist gut möglich, daß diese neue Auffassung von der Geometrie, vor allem vom theoretischen Teil, mit dem praktischen Sinn der Römer

---

70 Euklid-Kommentar zu Def. 4, ed. G. FRIEDLEIN: Procli Diadochi in primum Euclidis Elementorum librum commentarii, Leipzig 1873, S. 111.
71 Mathematik im Abendland (s. Anm. 3), S. 36.

und ihrem Streben nach Ordnung zusammenhängt. Denkbar war eine solche Umdeutung aber nur in Verbindung mit den geometrischen Leistungen, die Euklid, Archimedes und andere Griechen erzielt hatten.

Aus der theoretischen Geometrie, wie sie Heron und die Agrimensoren verstanden, entstand dann im Hochmittelalter die Geometria speculativa. Sie steht im Gegensatz zur Geometria practica[72], die schon bei Heron[73] in Streckenmessung (euthymetrikon), Flächenmessung (embadometrikon) und Körpermessung (stereometrikon) untergliedert wird. Diese Dreiteilung findet man auch im Corpus agrimensorum. Die Aufgaben, die ich Ihnen vorgestellt habe, beziehen sich naturgemäß fast alle auf das zweite Gebiet, die Flächenmessung; viel seltener wird die Strecken- oder Längenmessung behandelt, für die üblicherweise Visiereinrichtungen oder andere Geräte benutzt werden, und noch seltener beschäftigt man sich im Corpus agrimensorum mit der Körperberechnung. Die geometrischen Schriften der Agrimensoren haben die Geometrie im westlichen Mittelalter sehr stark beeinflußt; es gibt Traditionen, die ohne Bruch von der Römerzeit bis ins 16. und 17. Jahrhundert reichen. Aber dies näher auszuführen, wäre Aufgabe eines anderen Vortrags.

---

72 Siehe hierzu Stephen K. Victor: Practical geometry in the high Middle Ages. Artis cuiuslibet consummatio and the Pratike de geometrie. Transl. and commentary by the author. Philadelphia 1979.
73 Geometrica; Opera 4, S. 180.

Anhang I
Zusammenstellung der mathematischen Probleme

Bedeutung der Abkürzungen:
Varro (1): Fragment, ed. Bubnov 495,22–503,17. Kapitelzählung nach Bubnov
Varro (2): Fragment, ed. Bubnov 503–508. Kapitelzählung nach Bubnov
Pod. = Liber podismi, ed. Bubnov 510–516. Kapitelzählung nach Bubnov
Ep. = Epaphroditus und Vitruvius Rufus, ed. Bubnov 516–551. Buch- und Kapitelzählung nach Bubnov
Iug. met. = De iugeribus metiendis, ed. Lachmann 354,1–356,20. Zählung nach Absätzen bei Lachmann
Colum. = Columella, Buch 5, Kapitel 2. Die neun Aufgaben wurden numeriert

1. Terminologisches, Metrologisches

Benennung und Umrechnung von Maßen: Varro (1), 9–10
Einteilung und Umrechnung der Maße, Arten der Winkel: Ep. II 11–14
Namen der Dreiecke: Ep. I 1
Umrechnung Quadratfuß – *iugera*: Ep. II 20

2. Ebene Geometrie

2.1 Rechtwinkliges Dreieck
Beziehungen zwischen a, b, c (Satz des Pythagoras): Ep. I 5; Ep. II 1

$h = \frac{ab}{c}$: Pod. 6

$F = \frac{ab}{2}$: Ep. I 5; Ep. II 3; Colum. 5

Bestimmung von a, b aus c, F mit $a \pm b = \sqrt{c^2 \pm 4F}$: Pod. 2 a
Bestimmung von a, b aus $a + b, c, F$: Pod. 2 b

2.2 Gleichseitiges Dreieck
$F = \frac{a \cdot h}{2}$

(Bestimmung von h fehlerhaft oder fehlend): Varro (1), 5; Ep. I 7

$F = \frac{a^2}{3} + \frac{a^2}{10}$: Colum. 4

$F = a \cdot \frac{a}{2}$: Iug. met. 3

2.3 Sechseck
$F = \left(\frac{a^2}{3} + \frac{a^2}{10}\right) \cdot 6$: Iug. met. 10; Colum. 9

2.4 Gleichschenkliges Dreieck
Bestimmung der Höhe aus Basis und Schenkel mit Hilfe des Satzes des Pythagoras; Flächenbestimmung: Ep. I 2

2.5 *Spitzwinkliges Dreieck*
Bestimmung der *praecisura* x aus den Seiten *a, b, c*:

$$x = \frac{a^2 + b^2 - c^2}{2b} \; ;$$

Bestimmung der Höhe *h* mit Hilfe des Satzes des Pythagoras;

$$F = \frac{c \cdot h}{2} : \text{Pod.3}; \ \text{Ep.I4}; \ \text{Ep.I10}$$

2.6 *Stumpfwinkliges Dreieck*

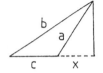

eiectura $x = \frac{b^2 - a^2 - c^2}{2ac} : \text{Pod.1}$

2.7 *Allgemeines Dreieck*

$$F = \frac{a \cdot h}{2} \ (\text{Bestimmung von } h \text{ fehlt}): \text{Varro (1), 6}$$

$$F = \sqrt{s(s-a)(s-b)(s-c)} : \text{Pod.5}$$

2.8 *Mondförmiges Feld*

$$F = \frac{a+c}{2} \cdot \frac{b}{2} : \text{Iug.met.6}$$

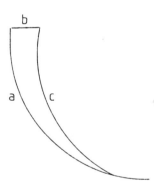

2.9 *Quadrat*
$F = a^2 : \text{Ep.II2}; \ \text{Iug.met.1}; \ \text{Colum.1}$

2.10 *Rechteck*
$F = ab : \text{Colum.2}$
Diagonale $d = \sqrt{a^2 + b^2} : \text{Ep.I3}$
Zahl der Bäume auf rechteckigem Acker: Ep.II 4–5
Rechteckige Felder: siehe unter 4.2

2.11 *Rechtwinkliges Trapez*
$F = \frac{a+b}{2} \cdot c : \text{Varro (1), 4; Ep.I6; Ep.II6; Colum.3}$

$s = \sqrt{c^2 + b^2 + a^2 - 2ab} : \text{Ep.I6}$

2.12 *Gleichschenkliges Trapez*
$F = \frac{a+b}{2} \cdot h : \text{Varro (1), 7}$

2.13 *Allgemeines Trapez*

$F = \frac{a+b}{2} \cdot h$: Varro (1), 8

2.14 *Rhombus*

$h = \sqrt{a^2 - (\frac{d}{2})^2}$; $F = d \cdot h$: Ep. II 7

$F = a^2$, wenn $2h = a$: Iug. met. 4

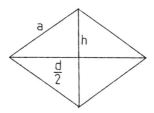

2.15 *Allgemeines Viereck*

$F = \frac{a+c}{2} \cdot \frac{b+d}{2}$: Iug. met. 5

2.16 *Achteck*
Fläche mit Bestimmung der Höhe $h = \sqrt{\frac{s^2}{2} + \frac{s}{2}}$: Varro (2), 3

2.17 *Flächenbestimmung der Polygone*
siehe auch unter 4.1

2.18 *Kreis*
$U = 3\frac{1}{7}d$: Varro (1), 1–2

$d = \frac{U}{22} \cdot 7$: Varro (1), 3

$F = d^2 \cdot \frac{11}{14}$: Ep. I 20; Iug. met. 8; Colum. 6

$F = \frac{U}{2} \cdot \frac{d}{2}$: Ep. I 21

$F = (\frac{U}{4})^2$: Iug. met. 2

$d = a + b - c$ (Bestimmung des Durchmessers d des in ein rechtwinkliges Dreieck einbeschriebenen Kreises): Ep. I 24

2.19 *Halbkreis*
$F = d \cdot r \cdot \frac{11}{14}$ (d = *basis*, r = *curvatura* oder *curvaturae latitudo*): Ep. I 22–23;
    Iug. met. 7; Colum. 7

2.20 *Kreisabschnitt*
$F = \frac{a+b}{2} \cdot b + \frac{(\frac{a}{2})^2}{14}$: Iug. met. 9;
    Colum. 8

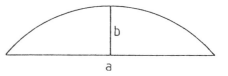

3. *Stereometrie*

3.1 *Kegelstumpf*
Bestimmung des Mantels durch Mittelbildungen (Oberfläche von Bergen):
    Ep. II 8–10

3.2 *Kugel*

Oberfläche $O = \frac{(2d)^2 \cdot 11}{14}$: Ep. I 19

4. *Arithmetik, Zahlentheorie*

4.1 *Polygonalzahlen, Pyramidalzahlen*

Bestimmung der Polygonalzahl p_n^a und Identifizierung mit der Fläche $F_n(a)$
des n-Ecks der Seite a

$$F_n(a) = p_n^a = \frac{(n-2)\,a^2 - (n-4)\,a}{2} :$$

Varro (1), 15–18 ($n = 3, 5, 6, \ldots$); Varro (2), 1 ($n = 3, 5, \ldots, 10$);
Ep. I 9 a ($n = 3$); Ep. I 11 a, b – 18 a, b ($n = 5, 6, \ldots, 12$)

Bestimmung der Seite a aus der Polygonalzahl $p_n^a = F_n(a)$

$$a = \frac{\sqrt{8\,(n-2)\,F_n(a) + (n-4)^2} + (n-4)}{2\,(n-2)} :$$

Varro (2), 2 ($n = 3, 5, \ldots, 10$); Ep. I 9 b ($n = 3$);
Ep. I 11 c – 18 c ($n = 5, 6, \ldots, 12$)

Bestimmung der Pyramidalzahlen aus den Polygonalzahlen

$$P_n^a = p_n^1 + p_n^2 + \ldots + p_n^a = \frac{(2 p_n^a + a)(a + 1)}{6} :$$

Ep. I 11 d – 18 d ($n = 5, 6, \ldots, 12$)

4.2 *Sonstiges*

Rechteckige Felder (Beziehungen zwischen Länge a, Breite b, Fläche F;
griechische Potenzbezeichnungen): Varro (1), 11–14, 19–23

Berechnung von 10^4 und 10^3; griechische Potenzbezeichnungen: Ep. II
16–17

Summe der Quadratzahlen: Ep. II 15 (siehe auch 4.1)

Summe der Kubikzahlen: Ep. II 18 (siehe auch 4.1)

Erzeugung pythagoreischer Zahlentripel: Pod. 4; Ep. I 5 a; Ep. I 8

5. *Sonstiges*

$(a + b)^2 = a^2 + b^2 + 2\,ab$ (Euklid II 4): Varro (2), 4

Höhenmessung ohne Benutzung der *umbra*: Ep. II 19

334

Anhang II
Text des Fragments über die Felder (Lachmann, S. 290, 6–16)

1. LACHMANNS Text (nach Handschrift E):

Actus tamen invaserunt a ‹   › in re iugera LXV. sic
querimus. semper in uad ‹   › duco quater id est LXV.
fit. CLX. huius sumo partem‹   ›XX. fit XIII. erit
CL. catectus trigoni actus‹   ›XIII. in sequentem autem iunctum trigono ad tra‹   ›cta qualiter erunt
CCCC. huius parte XX erit basis ‹   › uv. VII. erit
contraria basis actus VII. similiter in‹

2. Text der Handschrift F, f. 23 v:

Actus tamen in base sunt XX. sicut pata L fit.I.C quinque in re iugera LXV
sic quaerimus semper in vadum duco quater id est LXV. fit CLX. Huius sumo
partem XX fit XIII. erit CL catectus trigoni actus XIII. In sequentem autem
iunctum trigono ad trapizeo similiter quae si fuerint in trapizeo iugera C. iugera
dicta qualiter erint Huius parte XX erit basis. Deducto contrario idest XX. fit
reliquum VII. erit contraria basis actus VII. Similiter in reliquis pedibus fuerint
CC.

3. Text der Handschrift N, f. 37 r:

Actus tamen in base sunt XX. sic utputa in pentagono LIIS bis ducti. faciunt
CV. qui in se ducti. faciunt iugera LXV. cathetum querimus. Embadum duco
quater. id est LXV. fiunt CCLX. Huius summe pars XX^a fit XIII. erit cathetus.
In trigono sunt actus XLII. iugera CL. In sequentem actum iunctum trigono ac
trapezeo similiter. Si autem fuerint in trapezea iugera. ducta quater erunt
CCCC. Horum pars XXa. hoc est XX. erit basis. Deducto contrario idest XX.
fit reliquum VII. erit contraria basis actui ‹?› VII. Similiter in reliquis pedibus.
si fuerint CC.

In F und N steht die folgende Figur:

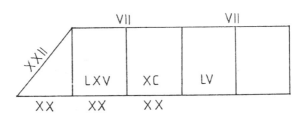

III

De arithmeticis propositionibus.
A Mathematical Treatise
Ascribed to the Venerable Bede[*]

Among the works which are ascribed to Beda Venerabilis (672/3 to 735) in the early printed editions[1] is the short treatise entitled *De arithmeticis propositionibus* (here abbreviated as *Ar. pr.*). Until now it has been neglected by mediaevalists and historians of mathematics, though only few mathematical works are known from 8th- and 9th-century Europe and the content of the work is of considerable interest. Of the four parts the first three give procedures to find numbers. The procedures outlined appear to be described in Western Europe for the first time in this treatise; afterwards they became a constant feature of recreational mathematics until early modern times. The fourth section gives rules for the addition of positive and negative numbers. There is no other text like this in Western Europe before the 15th century; comparable rules with negative numbers before this time are known only in Indian and Chinese sources. Parts 1 and 2 of this article are concerned with the transmission of the text and the affiliation of the manuscripts. In Part 3 the importance of this treatise for the history of mathematics is discussed. In Part 4 questions of authorship and the date of composition are treated.

1 Transmission

Nineteen manuscripts of the *Ar. pr.* are known[2]. They are described with some details of their contents in the following.

---

[*]This is an enlarged and updated English version of the article "Pseudo-Beda: De arithmeticis propositionibus. Eine mathematische Schrift aus der Karolingerzeit", *Sudhoffs Archiv* 56 (1972), 22–43.

[1]Beda 1563, cols. 133–135; Beda 1688, cols. 100–102; Beda 1850, cols. 665–668. I have not had access to the reprint of Hervagius' edition (Cologne 1612) mentioned by Jones, 'Pseudepigrapha', p. 18.

[2]Two further codices could be eliminated as being copies of existing manuscripts: New York, Columbia University, Plimpton 250, f. 159rv, 15th century (copy of V_4), and Vat. lat. 4539, ca. f. 80, 17th c. (copy of M or a sister of it). Since the text of the *Ar. pr.* is short and transmitted anonymously, other manuscripts may exist. Jones ('Pseudepigrapha', p. 51) mentions only three manuscripts.

2 Ps. Bede: *De arithmeticis propositionibus*

1. B = Berlin, Staatsbibliothek, Phillipps 1832, f. 55v, 9th c.

According to V. Rose, this manuscript was written in Metz in the 9th or 10th century[3]. Later owners: Collegium Parisiense Societatis Iesu (Collège de Clermont), no. 633 (up to 1764); G. Meerman, no. 718 (up to 1824); Th. Phillipps, no. 1832. The codex, which Rose described meticulously[4], principally contains Bede's writings on *computus* (finding the date of Easter and other calendrical matters), his well-known history of the church, and also Aratus' *De prognosticis* and Germanicus' reworking of Aratus' *Phaenomena*. Some texts in this manuscript belong to the so-called *Liber calculationis*, an encyclopaedic work on computistical matters in three books, which was written in 818 and set going by Arn, abbot of St. Amand (782) and later (785) archbishop of Salzburg[5]. To fill the space at the end of Bede's *De ratione temporum*, one of the *computus* writings, on f. 54v–55v three short texts are added in the same hand which also wrote *De ratione temporum*. They are: *Argumentum quomodo feria qua dominus passus est invenitur*; a note on *azymus panis*; and finally *Ar. pr.* 1–2, which breaks off incomplete at the end of page 55v.

2. B_1 = Bordeaux, Bibliothèque municipale, Ms. 11, f. 209v–210v, shortly after 1104

This codex was written in the Abbey La Sauve-Majeure (near Bordeaux), shortly after 1104[6]. It contains numerous theological and historical works. On f. 190–212 there are various treatises on computus and other calendrical matters. They are excerpts of the computistical-astronomical encyclopaedia of 809 in seven books, which was compiled at the behest of Charlemagne and might have been written in Aachen at his court[7]. In this part, on f. 209v–210v, is a copy of *Ar. pr.* 4 and 1–3.

---

[3] Rose, 'Meerman-Handschriften', p. 289. Borst considers that the extracts of the computistical encyclopaedia in three books (see note 5) were written in Laon about 873 (Borst, 'Naturgeschichte', p. 175, n. 23).

[4] Rose, 'Meerman-Handschriften', pp. 289–293.

[5] For this 3-book encyclopaedia see Borst, 'Naturgeschichte', pp. 170–175. Book I is based upon books I–IV of the 7-book Computus of 809 (see note 7) and is augmented by long citations from Bede; book II deals with astronomy, meteorology and metrology; book III contains Bede's *De natura rerum*.

[6] Description in *Catalogue général* 1894, pp. 7–16. See Borst, 'Naturgeschichte', p. 237, n. 73; Borst, 'Kalenderreform', p. XXV (g 3); Borst, 'Reichskalender', pp. 283f.

[7] This Carolingian handbook contains calendars, chronological writings, computistical texts, writings on astronomy, extracts from encyclopaedical works (Isidore, Macrobius, Martianus Capella) and, at the end, Bede's *De natura rerum*. For the origin and contents of the 7-book encyclopaedia see Borst, 'Naturgeschichte', pp. 156–165, and Borst, 'Kalenderreform', pp. 146f.

3. H = London, British Library, Harley 3017, f. 152r–153v, 180rv, between 862 and 864

This manuscript, which was written between 862 and 864 in Fleury[8], contains mainly texts on calendrical and computistical matters[9]. After a table on the Greek alphabet and the numerical value of its letters (f. 151v) follow, on f. 152r–153v, nos. 1–3 of the *Ar. pr.* Much later in this manuscript, on f. 180rv, we find *Ar. pr.*, no. 4. It is preceded by an arithmetical treatise which is in the printed editions attributed to Bede[10] and by a short text on geometry and geometrical figures[11].

4. L = London, British Library, Burney 59, f. 11rv, 1st half 11th c.

The codex was written in the first half of the 11th century in western Germany or eastern France[12]. According to a handwritten note in the catalogue of the British Library it belonged to the Benedictines in Dijon[13]. The manuscript contains on f. 2v–7v the Romulus text of Aesop's fables[14]. There follow the *Propositiones ad acuendos iuvenes* ascribed to Alcuin (f. 7v–11r)[15] and *Ar. pr.* 1–3 (f. 11rv). The manuscript ends with the following conundrum in verse:

> *Dum iuvenis fui, quattuor fontes siccavi.*
> *Cum autem senui, montes et valles versavi.*
> *Post mortem meam vivos homines ligavi.*

5. M = Montpellier, Bibliothèque de la Faculté de Médecine, H 491, f. 108r–110r, beginning 11th c.

The codex was written at the beginning of the 11th century in eastern France[12]. Later owners: Pierre Pithou (from Troyes, 1539–1596); Oratoire de Troyes; since the time of the French Revolution in Montpellier. There is a good description of the contents of the manuscript by N. Bubnov

---

[8]See Borst, 'Naturgeschichte', p. 117, n. 94; Borst, 'Kalenderreform', p. XX (C 5); Borst, 'Reichskalender', pp. 135f.

[9]See *Catalogue of the Harleian Manuscripts*, p. 725. Some of them are based upon the 7-book *computus* of 809 (note 7); see Borst, 'Naturgeschichte', p. 185, n. 48.

[10]It begins on f. 173r with the words "Arithmetica est disciplina numerorum". See Jones, 'Pseudepigrapha', p. 48, and Beda 1850, cols. 641–648.

[11]On f. 178v–180r. Incipit: "Geometriae disciplina primum ab egiptiis reperta".

[12]The late Professor Bernhard Bischoff, Munich, was so kind as to inform me about the date and provenience of the manuscripts $LM M_1 M_2 M_3 P_1 S V_1 V_2 V_3 V_4 W$.

[13]There is no printed description of the codex.

[14]The so-called "Recensio Gallicana"; see Thiele, 'Aesop'.

[15]Critical edition in Folkerts, 'Propositiones ad acuendos iuvenes'; reprinted in this volume, item V.

4   Ps. Bede: *De arithmeticis propositionibus*

('Gerberti opera', pp. XXXIXf.). It contains mostly writings on the abacus (by Gerbert, Bernelinus, Herigerus), the correspondence of Gerbert and Adelboldus and also extracts of the anonymous continuation of Gerbert's *Geometry*. After a text on music (Bernelinus [?], *Commensuralitas fistularum et monochordi*, f. 81r–93v) follow the *Propositiones ad acuendos iuvenes* ascribed to Alcuin (f. 94r–108r) and *Ar. pr.* 1–3 (f. 108r–110r). After this come, without title, some interesting texts on finding numbers, on perfect numbers and on the Josephus problem.

6. M_1 = Munich, Bayerische Staatsbibliothek, Clm 18765, f. 1r–2r, mid-10th c.

This manuscript, which was written in the mid-10th century in western Germany[12], came from Tegernsee (shelfmark: Teg. 765) to Munich. It starts with *Ar. pr.* 1–3 and continues with theological writings of Boethius and his *Consolatio*[16].

7. M_2 = Munich, Bayerische Staatsbibliothek, Clm 14689, f. 13rv, ca. 1100

The codex was written in the 11th to 12th century in Bavaria[12]. It was brought from St. Emmeram, near Regensburg (shelfmark: Em. G 73), to Munich. It contains, besides metrological and some astronomical texts, mostly writings on the abacus (Bernelinus, Gerbert, Herigerus, Gerlandus) and other arithmetical texts[17]. The *Ar. pr.* 1–3 lies between a fragment on music and the *Propositiones ad acuendos iuvenes* attributed to Alcuin.

8. M_3 = Munich, Bayerische Staatsbibliothek, Clm 14836, f. 78v–80r, 11th c.

This manuscript, from the 11th century, was also once in the monastery of St. Emmeram (Em. k. 6); perhaps it was written there[12]. Its contents are described in the Munich manuscript catalogue (*Catalogus* 1876, pp. 240f.) and by Bubnov ('Gerberti opera', pp. XLVI–XLVIII); there is also much information in a long article by Curtze that is devoted to this manuscript (Curtze, 'Handschrift'). Like M_2, this manuscript contains mostly arithmetic and geometric texts and (from f. 136r) metrological and astronomical excerpts[18]; there are also extracts (f. 83–107) from the *agrimensores* (Roman land surveyors). After the *Ar. pr.* 1–3 follow a letter-puzzle and

---

[16]See the description in *Catalogus* 1878, p. 206.

[17]See *Catalogus* 1876, pp. 217f., and Bubnov, 'Gerberti opera', pp. XLIII–XLV.

[18]Most of them come from the computistical encyclopaedia in seven books (see note 7): Borst, 'Naturgeschichte', p. 210, n. 4.

two problems in recreational mathematics. These items and the *Ar. pr.* are embedded in extracts from the anonymous continuation of Gerbert's *Geometry*.

9. P_1 = Paris, Bibliothèque nationale de France, lat. 3652, f. 10r, 12th c.

This manuscript is formed from five different parts[19]. The first ten folios, which formerly were a separate manuscript, are apparently from the 12th century and were written in south-western France[12]. They contain Boethius' *De fide catholica*, his *Liber contra Euthychen et Nestorium*, and at the end an extract from *Ar. pr.* (no. 1).

10. P_2 = Paris, Bibliothèque nationale de France, lat. 5543, f. 147v–148r, ca. 847

The main part of the manuscript (f. 1–157) was written in the 9th century (perhaps about 847), probably in Fleury – it belonged to the monastery there in the 10th century[20]. It contains various texts, mostly on astronomy. Some remarks and marginal notes refer to Fleury and Sens. This codex contains Bede's writings on chronology, his *Historia ecclesiastica* and numerous, mostly anonymous, tracts on computus[21]. The *Ar. pr.* 1–3 is between verses, mostly taken from the *Anthologia Latina*[22], and excerpts from Isidore (*De temporibus* and *De astronomia*).

11. P_3 = Paris, Bibliothèque nationale de France, lat. 7299A, f. 34r, before 1150

The manuscript was written before 1150 in Limoges or nearby[23]. It contains mainly texts on computus and astrology[24]. Among the computis-

---

[19] Description in *Bibliothèque Nationale* 1975, pp. 458–461.

[20] Samaran / Marichal, 'Catalogue', p. 275. See Borst, 'Naturgeschichte', p. 117, n. 94; Borst, 'Kalenderreform', p. XX (C 4); Borst, 'Reichskalender', pp. 134f.

[21] Some of them are based upon a Franconian schoolbook written in 760 (*Lectiones seu regula conputi*; see Borst, 'Naturgeschichte', pp. 116–118), others go back to the computistical encyclopaedia in 7 books (note 7; see Borst, 'Naturgeschichte', 185, n. 47).

[22] Buecheler / Riese, 'Anthologia Latina', nos. 639, 394, 395, 640.

[23] See Borst, 'Naturgeschichte', p. 118, n. 95.

[24] A short description of the contents is in *Catalogus* 1744, p. 338. The computistical texts and tables form the first part. On f. 56r begins an astrological text with the incipit "Quicumque vero desiderat". From f. 67r on there are excerpts from Bede's *De natura rerum*. On f. 70v starts Albumasar's *Introductorium in astrologiam*.

6 Ps. Bede: *De arithmeticis propositionibus*

tical texts are extracts of the *Lectiones seu regula conputi*[25] and of the computistical encyclopaedia in 7 books[26], as well as *Ar. pr.* (no. 1).

12. *S* = Strasbourg, Bibliothèque nationale et universitaire, Ms. 326, f. 110v–111r, ca. 900

This manuscript was written about 900, probably in western France[27]. The codex begins with Bede's writings on the computus and his chronicle[28]. There follow on f. 107r–110r two short pieces (*De ortu et obitu patrum* and *De alphabetis*), then *Ar. pr.* 1–3 (f. 110v–111r) and after that, from f. 112r on, short texts on astronomical questions – *inter alia* extracts from Macrobius' commentary to the *Somnium Scipionis*, from Martianus Capella and from Isidore; the *De prognosticis* attributed to Aratus and Germanicus' reworking of Aratus' *Phaenomena*. Many of the texts in *S* can be found in computistical collections of the 8th and 9th centuries[29].

13. *T* = Trier, Stadtbibliothek, Hs. 2500, f. 72rv, 1st half 9th c.

The main part of this famous manuscript[30] was written in north-western France (most probably in Reims or Laon), as it seems, in the first half of the 9th century[31]. It has notes by Manno (b. 843), who taught at the cathedral school of Laon and later at that of Trier. Until the beginning of the 19th century the manuscript belonged to the Benedictine abbey St. Maximin in Trier. Later owners: Joseph Görres (1776–1848), Th. Freiherr von Cramer Klett; Heinrich and Erwin Rosenthal; Harrison D. Horblit; Peter Ludwig; Getty Museum (Malibu; shelf-mark: Ludwig MS. XII.3). It returned to Trier in 1990. The codex contains the three

---

[25]Note 21; see Borst, 'Naturgeschichte', p. 118, n. 95.

[26]Note 7; see Borst, 'Naturgeschichte', p. 237, n. 70.

[27]Information from the late Professor Bischoff. Borst dates the calendar texts of this manuscript (f. 173v–179r, 186r–190v) to "um Angoulême um 1000" (Borst, 'Naturgeschichte', p. 118, n. 95; Borst, 'Kalenderreform', pp. XXf.: Ms. C 3 and c 3b; Borst, 'Reichskalender', pp. 133f.).

[28]The content of the manuscript is described in *Catalogue général* 1923, pp. 139–143.

[29]The *Lectiones seu regula conputi* from 760 (note 21; see Borst, 'Naturgeschichte', p. 118, n. 95); the 7-book encyclopaedia of 809 (note 7; see Borst, 'Naturgeschichte', p. 210, n. 4) and an addition to it (see Borst, 'Naturgeschichte', p. 169, n. 9).

[30]See de Hamel / Nolden, 'Beda-Handschrift', and the description in von Euw / Plotzek, 'Handschriften', vol. 3 (1982), pp. 145–153.

[31]There is some dispute whether it was written in the beginning or in the second half of the 9th century; see de Hamel / Nolden, 'Beda-Handschrift', p. 16. Borst gives the date "Reims um 840" (Borst, 'Kalenderreform', p. XX; siglum: c 2; Borst, 'Reichskalender', pp. 145f.) and "vermutlich Laon oder Reims nach 840" (Borst, 'Naturgeschichte', p. 151, n. 52), respectively.

most important scientific works by Bede (*De natura rerum, De temporibus, De temporum ratione*) and other works on *computus* and related
matters. Some of them belong to the Carolingian encyclopaedia of 809[32].
Between works on *computus* by Bede and Hrabanus Maurus is the main
part of *Ar. pr.* This text begins, on the top of f. 72r, in the mid of no. 2
and ends on f. 72v with no. 4.

14. *V* = Venice, Biblioteca Nazionale Marciana, Z. L. 497 (1811),
f. 169v–170r, ca. 1045–1063

This manuscript is a handbook of the liberal arts which is connected with
Lawrence of Amalfi. Lawrence was a monk of Monte Cassino who later
became archbishop of Amalfi; being forced to go into exile, he took refuge
in Florence; then he taught, in Rome, the boy Hildebrand, who was to
become pope as Gregory VII; and at the close of his life he was in affec-
tionate friendship with Odilo of Cluny[33]. This handbook was compiled
by Archbishop Lawrence or, at the very least, reflects his teaching; the
writing is typically Italian – in particular Beneventan –, and the contents
date the manuscript most probably to the years 1045–1063[34]. It contains
in its 202 leaves numerous treatises of the *artes liberales*, including the
original Boethian translation of Aristotle's *Categories*[35]. The mathematical part of the manuscript is on folios 156v–170r, between the sections on
music and on astronomy[36]. There are texts on geometry, most of which
being connected with the so-called *Geometria incerti auctoris*, and different treatises on the abacus and other arithmetic matter; on f. 164v–169r
there is the *Propositiones ad acuendos iuvenes* attributed to Alcuin. At
the end of the arithmetical section, on fol. 169v–170r, there is *Ar. pr.* 1–3.
– Later, the manuscript belonged to Cardinal Bessarion, from whom it
came into the Biblioteca Marciana.

15. *V₁* = Biblioteca Apostolica Vaticana, Reg. lat. 309, f. 1v–2r, 16r,
9th c.

This manuscript, written in the 9th century, is from the monastery of
St. Denis near Paris; the numerous early additions (10th to 11th century),

---

[32]See Borst, 'Naturgeschichte', p. 185, n. 48. To this encyclopaedia in seven books
see note 7.

[33]See Newton, 'Lawrence', p. 445.

[34]Newton, 'Tibullus', pp. 277–280.

[35]For the description of the Aristotelian texts, see Lacombe, 'Aristoteles Latinus',
pp. 1123f.; a very superficial description of the manuscript is in Zanetti, 'Bibliotheca',
p. 202.

[36]This part is described in Newton, 'Lawrence', pp. 446–449.

8   Ps. Bede: *De arithmeticis propositionibus*

too, are from there[37]. Later owners: Paul Petau (M 50), Alexander Petau (no. 397), Queen Christina of Sweden (no. 499)[38]. The main part of the manuscript contains the computistical-astronomical encyclopaedia of 809 in seven books[39]. On folios 1v–2r, 16rv, 3v–4r, which were empty, *Ar. pr.* 1–3 and excerpts of the *Propositiones ad acuendos iuvenes* attributed to Alcuin were added later in the 9th century .

16. V_2 = Biblioteca Apostolica Vaticana, Reg. lat. 755, f. 97v–98r, mid-9th c.

The older part of the manuscript (f. 1r–99v except f. 18r–21v) was written in mid-9th century, perhaps in Sens or Blangy (Artois)[40]. Extracts from the *Annales S. Columbae Senonensis* in the inner margin of f. 9r–13r show that the codex was in Sens in the 10th to 13th centuries. Later owners: Paul Petau (Y. 50), Queen Christina of Sweden. The codex contains large parts of the Carolingian encyplopedia of 809[41]: many texts on astronomy, metrology and *computus*; the *De cyclo paschali* of Dionysius Exiguus; Bede's three computistical writings. Directly after Bede's *Chronicon minus* follows *Ar. pr.* 1–4; no. 4 is written in another ink, perhaps somewhat later than nos. 1–3. Various tables of kindred and affinity close the older part of the manuscript.

17. V_3 = Biblioteca Apostolica Vaticana, Ottob. lat. 1473, f. 35v, 37rv, ca. 1000

The manuscript comprises three parts (f. 1–27: s. 14; f. 28–37: s. 10/11; f. 38–45: s. 14 in.). The oldest part was written by one hand of western German or eastern French origin[12]. Owners were: Paul Petau, Christina of Sweden and then Cardinal P. Ottoboni. On f. 28r are two short texts on measures of length and the conversion of the circumference of the earth[42].

---

[37] According to Borst ('Naturgeschichte', p. 162, with n. 95) the text on f. 2v–120v is from 859; additions were written in St. Denis about 885. Borst ('Kalenderreform', pp. XXIIf.; 'Reichskalender', pp. 184f. and 216–218) dates the calendar on f. 6r–12v to 859–860 (siglum: D 6) and the calendar addition on f. 128v–140r to "nach 1000" (siglum: e 4).

[38] All details are to be found in the exact description of A. Wilmart, 'Codices Reginenses Latini', pp. 160–174.

[39] See note 7.

[40] Date by the late Professor Bischoff. Borst has "vielleicht Sens um 850" (Borst, 'Naturgeschichte', p. 185, n. 48) and, for f. 2r–22v, "wohl Frankreich um 1100" (Borst, 'Kalenderreform', p. XXVII and p. 178; siglum: k1), respectively. To the texts on music, see Bannister, 'Monumenti Vaticani', pp. 45f.

[41] See note 7.

[42] This text is identical with M_3, f. 118r.

There follow on f. 28v–35v the *Propositiones* attributed to Alcuin and on f. 35v, 37rv *Ar. pr.* 1–3. On f. 36, which is inserted within these two treatises, there are various texts on stones, catching birds, medical recipes etc. The owner's mark on f. 37v is unfortunately no longer readable (*Iste liber est sancti ///*).

18. V_4 = Biblioteca Apostolica Vaticana, lat. 3123, f. 111rv, 12th c.

This codex, of the 12th century, comprises two parts: f. 1–64 appears to be of English origin; f. 65–112 was probably written in western Germany or eastern France[12]. The first part contains, besides some text on the calculation on the abacus (Gerbert, Turchillus), treatises on *computus*; in the second there are the *Geometry II* ascribed to Boethius, Franco of Liège's text on squaring the circle, a treatise *De mensura fistularum*, then *Ar. pr.* 1–2, a wind-rose and the poem *Anthologia Latina*, no. 484[43].

19. W = Würzburg, Universitätsbibliothek, Mp. th. 4° 60, f. 94v–96v, 13th c.

This codex is a theological-historical collection of texts and may be divided into four parts (part 1: f. 1r–7r, s. 13; part 2: f. 7v–78v, s. 11; part 3: f. 79r–121v, s. 13; part 4: f. 122r–181v, s. 15). It is of German origin[12] and belonged to the Cistercian monastery of Ebrach, from where it came to Würzburg[44]. On f. 79r–121v there are extracts from Latin theological literature, some of them with glosses in German. Within these texts appears, unannounced, *Ar. pr.* 1–4. It lies between texts with the titles *De filiis Israel* and *De vita Ade*.

The text of *Ar. pr.* has been edited three times, always as part of Bede's collected works:

20. h = edition by J. Herwagen, Basel 1563 (Beda 1563, cols. 133–135)

The Basel printer Johannes He(e)rwagen (Hervagius), who tried to give the first complete edition of Bede's writings[45], has included in the first volume the *Ar. pr.*, nos. 1–4, without discussing its authenticity. The *Ar. pr.* is preceded by the *De divisionibus temporum* ascribed to Bede[46]. They are followed by the *Propositiones ad acuendos iuvenes* attributed to Alcuin[47]. In Hervagius' edition the *Ar. pr.* and the *Propositiones ad*

---

[43] Description in Folkerts, 'Geometrie II', pp. 12–14.

[44] A description of the various texts is to be found in Thurn, 'Handschriften', pp. 84–86.

[45] To this edition and its contents see Jones, 'Pseudepigrapha', pp. 14–18.

[46] On this, see Jones, 'Pseudepigrapha', pp. 48–51. This text is on cols. 117–132.

[47] On cols. 135–146.

acuendos iuvenes are published as if they were one work; we only read at the beginning of the second treatise: "Incipiunt aliae propositiones ad acuendos iuvenes". Then we find an incomplete text of the *Calculus* by Victorius of Aquitaine, i. e. his multiplication tables for integers and Roman fractions[48].

21. f = edition by J. W. Friessem, Cologne 1688 (Beda 1688, cols. 100–102)

The second edition of the *Ar. pr.* was published by Johannes Wilhelm Friessem in Cologne in 1688. It is part of the first volume of his Bede edition. The order of the writings published in this volume correspond to that in h; and indeed, Friessem's edition is nearly a reprint of h.

22. m = edition by J. P. Migne, Paris 1850 (Beda 1850, cols. 665–668)

Volume 90 of J. P. Migne's monumental *Patrologia Latina* is devoted to Bede's "Opera Didascalica"[49]. Migne's edition depends upon the folio editions h and f and on the edition of Bede's authentic works by J. A. Giles (1843). Writings which seemed to be spurious were collected in the "Sectio secunda" which had the subtitle "Dubia et spuria". In this part we find *Ar. pr.* 1–4 (on cols. 665–668).

2 Filiation of the manuscripts and early printed editions

On the basis of the variant readings it is possible to find special groups of manuscripts which are close to each other, though no stemma can be given.

First, we find that none of the extant manuscripts was copied from any other: in every manuscript there are mistakes which would have been transmitted to any copy of it.

One clearly defined group is represented by LMM_1VV_3, which transmit a specific text of problem 1. We shall call this group the "β-recension", all the other manuscripts belonging to the "α-recension". Typical for the β-recension in problem 1 are: a more succinct mode of expression (in general, the α-recension is long-winded) and the insertion of the first numerical example[50]. This addition is necessary and it makes the

---

[48] On cols. 147–158. The title in h is: "Praefatio in libellum de ratione calculi". See Jones, 'Pseudepigrapha', p. 53.

[49] To this edition see Jones, 'Pseudepigrapha', pp. 18f.

[50] The case that 1 is the number thought of (section 1.3′ of the critical edition at the end of this article); it has no equivalent in α. The second case – that 2 is the number thought of – is present in both recensions α and β (section 1.4 of this edition).

sense clear, since the main characteristic of problem 1 is the distinction between the two possible cases of an even or an odd number. None the less, recension β seems to have been written after α, since the oldest manuscripts belong to the α tradition. It seems that β was an attempt to shorten and to clarify the diffuse text in α. The most carefully written manuscript in β is M_1.

Within the α-recension manuscripts BP_2SV_2 form a subgroup (γ): this is indicated by several mistakes common to these manuscripts, but not to others. It may suffice to mention the variants 2,25; 2,64; 2,96 where these manuscripts have the same misreading[51]. The lacuna 2,20, common to P_2 and S, shows that P_2 and S form a subgroup within γ.

Three manuscripts of the α-recension, M_2M_3W, are also closely related to each other and thus form another subgroup (δ). This is sufficiently shown by the variants 1,114 and 2,108 and by the lacuna 2,116, which occur only in these manuscripts. The lacuna 1,51, which is common to δ and $P_1P_3V_1V_4$, shows that these four manuscripts are close to group δ.

The texts in the three early printed editions, Hervagius (h), Friessem (f) and Migne (m), are almost identical and have almost identical mistakes. In some places f and m corrected evident mistakes in h, but not by reference to an extant manuscript; at the same time they have introduced new errors. It may therefore be supposed that both f and m depend ultimately on h and did not use a manuscript. Hervagius had clearly used a manuscript close to M_3[52], but not M_3 itself[53].

For the establishment of the text the oldest group of manuscripts, the recension γ $(= BP_2SV_2)$, was taken as a basis; and of this group the manuscripts P_2S present the best text. The best β manuscript is M_1. The codices of the later group δ $(= M_2M_3W)$ contain variants that appear to be deliberate alterations of the transmitted text; so the value of this group is limited.

---

[51] The first number of the variant refers to the problem, the second to the variant number within the problem; see the critical apparatus of the edition at the end of this article. In 2,25 and 2,96 the misreading is common with H and L or T. B does not transmit the text to which the variant 2,96 belongs.

[52] This is evident from the big lacuna (variant 3,44) which hfm share with M_3. Already in 1939 Jones suggested that Hervagius used a later manuscript from southern Germany (Jones, 'Pseudepigrapha', p. 15).

[53] Problem 4 is missing from M_3, but is present in h; M_3 has many individual readings not shared by h or any other manuscript.

3   Mathematical content

Problems 1–3 give various procedures for one person to find the number thought of by another.

Problem 1: One person thinks of a number x, triples it and then halves the result. If there is a remainder, the result is increased by 1. Whatever comes out is tripled. The person is then asked how many nines (n) there are in the result and if there is a remainder. If there is no remainder, $2n$ is the number first thought of; if there is the remainder 6, $2n + 1$ is the number first thought of.

In both cases (x even or odd) the correctness of the procedure is evident:

a) $x = 2n$: One forms $2n$; $2n \cdot 3 = 6n$; $6n = 3n + 3n$; $3n \cdot 3 = 9n$; $9n \div 9 = n$ with remainder 0; thus $2n = x$.

b) $x = 2n + 1$: One forms $2n + 1$; $(2n + 1) \cdot 3 = 6n + 3$; $6n + 3 = (3n + 1) + (3n + 2)$; take $3n + 2$; $(3n + 2) \cdot 3 = 9n + 6$; $(9n + 6) \div 9 = n$ with remainder 6; thus $2n + 1 = x$.

Problem 2: This begins in the same way as problem 1, but after the second multiplication by 3 the result is again halved. After each halving one asks whether there is a remainder, and, if so, the result is raised to the next integer ("maior pars triplicanda est"). The questioner keeps in mind 1, if the first halving produces a remainder; he keeps 2, if the second produces a remainder. As before, one asks how many nines (n) there are in the result. The questioner multiplies this by 4 ("quot novenarii in ea inveniuntur, tot quaternos divinator sumere debet") and adds in the sum of the numbers he has kept. The result is the number x that the other person first thought of.

Explanation:

a) $x = 4n$: the series is $4n$, $12n$, $6n$ (note 0), $18n$, $9n$ (note 0), n with remainder 0, thus $4n + 0 + 0 = x$.

b) $x = 4n + 1$: the series is $4n + 1$, $12n + 3$, $6n + 2$ (note 1), $18n + 6$, $9n + 3$ (note 0), n with remainder 3, thus $4n + 1 + 0 = x$.

c) $x = 4n + 2$: the series is $4n + 2$, $12n + 6$, $6n + 3$ (note 0), $18n + 9$, $9n + 5$ (note 2), n with remainder 5, thus $4n + 0 + 2 = x$.

b) $x = 4n + 3$: the series is $4n + 3$, $12n + 9$, $6n + 5$ (note 1), $18n + 15$, $9n + 8$ (note 2), n with remainder 8, thus $4n + 1 + 2 = x$.

Problem 3: Finding the day of a week. If x is the number of the day, the questioner asks the other man step by step to form $(2x + 5) \cdot 5 \cdot 10$ and tell him the result. When 250 is subtracted, he asks himself how many hundreds there are in this.

The procedure follows from the identity $[(2x+5)\cdot5\cdot10-250]/100 = x$.

All three problems belong to recreational mathematics. Finding numbers thought of by someone else is one of the oldest mathematical problems[54]. Such simple problems could be solved without algebraical procedures, by trial and error or working backwards. Our text seems to be the earliest Western text in which the three procedures appear[55]. In the following centuries we often come across these three problems: no. 1 is to be found in the arithmetic part of the *Secretum philosophorum*[56], in 15th-century collections of problems[57] and in later collections[58]. Problem 2 appears (presumably from an Arabic source) about 1140 in a work of Johannes Hispalensis[59], again about 1240 in the *Annales Stadenses*[60], in Leonardo Fibonacci's *Liber abbaci* (1202, reworked 1228)[61], in a Byzantine work on practical arithmetic of the 15th century[62] and in Latin collections of problems of the 14th/15th centuries[63]. Just as often is problem 3 to be found in the later literature: e. g. in Leonardo Fibonacci's *Liber abbaci*[64], in the *Algorismus Ratisbonensis* (about 1450)[65] and, sometimes slightly varied, in numerous Latin collections of about the same time[66].

Problem 4 is of particular interest. Here there are detailed instructions for the addition of positive and negative numbers. The term for "positive number" is *verum* and the term for "negative number" is *minus*; other terms used are *essentes numeri* or *existentes numeri* for positive numbers and *non essentes* or *non existentes numeri* for negative numbers. After the lapidary statements "Verum cum vero facit verum. Minus cum vero facit verum. Verum cum minus facit minus. Minus cum minus facit minus" and the definition of *verum* and *minus*[67] it is shown, by

---

[54]See Tropfke, 'Elementarmathematik', pp. 642–651.

[55]For the history of these procedures see Tropfke, 'Elementarmathematik', pp. 643f.

[56]This treatise was presumably written in England in the 13th century; see Thorndike, 'History', pp. 788–791, 811f.

[57]Not seldom for calculating the contents of a purse. Latin manuscripts containing this problem are given in Folkerts, 'Aufgabensammlungen', p. 62.

[58]See Tropfke, 'Elementarmathematik', p. 644.

[59]Edited in Boncompagni, 'Trattati', pp. 125f..

[60]Edited in Lappenberg, 'Annales Stadenses', p. 334.

[61]Boncompagni, 'Scritti', pp. 303f.

[62]Hunger / Vogel, 'Rechenbuch', pp. 38f. (no. 44).

[63]E. g. in London, British Library, Cotton Cleop. B IX, f. 20v.

[64]Boncompagni, 'Scritti', p. 304.

[65]Vogel, 'Practica', nos. 270, 310.

[66]See Folkerts, 'Aufgabensammlungen', pp. 61f.

[67]"Verum essentiam, minus nihil significat."

14   Ps. Bede: *De arithmeticis propositionibus*

means of four examples[68], how a negative number and a positive number of different values can be combined[69]. In the text negative numbers are clearly treated as independent objects. This shows essential advance over Diophantus who is known as the only mathematician of antiquity to give rules for operations with quantities to be subtracted[70], but he did not have the abstract conception of negative number: his rules refer to the coefficients of subtracted quantities which can occur in polynomials[71]. To our present knowledge, the conceptual step from subtractive coefficients to negative numbers was taken before the 15th century only by Chinese and Indian mathematicians[72]: we first meet negative numbers in book 8 of the *Mathematics in nine books*, the central work of early Chinese mathematical literature. This book was reworked in the first century B.C., but goes back to earlier mathematical writings[73]. In this book the rules for the addition and subtraction of positive and negative quantities are formulated in general form; negative numbers are introduced to extend the algorithm for formally solving linear equations to problems with negative coefficients. Somewhat later the rules for calculating with positive and negative numbers, including multiplication, division, squaring and finding roots, can be found in Indian texts – beginning with Brahmagupta (ca. 628)[74]. The Indian terms for positive and negative numbers have the meaning of "property" and "debt", respectively. It should not be supposed that there is a connection between these Chinese or Indian works and our text. Our text appears to be an independent achievement and one without direct influence: not before the 16th century, in the works of the mathematician Michael Stifel (1487?–1567), were negative numbers

---

[68] $(+7) + (-3)$, $(+3) + (-7)$, $(-3) + (-7)$, $(+3) + (+7)$.

[69] The terms for "combine" are: *iungere, ponere, conferre*.

[70] Tropfke, 'Elementarmathematik', pp. 144f.

[71] A. P. Juschkewitsch writes ('Geschichte', p. 37): "Dessen ungeachtet kannte Diophantos die negativen Zahlen nicht. Die subtraktiven Zahlen waren bei ihm keine selbständigen Objekte, und die Regeln für das Operieren mit Vorzeichen bezogen sich lediglich auf die Glieder von Differenzen, wie etwa die bei uns gebräuchlichen Ausdrücke $a - b$ oder $ax^2 - bx$ usw., bei denen der Minuend größer ist als der Subtrahend."

[72] See Tropfke, 'Elementarmathematik', pp. 145f.; Juschkewitsch / Rosenfeld, 'Mathematik', pp. 126–129; Sesiano, 'Negative Solutions', pp. 106–108. "There is at present no indication that Arabic mathematicians ever considered negative solutions acceptable" (Sesiano, 'Negative Solutions', p. 108).

[73] See, e. g., Juschkewitsch, 'Geschichte', pp. 23–26, 36–39.

[74] In chapter 18, stanzas 30–35. See Juschkewitsch, 'Geschichte', pp. 126–129, and Datta / Singh, 'Hindu Mathematics', pp. 21f.

accepted as having the same status as positive numbers[75]. We meet the term *numeri veri* for positive numbers again in texts of the 17th century, negative numbers being called *numeri falsi* or *negati*[76]. In his edition of a Byzantine book on practical arithmetic, Kurt Vogel, one of the best historians of mediaeval mathematics, writes[77]: "The calculations with negative numbers treated there (i. e., in Friessem's edition of our text) are scarcely possible in his (i. e. Bede's) time." Vogel accordingly doubted that the text was written by Bede. After inspection of the manuscript tradition rather than Friessem's text, this judgement should be modified: as will be seen in the next section, it is clear that, though this part of the text was probably not written in Bede's time, it was certainly written before the 10th century.

4 Authorship and date of the text

Of the 19 manuscripts only one (W) has the complete text as it appears in the early editions; the others stop at various places or omit parts of the text given in the editions. In detail:

P_1 and P_3 contain only problem 1 $(1.1–1.5)$[78]

B: problems $1 – 2$ middle $(1.1–2.6)$

V_4: problems $1 – 2$ end $(1.1–2.8)$

$LMM_1P_2SVV_1V_3$: problems $1 – 3$ middle $(1.1–3.3)$

M_2M_3: problems $1 – 3$ end $(1.1–3.9)$

HV_2: problems $1 – 3$ middle, 4 $(1.1–3.3, 4.1–4.6)$

B_1: problems 4, $1 – 3$ middle $(4.1–4.6, 1.1–3.3)$

T: problems 2 middle $– 3$ middle, 4 $(2.7–3.3, 4.1–4.6)$

$Whfm$: problems 1–4 $(1.1–4.6)$.

Accordingly, four groups may be distinguished[79]:

(1) The majority of the manuscripts – i. e. all except $B_1HM_2M_3TV_2W$ – end in the middle of problem 3 (3.3).

---

[75]But it should be said that negative solutions of linear equations or of systems of linear equations did occur in some mediaeval texts, e. g. by Leonardo Fibonacci (13th c.), in two arithmetics of the 15th century written in Provençal, and in Nicolas Chuquet's *Triparty* (1484); see the detailed analysis in Sesiano, 'Negative Solutions', pp. 109–142.

[76]To the terminology of negative numbers see Tropfke, 'Elementarmathematik', pp. 150f.

[77]Hunger / Vogel, 'Rechenbuch', p. 98, n. 5.

[78]The numbers in brackets refer to the paragraphs of the edition at the end of this article.

[79]The shortened texts in $BP_1P_3V_4$ can be taken out of consideration here.

(2) Three codices (M_2M_3W) and the printed editions hfm contain in addition the end of problem 3 (3.4–3.9).

(3) Only B_1HTV_2W and the printed editions hfm have the text of problem 4.

(4) B_1 has problem 4 first and then problems 1 – 3 middle (1.1–3.3); in H, problem 4 is found at a much later place than problems 1–3; in V_2, problem 4 seems to have been added later.

It is therefore probable that our text originally went as far as the middle of problem 3 (3.3), i. e. the place where group 1 ends, and that subsequently two separate extensions, (2) and (3), were made. There is support for this suggestion in the content of the text: the end of problem 3 (i. e. 3.4–3.9) is superfluous and inappropriate. In the first part (i. e. 3.1–3.3) the author says everything that is necessary: he gives an example (the first case) and then the general rule. In the second part (3.4–3.9) the text takes each of the other cases and works through the procedure of the rule. The supposition of a later addition is supported by the word *Item* at the beginning of the passage.

Problem 4 has nothing in common with problems 1 to 3 – so it is probable that this passage, too, is a later addition to the text. It seems very probable that problem 4 was an isolated short text, without title, that was by chance near the text of problems 1–3.3 and was transcribed either before (B_1) or after (HTV_2W) the main text. We find just such a situation in manuscript M: after our text there follow five individual mathematical problems without pause or separate title; a reader might very well think that the five problems belonged to the preceding text. H, T and V_2 are the only old manuscripts in which the scribes have included problem 4. But it must also have been present in δ since it is present in W and in the editions hfm; we note that it was left out in M_2 and M_3.

Thus there were two extensions to the basic text (1.1–3.3). One of them (problem 4) was made at the latest in the 9th century, the date of manuscripts H, T and V_2, and the other (the end of problem 3) in the 11th century, the date of M_3, or earlier. The first extension was perhaps made in France, since B_1HTV_2 were written there, and the second with great probability in Bavaria, perhaps in St. Emmeram, because M_2M_3W are all of Bavarian origin.

As for the authorship of the original text (i. e. problems 1 to 3.3): the printed editions treat the text as part of the Bede corpus. But this ascription has no manuscript support; nor is the content redolent of Bede's other works. Bede does not mention the work in his list of his own books, which is found at the end of his *Historia ecclesiastica*, finished in 731.

Further, Bede is not given as author in any of the 19 manuscripts: the typical title is "Incipiunt numeri, per quos potest qui voluerit alterius cogitationes de numero quolibet, quem animo conceperit, explorare", which describes the content without giving any hint to the author. No doubt Hervagius put the work in his edition of Bede because he found it in a codex which contained other works by him: in most of the older extant manuscripts ($BHP_2STV_1V_2V_4$) the text is to be found together with texts on chronology, computus and astronomy, some of which are by Bede. The title in the edition, *De arithmeticis propositionibus*, was evidently Hervagius' own; it is used in this article only to keep to the tradition.

We may note that our text appears in six codices ($LMM_2VV_1V_3$) before or after the *Propositiones ad acuendos iuvenes* attributed to Alcuin. This circumstance suggests that the two texts came from the same area and were written at much the same time. Whether they were written by the same author is another question. As for the place of origin, western Germany or eastern France is suggested since this is where most of the older manuscripts were written. It should be mentioned that in ten manuscripts ($BB_1HM_3P_2P_3STV_1V_2$) the *Ar. pr.* is transmitted in connection with excerpts of the two computistical encyclopaedias of 809 and 818 respectively, the first of which probably originated at the court of Charlemagne in Aachen. The *Ar. pr.* must have been written at least as early as the oldest manuscripts, which are of the 9th century, and may well have been somewhat older.

5 Edition of the text

The edition is based on all 19 manuscripts and the three printed editions. In the apparatus all variants are noted. The only exceptions are trivial variants in spelling, e.g.: *-ti- – -ci-*; *quoties – quotiens*; *habuerit – abuerit*; *hiis – his*; *nihil – nichil, nil*; *assumptus – adsumptus*; *duplicare – dupplicare*; *Quodsi – Quod si* (in all of these cases, the first spelling is used in the text). Variants in writing the numbers (e. g. *X et VIII* instead of *XVIII* or the use of Arabic numerals in the editions *hfm*) are only mentioned if the number given in the manuscript or early edition has another value than the number given in the text presented here. In the apparatus, the following abbreviations are used: *add.* = has added, *om.* = has left out; *tr.* = has transposed; *corr. ex* = has corrected from; *supr.* = has superscribed.

Bibliography

H. M. Bannister, *Monumenti Vaticani di paleografia musicale latina*, Leipzig 1913.

Beda 1563: *Opera Bedae Venerabilis presbyteri Anglosaxonis* ... Basileae, per Ioannem Heruagium. [Tom. I.] 1563.

Beda 1688: *VENERABILIS BEDAE PRESBYTERI ANGLO-SAXONIS viri sua ætate doctissimi, OPERA quotquot reperiri potuerunt omnia.* Coloniae Agrippinae, Sumptibus Joannis Wilhelmi Friessem. [Vol. 1.] 1688.

Beda 1850: J. P. Migne (ed.), *Patrologia Latina*, vol. 90: *Venerabilis Bedae Anglosaxonis presbyteri opera omnia. Tomus primus*, Paris 1850.

Bibliothèque Nationale. Catalogue général des manuscrits latins. Tome VI (N°s 3536 à 3775^B), Paris 1975.

B. Boncompagni (ed.), *Scritti di Leonardo Pisano.* Vol. 1: *Il Liber abbaci*, Rome 1857.

B. Boncompagni (ed.), *Trattati d'aritmetica. II: Ioannis Hispalensis liber algorismi de pratica arismetrice*, Rome 1857.

A. Borst, *Das Buch der Naturgeschichte. Plinius und seine Leser im Zeitalter des Pergaments*, Heidelberg 1994.

A. Borst, *Die karolingische Kalenderreform*, Hannover 1998.

A. Borst (ed.), *Der karolingische Reichskalender und seine Überlieferung bis ins 12. Jahrhundert*, 3 parts, Hannover 2001 (Monumenta Germaniae Historica, *Libri memoriales*, II).

N. Bubnov, *Gerberti postea Silvestri II papae opera mathematica (972–1003)*, Berlin 1899.

F. Buecheler, A. Riese, *Anthologia Latina sive poesis Latinae supplementum. Fasciculus I/II*, Leipzig 1894–1906.

A Catalogue of the Harleian Manuscripts, in the British Museum. Vol. II, London 1808.

Catalogue général des manuscrits des bibliothèques publiques de France. Départements. Tome XXIII: Bordeaux. (Octavo Series), Paris 1894.

Catalogue général des manuscrits des bibliothèques publiques de France. Départements. Tome XLVII: Strasbourg. (Octavo Series), Paris 1923.

Catalogus codicum manuscriptorum bibliothecae regiae. Pars tertia. Tomus quartus, Paris 1744.

Catalogus codicum manu scriptorum bibliothecae regiae Monacensis. Tomi IV pars II, Munich 1876.

Catalogus codicum manu scriptorum bibliothecae regiae Monacensis. Tomi IV pars III, Munich 1878.

M. Curtze, "Die Handschrift No. 14836 der Königl. Hof- und Staatsbibliothek zu München", *Abhandlungen zur Geschichte der Mathematik* 7 (= *Supplement zur Zeitschrift für Mathematik und Physik* 40) (1895), 75–142.

B. Datta, A. N. Singh, *History of Hindu Mathematics. Part 2*, Lahore 1938.

A. von Euw, J. M. Plotzek, *Die Handschriften der Sammlung Ludwig*, 4 vols., Cologne 1979–1984.

M. Folkerts, *"Boethius" Geometrie II, ein mathematisches Lehrbuch des Mittelalters*, Wiesbaden 1970.

M. Folkerts, "Mathematische Aufgabensammlungen aus dem ausgehenden Mittelalter. Ein Beitrag zur Klostermathematik des 14. und 15. Jahrhunderts", *Sudhoffs Archiv* 55 (1971), 58–75.

M. Folkerts, "Die älteste mathematische Aufgabensammlung in lateinischer Sprache: Die Alkuin zugeschriebenen Propositiones ad acuendos iuvenes. Überlieferung, Inhalt, Kritische Edition", *Denkschriften der Österreichischen Akademie der Wissenschaften, mathematisch-naturwissenschaftliche Klasse*, 116. Band, 6. Abhandlung, pp. 13–80, Vienna 1978. (Reprinted in this volume, item V.)

Ch. de Hamel, R. Nolden, "Die neuerworbene Beda-Handschrift aus St. Maximin", in: *Stadtbibliothek Trier. Karolingische Beda-Handschrift aus St. Maximin*, Trier 1990, pp. 14–28.

H. Hunger, K. Vogel, *Ein byzantinisches Rechenbuch des 15. Jahrhunderts. 100 Aufgaben aus dem Codex Vindobonensis Phil. Gr. 65. Text, Übersetzung und Kommentar*. (Österreichische Akademie der Wissenschaften, Philosophisch-historische Klasse, Denkschriften, 78. Band, 2. Abhandlung), Vienna 1963.

Ch. W. Jones, *Bedae Pseudepigrapha. Scientific Writings Falsely Attributed to Bede*, Ithaca 1939.

A. P. Juschkewitsch, *Geschichte der Mathematik im Mittelalter*, Leipzig 1964.

A. P. Juschkewitsch, B. A. Rosenfeld, "Die Mathematik der Länder des Ostens im Mittelalter", in: G. Harig (ed.), *Sowjetische Beiträge zur Geschichte der Naturwissenschaft*, Berlin 1960, pp. 62–160.

G. Lacombe, *Aristoteles Latinus. Pars posterior*, Cambridge 1955.

J. M. Lappenberg (ed.), "Annales Stadenses auctore Alberto", in: *Monumenta Germaniae Historica, Scriptores* [5, in folio], tom. 16: *Annales aevi Suevici*, Hannover 1859, pp. 271–379.

F. L. Newton, "Tibullus in Two Grammatical *Florilegia* of the Middle Ages", *Transactions and Proceedings of the American Philological Association* 93 (1962), 253–286.

F. L. Newton, "Lawrence of Amalfi's Mathematical Teaching", *Traditio* 21 (1965), 445–449.

III

V. Rose, *Die lateinischen Meerman-Handschriften des Sir Thomas Phillipps in der Königlichen Bibliothek zu Berlin*, Berlin 1892.

Ch. Samaran, R. Marichal, *Catalogue des manuscrits en écriture latine portant des indications de date, de lieu ou de copiste. Tome II: Bibliothèque Nationale, fonds latin*, Paris 1962.

J. Sesiano, "The Appearance of Negative Solutions in Mediaeval Mathematics", *Archive for History of Exact Sciences* 32 (1985), 105–150.

G. Thiele, *Der lateinische Aesop des Romulus und die Prosa-Fassungen des Phaedrus*, Heidelberg 1910.

L. Thorndike, *A History of Magic and Experimental Science During the First Thirteen Centuries of Our Era*. Vol. 2, New York / London 1923.

H. Thurn, *Die Handschriften der Universitätsbibliothek Würzburg. Erster Band: Die Ebracher Handschriften*, Wiesbaden 1970.

J. Tropfke, *Geschichte der Elementarmathematik. 4. Auflage. Band 1: Arithmetik und Algebra. Vollständig neu bearbeitet von Kurt Vogel, Karin Reich, Helmuth Gericke*, Berlin / New York 1980.

K. Vogel, *Die Practica des Algorismus Ratisbonensis*, Munich 1954.

A. Wilmart, *Codices Reginenses Latini. Tomus II: Codices 251–504*, Vatican 1945.

A. Zanetti, *Latina et Italica D. Marci Bibliotheca codicum manu scriptorum Per titulos digesta*, Venice 1741.

Sigla codicum editionumque

B	=	Berlin, SB, Phillipps 1832, f. 55v, s. 9
B_1	=	Bordeaux, BM 11, f. 209v–210v, post 1104
H	=	London, BL, Harley 3017, f. 152r–153v, 180rv, a. 862–864
L	=	London, BL, Burney 59, f. 11rv, s. 11[1]
M	=	Montpellier, H 491, f. 108r–110r, s. 11 in.
M_1	=	München, BSB, Clm 18765, f. 1r–2r, ca. 950
M_2	=	München, BSB, Clm 14689, f. 13rv, ca. 1100
M_3	=	München, BSB, Clm 14836, f. 78v–80r, s. 11
P_1	=	Paris, BnF, lat. 3652, f. 10r, s. 12
P_2	=	Paris, BnF, lat. 5543, f. 147v–148r, ca. 847
P_3	=	Paris, BnF, lat. 7299A, f. 34r, ante 1150
S	=	Strasbourg, BU, Ms. 326, f. 110v–111r, ca. 900
T	=	Trier, StB, Hs. 2500, f. 72rv, s. 9[1]
V	=	Venezia, BNM, Z. L. 497 (1811), f. 169v–170r, ca. 1045–1063
V_1	=	Vat. Reg. lat. 309, f. 1v–2r, 16r, s. 9
V_2	=	Vat. Reg. lat. 755, f. 97v–98r, ca. 850
V_3	=	Vat. Ottob. lat. 1473, f. 35v, 37rv, ca. 1000
V_4	=	Vat. lat. 3123, f. 111rv, s. 12
W	=	Würzburg, UB, Mp. th. 4° 60, f. 94v–96v, s. 13
h	=	ed. Herwagen (1563)
f	=	ed. Friessem (1688)
m	=	ed. Migne (1850)

Edition

[1]INCIPIUNT NUMERI, PER QUOS POTEST QUI VOLUERIT ALTERIUS CO-
GITATIONES DE NUMERO QUOLIBET, QUEM ANIMO CONCEPERIT[2], EX-
PLORARE[3].

<PROPOSITIO 1>

1.1–1.5: recensio α [γ (= BP$_2$SV$_2$) δ (= M$_2$M$_3$W) B$_1$HP$_1$P$_3$V$_1$V$_4$ hfm]
1.1, 1.2′, 1.3′, 1.4, 1.5′: recensio β [= LMM$_1$VV$_3$]

QUOMODO[4,5] NUMERUS A QUOLIBET ANIMO CONCEPTUS, QUIS SIT,
POSSIT AGNOSCI.

<1.1> Assumatur numerus quilibet[6,7] ac triplicetur[8]. Triplicatus[9] di-
vidatur[10] in duas partes[11], et si ambae aequales extiterint[12,13], qualem
volueris absque[14] ulla differentia iterum[15] triplicabis. Quodsi inaequales
fuerint, quae[16] maior fuerit[17,18] triplicetur[19], quotiesque in ea, postquam
triplicata fuerit[20,21], VIIII[22] inveniri possint[23], consideretur[24]; quia[25] quoties VIIII[26] habuerit[27], toties duo sumendi sunt[28].

---

[1] *inscr. om.* B$_1$LMM$_1$P$_1$P$_3$V$_1$V$_3$V$_4$, ALIA RATIO V, De numero W, BEDA (BEDA
om. m) DE ARITHMETICIS PROPOSITIONIBUS. Divinatio numeri ab aliquo in
animo concepti hfm   [2] quem animo conceperit] animo concepto V$_2$ [3] explorare] co-
gnoscere (*post* cogitationes) M$_3$ [4] Quomodo ... agnosci *om.* LMM$_1$VV$_3$V$_4$ [5] *ante*
Quomodo *add.* P$_1$ ARGUMENTUM   [6] numerus quilibet *tr.* MVW   [7] quislibet
M$_1$V$_3$L   [8] tripliciter P$_1$ [9] inplicitus P$_1$, triplicatur V$_3$hf, Qui triplicatus M$_3$
[10] dividat H   [11] dividatur ... partes] in duas partes dividatur MV   [12] aequales
extiterint] fuerint equales P$_3$ [13] extiterint] existerint B$_1$, constiterint V$_3$ [14] absque]
sine MV   [15] iterum *om.* MV   [16] quae] qui V$_1$ [17] quae maior fuerit] maior MV
[18] fuerit] est LM$_1$V$_3$ [19] triplicetur] iterum triplicabis (*ante* quae maior fuerit) M$_3$,
triplicatus P$_3$ [20] postquam triplicata fuerit *om.* LMM$_1$V   [21] triplicata fuerit] triplicatur M$_3$ [22] novenarius MV   [23] possunt M$_3$P$_3$V$_3$, potest MV   [24] considerentur
M$_2$M$_3$hfm   [25] quia *om.* VV$_4$ [26] novenarium M, novenarius V   [27] fuerit V$_4$,
habuerint W   [28] duo sumendi sunt] in ea binarius sumendus est MV, binarius est
sumendus V$_4$

recensio α

<1.2> Observandum est autem[29], ut[30], postquam numerus triplicatus[31] et[32] divisus est iterumque[33] medietas fuerit triplicata[34] et de novenario fuerit interrogatum[35], hoc etiam adiungatur, si aliquid[36] supra[37] unum vel[38] duos[39] vel quot[40] fuerint[41,42] novenarios[43,44] remansisset[45]. <1.3> Quia si aliquid[46] remansit[47], sex[48] fuerunt, et de his[49] unus sumendus est[50]; si[51] nihil remansisse responderit, nihil sumendum est. Hisque[52] in summam[53] redactis pronuntiandus[54] numerus, qui primus[55] fuerat[56] mente[57] conceptus.

recensio β

<1.2'> Si autem usque ad VIIII non pervenerit, senario[58] numero summa concluditur, de quo unus[59] sumendus[60] est.

<1.3'> Ut verbi gratia[61]: Si unum fuerit mente conceptum[62] et triplicatum, tres efficiuntur, qui[63] divisi in duos[64] et unum resolvuntur. Duo vero, qui[65] huius divisionis maiorem obtinent[66] portionem[67], iterum triplicati VI tantum efficiunt, nec remanet aliquid[68]. Unum[69,70] ergo fuit, quod[71] prius[72] mente[73] conceptum[74] est[75].

<1.4> Verbi gratia[76] si duo fuerint animo[77] concepti[78], cum triplicabuntur[79], VI faciunt[80,81]; sex[82] divisi in tres[83], et[84,85] tres resolvuntur[86].

---

[29] est autem] autem P_1, *tr.* V_4 [30] ut] quod M_3, possint *add.* hfm, *om.* P_1 [31] triplicatur $M_2P_1P_2P_3V_1$hfm, triplicetur M_3 [32] et] ac W   [33] iterumque] et iterum V_4 [34] fuerit triplicata *tr.* W   [35] interrogandum P_1 [36] aliquod $M_2P_1P_3$W   [37] super SV_4 [38] vel] aut V_2 [39] duo B_1 [40] quod $BHP_1P_2P_3V_1V_2$W, *corr.* P_2^1, quotquot B_1 [41] quot fuerint] plures V_4 [42] fuerint] fuerit BM_3, *om.* P_3 [43] fuerint novenarios] novenarii fuerint V_2 [44] novenarios] novenarius $B_1M_2M_3$Whfm, novenarii S   [45] remansissent M_3, remansisse P_3 [46] aliquod $M_2M_3P_1$, aliqui B, aliquit H, supra *add.* M_3 [47] remansisset fm   [48] VII M_3, septem hfm   [49] is P_3 [50] unus sumendus est] sumendus est unus $P_1P_3V_1$, sumendus unus V_4 [51] si ... sumendum est *om.* $M_2M_3P_1P_3V_1V_4$Whfm   [52] Hisque] His in P_1 *et del.* in   [53] summa P_1V_1, su$\overline{\text{ma}}$ B_1 [54] pronuntiandus] est *add.* $B_1M_3P_1$, est *supr.* V_4, est *(?) supr.* M_2 [55] prius $P_1P_3V_1$, *om.* V_4 [56] fuerit $P_1P_2SV_2$ [57] fuerat mente *tr.* M_3 [58] senariorum L   [59] unius M   [60] sumendum M   [61] Ut verbi gratia] Verbi causa MV   [62] fuerit ... conceptum] mente fuerit conceptus (*corr. in* conceptum) V   [63] qui] si *add.* V_3 [64] duo MV   [65] quae LM_1V_3, simul *add.* V   [66] optinet MVV_3, continent L   [67] partem MV   [68] aliquid] nec superhabundat *add.* MV   [69] Unum ... <1.4> nec remanet aliquid *om.* V_4 [70] Unum] Unus V_3 [71] quod *om.* MV   [72] primis LM_1 [73] mente *om.* L   [74] concentum V_3 [75] est *om.* MV   [76] Verbi gratia] Item LMM_1V, Iterum V_3 [77] animo] mente MVV_2 [78] fuerint animo concepti] animo concepti sunt P_1 [79] triplicantur $LMM_1P_1VV_3$ [80] VI faciunt] fiunt VI M_3 [81] faciunt] fiunt M_1W   [82] sex] VIII P_3, Sed P_1, *om.* MV, vero *add.* M_3 [83] tres, et tres resolvuntur] duas partes utraque pars habet tres W   [84] in tres, et] per binarium M_3 [85] et *om.* P_3 [86] resolvuntur] fiunt M_3

24   Ps. Bede: *De arithmeticis propositionibus*

Tres vero[87,88] triplicati VIIII tantum efficiunt[89], nec remanet[90] aliquid[91]. Ergo duo[92] fuerunt, qui[93] prius[94] mente[95] comprehensi sunt[96].

recensio α

<1.5> Et ideo in omni huius supputationis[97] ratione novenarius[98] duos[99] semper significat[100] et[101] senarius[102] unum[103]; sciendumque[104], quod omnis impar numerus, qui assumptus[105] fuerit[106] et[107] ordine quo praedictum est triplicatus[108] atque[109] divisus[110], in[111] senarium[112] terminatur, par vero numerus in novenarium, quia unitas secundum supradictum modum triplicata atque[113] divisa senarium procreat, binarius vero[114] simili modo[115] triplicatus[116] atque[117] divisus[118] novenarium, quia hi[119] duo numeri, id est[120] unus et[121] duo, paritatis et imparitatis[122] dicuntur esse principia.

recensio β

<1.5'> Ideo[123] sciendum est[124], quod in hac supputatione unitas, quae est imparitatis[125] exordium[126], senarium procreat et binarius, qui est paritatis principium, novenarium generat. Ac propter hoc quoties novem[127] conceptor numeri se habere responderit[128], toties duo a divinante[129] colligendi[130] sunt. Et si supra VIIII aliquid superesse responderit[131], VI esse non dubium est; de quibus unitas, quae ipsos generare solet[132], sumenda est.

* * *

---

[87] Tres vero] Qui M₃   [88] vero] autem MV   [89] VIIII tantum efficiunt] VIIII faciunt MV, in VIIII consurgunt M₃, fiunt tantum VIIII W   [90] remanet] superest LMM₁VV₃   [91] remanet aliquid *tr.* M₃   [92] Ergo duo] Duo igitur LMM₁VV₃   [93] qui *om.* MVV₄   [94] prius LMM₁P₁P₃VV₃] primis BB₁HP₂SV₁, primitus V₂, primi M₂hfm, primum M₃, primo W, *om.* V₄   [95] mentem S, *om.* P₁V₂   [96] comprehensi sunt] concepti MVV₄   [97] supplicationis P₃, *corr.* P₂ *ex* supputationibus   [98] novenarius] .IX. V₄   [99] duo M₂M₃P₁W   [100] significant V₄   [101] et *om.* M₃W   [102] senarius] .VI. V₄   [103] unus M₂   [104] sciendumque] Sciendum quoque W, est *add.* V₄   [105] qui assumptus] quassumtus B   [106] fuerit] est M₃   [107] et] in *add.* P₃   [108] triplicatur P₁   [109] ac M₂W   [110] divisis M₃   [111] in ... secundum supradictum *om.* M₃   [112] senario B₁   [113] ac M₂W   [114] vero] numerus *add.* M₂M₃Whfm, *om.* P₃   [115] simili modo *om.* V₄   [116] triplicatur M₂P₂   [117] ac W   [118] divisus *corr. in* dividitur M²₁, in *add.* M₂W   [119] quia hi] Qui B₁   [120] id est *om.* M₂M₃, idē B₁P₂P₃SV₂, id' V₄, idem hfm   [121] unus et *om.* P₃   [122] imparatis V₄   [123] Ideo] Id V₃   [124] est *om.* MV   [125] inparietatis V₃   [126] exordium] initium MV   [127] novem *post* numeri MV   [128] responderit] dixerit MV   [129] a divinante M₁V] a divinitate L, ad unitatem V₃, adiuvante M   [130] collegendi M₁   [131] responderit] dixerit MV   [132] generare solet] generat MV

<PROPOSITIO 2>

2.1–2.8: γ (= BP_2SV_2; B *deest post 2.6*) δ (= M_2M_3W) B_1HT(*in 2.7 incipiens*)V_1V_4 β (= LMM_1VV_3) hfm

ITEM ALITER.[1]

<2.1> Assumatur[2] numerus quilibet[3] ac[4] triplicetur[5]. Triplicatus[6] dividatur in duas partes[7], et[8] tunc ille, qui numerum[9] mente concepit[10], interrogetur, si ipsius numeri[11] sit aequa[12] divisio. <2.2> Quodsi pares[13] ambas[14] esse[15,16] partes[17,18] responderit[19], nihil[20] sumatur; si autem[21] impares esse dixerit[22,23], unum sumatur[24] in hac prima divisione ad memoriam divinantis[25], atque iterum[26] una pars divisionis[27], quando[28] ambae aequales fuerint[29], absque differentia triplicetur[30]. <2.3> Si vero impares fuerint, maior pars[31] triplicanda est[32] atque dividenda[33], sicut superius, in duas partes iterumque[34] interrogandum[35], si aequalis aut inaequalis[36] sit facta[37] divisio. <2.4> Et si aequalis[38] quidem facta est, nihil[39] sumendum[40] est[41], si autem inaequalis, duo sumendi sunt[42] in secunda divisione. <2.5> Et in medietate divisionis[43] si[44] aequalis fuit[45], absque ulla partium differentia, quot[46] novenarii[47] contineantur[48], interrogandum[49]. <2.6> Quodsi inaequalis[50] fuit[51,52], in maiori parte quaerendum est, quia quot[53] novenarii in ea[54] inveniuntur[55], tot quaternos[56] divinator sumere[57] debet[58].

---

[1] ITEM ALIA HV_3, ITEM L, ITEM VNDE SVPRA V, *om.* MM_1V_4W [2] Assumitur M_2M_3m, Adsumitur hf   [3] quislibet LM_1, *om.* M_3 [4] ac] et M_3, aut V_4 [5] triplicatur fm   [6] Triplicatur B_1h, triplicetur fm, *corr.* B_1^1, vero *add.* M_3 [7] partes *om.* BHP_2S [8] et *om.* MV   [9] numerum] in *add.* W   [10] concipit V   [11] numerus W, *om.* M_3 [12] sit aequa *tr.* MV   [13] pares ... partes] ambas partes esse pares V_3 [14] pares ambas] ambas pares BHP_2SV_2, aequa V, aequas MV^1 [15] ambas esse *tr.* M_3 [16] esse *om.* V_4 [17] esse partes *tr.* W   [18] partes *om.* B_1MV, *add.* B_1^1 [19] responderit] dixerit MV   [20] nihil ... esse dixerit *om.* P_2S [21] autem] vero MV   [22] esse dixerit *om.* MV   [23] dixerit] responderit BHV_2 [24] unum sumatur] unus adsumatur M_3 [25] divinitatis $BHLP_2SV_2$ [26] iterum *om.* V_4 [27] divisionis *corr.* M_3 *ex* divisiones   [28] quando] cum MV   [29] fiunt V_4 [30] triplicetur] ///cetur M_3, iterum triplicetur V_2 [31] pars *om.* L   [32] est *om.* V_2 [33] dividendi V_3 [34] iterumque] Iterum quae M_3h [35] interrogandum] est *add.* M_2W [36] aut inaequalis *om.* B   [37] sit facta *tr.* V_2 [38] aequales B   [39] nihil] unum MV, *corr.* V^1 [40] sumendus V_1 [41] est *om.* V_4 [42] sumendi sunt *om.* MV   [43] divisiones M_1 [44] si *ante* in medietate M_2M_3Whfm   [45] fuerit MVW   [46] quod $BB_1HP_2V_1V_2W$, *corr.* B_1^1, qui hfm   [47] novenari B_1, in eis *add.* M_3 [48] contineantur] sint V_4 [49] interrogandum] interrogetur M_3, est *add.* W   [50] inaequalis] aequalis B   [51] inaequalis fuit] inaequales fiunt MV   [52] fuit] fuerit W   [53] quia quot $BB_1M_1V_3V_4$] quia quod HV_1, quot $MM_2M_3VV_2$, quod W, quodquod P_2, quotquot P_2^1Shfm   [54] in ea *om.* MVV_4 [55] inveniantur BHP_2SV_2W, et *add.* MV   [56] quaternarios MM_2, novenarios M_3 [57] summere SV_2 [58] debet] *hic desinit* B

26   Ps. Bede: *De arithmeticis propositionibus*

<2.7> Ut[59] verbi gratia: si sex fuerint mente concepti, cum triplicati fuerint, XVIII faciunt[60]; XVIII[61,62] divisi in VIIII et VIIII partiuntur, et[63] quia aequalis est divisio, nihil ibi[64] sumendum est[65]. Novem iterum[66], quae[67,68] est[69] medietas huius divisionis, triplicati faciunt[70] XXVII[71]. Qui[72] XXVII[73] divisi in XIII et XIIII[74,75] resolvuntur, et quia ista[76] est[77] secunda divisio et inaequalis[78], duo[79] sumendi sunt[80]. Tum[81] in maiori parte ipsius[82,83] divisionis, hoc est[84] in[85] XIIII, quaerendum est, quoties[86] VIIII possint[87] inveniri[88]. In[89] XIIII[90] semel VIIII[91] sunt[92,93,94]. De his VIIII[95] IIII a divinante[96] colligendi sunt, qui[97] duobus, qui in secunda divisione[98] collecti sunt, adiuncti VI[99] faciunt[100]. Senarius ergo[101] numerus[102] primus[103] mente[104] comprehensus[105] est. <2.8> Et[106] hoc[107] in hac ratione[108] notandum est, quod[109], si[110] ambae divisiones[111] paritati[112] responderint[113], nihil ex his[114] sumendum est[115]. Si[116] vero prima impar[117] fuerit, unus sumendus[118] est; si secunda, duo sumendi[119] sunt[120]. Novenarius vero quaternarii[121] significationem continet[122]. Quod ideo in hac supputatione aliter quam[123] in[124] superiori accidere[125] videtur[126], quia[127] haec[128] bis triplicatur[129,130] et bis dividitur[131], illa[132] vero

---

[59] Ut *om.* $B_1MM_3VV_4W$ [60] faciunt] fiunt MVV_3W [61] XVIII ... partiuntur] quia in IX et XI dividuntur V_4 [62] XVIII] Qui M_3, et L, *om.* MV   [63] et *om.* MV   [64] ibi] sibi P_2SV_2(*post* sumendum est), *om.* MM_3VV_4 [65] est *om.* V_4 [66] iterum] igitur MV, vero M_3 [67] iterum, quae] iterumque LV_1V_3 [68] quae] qui M, quod V   [69] est *om.* L   [70] faciunt] fiunt W   [71] XXVII] XXVI M_3 [72] Qui *om.* $B_1HLM_1P_2SV_1V_2V_3$Whfm   [73] XXVII] XIII M_2, *om.* MM_3VV_4W [74] XIII et XIIII] XIIII et XIII P_2 [75] XIIII] XIII $M_2M_3V_1$ [76] ista *om.* M_3 [77] est *om.* W   [78] in(a)equalis MM_2W] in(a)equa $B_1HM_1M_3P_2V_1V_2V_3$hfm, in equa LS, inequalis V_4 *et corr. in* inequa   [79] duo sumendi sunt] *hic incipit* T *inscriptione* DE NVMERO *verbisque* Sunt vero duo sumendi   [80] sunt *om.* M_3 [81] cum M_2Whfm, tunc M_3V_4, *om.* MV   [82] parte ipsius *tr.* V_2 [83] ipsius] huius W   [84] hoc est] id est MV   [85] in *om.* L   [86] quoties] utrum *supr.* L^1 [87] possunt MV, possit M_3 [88] possint inveniri *tr.* V_3 [89] In] Sed in MV, *om.* hfm   [90] XIIII] .XVIII. W   [91] semel VIIII] .IX. semel V_4 [92] semel VIIII sunt] novem semel est MV   [93] VIIII sunt *tr.* M_2 [94] sunt *om.* L   [95] VIIII *om.* M_3V_4 [96] a divinante $LM_1M_3VV_1V_3V_4$] a divinitate HP_2TV_2, ad divinitatem S, ad unitatem M_2W, *corr.* T^1 [97] qui] Quibus W   [98] secunda divisione *tr.* T   [99] VII M_2 [100] VI faciunt] fiunt VI M_3, faciunt .VII. W   [101] ergo] igitur MVV_4 [102] numerus *om.* V_4 [103] primis $B_1LM_1V_1$, primum M_3, in V_4, *om.* V_2 [104] mentem M_3 [105] conceptus MP_2VV_4 [106] Et hoc ... ratione] Qua in re V_4 [107] hoc *om.* MV   [108] oratione M_2M_3Whfm   [109] quod] quia V_4 [110] si *om.* S   [111] divisionis S   [112] paritate H, *om.* M_3 [113] responderunt V_3, responderit T, *corr.* T^1 [114] his] eis $B_1LMVV_1V_3V_4$ [115] est *om.* V_4 [116] Si ... sumendus est *om.* M_2M_3Whfm   [117] prima impar *tr.* V_1 [118] unum sumendum V_1 [119] sumendi] simili M_2 [120] sumendi sunt *om.* V_4 [121] quaternari M_3, quartam M, .IIII. V   [122] optinet MV   [123] quam] quidem (?) B_1, *corr.* B_1^1 [124] in] alia *add.* V_4 [125] accideri H, contingere V_4, accedere W   [126] videretur V_3 [127] quia *om.* M_2Whfm   [128] haec *om.* M_2 [129] triplicatur et bis dividitur] dividitur et triplicatur M_3 [130] triplicatur] triplicatus H   [131] bis dividitur ... triplicatur et *om.* V_4 [132] illa ... semel dividitur *om.* S

bis[133] triplicatur et semel dividitur. Quapropter[134] in hac[135] novenarius quaternarium[136], in illa[137] vero[138] binarium significat.

* * *

<PROPOSITIO 3>

3.1–3.3: γ (= P_2SV_2) δ (= M_2M_3W) β (= LMM_1VV_3) B_1HTV_1 hfm
3.4–3.9: δ (= M_2M_3W) hfm

ITEM ALITER.[1]

QUOMODO DIVINANDUM[2] SIT, QUA[3] FERIA SEPTIMANAE QUISLIBET[4] HOMO REM QUAMLIBET[5],[6] FECISSET[7].

<3.1> Quemcumque[8] numerum cuiuslibet[9] feriae nomen continentem[10],[11] animo conceperit[12], primo debet duplicare[13], deinde illi[14] numero duplicato[15] quinque adiungere[16],[17] ipsamque[18] summam, quae de[19] his collecta est[20],[21], quinquies multiplicare[22], deinde totum[23] decies[24] ducere[25], post haec[26] ex toto[27] ducentos L[28] tollere[29] et hoc[30], quod remanet, pro feriae numero[31] tenere[32]. <3.2> Ut[33] verbi gratia: si de prima feria ratio habeatur[34],[35], unum duplicetur[36], fiunt duo; his V adiungantur[37], fiunt VII; qui VII[38] quinquies[39] multiplicati[40] fiunt XXXV[41]; qui XXXV[42] decies ducti fiunt[43] CCCL. De quibus si CCL tollantur[44],[45], remanent C, qui pro monade[46],[47], id est uno, qui primam feriam[48] significat, sumendi sunt[49]. <3.3> Hoc quoque[50] in[51] ista[52] supputatione[53] servandum[54] est,

---

[133] bis *om.* B₁, *supr.* B₁¹   [134] Quopropter V₂   [135] in hac] in hoc W, hoc HP₂SV₂, haec T   [136] quaternarius M₃   [137] in illa] illa HP₂STV₂, nulla hfm   [138] vero *om.* M₃
[1] ITEM ALIA HP₂ST, ITEM ALIA CAPITULUM V₃, ITEM UNDE SUPRA V₂, De divinatione W, *om.* B₁MM₁M₂VV₁   [2] dividendum V₃   [3] qua] QVACVMQVE L   [4] quilibet Whfm   [5] rem quamlibet *tr.* MM₂M₃VWhfm   [6] quamlibet] qualibet H   [7] fecisse L   [8] Quem cum V₂, Quicumque Wfm, Quaecunque h   [9] cuiusque M₂Whfm, cuius M₃   [10] continentem … conceperit] tenuerit animo V₂   [11] continentem] tenentem P₂S   [12] concepit W, quis *add.* B₁   [13] duplicari W, *ex* triplicare *corr.* M₃   [14] illo LM₃P₂SV₂W   [15] *ex* triplicato *corr.* M₃   [16] quinque adiungere] adice quinque V₂   [17] adiungere] adiunges W, debet *add.* M₃   [18] ipsam P₂S   [19] de] ex MV   [20] quae de his collecta est *om.* V₂   [21] est *om.* M₃   [22] multiplica W   [23] totam MV   [24] totum decies *tr.* M₃   [25] ducito V₂, duc W   [26] post haec] postea M₂M₃Whfm   [27] toto] octo W   [28] ducentos L] CCL. CCI M₂, CL M₃   [29] tollero B₁, tolle W   [30] hoc *om.* MVV₂   [31] feriae numero *tr.* W   [32] teneto M₃V₂, tene W, computare V   [33] Ut *om.* M₃VW   [34] ratio habeatur] agitur V₂   [35] habeatur] habebatur V₃, habetur W   [36] dupliciter B₁, et *add.* LM₂W   [37] adiungantur] adiuncti MV, adiunctis M₂, iunctis V₂, et *add.* W   [38] VII *om.* MVV₂   [39] quinquies *om.* V   [40] multiplicati] ducti M₃   [41] fiunt XXXV *tr.* V₁   [42] qui XXXV] qui MM₃VV₂, *om.* V₁   [43] fiunt] faciunt S   [44] tollantur … <3.3> ut semper CCL *om.* M₃hfm   [45] tollantur] tolluntur M₂   [46] pro monade] promade V   [47] modade S   [48] prima feria V   [49] sumendi sunt] accipitur V₂, *ante* significat S   [50] quoque] inquam V₃   [51] in ista … servandum est] observandum est ista supputatione M₂W   [52] ista] hac MVV₂   [53] suppositione B₁, *corr.* B₁¹   [54] servandum] considerandum V₂

ut semper CCL[55] de totius summae collectione[56,57] auferantur[58] et[59,60] quot[61] centenarii[62] remanserint[63] diligenter consideretur[64], quia, sicut[65] supradictum[66] est, semel centeni primam[67] feriam[68] significant[69], bis[70] centeni secundam[71], ter centeni tertiam, quater centeni quartam, quinquies centeni quintam, sexies centeni sextam, septies centeni septimam[72].

[<3.4> Item si de[73] secunda feria[74] ratio habeatur, II duplicentur, et[75] fiunt IIII. His V adiungantur, et[76] fiunt VIIII[77]. Qui VIIII quinquies multiplicati[78] fiunt XLV. Qui[79] XLV decies ducti[80] fiunt[81] CCCCL. De his tolle CCL[82], remanent CC, qui secundam feriam significant. <3.5> Si[83] de tertia feria ratio habeatur, III duplicentur, et[84] fiunt VI; his V adiungantur, et[85] fiunt XI; qui XI quinquies multiplicati fiunt[86] LV; qui LV decies ducti[87] fiunt DL[88]; de his tolle CCL et[89] remanent[90] CCC, qui tertiam feriam significant[91]. <3.6> Si de quarta feria ratio habeatur[92], IIII duplicentur, et[93] fiunt[94] VIII; his V[95] adiungantur[96], et[97,98] fiunt XIII; qui XIII[99] quinquies multiplicati fiunt LXV[100]; qui LXV decies ducti fiunt DCL; de his tolle CCL, et[101] remanent CCCC[102], qui quartam feriam significant. <3.7> Si de quinta feria ratio habeatur, V duplicentur, et fiunt X; his V adiungantur, et[103] fiunt XV, qui quindecim[104] quinquies multiplicati fiunt LXXV; qui LXXV decies ducti fiunt septingenti L; de his tolle CCL, et[105] remanent quingenti, qui quintam feriam significant. <3.8> Si de sexta feria[106] ratio habeatur[107], VI duplicentur, et[108] fiunt XII; his V adiungantur[109], et[110] fiunt[111] XVII; qui XVII[112] quinquies multiplicati fiunt[113] LXXXV; qui LXXXV decies ducti fiunt[114] DCCCL; de his tolle

---

[55] CCL *post* auferantur M$_2$ [56] de totius summae collectione *om.* M$_3$ [57] collectione] collectionis V$_2$, *ex* collatione *corr.* T   [58] auferantur] tollantur (*ante* de totius) V   [59] et quot ... septimam *om.* H *(rasura?)*   [60] et *om.* M$_3$ [61] quod M$_1$V$_1$W   [62] centum M, centeni V   [63] remanserunt Whfm   [64] considerentur V   [65] ut W, *om.* P$_2$S   [66] supradictum] superius dictum M$_3$, praedictum B$_1$LM$_1$V$_1$hfm   [67] primam ... quater centeni *om.* S   [68] feriam *om.* L   [69] significat P$_2$TV$_2$, *corr.* T^1 [70] bis ... septimam *om.* M$_1$ [71] secundam] II. feriam M$_3$, feriam significant *add.* V$_3$ [72] septimam] VII. M, septem h, Haec ratio minime fallit *add.* MV, EXPLICIT *add.* V$_3$ [73] de *om.* M$_2$h   [74] secunda feria] secundam M$_3$ [75] et *om.* M$_2$M$_3$hfm   [76] et *om.* M$_3$hfm   [77] VIII M$_3$ [78] multiplicati *om.* M$_3$ [79] fiunt XLV. Qui M$_2$W] fiunt M$_3$, *om.* hfm   [80] ducti] multiplicati W   [81] fiunt *om.* M$_3$ [82] CCL] et *add.* W   [83] Si ... <3.7> qui quintam feriam significant *om.* M$_3$ [84] et *om.* M$_2$hfm   [85] et *om.* M$_2$hfm   [86] fiunt *om.* hfm   [87] ducti] multiplicati W   [88] DL] 5500 h   [89] et *om.* M$_2$hfm   [90] remanet f   [91] significat hf   [92] habeatur] qu(a)eritur M$_2$hfm   [93] et *om.* hfm   [94] fient W   [95] V] 7 m   [96] *post* adiungantur *signo* * *lacunam indic.* hf   [97] et ... <3.7> his V adiungantur *om.* mh   [98] et *om.* f   [99] XIII] XV M$_2$ [100] LXV ... <3.7> multiplicati fiunt *om.* f   [101] et *om.* M$_2$ [102] CCCC ... <3.7> tolle CCL, et remanent *om.* M$_2$ [103] et *om.* hm   [104] quindecim] 12 h   [105] et *om.* hfm   [106] feria *om.* W   [107] habeatur] inquiritur M$_2$M$_3$hfm   [108] et *om.* M$_3$hfm   [109] V adiungantur] adde V W   [110] et *om.* M$_3$hfm   [111] fient W   [112] XVII *om.* M$_3$ [113] fiunt *om.* M$_3$ [114] fiunt *om.* W

CCL; remanent DC, qui sextam feriam significant[115,116]. <3.9> Si[117] de septima feria[118] ratio habeatur[119], VII duplicentur, et[120] fiunt XIIII[121]; his V[122] adiungantur, et[123] fiunt XVIIII; qui XVIIII[124] quinquies multiplicati fiunt LXXXXV; qui[125] decies ducti[126,127] fiunt DCCCCL[128]; de his tolle CCL[129]; remanent DCC, qui septimam feriam significant. Si huic numero animum tuum[130] diligenter commendaveris, omnes ferias absque errore explorare poteris.]

* * *

<PROPOSITIO 4>

4.1–4.6: B₁HTV₂W hfm

ITEM.[1]

<4.1> Verum cum vero facit[2] verum. Minus cum vero facit verum. Verum cum minus facit minus. Minus cum minus[3] facit minus. Verum[4] essentiam, minus nihil[5] significat. <4.2> Pone summam numeri quam volueris in veri, hoc est[6] essentiae, nomine, et pone aliam summam cuius volueris numeri in[7] adverbii[8], quod minus dicitur, nomine, quod[9] nihil[10] significare dixi, et confer illas duas summas. Quae maior fuerit, vincit minorem et consumit eam iuxta quantitatem magnitudinis[11] suae. <4.3> Ut[12] verbi gratia: si iungantur[13] duae summae numerorum, una, quae[14] veri nomine, id est essentiae, appellata sit[15], ut sunt VII, alia[16], quae minus adverbii nomine vocetur[17], ut sunt III, hae duae summae numerorum[18] sibi[19] collatae, hoc est essentis et non essentis, quia[20] maior est summa veri quam illius, quae dicitur minus, sicut[21] plus sunt VII quam III, vincit[22] VII verum III[23,24] minus, sed[25] non maiori[26] parte quam vincere potest, hoc est[27] IIII; valet enim III minus inde[28] VII tollere, VII autem tribus sublatis IIII superat, et ideo quando iunguntur III minus et

---

[115] qui sextam feriam significant *om.* M₃   [116] significant] significent fm   [117] Si] autem *add.* M₃   [118] feria *om.* M₃   [119] habeatur] inquiritur M₂hfm   [120] et *om.* M₃hfm   [121] XIIII] 19 hf   [122] his V … fiunt XVIIII *om.* W   [123] et *om.* M₂M₃hfm   [124] XVIIII *om.* M₃   [125] qui] Hi vero M₃, *om.* hf, 95 *add.* m   [126] qui decies ducti] ducti decies M₂   [127] ducti] multiplicati M₃   [128] DCCCCL] 905 hf   [129] CCL] et *add.* M₃   [130] animum tuum] animum M₃, *om.* W
 [1] Solutio de ea propositione qua dicitur verum cum vero facit verum et reliqua. B₁, Item aliud argumentum hfm, *om.* HW   [2] faciat f   [3] cum minus *bis habet* V₂   [4] Verus TV₂   [5] nihil *om.* W   [6] est] in *add.* B₁W   [7] in] ad Whfm   [8] adverbium hfm   [9] quid W   [10] ni(c)hilum B₁HTWhfm   [11] magnitudinem quantitatis Whfm   [12] Ut *om.* W   [13] iungatur T   [14] una que B₁, unaque HWhfm, unaquę T   [15] appellati sint Whf, appellata sint (sint *corr. in* sit) T   [16] ali(a)e HTV₂   [17] vocetur *om.* V₂   [18] numero hfm   [19] numerorum sibi] sibi numero W   [20] quae V₂   [21] sicut] minus *add.* W   [22] vincit] numerum *add.* HTV₂   [23] vincit VII verum III *om.* Whfm   [24] III *om.* HTV₂   [25] sed] si hfm   [26] maiore B₁   [27] est *om.* Whfm   [28] inde … iunguntur III minus *om.* HTV₂Whfm

VII, remanent IIII. <4.4> Similiter si iungantur[29] III veri nomine et VII minus, quia maior est[30] nihili[31] quam essentiae summa, vincit septenarius[32] non existens ternarium subsistentem et consumit eum[33,34] sua non essentia, et remanent de ipso sibi IIII numeri non existentes[35]; et hoc est quod dicitur: iunge III et VII minus, faciunt IIII minus. <4.5> Si autem III non existentes simul[36] et VII similiter non existentes numeros[37] iunxeris, decem non[38] esse monstrabis. Sicut enim duo veri, hoc est[39] existentes, numeri, ut sunt VII et III, verum[40], id est existentem, numerum efficiunt, hoc est denarium, sic duae[41] non existentes[42] numerorum summae denominatae[43], ut sunt III minus[44] et VII minus, X minus faciunt[45]. <4.6> Valet enim verum efficere et minus non efficere. Iunge[46] III[47] et VII, fiunt[48] X; iterum[49] iunge III minus et VII, fiunt[50] IIII; iunge III et VII minus, fiunt IIII minus[51,52,53]; iunge III minus[54] et VII minus[55], fiunt X minus[56].

---

[29] si iungantur *om.* V$_2$ [30] maior est *om.* B$_1$, *add.* B$_1^1$ [31] nichili B$_1$] ni(c)hilis HTV$_2$W   [32] septenari HT, septenarii V$_2$ [33] consumit eum] consumendum W   [34] eum] eam B$_1$ [35] existens T   [36] similiter W   [37] numeros … hoc est existentes *om.* Whfm   [38] non *om.* T   [39] hoc est … X minus] & cetr. H   [40] veri Whfm   [41] due B$_1$] duo TV$_2$hfm, v° (= vero *(?)*) W   [42] existes V$_2$ [43] denominatae] de nomine W   [44] minus *om.* TV$_2$ [45] efficiunt B$_1$TV$_2$ [46] Iunge … VII, fiunt IIII *om.* B$_1$, *in marg. add.* B$_1^1$ [47] III] minus *add.* Whfm   [48] fient W   [49] Item B$_1^1$ [50] fient W   [51] iunge III et VII minus, fiunt IIII minus *om.* W   [52] fiunt IIII minus *om.* hfm   [53] IIII minus *om.* B$_1$, *supr.* B$_1^1$ [54] minus *om.* TV$_2$ [55] minus *om.* W   [56] minut f

IV

The *Propositiones ad acuendos iuvenes* Ascribed to Alcuin[*]

Among the writings attributed to Alcuin the *Propositiones ad acuendos iuvenes* ("Problems to sharpen the minds of youths") has a special place. In this article first the transmission of this collection, then the question of authorship and finally its relation to other collections of recreational mathematics will be discussed[1].

1 Transmission

Fourteen manuscripts of the *Propositiones*, either in whole or in part, are known. The oldest, Vat. Reg. lat. 309 (R_1), was written at the end of the 9th century in the monastery of St. Denis near Paris. The codex comprises principally Bede's writings on *computus*, his chronicle and extracts from Isidore, Macrobius, Martianus Capella and Aratus. On some folios originally left blank (f. 1v–2r, 16rv, 3v–4r) the *De arithmeticis propositionibus* attributed to Bede[2] and without break 21 of the 56 problems of the *Propositiones* were added, also in the 9th century. For most of the problems solutions are given, and there is one solution to a problem not in the manuscript.

Seven of the fourteen manuscripts were written about 1000 or a little later; two other manuscripts are from the 11th century and there is one for each of the 12th, 13th and 15th centuries. The following remarks are about the origins of these manuscripts and about the texts that are found in the manuscripts in the vicinity of the *Propositiones*.

Shortly before 1000 manuscript Karlsruhe, Augiensis 205 (A) was written in the monastery at Reichenau. It is often mentioned in the lit-

---

[*]This is the English translation, with some changes, of the article "Die Alkuin zugeschriebenen *Propositiones ad acuendos iuvenes*", in P. L. Butzer, D. Lohrmann (ed.), *Science in Western and Eastern Civilization in Carolingian Times*, Basel 1993, pp. 273–281.

[1]For more details, see Folkerts, 'Aufgabensammlung', pp. 15–41. Three translations, into German, English and Czech, have recently been published: Folkerts / Gericke, 'Propositiones'; Hadley / Singmaster, 'Problems'; Mačák, 'Sbírky', pp. 11–36 (Czech translation of the propositions, but not of the proofs).

[2]For this, see Folkerts, 'De arithmeticis propositionibus'.

erature on history of science, mostly as "the" manuscript of the *Propositiones*. The Munich manuscript Clm 14272 (M_2), which only transmits five problems, was copied shortly after 1000 by the monk Hartwic of St. Emmeram during his stay in Chartres and is connected with the teaching of Fulbert, who was Hartwic's teacher in Chartres; Hartwic brought back the texts when he returned to St. Emmeram. The Leiden codex Voss. lat. oct. 15 (V) was written in the years 1023–1025 in the monastery of St. Martial in Limoges by Ademar of Chabannes (988–1034). Also shortly after 1000 the Vienna manuscript ÖNB 891 (W) was copied in southern Germany, possibly in Füssen. On palaeographical grounds three manuscripts may be ascribed to western Germany or eastern France: Vat. Ottob. lat. 1473 (O; about 1000); London, BL, Burney 59 (B; first half of the 11th century, perhaps Dijon); Montpellier 491 (M; beginning of the 11th century).

Manuscript Vatican, Reg. lat. 208 (R), which only transmits 24 problems in its present form (the three following folios are missing), is from the 11th century and belonged about this time to the monastery of St. Mesmin near Orléans. From the middle of the 11th century is the manuscript Venice, Z. L. 497 (1811) – here abbreviated as V_1 –, a handbook of the liberal arts which is connected with Lawrence of Amalfi[3]. In the 12th century the manuscript Munich, Clm 14689 (M_1), was written in Bavaria; like Hartwic's codex M_2, it also came from St. Emmeram to Munich. Of English origin are the two London manuscripts BL, Cotton Iulius D. VII (C), and Sloane 513 (S). C, which is incomplete (only 12 problems), was written in St. Albans in the 13th century, and S in Bukfastleigh (Devon) in the 15th century. To these manuscripts we should add a single folio (f. 229), probably of the 13th century, in the collection of fragments in manuscript BL, Sloane 1044[4]: on the recto side are the letters of the Greek alphabet, their names and the corresponding numerical value of them (given in Greek words and in Roman numerals). On the verso side are problems 3, 4, 32, 33 of the *Propositiones*. There

---

[3]For this manuscript, which I could not use for my edition in Folkerts, 'Aufgabensammlung', see Folkerts, 'De arithmeticis propositionibus', p. 7. The manuscript contains on f. 164v–169r the text of the *Propositiones* in the following order (the numbering according to the edition in Folkerts, 'Aufgabensammlung'): problems 7–11, 11b, 11a, 12, 13, 15–19, 21–33, 33a, 35–48, 51–53, 1 (with solution), 49; solutions of problems 7–10, 12, 13, 16–19, 21–33, 33a, 34 (problem with solution), 35–38, 51–53, 1, 49. This manuscript is in the order of the problems and in the wording of the text very close to manuscript M.

[4]This fragment could not be used for my edition in Folkerts, 'Aufgabensammlung'.

follow – written by another hand – four other problems of recreational mathematics which have no direct equivalent in the *Propositiones.*

In six manuscripts (B, M, M_1, O, R_1, V_1) the *Propositiones* comes immediately before or after the *De arithmeticis propositionibus* ascribed to Bede. In three cases the text of the *Propositiones* is similarly close to collections of fables (B, V: the Latin Aesop; R: Avian). In M, M_1 and V_1 the *Propositiones* is transmitted together with mathematical writings from the Gerbert circle, particularly texts on the abacus. We may also mention that in two manuscripts the *Propositiones* is transmitted with philological treatises (W: *Cena Cypriani* and excerpts from Isidore and Solinus; R: Servius and Fulgentius). In two cases (M_2, R_1) the *Propositiones* was added to the codex to fill blank pages. Later, individual problems from the *Propositiones* are found in collections of recreational mathematics, of which there was an increasing number since the 11th century: as examples we will mention those in manuscripts London, BL, Cotton Cleopatra B. IX (about 1270; from Abbotsbury Abbey, Dorset), and Oxford, BL, Digby 98 (15th century).

There were two independent printed editions of the *Propositiones*: as parts of the *Opera omnia* of Bede and Alcuin respectively. In 1563 the *Propositiones* was published in the Bede edition by Herwagen in Basel from a manuscript similar to the Emmeram manuscript M_1. In M_1 the *Propositiones* follows the *De arithmeticis propositionibus*, which could be considered as written by Bede; and this seems to be the reason why Herwagen ascribed the *Propositiones* to Bede. In later editions of Bede (Friessem, 1688; Migne, 1844) the *Propositiones* was reprinted practically unchanged. The *Propositiones* was also published as part of Alcuin's work, for the first time in 1777 in the edition by the Emmeram monk Forster (Frobenius). Astonishingly, Forster used for his edition neither of the two manuscripts which were at this time in St. Emmeram, but manuscript A. He apparently included the text in the edition of Alcuin's writings because in A the *Propositiones* follows Alcuin's (genuine) commentary on Genesis and the text of the *Propositiones* begins simply with "Incipiunt capitula propositionum ad acuendos iuvenes" – with no new author mentioned. After Forster's edition, which was reprinted by Migne in 1851[5], the *Propositiones* was normally ascribed to Alcuin.

---

[5] The text thus appears twice in the *Patrologia Latina.*

2   Authorship

To answer the question of authorship, one must first collect some information from the manuscripts. In no case is an author mentioned; and where there is a title, it usually runs "Incipiunt propositiones ad acuendos iuvenes". For its provenance the origins of the oldest manuscripts should be considered. All the older manuscripts came from east or north France or from west or south Germany. Thus this text was clearly well known in this area in the 10th and 11th centuries. Again, the *leuga* (*leuva*), a measure of length known in parts of what is now France, appears in two problems (nos. 1 and 52): this fits in well with the hypothesis of a French origin. The time of writing may be assumed to be at the latest in the 9th century, since the oldest manuscript, R_1, was written at the end of this century. In two problems (nos. 39 and 52) camels are mentioned: in no. 39 as being bought in the Orient, in no. 52 to bring grain from one place to another. Since there were diplomatic contacts between the Carolingian court and the Baghdad Caliphate at least by the time of Charlemagne, Alcuin's authorship appears more likely. There are other indications of his authorship:

1. Alcuin loved intellectual puzzles. Some are presented in his poems or in the *Disputatio Pippini cum Albino scholastico*. Besides, he had great interest in mathematics and its applications. In a letter that he wrote to Charlemagne in 800 Alcuin says that he sends him "puzzles of arithmetic subtlety for delight"[6]. These puzzles could well have been our *Propositiones*.

2. We may consider texts of mathematical content that are attributed to Alcuin in manuscripts or medieval library catalogues. In the editions of Alcuin's works, between *De rhetorica* and *De dialectica* are diagrams and short texts which include mathematical pieces[7]. Bernhard Bischoff has shown that this material constitutes an appendix that a scholar of the early Carolingian time – probably not Alcuin himself – inserted between these two writings in a manuscript of Alcuin's dialogues in order to unify different material of the *artes liberales*[8]. Manuscript Vienna 2269, which was written in France in the first half of the 11th century, begins with writings on dialectic, rhetoric, arithmetic, music and astronomy which are ascribed to Alcuin. But of geometry there is only a title (the following five columns

---

[6] "Misi ... et aliquas figuras arithmeticae subtilitatis laetitiae causa".
[7] Migne, *Patrologia Latina* 101 (1852), cols. 945–950.
[8] Bischoff, 'Einteilung', pp. 275f., 286f.

are left empty), and the arithmetic is not by Alcuin, but is an excerpt from the *Etymologiae* of Isidore of Seville (book 3, chapters 1–9). In medieval library catalogues there are at least three mentions of mathematical writings ascribed to Alcuin: in the cathedral library of Puy (11th century), in Christ Church, Canterbury (about 1300) and in Fulda (about 850). In the first two cases the descriptions are too vague – a "de arimetica" or "de arismetrica" is named. In the Fulda catalogue, among the writings of Alcuin there is a manuscript characterized by "eiusdem quaestiones in genesim. eiusdem de formulis arithmeticae artis. eiusdem de grammatica in I. vol.". This manuscript is lost. But if the attribution in the Fulda catalogue is correct, we have here a strong indication that Alcuin is the author. It should be mentioned that in the Reichenau manuscript *A* the *Propositiones* comes directly after the commentary on Genesis by Alcuin, as in the manuscript mentioned in the Fulda catalogue, and that the characterization "de formulis arithmeticae artis" in the Fulda catalogue is reminiscent of the "figurae arithmeticae subtilitatis" of Alcuin's letter.

There are, therefore, strong indications, but no definite proof that Alcuin wrote the *Propositiones*.

3   Content of the *Propositiones*

The *Propositiones* transmits 56 problems. Their outward form point to the monastic life: in a cathedral a floor is paved with stones; a bishop distributes bread among his clergy; an abbot distributes eggs to his monks. Manual work and trade play a minor role, *motifs* from rural life predominate. Among the objects mentioned are: bowls, barrels, wine jugs, oil bottles, ladders, wagon wheels. The animals that we meet are: donkey, hare, dog, hedgehog, camel, horse, cow and ox, sheep, snail, swallow, pig, stork, pigeon, wolf, goat – apart from the camel, all animals common in Central Europe.

Which problems are treated in the *Propositiones*? With the exception of a few single problems, we find the following groups[9]:

1. Linear problems with one unknown (13 items). Most of these problems belong to the so-called "*hau* problems", named after similar problems in the Papyrus Rhind where a *hau* (*aha*) or "heap" has to be found. The problems are to find the number of a group or the age of a person

---

[9]For details, see Folkerts, 'Aufgabensammlung', pp. 35–41. For the history of these types of problems, see Tropfke, 'Elementarmathematik', chapter 4 (pp. 513–660).

and lead to a linear equation $nx + p = 100$, where n is usually the sum of natural numbers or fractions and p is an integer or zero. To this group we can count a problem of speeds: a dog pursues a hare, which has a given start. Another problem is to determine the weights of the constituents of a discus, when the ratio of the different constituents is known. Also to this group belongs a vessel problem: a barrel is filled through three pipes. But in this case simple division suffices for the solution, in contrast to the normal problems of this class of "cistern problems".

2. Linear problems with several unknowns (9 items). Eight of the nine problems belong to the type usually known as "The hundred fowls". Here among a certain number of persons the same number of objects is distributed, though unevenly; or 100 animals of different value cost 100 *solidi*. In each case whole-number solutions of a system of two simultaneous equations with three unknowns are sought. The oldest known problems of this sort were formulated in China in the 5th century A.D.; most of such problems involve 100 birds of various species. Among the Arabs Abū Kāmil (ca. 850–930) treated several such problems and considered all possible solutions. It is fully possible that knowledge of such puzzles found the way to Central Europe through the Arabs; we note that Charlemagne was in diplomatic contact with Hārūn al-Rashīd.

3. Problems with sequences and series (3 items). Two of the problems treat geometrical progressions and the third an arithmetic progression. In one case a king collects together an army by going into 30 houses and from each brings out as many soldiers as went in to collect them. The sum of the geometric progression with quotient 2 is found by simple addition. In the problem with the arithmetic progression the number of pigeons on a staircase with 100 steps is to be determined: on the first step there is one pigeon, on the second there are two, on the third three, and so on. In this case, exceptionally, a method of solution is given: one adds together the pairs $1 + 99$, $2 + 98$, ..., $49 + 51$; since there are 49 such pairs, this makes 4900 pigeons; to these one must add 50 and 100. The idea of the solution reminds us of the procedure of the young Gauss, who in the primary school found the sum of the first 100 integers in a similar way.

4. Problems of ordering in the widest sense (11 items). To this category belong the four transport problems. The best-known is that of the wolf, the goat and the cabbage which have to be transported across a river without the goat's eating the cabbage or the wolf's eating the goat. Two other examples belong to the "barrel-sharing problems"; in one, 10 full, 10 half-full and 10 empty glass flasks with oil have to be divided

among three persons so that each of them receives equally of both glass and oil. Another problem concerns the inheritance of twins when the portion for each child and for the mother have definite ratios. One problem is about buying and selling pigs at the same total price, but with a profit for the dealers.

5. Geometrical problems (12 items). In these areas are to be determined or transformed into other areas. These are not questions of recreational mathematics, but show geometric rules by means of examples. In contrast to most of the other problems, the method of solution is here usually described. The form of the problems and the methods of solution show that these questions stand in the tradition of the Roman land surveyors (*agrimensores*): there are similar problems to be found in Heron, in Columella's *De re rustica* and in many writings of the *Corpus agrimensorum*. There is a particularly strong similarity to the so-called *Geometria incerti auctoris*, an anonymous geometrical writing which is connected with Gerbert's *Geometry*, but is older. As Bubnov showed some hundred years ago, this treatise is of the 10th century at the latest and was probably written in the 9th; but a rigorous investigation of this important text has still to be made. The geometrical problems in the *Propositiones* are in the same order as in the *Geometria incerti auctoris*, the numbers in the examples and the methods of solving the problems are mostly the same. Where there are differences, the method in the *Geometria incerti auctoris* is always better and the solution is more exact. Its author knew the basic rules for geometry and calculation with fractions, whereas the solutions in the *Propositiones* often seem clumsy and even faulty. It seems that both texts were derived from a lost source. In many of the geometrical problems in the *Propositiones* there are two redactions of the text, the first being better and closer to the *Geometria incerti auctoris*[10].

6. There are a few problems that have nothing to do with mathematics: three problems appear to be spoof questions and three have to do with complicated family relationships.

All items of the collection are put in an attractive form. The method of solution is given only in the geometric problems, in the calculation of the number of pigeons on the staircase and of the number of soldiers to be collected in the town: in all other cases the text is limited to giving the solutions and checking that they are correct.

---

[10]See Folkerts, 'Aufgabensammlung', pp. 25–30.

4   Sources and influence

The *Propositiones* stands in a long tradition of recreational mathematics, which stretches back at least to about 1800 B.C. among the ancient Egyptians and Babylonians. The Greeks and Romans, Chinese and Indians also had similar problems. We find the so-called *"hau* problems" in ancient Egypt in the Rhind mathematical Papyrus and also in the arithmetical epigrams in the *Anthologia Graeca* from late Greek or Byzantine time, most of which are supposed to have been formulated by Metrodorus (4th to 5th century A.D. ?). The Greek problems were possibly transmitted to the West by Byzantine diplomats: we know that in 781 A.D. there was an embassy from Byzantium to Charlemagne and that from this legation a Greek named Elissaios was left behind as a teacher of Princess Rotrud, but, of course, transmission through less official channels is always possible. Further, there are clear influences on the *Propositiones* from the Roman tradition: already mentioned are the geometrical problems similar to those in the *agrimensores* tradition. To these we may add the three inheritance problems and the four items on family relationships, but it should be noted that the author of the *Propositiones* understood very little of Roman legal practice[11]. Finally, some items, such as the wolf-goat-cabbage puzzle and the other transport problems, seem to have first appeared in the *Propositiones*.

The *Propositiones* seems to have had great influence. Items from the collection reappear in collections of problems which were compiled, largely from mathematical material, in the later Middle Ages. They are also found – together with other problems from the Arabic tradition – in the writings of Leonardo Fibonacci at the beginning of the 13th century; from there they were transmitted to the Italian *maestri d'abbaco* and to the German *Rechenmeister* in the late Middle Ages and the Renaissance. Some of them were intensely studied by Leonard Euler in the 18th century and subjected to an exact mathematical treatment. Thus the *Propositiones* forms part of the tie that binds Antiquity, the oriental and occidental Middle Ages, the Renaissance and modern mathematics. From this story one is reminded that mathematics has not only been a strong logico-deductive science, but also a source of fun and entertainment: the mathematician is and was also a *homo ludens*.

---

[11]See Folkerts, 'Aufgabensammlung', p. 33.

Bibliography

B. Bischoff, "Eine verschollene Einteilung der Wissenschaften", in B. Bischoff, *Mittelalterliche Studien*, vol. 1, Stuttgart 1966, pp. 273–288.

M. Folkerts, "Die älteste mathematische Aufgabensammlung in lateinischer Sprache: Die Alkuin zugeschriebenen Propositiones ad acuendos iuvenes. Überlieferung, Inhalt, Kritische Edition", *Denkschriften der Österreichischen Akademie der Wissenschaften, mathematisch-naturwissenschaftliche Klasse*, 116. Band, 6. Abhandlung, pp. 13–80, Vienna 1978. (Reprinted in this volume, item V.)

M. Folkerts, "*De arithmeticis propositionibus*. A Mathematical Treatise Ascribed to the Venerable Bede" (In this volume, item III).

M. Folkerts, H. Gericke, "Die Alkuin zugeschriebenen Propositiones ad acuendos iuvenes (Aufgaben der Schärfung des Geistes der Jugend)", in P. L. Butzer, D. Lohrmann (ed.), *Science in Western and Eastern Civilization in Carolingian Times*, Basel etc. 1993, pp. 283–362.

J. Hadley, D. Singmaster, "Problems to sharpen the young. An annotated translation of *Propositiones ad acuendos juvenes*, the oldest mathematical problem collection in Latin, attributed to Alcuin of York", *The Mathematical Gazette* 76 (1992), 102–126.

K. Mačák, *Tři středověké sbírky matematických úloh. Alkuin, Métrodóros, Abú Kámil*, Prague 2001.

J. Tropfke, *Geschichte der Elementarmathematik. 4. Auflage. Band 1: Arithmetik und Algebra. Vollständig neu bearbeitet von K. Vogel, K. Reich, H. Gericke*, Berlin / New York 1980.

V

Die älteste mathematische Aufgabensammlung in lateinischer Sprache:

Die Alkuin zugeschriebenen

PROPOSITIONES AD ACUENDOS IUVENES

Überlieferung, Inhalt, Kritische Edition

INHALTSVERZEICHNIS

Denkschrift math.-nat. Kl. 116. Band, 6. Abhandlung

1. EINLEITUNG

Eine sehr reizvolle Aufgabe mathematikhistorischer Forschung besteht darin, die Geschichte bestimmter mathematischer Aufgabentypen und Lösungsmethoden zu erforschen. Es ist schon lange bekannt, daß oft dieselben Probleme zu verschiedenen Zeiten und in voneinander weit entfernten Kulturkreisen behandelt wurden. Dabei nimmt man an, daß manche Probleme des angewandten Rechnens Bestandteil der Literatur vieler Völker sind, ohne daß man eine gegenseitige Beeinflussung vermuten darf. Wenn allerdings eine Aufgabe mit denselben nicht zu einfachen Zahlenwerten in verschiedenen Quellen überliefert wird, muß man an eine Abhängigkeit denken. Es ist jedoch auch in diesen Fällen gegenwärtig noch nicht möglich, zu sicheren Erkenntnissen über den Weg eines Problems zu gelangen; dazu sind die kulturellen Beziehungen zwischen den Völkern zu komplex und in den Einzelheiten zu wenig geklärt. Gemeinsam mit Mathematikhistorikern müßten hier Vertreter anderer historischer Disziplinen wie Wirtschafts- und Sozialgeschichte, aber auch die Philologen mitarbeiten. Eine solche Arbeit könnte dazu beitragen, die kulturellen Leistungen der beteiligten Völker, die Gemeinsamkeiten, aber auch die Unterschiede ihrer wissenschaftlichen Entwicklung herauszuarbeiten und dabei insbesondere den europazentrischen Standpunkt zu überwinden, der immer noch viele wissenschaftshistorische Darstellungen beherrscht.

Als Vorarbeit für eine derart anspruchsvolle Untersuchung stellt sich dem Mathematikhistoriker zunächst die Aufgabe, die zahlreichen Sammlungen praktischer Mathematik zu untersuchen, festzustellen, wo das einzelne Problem oder die verwendete Methode sich erstmals findet, und — wenn möglich — Aussagen über Entstehung und Einfluß der betreffenden Sammlung zu machen. Gerade in den letzten Jahrzehnten sind hier neue Untersuchungen erschienen. So hat K. VOGEL verschiedene wichtige Texte aus dem chinesischen[1]), byzantinischen[2]) und süddeutschen Raum[3]) ediert und zum Teil kommentiert, und sowjetische Gelehrte, insbesondere A. P. JUSCHKEWITSCH und B. A. ROSENFELD, haben in zahlreichen grundlegenden Arbeiten neue Erkenntnisse über die Leistungen der arabischen Mathematiker gewonnen und zusammenfassende Darstellungen über die Mathematik der Chinesen, Inder und Araber im Mittelalter vorgelegt[4]).

Der Erkenntnis, daß erst einmal Klarheit über die wesentlichen Sammlungen mathematischer Aufgaben bestehen muß, bevor eine endgültige Geschichte des angewandten Rechnens geschrieben werden kann, verdankt die vorliegende Arbeit ihre Entstehung. Sie beschäftigt sich mit der ältesten und vielleicht auch wichtigsten Sammlung der Unterhaltungsmathematik in lateinischer Sprache, nämlich mit den ALKUIN zugeschriebenen *Propositiones ad acuendos iuvenes* (künftig abgekürzt als *Propositiones*). Da diese Schrift im Gegensatz zu anderen mehrfach ediert ist[5]) und in der wissenschaftlichen Literatur oft erwähnt wird,

---

[1]) VOGEL (4).
[2]) VOGEL (2), VOGEL (3).
[3]) VOGEL (1).
[4]) Vor allem A. P. JUSCHKEWITSCH/B. A. ROSENFELD, Die Mathematik der Länder des Ostens im Mittelalter, in: *Sowjetische Beiträge zur Geschichte der Naturwissenschaft*, Berlin 1960, S. 62—160; A. P. JUSCHKEWITSCH, *Geschichte der Mathematik im Mittelalter*, Leipzig 1964.
[5]) Seit 1563; siehe Seite 21f.

bedarf es einer Begründung, warum gerade die *Propositiones* zum Thema einer Arbeit genommen werden.

Die maßgebliche Edition der *Propositiones*, die FORSTER 1777 veröffentlichte, benutzt, wie wir unten sehen werden, nur eine Handschrift, nämlich den bekannten Karlsruher Augiensis 205 von der Reichenau, der fast unverändert abgeschrieben wurde. Bei FORSTER und in späteren Arbeiten wird der Eindruck erweckt, als gebe es nur diesen Codex. Merkwürdigerweise hat in der Folgezeit keiner die Überlieferung dieser Schrift untersucht, sondern man betrachtete FORSTERS Ausgabe, die durch den Wiederabdruck in der *Patrologia Latina* leicht greifbar war⁶), gleichsam als maßgeblichen und endgültigen Text. Selbst in MAX MANITIUS' *Geschichte der lateinischen Literatur des Mittelalters* wird außer dem Augiensis nur noch die Handschrift *W* erwähnt⁷). Zu dieser unzulässigen Vereinfachung hat wohl die Tatsache geführt, daß M. CANTOR an verschiedenen Stellen⁸) ausführlich auf die *Propositiones* eingegangen ist und sich daher kein Mathematikhistoriker oder Philologe veranlaßt sah, selbständige Forschungen anzustellen. Ich kenne nur eine nicht in der Tradition CANTORS stehende Arbeit⁹), die aber von Historikern und Philologen kaum beachtet wurde, da sie an versteckter Stelle erschien¹⁰).

So gelten CANTORS Erörterungen noch immer als die maßgeblichen Aussagen zu den *Propositiones*. Dies ist bedauerlich, da CANTOR die Schrift nur nach FORSTERS Ausgabe und der Reichenauer Handschrift, die mit jener Edition fast identisch ist¹¹), kannte und er nur darauf Wert legte, die Aufgaben zu römischen Texten, insbesondere zu den Agrimensoren, in Beziehung zu setzen. Daher wissen wir durch CANTORS Arbeiten zwar einiges über den Inhalt der Schrift, jedoch nur wenig über benutzte Vorbilder und nichts über die Entstehungsgeschichte des Traktats sowie die Einflüsse auf spätere Sammlungen der praktischen Mathematik.

Hieraus ergibt sich das bescheidene Ziel, das sich die vorliegende Arbeit gesetzt hat: Vor allem durch Untersuchung der erhaltenen Handschriften (Kapitel 2 und 3) soll etwas Licht in die Frage der Entstehung und Verbreitung der *Propositiones* gebracht werden (Kapitel 4 und 5). Hierbei wird auch die Verfasserfrage behandelt. Die Analyse der Texte, die in den verschiedenen Codices überliefert werden, wird auch zu neuen Aussagen in der Quellenfrage führen (Kapitel 6): Zumindest einige vom Autor verwertete Bausteine kristallisieren sich heraus. Das Kapitel 7 beschränkt sich darauf, die Problemgruppen zu nennen, die in den *Propositiones* behandelt werden. Dagegen war es aus den im Anfang genannten Gründen nicht möglich, die Wirkungsgeschichte dieser Schrift schon jetzt exakt zu beschreiben. Es steht zwar außer Zweifel, daß die *Propositiones* die mittelalterlichen Aufgabensammlungen in lateinischer Sprache und in den Nationalsprachen direkt oder indirekt stark beeinflußten¹²), doch ist es zu früh, gegenwärtig auf Einzelheiten einzugehen. Den umfangreichsten Teil der Arbeit bildet eine kritische Edition des lateinischen Textes, die alle Handschriften berücksichtigt.

---

⁶) Siehe Seite 22.

⁷) 1. Band, 1911 = München 1965, S. 286. Zur Handschrift *W* siehe Seite 18.

⁸) CANTOR (1), S. 286—288; CANTOR (2), S. 139—150; CANTOR (3), S. 834—839.

⁹) THIELE (1), S. 22—25.

¹⁰) Immerhin erwähnt D. E. SMITH (*History of mathematics*, Band 1, 1923 = New York 1958, S. 186f.) THIELES Theorie.

¹¹) Siehe S. 22.

¹²) So etwa die Sammlungen der Klostermathematik, die Schriften LEONARDOS von Pisa und die Practica des *Algorismus Ratisbonensis*, wobei in den beiden letzten Fällen auch die kaufmännische Mathematik stark berücksichtigt ist. Siehe FOLKERTS (1); VOGEL (1); B. BONCOMPAGNI, *Scritti di Leonardo Pisano*, Rom 1857/62.

2. DIE HANDSCHRIFTEN UND DRUCKE

Nicht nur eine oder zwei Handschriften der *Propositiones* sind erhalten, sondern mindestens zwölf. Sie sollen im folgenden kurz charakterisiert werden, wobei jeweils Inhalt, Entstehungszeit und -ort und gedruckte Beschreibungen erwähnt werden. Die Reihenfolge ergibt sich aus dem Alter der Codices. Die Numerierung der *Propositiones* stimmt mit FORSTERS Ausgabe[13]) überein; die Nummern 11a, 11b und 33a bezeichnen diejenigen Zusätze, die in den BEDA-Ausgaben[14]) an den entsprechenden Stellen abgedruckt sind.

R_1 = VAT. REGIN. LAT. 309, f. 16rv. 3v—4r

Die Handschrift entstand vermutlich s. 9 ex. im Kloster St. Denis bei Paris; mit Sicherheit stammen die zahlreichen frühen Zusätze (s. 10/11) von dort. Spätere Besitzer: PAUL PETAU (M. 50), ALEXANDER PETAU (Nr. 397), CHRISTINE von Schweden (Nr. 499). Alle Einzelheiten sind in der sehr genauen Beschreibung von A. WILMART[15]) nachzulesen. Die Handschrift enthält u. a. BEDAS Schriften zur Zeitmessung, seine Chronik, Auszüge aus ISIDOR, MACROBIUS, MARTIANUS CAPELLA und ARATOS. Auf den vorher leeren Blättern 1v—2r, 16rv, 3v—4r wurden noch im 9. Jahrhundert zunächst Ps. BEDAS *De arithmeticis propositionibus* und dann unmittelbar anschließend Auszüge aus den *Propositiones ad acuendos iuvenes* nachgetragen. Es handelt sich um die Aufgaben 4, 5, 8, 11, 11a, 11b, 22—26, 32, 33a, 36, 40, 42, 48, 52, 49, 47, 45 und die Lösungen von Nr. 2, 5, 8, 11, 11a, 11b, 22—26, 32, 33a, 36, die eingeleitet werden durch die Überschrift *Proposi/ones* (!) *ad acuendos iuvenes.*

O = VAT. OTTOBON. LAT. 1473, f. 28r—35v

Die Handschrift besteht aus drei Teilen (f. 1—27: s. 14, f. 28—37: s. 10/11, f. 38—45: s. 14 in.). Beim ältesten Teil, der von nur einer Hand geschrieben wurde, deutet der Schriftcharakter auf westdeutschen oder ostfranzösischen Ursprung hin. Vorbesitzer: PAUL PETAU (P. 29), CHRISTINE von Schweden (1994), Kardinal P. OTTOBONI. Auf f. 28r befinden sich zwei kurze Texte über Längenmaße und die Umrechnung des Erdumfangs. Es folgen auf f. 28r—35v die *Propositiones* und anschließend (f. 35v. 37rv) Ps. BEDAS *De arithmeticis propositionibus.* Der Besitzvermerk auf f. 37v ist leider nicht mehr vollständig lesbar[16]). Die inscriptio der *Propositiones* lautet: *Incipiunt propositiones adecuendos* (!) *iuvenes.* Voraus gehen die Überschriften I—LVI; es folgen die Aufgaben 1—11, 11a, 11b, 12, 13[17]), 33a, 34—52, 14—33, 53 mit den Lösungen (außer zu 11, 11a, 11b), wobei sich diese unmittelbar an die Aufgaben anschließen.

A = KARLSRUHE, BADISCHE LANDESBIBLIOTHEK, AUGIENSIS 205, f. 54r—70r

Die Handschrift wurde Ende des 10. Jahrhunderts auf der Reichenau geschrieben und blieb dort, bis die dortigen Codices Anfang des 19. Jahrhunderts nach Karlsruhe geschafft wurden. Der Codex beginnt mit folgenden Texten: f. 1r—54r: ALKUIN, *Quaestiones in genesin;* f. 54r—70r: die *Propositiones;* f. 70rv: Scherzrätsel unter dem Titel *Enigmata rkskbklkb* (= *risibilia*). Es folgen Chroniken und Berichte, die die Reichenau betreffen, darunter auch

---

13) Siehe S. 22.
14) Siehe S. 21f.
15) *Codices Reginenses Latini,* Band 2, Vatikan 1945, S. 160—174.
16) *Iste liber est sancti ////.*
17) In der Vorlage von O muß eine Blattversetzung stattgefunden haben: Auf f. 30r bricht der Text mitten in der Lösung zu Aufgabe 13 ab; es folgen die letzten beiden Worte der Lösung zu Aufgabe 33 (*modia XXX*) und dann auf f. 30r—33r die Aufgaben 33a, 34—52 bis zum Anfang der Lösung (*In prima subvectione:* Zeile 582 der Edition). Dann schließt sich unmittelbar das Ende der Lösung 13 an und die Aufgaben 14—33 mit Lösungen (bis f. 35v). Das Ende bildet die Aufgabe 53 mit Lösung (f. 35v). Die ursprüngliche Reihenfolge war also 13—33, 33a, 34—52; durch die Versetzung ist die Lösung der Aufgabe 52 bis auf die ersten drei Worte verlorengegangen.

PURCHARTS Gedicht über die Taten des Abts WITIGOWO, das zwischen 994 und 996 verfaßt wurde. Der Codex schließt mit *Expositiones in parabolas Salomonis* (f. 84v—115v) und *in Ecclesiasten* (f. 115v—137r). Die Abschrift der *Propositiones* stammt wohl aus demselben Jahrzehnt wie das Original der *Gesta Witigowonis*. In der Sekundärliteratur wird die Handschrift öfter erwähnt und i. a. als „der" Codex der *Propositiones* bezeichnet, so etwa in dem ausführlichen Bericht von MORITZ CANTOR[18]). Der Codex ist genau beschrieben von ALFRED HOLDER[19]). Die *Propositiones* beginnen im Anschluß an ALKUINS Genesiskommentar ohne besondere Überschrift mit den Worten *Incipiunt capitula propositionum ad acuendos invenes*. Es folgen die Überschriften der Aufgaben 1—52 und dann nach der inscriptio *Incipiunt propositiones ad acuendos iuvenes* die Aufgaben 1—53 mit den Lösungen unmittelbar nach der jeweiligen Aufgabe.

W = WIEN, ÖSTERREICHISCHE NATIONALBIBLIOTHEK, MS. LAT. 891, f. 4v—27v

Der Codex wurde ganz kurz nach 1000 in Süddeutschland, möglicherweise in Füssen, geschrieben[20]). Er gehörte dem Benediktinerkloster St. Mang, Füssen, an[21]) und befand sich schon 1576 in der Wiener Hofbibliothek (alte Signatur: Philol. 425). Die Handschrift enthält auf f. 1r—4r die *Cena Cypriani*[22]). Es folgen auf f. 4v—27v die *Propositiones ad acuendos iuvenes* und auf f. 27v—66v Exzerpte aus ISIDOR und SOLIN. Zwei *Carmina de destructione Aquilegiae* mit einigen Anhängen beschließen die Handschrift. Sie ist kurz beschrieben in den *Tabulae*[23]) sowie bei HERMANN[24]) und ausführlicher bei ENDLICHER[25]). Die *Propositiones* tragen die Überschrift *Incipiunt propositiones ad exertitium acuendorum iuvenum*. Vorhanden sind die Aufgaben 1—10, 12—18, 21, 19, 20, 27, 28, 30—42, 44—49, 51—53 mit Lösungen, die jeweils direkt auf die Aufgaben folgen.

M₂ = MÜNCHEN, BAYERISCHE STAATSBIBLIOTHEK, Clm 14272, f. 181v

Die meisten Texte der Handschrift wurden kurz nach 1000 vom Emmeramer Mönch HARTWIC während seines Aufenthaltes in Chartres gesammelt und hängen mit dem Unterricht FULBERTS (1007—1029) zusammen, bei dem HARTWIC lernte[26]). HARTWIC brachte die Texte bei seiner Rückkehr nach St. Emmeram mit. Der Codex (alte Signatur: Em. C 91) enthält überwiegend philosophische Schriften (u. a. BOETHIUS' Kommentar zu CICEROS *Topica*, seine *Introductio in cathegoricos syllogismos*) und Traktate zur Musik (u. a. von BOETHIUS und HUCBALDUS). Nach einer musikalischen Abhandlung hat HARTWIC auf f. 181v Exzerpte aus den *Propositiones* eingetragen, und zwar die Aufgaben 1—4, 8 mit den jeweils folgenden

---

¹⁸) CANTOR (2), S. 141—150.219.

¹⁹) *Die Handschriften der Großherzoglich Badischen Hof- und Landesbibliothek in Karlsruhe. Band 5: Die Reichenauer Handschriften*, 1. Band, Leipzig 1906, S. 466—469.

²⁰) Briefliche Mitteilung von Prof. B. BISCHOFF, München. Nach HERMANN J. HERMANN, *Beschreibendes Verzeichnis der illuminierten Handschriften in Österreich. Neue Folge: Die illuminierten Handschriften und Inkunabeln der Nationalbibliothek in Wien, Band 1: Die frühmittelalterlichen Handschriften des Abendlandes*, Leipzig 1923, S. 206, entstand dieser Codex dagegen in der 2. Hälfte des 10. Jahrhunderts in der Diözese Aquileia. HERMANNS Vermutung beruht wohl darauf, daß der Codex auch ein Gedicht über die Geschichte Aquileias zur Zeit Kaiser Ludwigs enthält; vergleiche hierzu *MGH Poetae* 2, 150.

²¹) Auf der Versoseite des vorderen Blattes befindet sich der Besitzvermerk von einer Hand des 15. Jahrhunderts: *Sancti Magni est confessoris*. P. RUF (*Mittelalterliche Bibliothekskataloge Deutschlands und der Schweiz*, 3. Band, München 1932/39) erwähnt auf S. 112—120 auch das Benediktinerkloster St. Mang in Füssen, ohne die Wiener Handschrift zu nennen.

²²) *MGH Poetae* 4, 872ff.

²³) *Tabulae codicum manu scriptorum praeter Graecos et orientales in Bibliotheca Palatina Vindobonensi asservatorum*, Band 1—2, Wien 1864 = Graz 1965, S. 151.

²⁴) HERMANN²⁰), S. 206.

²⁵) STEPHAN ENDLICHER, *Catalogus codicum manuscriptorum Bibliothecae Palatinae Vindobonensis, pars I: Codices philologici Latini*, Wien 1836, S. 296—302 (dort unter Nr. CDXX).

²⁶) B. BISCHOFF, *Mittelalterliche Studien*, Band 2, Stuttgart 1967, S. 80f.

Lösungen zu 1—4, alles ohne Überschriften. Die Handschrift ist beschrieben im Katalog der Münchner Handschriften[27]).

V = Leiden, Bibliotheek der Rijksuniversiteit, Voss. lat. oct. 15, f. 203 v—205 v. 206 v—210 r

Der Codex wurde von Ademar von Chabannes (988—1034) und anderen Schreibern in den Jahren 1023—1025 im Kloster St. Martial zu Limoges geschrieben und verblieb nach Ademars Tod dort[28]). Spätere Besitzer: Paul Petau (H. 13), Alexander Petau (1028), Königin Christine von Schweden, Isaak Voss, Gerard Voss. L. Delisle geht neben anderen Handschriften des Ademar auch auf diese ein[29]). Der Codex enthält verschiedenartige Texte, darunter Symphosius, Avians Fabeln, Gedichte aus der *Anthologia Latina* und Glossen zu römischen Dichtern, Prudentius' *Psychomachia*, zwei Werke Priscians, Hygins *Astronomica*[30]). Den letzten Teil der Handschrift (ab f. 195) schrieb Ademar selbst. Hier finden wir Äsops Fabeln in lateinischer Übersetzung, an die sich ohne Trennung und Überschrift unmittelbar Teile von Alkuins (?) *Propositiones* anschließen[31]), und zwar befinden sich auf f. 203 v—205 v. 206 v—210 r die Aufgaben 1—11[32]), 11 a, 11 b, 12—20, 22, 21, 23—33, 33 a, 34—42, 44—53, 43 mit Lösungen (außer zu 11, 11 a, 11 b) direkt nach den Aufgaben. Der Text ist gegenüber der üblichen Fassung in vielen Kleinigkeiten überarbeitet.

B = London, British Museum, Burney 59, f. 7 v—11 r

Der Codex wurde s. 11[1] in Westdeutschland oder Ostfrankreich geschrieben; die Zuweisung nach St. Bénigne, Dijon, ist ungewiß[33]). Die Handschrift gehörte vielleicht der Sammlung des Grafen McCarthy-Reagh an[34]) und wurde im Jahre 1818 zusammen mit den übrigen Handschriften des Charles Burney († 1817) vom British Museum übernommen. Der Codex, von dem keine ausführliche gedruckte Beschreibung existiert[35]), enthält auf f. 2 v—7 v den Romulus-Text von Äsops Fabeln *(recensio Gallicana)* und unmittelbar anschließend auf f. 7 v—11 v die *Propositiones*. Der Beda zugeschriebene Traktat *De arithmeticis*

---

[27]) *Catalogus codicum manu scriptorum Bibliothecae Regiae Monacensis* IV, 2, München 1876, S. 152 f.

[28]) Vermerk in der Handschrift auf f. 141 v: *hic est liber sanctissimi domini nostri Marcialis Lemouicensis ex libris bonae memoriae ademari grammatici. Nam postquam idem multos annos peregit in domini seruicio ac simul in monachico ordine in eiusdem patris coenobio, profecturus hierusalem, ad sepulchrum domini nec inde reuersurus, multos libros in quibus sudauerat eidem suo pastori ac nutritori reliquid, ex quibus hic est unus.* Dieser Vermerk stammt aus dem 11. Jahrhundert.

[29]) Notice sur les manuscrits originaux d'Adémar de Chabannes, in: *Notices et Extraits des Manuscrits de la Bibliothèque Nationale*, 35, 1, Paris 1896, S. 241—358, vor allem S. 301—319.

[30]) Beschreibung der Handschrift durch K. A. de Meyier: *Bibliotheca Universitatis Leidensis, Codices manuscripti XV*: Codices Vossiani Latini 3, Leiden 1977, S. 31—40.

[31]) Die Fabeln und der Anfang der *Propositiones* (bis f. 205 v) sind im Faksimile wiedergegeben bei Thiele (1). Man vergleiche auch Thiele (2), S. 154 f.

[32]) Nach Aufgabe 2 ist das folgende sonst unbekannte Rätsel eingefügt: *Supra vivum sedebat mortuus. Risit inquit mortuus. mortuus est et (?) vivus. Quomodo hoc esse potest?* Zur Auflösung siehe Thiele (1), S. 62.

[33]) *Catalogue of Manuscripts in the British Museum*, New Series, Bd. 1, Teil 2: The Burney Mss., London 1840, S. 21: *... quondam, ut liquet ex collatione editionis Gudianae, monachorum Benedictinorum Divionensium.* In Dijon befand sich tatsächlich eine lateinische Aesop-Handschrift (Montfaucon, *Bibliotheca Bibliothecarum Manuscriptorum*, Paris 1739, Band 2, S. 1284, erwähnt unter den Handschriften von St. Bénigne, Dijon: *Aesopi Fabularum Libri Quattuor*), und Gudius benutzte für seine Aesop-Ausgabe (1698) eine Handschrift von St. Bénigne, aber G. Thiele vermutete, daß Burney 59 nicht die von Gudius benutzte Handschrift aus Dijon, sondern eine Schwesterhandschrift ist (Thiele [2], S. CLXXXVI). Allerdings ist Thieles Argumentation nicht unbedingt zwingend, da er einzig und allein aus der Qualität der von Gude angefertigten Abschrift (heute in Wolfenbüttel) Rückschlüsse zieht.

[34]) Der Einband von Burney 59 ähnelt sehr demjenigen von Egerton 3055, einer Sueton-Handschrift aus dem 12. Jahrhundert, die mit Sicherheit aus St. Bénigne stammt und McCarthy-Reagh gehörte. Beide Einbände sind relativ jung.

[35]) Erwähnt bei Thiele (2), S. CL f.

propositionibus[36]) beschließt die Handschrift (f. 11 rv). Die *Propositiones* beginnen mit *Incipiunt propositiones ad acuendos iuvenes* und enthalten den vollständigen Text in der Reihenfolge Nr. 1—11, 11a, 11b, 12—32, 33a, 34—53; Lösungen zu 1—10, 12—33, 33a, 34—53.

M = Montpellier, Bibliothèque Universitaire, H. 491, f. 94r—108r

Der Codex entstand Anfang des 11. Jahrhunderts in Ostfrankreich. Spätere Besitzer: Pierre Pithou (1539—1596); Oratoire de Troyes. Die Handschrift ist (im großen und ganzen zutreffend) von N. Bubnov beschrieben[37]). Sie enthält hauptsächlich Schriften zum Abakus und andere mathematische Traktate des Kreises um Gerbert († 1003). Nach einem musikalischen Text (Bernelinus ?, *Commensuralitas fistularum et monochordi,* f. 81r—93v) folgen die *Propositiones* und Ps. Beda, *De arithmeticis propositionibus*[36]) (f. 108r—110r). Daran schließen sich ohne Überschrift fünf interessante Texte über das Erraten von Zahlen, über vollkommene Zahlen und über das Josephspiel an. Die inscriptio der *Propositiones* lautet: *De coniecturis diligentibus oppositis.* Kapitelüberschriften fehlen. Die Aufgaben sind folgendermaßen angeordnet: Nr. 7—11, 11b, 11a, 12, 13, 15—19, 21—33, 33a, 35—48, 51—53, 1; Lösungen zu 1, 7—10, 12, 13, 16—19, 21—33, 33a, 34—49, 51—53. Offenbar eine Kopie von *M* oder einer Schwesterhandschrift ist der Codex Vat. lat. 4539 aus dem 17. Jahrhundert, der auf f. 72—83 die *Propositiones* in derselben Reihenfolge wie *M* enthält[38]).

R = Vat. Regin. lat. 208, f. 57v—61v

Der Codex stammt aus dem 11. Jahrhundert und gehörte nach einem Besitzvermerk auf f. 61v[39]) damals dem Kloster St. Mesmin bei Orléans an. Spätere Besitzer: Paul Petau (K. 20), Alexander Petau (1553), Christine von Schweden (1595). Die Handschrift, die von André Wilmart ausführlich beschrieben ist[40]), enthält theologische Schriften des Boethius, Servius' *Centimeter,* Avians Fabeln und zwei Schriften des Fulgentius[41]). Daran schließen sich ohne inscriptio[42]) die *Propositiones* an (Nr. 1—10, 12—19, 21[43]), 20, 22—25), wobei die Lösungen direkt nach den Aufgaben folgen. Die weiteren Aufgaben standen vielleicht auf den letzten drei Blättern, die später abgeschnitten worden sind.

M₁ = München, Bayerische Staatsbibliothek, Clm 14689, f. 13v—20r

Die Handschrift entstand im 12. Jahrhundert in Bayern[44]) und gelangte aus St. Emmeram nach München (Signatur: Em. G 73). Sie enthält neben metrologischen und einigen astronomischen Texten überwiegend Schriften zum Abakus und andere arithmetische Texte, die im Münchner Katalog[45]) und bei Bubnov[46]) beschrieben sind. Nach einem musikalischen Fragment folgen auf f. 13rv Ps. Bedas *De arithmeticis propositionibus* und dann die *Propositiones* unter dem Titel *Incipiunt propositiones ad acuendos iuvenes* in der Anordnung: Nr. 1—11, 11a, 11b, 12—32, 33a, 35, 33, 36, 34, 37—53; Lösungen zu 1—10, 12—33, 33a, 34—53 und fünf zusätzliche Lösungen[47]).

---

[36]) Siehe Folkerts (2).
[37]) Bubnov, Einl. S. 39f.
[38]) Bubnov, Einl. S. 79f.
[39]) *iste Liber est b. maximini (miciacensis) q(ui eum ab)s(tul)erit hanathema sit.*
[40]) *Codices Reginenses Latini,* Band 1, Vatikan 1937, S. 492—494.
[41]) *Mythologiae; Expositio Virgilianae continentiae moralis.*
[42]) Der Titel *De coniecturis diligentibus oppositis* (vgl. *M*) ist von einer späten Hand nachgetragen.
[43]) Vor Aufgabe 21 befindet sich in *R* ein Verweisungszeichen, das jedoch keine Entsprechung in der Handschrift hat.
[44]) Siehe Bischoff[26]), S. 94, Anm. 68.
[45]) *Catalogus . . .*[27]), S. 217f.
[46]) Bubnov, Einl. S. 43—45.
[47]) Zu diesen für die Textgeschichte der *Propositiones* wichtigen Ergänzungen siehe S. 26—29.

C = London, British Museum, Cotton Iulius D. VII, f. 132r

Die Handschrift stammt aus der St. Albans Abbey[48]); Teile davon wurden von Matthew Paris († 1259) geschrieben. Vermutlich gehörte sie zur Wymondham Priory in Norfolk, einer Abteilung von St. Albans, und später Philip Howard, Earl of Arundel († 1595). Die Handschrift, die im Katalog der Cotton-Manuskripte recht summarisch beschrieben ist[49]), enthält auf f. 132r ohne gesonderte Überschrift die Aufgaben 1, 2, 4, 5, 16, 18, 32, 33, 36, 44, 45, 47 mitsamt den Lösungen, die stets unmittelbar folgen.

S = London, British Museum, Sloane 513, f. 43v—48r. 52r—56v

Der größte Teil der Handschrift (ab f. 11) wurde im 15. Jahrhundert von Richard Dove, Mönch in Bukfastleigh (Devon), geschrieben; später gehörte der Codex John Shaxton. Die Handschrift enthält über 40 Traktate überwiegend astronomischen, mathematischen und alchemistischen Inhalts, aber auch Abhandlungen über die Chiromantie und Bemerkungen zur französischen Sprache[50]). Den *Propositiones* geht ein sonst unbekannter Algorismus-Traktat voraus; es folgen auf f. 57v Rätsel in Versform, die an den Codex Cotton Cleop. B. IX, f. 10v, erinnern, und eine anonyme Abhandlung über die arabischen Ziffern. Innerhalb des Textes der *Propositiones* (zwischen den Aufgaben und den Lösungen) begegnen uns auf f. 48r—51v Aufgaben zur Unterhaltungsmathematik, die aus einer anderen Quelle stammen. Vorhanden sind von den *Propositiones* die Aufgaben 1—9, 12—24, 26—33, 33a, 34—53, 11, 11a, 11b und im Anschluß daran geschlossen die Lösungen (Nr. 1—10, 12—33, 33a, 34—44). Eine inscriptio fehlt.

Einzelne Aufgaben der *Propositiones* sind zusammen mit anderen Problemen der Unterhaltungsmathematik vorhanden in den beiden Handschriften London, British Museum, Cotton Cleopatra B. IX, f. 17v—21r (geschrieben um 1270 in der Abbotsbury Abbey, Dorset)[51]) und in Oxford, Bodleian Library, Digby 98, f. 34r—39v (15. Jahrhundert).

Editionen:

Die *Propositiones* wurden erstmals in der Beda-Ausgabe gedruckt, die 1563 bei Herwagen in Basel erschien[52]). Sie folgen dort im Kapitel mit der Überschrift *Beda de arithmeticis propositionibus* unmittelbar nach dem 4. Problem jener kleinen Schrift[36]) ohne besondere Trennung vom Vorhergehenden. Eingeleitet von der im Druck nicht besonders hervorgehobenen Überschrift *Incipiunt aliae propositiones ad acuendos iuvenes*, sind auf den Spalten 135—146 folgende Aufgaben abgedruckt: Nr. 1—11, 11a, 11b, 12—32, 33a, 35, 33, 36, 34, 37—53; Lösungen zu Nr. 1—10, 12—33, 33a, 34, 35 Anfang. Der Text endet am Fuß der Spalte 146 abrupt inmitten der Lösung zu Nr. 35; es folgt die lapidare Bemerkung *Reliquae solutiones desiderantur. Potest autem quisque ratione arithmetica Propositiones illas solvere. Ita ad exercendum ingenium, omissa valebunt.* Der Text ähnelt stark der Handschrift M_1, jedoch steht fest, daß Herwagen nicht diesen Codex direkt benutzte. Wieso Herwagen die Schrift unter die Werke des Beda einreihte, läßt sich heute nicht mehr ausmachen. Vermutlich gingen in der Druckvorlage (ebenso wie in M_1) die *De arithmeticis propositionibus* voraus, die man bei oberflächlicher Betrachtung dem Beda zuschreiben konnte[53]), und Herwagen vermutete, daß die anonymen *Propositiones* Teil derselben Schrift seien.

---

[48]) Siehe N. R. Ker, *Medieval libraries of Great Britain*, ²London 1964, S. 166.

[49]) *Catalogue of the manuscripts in the Cottonian Library*, London 1802, S. 15 f.

[50]) Eine Beschreibung befindet sich im gedruckten, aber nicht publizierten *Catalogus librorum manuscriptorum bibliothecae Sloanianae* (1837—40; Exemplar im British Museum).

[51]) *Catalogue* . . .[49]), S. 578 f.; Ker[48]), S. 1.

[52]) *Opera Bedae Venerabilis Presbyteri, Anglosaxonis: viri in divinis atque humanis literis exercitatissimi: omnia in octo Tomos distincta, prout statim post Praefationem suo Elencho enumerantur. Addito Rerum et Verborum Indice copiosissimo.* Basileae, per Ioannem Hervagium, Anno M. D. LXIII. Band 1, Spalte 135—146.

[53]) Siehe die Bemerkungen zur Verfasserfrage bei Folkerts (2), S. 35 f.

HERWAGEN benutzte also für den Text der *Propositiones* eine mit Clm 14689 verwandte Handschrift und für Ps. BEDAS *De arithmeticis propositionibus* einen Codex, der der ebenfalls aus St. Emmeram stammenden Handschrift Clm 14836 nahestand [54]. Da durch scharfsinnige Untersuchungen B. BISCHOFFS [55]) feststeht, daß HERWAGEN für die Herstellung seiner BEDA-Ausgabe auch sonst die Emmeramer Bibliothek ausbeutete, ist es sehr wahrscheinlich, daß im 16. Jahrhundert in St. Emmeram eine heute verlorene Handschrift vorhanden war, bei der ähnlich wie in BMM_1OR_1 die *Propositiones* und Ps. BEDAS *De arithmeticis propositionibus* aufeinander folgten, wobei die Lesarten dieser Texte mit den beiden oben erwähnten erhaltenen Handschriften im großen und ganzen übereinstimmten. Dieser Codex wurde als Druckvorlage benutzt und ist heute wohl verschollen. Da somit die Entstehung der HERWAGENschen Edition weitgehend geklärt ist, kann sie im folgenden vernachlässigt werden.

Die *Propositiones* wurden nach HERWAGENS Text noch zweimal gedruckt, und zwar in den BEDA-Ausgaben von FRIESSEM (1688) [56]) und MIGNE (1844) [57].

Eine weitere Edition erschien 1777 durch FORSTER (FROBENIUS), Abt von St. Emmeram in Regensburg [58]). Dieser hat in seine ALKUIN-Gesamtausgabe unter die *Opuscula dubia* auch die *Propositiones* aufgenommen [59]). FORSTER berichtet zwar im Vorwort zu den zweifelhaften Werken, er habe die Schrift einer Einsiedler Handschrift entnommen [60]), doch scheint *Einsiedlensis* ein Versehen für *Augiensis* zu sein, da sich in Einsiedeln keine Handschrift der *Propositiones* nachweisen läßt und FORSTER unmittelbar vor Abdruck des Textes schreibt, er habe eine Abschrift der Reichenauer Handschrift erhalten [61]). Tatsächlich beruht FORSTERS Edition auf der Handschrift A, die dieser mit der Ausgabe der *Propositiones* in BEDAS Werken [52]) verglich. FORSTER sagt (S. 440), er habe Lücken der Handschrift mit Hilfe der BEDA-Ausgabe ergänzt und dies im Text durch eckige Klammern angegeben. Einige Varianten der Handschrift und der BEDA-Edition sowie wenige Erläuterungen hat FORSTER zusätzlich als Fußnoten aufgenommen. Es nimmt wunder, daß FORSTER keine der beiden Handschriften benutzte, die sich zu seiner Zeit in St. Emmeram befanden (M_1 und M_2; siehe S. 20. 18f.).

FORSTERS Ausgabe wurde mit dessen Bemerkungen von MIGNE als Bestandteil von ALKUINS Werken wiederabgedruckt [62]). Beide Editionen unterscheiden sich praktisch nicht.

3. DER WERT DER EINZELNEN HANDSCHRIFTEN

Die Beschreibung der Handschriften zeigt, daß Zahl und Abfolge der Aufgaben von Codex zu Codex stark variieren. Bevor man hieraus Folgerungen zieht, muß man auf die Qualität der einzelnen Handschriften eingehen. Die folgenden Beobachtungen beruhen auf

---

[54]) FOLKERTS (2), S. 29.

[55]) Zur Kritik der Heerwagenschen Ausgabe von Bedas Werken (Basel 1563), in: *Mittelalterliche Studien*, Band 1, Stuttgart 1966, S. 112−117.

[56]) *Venerabilis Bedae Presbyteri Anglo-Saxonis viri sua aetate doctissimi opera quotquot reperiri potuerunt omnia Hac ultima impressione ornatius in lucem edita.* Coloniae Agrippinae, Sumptibus Ioannis Wilhelmi Friessem, Anno MDCLXXXVIII, Band 1, Spalte 102−110.

[57]) *Patrologia Latina* 90 (1844), Sp. 667−676.

[58]) (FROBENIUS) FORSTER lebte von 1709 bis 1791. Siehe ADB 7, 163, und NDB 5, 302f. Der Name FROBENIUS wurde ihm wohl erst im Orden verliehen.

[59]) *Beati Flacci Albini seu Alcuini abbatis, Caroli Magni regis ac imperatoris, magistri opera. Post primam editionem, a viro clarissimo D. Andrea Quercetano curatam, de novo collecta, multis locis emendata, et opusculis primum repertis plurimum aucta, variisque modis illustrata cura ac studio Frobenii, S. R. I. principis et abbatis ad S. Emmeramum Ratisbonae.* Tomi secundi volumen secundum. S. Emmerami, M.DCC.LXXVII. Seite 440−448.

[60]) S. 369: *Tertio recensemus Propositiones ad acuendos juvenes, ad fidem Cod. Ms. Illustrissimi Monasterii Einsiedlensis.*

[61]) S. 440: *Extat vero sub nomine Alcuini in pervetusto Codice MS. Monasterii Augiae Divitis, unde descriptum ad nos pervenit.*

[62]) *Patrologia Latina* 101 (1851), Sp. 1143−1160.

einer vollständigen Kollation aller Handschriften[63]). A vor einer Zahl bedeutet: Aufgabe, L vor einer Zahl: Lösung der Aufgabe. Die Numerierung stimmt mit derjenigen in der kritischen Edition überein. Auch die Zeilenzählung bezieht sich auf diese Ausgabe.

HANDSCHRIFT A: Bei genauer Betrachtung findet man so viele Sonderfehler, daß eine Bevorzugung dieser Handschrift keineswegs gerechtfertigt ist. Die wichtigsten sind längere Lücken in L 28 und A 53, die bewußte Auslassung von A 11a, A 11b, A 33a, L 33a sowie die zusätzliche Lösung L 11. Andrerseits hat A den Text einigermaßen treu kopiert.

HANDSCHRIFT B: Der Schreiber von B gibt seine Vorlage offenbar noch getreuer wieder. Absichtliche Änderungen sind nicht festzustellen. Allerdings wurden aus Unachtsamkeit an mehreren Stellen längere Abschnitte ausgelassen (z. B. L 2, A 23, L 27, A 28, L 29, A 30, A 35, L 36, L 37, L 47, L 52).

HANDSCHRIFT C: Trotz des kurzen Textes weist C eine Menge Fehler auf. Auffällig sind vor allem die Lücken in L 16, L 32, L 33, L 36. Einige Versehen, die C und S gemeinsam enthalten, deuten auf eine Verwandtschaft dieser beiden Handschriften hin.

HANDSCHRIFT M: Der Schreiber geht recht eigenwillig mit dem Text um: Ganze Aufgaben bzw. Lösungen weichen von dem herkömmlichen Text ab (A 1, L 1, L 13, L 41, L 42, L 44, L 49); dazu kommen die üblichen Versehen (etwa die Lücken in L 12, L 24).

HANDSCHRIFT M_2: Wegen des kurzen Textes ist eine eindeutige Einordnung nicht möglich. Vermutlich ist M_2 mit M_1 verwandt.

HANDSCHRIFT O: Dieser Codex bietet eine der besten Kopien, deren Wert durch mehrere längere Auslassungen (in L 17, A 23, L 23, L 24, L 32, A 52) nur unwesentlich beeinträchtigt wird. Änderungen sind nicht festzustellen. Der Text ist durch Blattversetzung der Vorlage etwas in Unordnung geraten[17]).

HANDSCHRIFT V: ADEMAR hat, ähnlich wie beim ÄSOP-Text, auch die *Propositiones* außerordentlich frei bearbeitet[64]), wobei er in diesem Teil den Text wohl nicht kontaminiert hat. Aus diesem Grunde ist die Handschrift V, die nicht der Gruppe M_1RR_1S nahesteht, für die Konstitution des Textes nicht wichtig.

HANDSCHRIFT W: Dieser Codex ist außerordentlich sorgfältig geschrieben. Sonderfehler sind kaum vorhanden[65]). Im großen und ganzen stimmt der Text gut mit AO überein.

HANDSCHRIFTEN M_1RR_1S: Diese Handschriften repräsentieren eine von den übrigen Codices teilweise abweichende Textrezension (siehe Kapitel 4 und 5). Dabei sind die älteren, aber unvollständigen Handschriften R_1R den vollständigeren jüngeren Codices M_1S vorzuziehen. M_1, der nur wenige Sonderfehler aufweist (etwa Lücken in L 17 und A 31), wurde durchgehend nach einer mit ABO verwandten Handschrift verbessert (M_1^1). R hat einen längeren Text der Lösungen L9 und L10 und weist Lücken (in L 19 und A 23) und eigene Konjekturen (z. B. in L 21) auf. R_1, dessen Text im übrigen der beste dieser Gruppe ist, zeichnet sich durch zusätzliche Lösungen zu 11, 11a, 11b sowie durch die bewußte Auslassung der Aufgabe 33 aus[66]). S schließlich weicht in vielen Kleinigkeiten vom üblichen Text ab. Sein Schreiber bricht bei den Lösungen mitten in L 44 ab und schreibt statt des Schlusses nur: *etc.*

Die Kollation der Handschriften ergibt also, daß kein Codex Abschrift eines erhaltenen anderen ist. Der Text der üblichen Fassung (Rezension I) wird am zuverlässigsten durch die Handschriften $ABOW$ wiedergegeben. Eine Gruppe von Handschriften (M_1RR_1S) weicht an zahlreichen Stellen stark von der Rezension I ab. Diese im folgenden als Rezension II bezeichnete Fassung ist zu großen Teilen nur durch die relativ junge Handschrift M_1 vertreten, da die älteren Codices RR_1 einen fragmentarischen Text aufweisen.

---

[63]) Ausgenommen ist der Codex V, von dem nur Stichproben genommen werden konnten, da die Schrift auf dem Mikrofilm, der mir zur Verfügung stand, oft kaum lesbar ist.

[64]) Über ADEMARS Arbeitsweise siehe THIELE (2), S. CLXXXI–CLXXXV.

[65]) Ausgenommen Lücken Z. 124, 125, 130f., 296, 585f.

[66]) Nach A 32 fügt R_1 hinzu: *sic et de XXX*.

4. FOLGERUNGEN FÜR DEN TEXT DER PROPOSITIONES

Stellen wir einmal die in den einzelnen Handschriften vorhandenen Probleme in einer Übersicht zusammen, wobei A wieder „Aufgabe", L „Lösung" bezeichnet:

R_1 (s. 9 ex.): A 4, 5, 8, 11, 11a, 11b, 22—26, 32, 33a, 36, 40, 42, 48, 52, 49, 47, 45; L 2, 5, 8, 11, 11a, 11b, 22—26, 32, 33a, 36

O (ca. 1000): Überschriften; A+L 1—11, 11a, 11b, 12, 13, 33a, 34—52, 14—33, 53 (keine Lösungen zu 11, 11a, 11b)

A (ca. 1000): Überschriften; A+L 1—53

W (ca. 1010): A+L 1—10, 12—18, 21, 19, 20, 27, 28, 30—42, 44—49, 51—53

M_2 (ca. 1020): A+L 1—4, A 8

V (ca. 1025): A+L 1—11, 11a, 11b, 12—20, 22, 21, 23—33, 33a, 34—42, 44—53, 43 (keine Lösungen zu 11, 11a, 11b)

B (s. 11[1]): A 1—11, 11a, 11b, 12—32, 33a, 34—53; L 1—10, 12—33, 33a, 34—53

M (s. 11[1]): A 7—11, 11b, 11a, 12, 13, 15—19, 21—33, 33a, 35—48, 51—53, 1; L 1, 7—10, 12, 13, 16—19, 21—33, 33a, 34—49, 51—53

R (s. 11): A+L 1—10, 12—19, 21, 20, 22—25

M_1 (s. 12): A 1—11, 11a, 11b, 12—32, 33a, 35, 33, 36, 34, 37—53; L 1—10, 12—33, 33a, 34—53

C (ca. 1250): A+L 1, 2, 4, 5, 16, 18, 32, 33, 36, 44, 45, 47

S (s. 15): A 1—9, 12—24, 26—33, 33a, 34—53, 11, 11a, 11b; L 1—10, 12—33, 33a, 34—44.

Man sieht, daß Auswahl und Reihenfolge der Aufgaben stark voneinander abweichen. In manchen Codices folgt die Lösung stets unmittelbar auf die jeweilige Aufgabe (in ACM_2 $ORVW$), in anderen stehen alle geschlossen am Schluß nach den Aufgaben (in BMM_1R_1S). Vermutlich ist die erste Anordnung die ursprüngliche, da sie von der Mehrzahl der besseren Codices vertreten wird. Ob im Original wie in den Handschriften A und O vor den Aufgaben die Überschriften aller Probleme in einem Block zusammengefaßt waren, ist ungewiß, allerdings nicht sehr wahrscheinlich.

Die Zusammenstellung zeigt, daß auch Anzahl und Anordnung der Aufgaben divergieren. Einige Handschriften enthalten nur wenige Aufgaben (CM_2RR_1). Mehrere Schreiber, darunter auch der Kopist der ältesten Handschrift R_1, verkürzten offenbar bewußt vollständigere Vorlagen: Bisweilen wurden Lösungen von einigen Aufgaben abgeschrieben, die überhaupt nicht in der Handschrift vorhanden waren[67]). Oft stimmt die Reihenfolge der Aufgaben nicht mit derjenigen der Lösungen überein, oder einzelne Lösungen fehlen[68]). Hierbei handelt es sich offenbar um individuelle Eigenheiten der Handschriften, da dieselbe Verkürzung oder Umordnung nicht in verschiedenen Codices anzutreffen ist. Sieht man von diesen Besonderheiten ab, so zeigen die erhaltenen Handschriften, daß folgende Reihenfolge mit großer Wahrscheinlichkeit die ursprüngliche ist: Aufgaben 1—11, 11a, 11b, 12—33, 33a, 34—53 mit Lösungen (außer zu 11, 11a, 11b)[69]).

Es hat sich also herausgestellt, daß die Zahl der Aufgaben in den gedruckten ALKUIN-Ausgaben nicht mit dem ursprünglichen Bestand übereinstimmt: Drei Aufgaben (11a, 11b, 33a) und eine Lösung (zu 33a) sind einzufügen, dafür ist eine Lösung (zu 11) zu streichen. Schon hieraus folgt die Notwendigkeit einer kritischen Ausgabe, die alle Handschriften berücksichtigt. Mehr noch: Abgesehen von diesen besonders gravierenden Ergänzungen bzw. Streichungen, gibt es ungefähr 300 Stellen, an denen die herkömmlichen gedruckten Texte der *Propositiones* aufgrund der handschriftlichen Lesarten geändert werden müssen.

---

[67]) In B: L33; in M: L34, L49; in R_1: L2; in S: L10, L25.
[68]) In M: L15; in M_2: L8; in R_1: Lösungen 4, 40, 42, 48, 52, 49, 47, 45; in S: Lösungen 45—53.
[69]) Zu A11 gibt es nur in A und R_1 Lösungen, zu 11a und 11b nur in R_1. Da die Lösungen in A und R_1 abweichen und inhaltlich wenig zufriedenstellen, handelt es sich um Erfindungen der Kopisten.

5. DIE BEIDEN REZENSIONEN I UND II

Eine wesentliche Frage ist freilich noch zu klären: Wir haben bereits erwähnt, daß die Handschriften zwei Redaktionen der *Propositiones* verkörpern, die sich allerdings nur bei bestimmten Aufgaben stark voneinander unterscheiden. Es wird nötig sein, diese beiden Fassungen gegeneinander abzugrenzen. Hierbei ergibt sich eine gewisse Schwierigkeit dadurch, daß die besseren Handschriften der Rezension II nur Auszüge aus den *Propositiones* aufweisen.

Es gibt acht Stellen, an denen die Handschriften $M_1 RR_1 S$ (oder wenigstens $M_1 S$, wenn RR_1 diesen Text nicht aufweisen) stärker von der Lesart der übrigen Codices abweichen:

1. Am Ende von L 17 haben $M_1 RS$ (R_1 fehlt) den Satz: *Tali igitur sicque sollicitante studio facta est navigatio nullo fuscante inquinationis contagio*, während es in den übrigen Handschriften heißt: *Et fieret expleta transvectio nullo maculante contagio*. Beide Formulierungen sind möglich.

2. Die Lösung 20 endet in $M_1 RS$ (R_1 fehlt) mit den Worten *expleta salubris transvectio nullo formidante mortis naufragio*, während überall sonst die Worte *salubris* und *mortis* fehlen. Auch hier erscheint der Satz also etwas aufgebläht.

3. Wesentlicher sind die Unterschiede bei der Lösung 24. Dabei geht es darum, die ermittelte Fläche von 270 *perticae* in *aripenni* zu verwandeln (1 *aripennus* = Fläche eines Quadrats mit der Seitenlänge 12 *perticae*, also 144 Flächeneinheiten der Länge 1 *pertica*). Statt nun $270 : 144 = 1\frac{1}{2}$ *(aripenni)* $+ 54$ *(perticae)* zu rechnen, wie es bei derselben Aufgabe in der mit Gerbert verbundenen, aber sicherlich früheren anonymen Geometrie (in Zukunft zitiert als: *Geometria incerti auctoris*)[70]) geschieht[71]), wird in den *Propositiones* folgendermaßen vorgegangen:

Handschriften *ABMO*:

. . . fiunt CCLXX. Fac exinde bis XII, id est, divide CCLXX per duodecimam: fiunt XXII et semis. Atque iterum XXII et semis per duodecimam divide partem: fit aripennus unus et perticae X ac dimidia.

Die Handschriften $M_1 RR_1 S$ haben denselben Text bis *partem* und dann folgendes:

Et fiunt II et remanent IIII, quae est tertia pars (de) XII. Sunt ergo aripenni in hoc numero II et tertia pars de aripenno tertio. Der Schlußsatz *fit . . . dimidia* fehlt in RR_1, nicht aber in M_1; S hat ihn verändert in: *fit aripennus unus et per terciam XII.* M_1^1 hat über *et fiunt duo et remanent IIII* hinzugefügt: *vel fit aripennus unus et pertice X ac dimidia*, also auch hier den ursprünglichen Text nach einer mit *ABO* verwandten Handschrift korrigiert.

In der Version I (Handschriften *ABMO*) wird also zum Zweck der Umrechnung nicht gleich durch 144 geteilt, sondern zweimal durch 12:

$270 : 12 = 22\frac{1}{2}$; $22\frac{1}{2} : 12 = 1$ *(aripennus)* $+ 10\frac{1}{2}$ *(perticae)*.

Bei der zweiten Division wird nicht berücksichtigt, daß es sich um Flächeneinheiten handelt, so daß der Rest falsch ist.

Noch konfuser ist die Rechnung in der Version II (Handschriften $M_1 RR_1 S$). Offenbar wird hier gerechnet: $22\frac{1}{2} = 22 + 6$ (die Hälfte von 12) $= 28$, $28 : 12 = 2\frac{1}{3}$ *(aripenni)*. M_1 und S kennen vermutlich den Text der Version I, den M_1 wörtlich übernimmt und damit die Widersprüche noch vergrößert, während S — erfolglos — versucht, den Unsinn zu heilen. Diese Stelle wirft ein bezeichnendes Licht auf die Arbeitsweise der Schreiber von M_1 und S; gleichzeitig erkennt man, daß hier die Version I weniger schlecht als die Version II ist.

---

[70]) Zu dieser Geometrie, die vor Gerbert entstand, aber lange Zeit als Teil seiner Geometrie galt, siehe S. 39. Sie ist ediert von Bubnov, S. 310—365.

[71]) IV 33 = S. 354, 1—8 Bubnov.

V

4. In Aufgabe 28 heißt es in der Rezension I folgendermaßen:
   volo enim ibidem aedificia domorum construere. Die Handschriften M_1S, die hier als einzige die Rezension II repräsentieren, schreiben stattdessen: *volo, ut fiat ibi domorum constructio.*

5. Bei der Lösung derselben Aufgabe ist es erforderlich, festzustellen, wie oft 20 in 100 und 10 in 45 enthalten ist. Dies wird folgendermaßen ausgedrückt:
   Rezension I (*A* fehlt): *itaque in C quinquies XX et in XL quater X sunt.*
   Rezension II (nur M_1S): *duc XX. partem de C, fiunt V. Et pars decima quadragenarii IV sunt.*
   Beide Formulierungen sind gleichermaßen möglich. Man sollte bemerken, daß dieselbe Aufgabe mit denselben Zahlen und ähnlichen Worten auch in der *Geometria incerti auctoris* vorhanden ist[72]). Dort heißt es ähnlich wie in der Rezension II: *... duc vigesimam de C, fiunt V, et decimam de XL, fient IV.*

6. Am Schluß der Aufgabe 31 haben M_1S (RR_1 fehlen) den Zusatz:
   et unaquaeque cupa habeat pedes VII. Diese Ergänzung wiederholt nur einen fast gleichlautenden Satz zuvor und ist daher überflüssig.

7. In Aufgabe 46 wird ein Spaziergänger überfallen. Man stiehlt ihm die Börse und teilt den Betrag unter sich gleichmäßig auf. Jeder bekommt 50 *solidi* von den 2 *talenti.* Gefragt wird nach der Anzahl der Personen. Dieser einfache Sachverhalt wird von den Handschriften der Rezension II (M_1S; RR_1 fehlen) klar ausgedrückt:
   Ipsi vero irruentes diripuerunt sacculum, et tulit sibi quisque solidos quinquaginta. Dicat qui vult, quot homines fuerunt. Die übrigen Codices fügen nach *quinquaginta* folgenden Satz ein: *Et ipse postquam vidit se resistere non posse, misit manum et rapuit solidos quinquaginta.*
   Durch diese schon aus Gründen der Logik nicht sehr wahrscheinliche Ergänzung wird der Sachverhalt unnötig kompliziert. Die Lösung geht von der einfacheren Aussage der Rezension II aus. Es handelt sich hier also nicht um das versehentliche Auslassen eines Satzes, wie man zunächst annehmen würde[73]), sondern entweder ist in der Rezension II der ursprünglich längere Text, wie ihn die Rezension I aufweist, sinnvoll verkürzt worden, oder der Text der Rezension II ist der ursprüngliche, und in der Rezension I hat man aus irgendwelchen Gründen einen Satz sehr ungeschickt eingefügt.

8. Der Text der Aufgabe 52 weicht in den beiden Fassungen stark voneinander ab:
   Rezension I:
   Quidam paterfamilias iussit XC modia frumenti de una domo sua ad alteram deportari, quae distabat leuvas XXX, ea vero ratione, ut uno camelo totum illud frumentum deportaretur in tribus subvectionibus et in unaquaque subvectione XXX modia portarentur, camelus vero in unaquaque leuva comedat modium unum. Dicat, qui velit, quot modii residui fuissent.
   Rezension II (nach M, R_1S; *R* fehlt):
   Quidam paterfamilias habebat de una domo sua ad alteram domum leugas XXX et habens camelum, qui debebat in tribus subvectionibus ex una domo sua ad alteram de annona ferre modia XC et in unaquaque leuga isdem camelus comedebat semper modium I. Dicat, qui valet, quot modia residua fuerunt.
   Beide Formulierungen geben das Problem richtig wieder. Vielleicht ist die kürzere Fassung der Rezension II vorzuziehen, da im längeren Text zumindest der triviale Zusatz *et in unaquaque subvectione XXX modia portarentur* fehlen könnte.

Um zu klären, wie sich beide Rezensionen zueinander verhalten, muß man noch fünf weitere Stellen betrachten, die sich von den erwähnten dadurch unterscheiden, daß die Handschrift *S* hier nicht mehr den abweichenden Text der Rezension II vertritt, sondern mit den

---

[72]) IV 36 = S. 354, 30—355, 8 Bubnov.
[73]) Zwei aufeinanderfolgende Sätze enden mit demselben Wort: *quinquaginta.*

übrigen Codices übereinstimmt[74]). Es handelt sich um die Lösungen zu den Aufgaben 22, 25, 29, 30, 31. Diese Probleme gehören zur Gruppe der geometrischen Aufgaben; sie betreffen Flächenverwandlungen und werden ebenfalls in der *Geometria incerti auctoris* behandelt[75]). In den *Propositiones* stimmt der Text der Aufgaben in beiden Rezensionen überein, die Beweise jedoch unterscheiden sich grundsätzlich voneinander. Die Aufgaben werden in der Rezension II nicht einheitlich überliefert: Die beiden ersten Lösungen stehen in M_1RR_1 (in M_1 zusätzlich zur üblichen Fassung am Ende der Lösungen); die letzten drei Lösungen sind in R und R_1 nicht mehr vorhanden, jedoch in M_1, dort wiederum als Zusatz am Schluß. Es sprechen gewichtige Gründe dafür, daß alle fünf abweichenden Lösungen der Rezension II angehören[76]). Im folgenden sollen die Unterschiede der beiden Fassungen bei diesen fünf Aufgaben besprochen werden. Dabei wird es bisweilen nützlich sein, die Lösungsverfahren in der *Geometria incerti auctoris* mit heranzuziehen.

Aufgabe 22:

Es soll die Fläche eines schrägen (viereckigen) Feldes mit folgenden Maßen bestimmt werden: Länge 100 *perticae*, Breite oben und unten: 50 *perticae*, in der Mitte: 60 *perticae*. In beiden Rezensionen wird ähnlich gerechnet: Man multipliziert die Länge (100) mit der gemittelten Breite $\frac{1}{3} \cdot (50 + 50 + 60) = 53$ und erhält als Fläche 5300 *perticae*[77]). Bei der Verwandlung der *perticae* in *aripenni* wird auch hier[78]) nicht gleich durch 144, sondern zweimal durch 12 dividiert, wobei die Reste vernachlässigt werden: $\frac{5300}{12} = 441$, $\frac{441}{12} = 37$ *(aripenni)*. Das Verfahren in den *Propositiones* weicht von dem in der *Geometria incerti auctoris* ab[79]).

Aufgabe 25:

Wie groß ist die Fläche eines kreisförmigen Ackers mit dem Umfang $U = 400$ *perticae*? Die in der Tradition der römischen Agrimensoren stehenden Mathematiker des frühen Mittelalters lösten dieses Problem üblicherweise, indem sie zunächst den Durchmesser d bestimmten $\left(d = \frac{1}{3} \left(U - \frac{U}{22} \right) \right)$ und dann $F = \frac{d}{2} \cdot \frac{U}{2}$ berechneten[80]). Genauso wird in

---

[74]) Hieraus lassen sich keine weiterreichenden Schlüsse ziehen, da S oft frei den Text bearbeitete und vermutlich mehrere Vorlagen benutzte.

[75]) Zu dieser Problemgruppe vergleiche die Bemerkungen im Kapitel 7 (S. 39—41).

[76]) In M_1 stehen also an den „richtigen" Stellen die üblichen Lösungen zu 22, 25, 29, 30, 31 und zusätzlich geschlossen am Schluß die abweichenden Lösungen jener fünf Aufgaben. Daß M_1 beide Arten der Lösungen aufweist, ist nicht verwunderlich, da M_1 zwar der Rezension II angehört, aber auch eine Vorlage der Rezension I benutzte (siehe S. 23). Wie aus der Übereinstimmung in M_1RR_1 folgt, entsprechen die zusätzlichen Lösungen zu 22 und 25 der Rezension II. Daß dies auch für die übrigen drei zusätzlichen Lösungen zutrifft, ist sehr wahrscheinlich, da sie in M_1 gemeinsam mit den beiden anderen am Schluß stehen und die Art der Lösung in L25 und L29 übereinstimmt.

[77]) In der Rezension II wird überflüssigerweise auch das Mittel aus den beiden gleichen Längen gebildet: $\frac{1}{2}(100 + 100) = 100$. Hier schwebte wohl die uralte Rechtecksformel $F = \frac{a+c}{2} \cdot \frac{b+d}{2}$ vor, die mechanisch angewandt wurde.

[78]) Man vergleiche dasselbe Verfahren in Aufgabe 24; siehe S. 25.

[79]) IV 31 = S. 352, 22—353, 19 BUBNOV: Dort rechnet man $\frac{50+60}{2} \cdot 100 = 5500$ *(perticae)* $= 38$ *(agripenni)* $+ 28$ *(perticae)*. Ein zweites Verfahren wird als „Probe" angegeben: Fläche des Ackers = Rechteck + 2 aufgesetzte gleichschenklige Dreiecke $= 50 \cdot 100 + 2 \cdot \frac{100 \cdot 5}{2} = 5500$ *(perticae)*.

[80]) So etwa VARRO (?): BUBNOV 496, 4—6; EPAPHRODITUS und VITRUVIUS RUFUS: BUBNOV 546, 10—14 ($= \S 31$); *Geometria incerti auctoris*: BUBNOV 346, 9—16 ($=$ IV 18); ADELBOLD, *Epistula ad Gerbertum*: BUBNOV 304, 17—305, 3.

der *Geometria incerti auctoris* IV 34 (S. 354,9—17 BUBNOV) verfahren. Hier soll ein Acker des Umfangs $U = 418$ bestimmt werden, und es begegnen uns die Zahlen:

$$U - \frac{U}{22} = 399, \quad 399 : 3 = 133 = d, \quad \frac{d}{2} \cdot \frac{U}{2} = 66\tfrac{1}{2} \cdot 209 = F.$$

Ganz anders rechnet man in den *Propositiones*, Rezension I:

$\frac{U}{4} = 100$, $100 \cdot 100 = 10\,000$ *(perticae)* $= 69$ *(aripenni)*, da $10\,000 : 12 = 833$ und $833 : 12 =$

$= 69$ ist. Hier wird also die Fläche eines Kreises mit demjenigen Quadrat gleichgesetzt, dessen Seite gleich dem Viertelumfang des Kreises ist. Diese Methode der Flächenberechnung, die auch in dem Agrimensoren-Text *De iugeribus metiundis* überliefert wird, ist von A. J. E. M. SMEUR analysiert worden [81]). In der Rezension II geht man wiederum ganz anders vor:

Man rechnet $\frac{U}{4} = 100$, $\frac{U}{3} = 133$, $100 : 2 = 50$, $133 : 2 = 66$, $50 \cdot 66 = 3151$ [82]), $3151 : 12 =$

$= 280$, $280 : 12 = 24$, $24 \cdot 4 = 96$ *(aripenni)*. Sieht man einmal von den beiden Divisionen durch 12 ab, die notwendig werden, sobald man *perticae* in *aripenni* umrechnet, so wird hier

also gerechnet: $F = \frac{U}{4 \cdot 2} \cdot \frac{U}{3 \cdot 2} \cdot 4$. Diese überraschend komplizierte Formel ist mir aus der abendländischen Mathematik des frühen Mittelalters sonst nicht bekannt; möglicherweise beruht sie auf der Vorstellung, das Verhältnis von Kreisumfang und Durchmesser

sei gleich 3. Dann würde sie wegen $d = \frac{U}{3}$ der obigen Formel $F = \frac{U}{2} \cdot \frac{d}{2} = \frac{U}{4} \cdot d = \frac{U^2}{12}$

entsprechen, die auch bei den Ägyptern und Babyloniern [83]) sowie im alten China belegt ist [84]). Warum allerdings zunächst zweimal durch 2 dividiert und am Schluß wieder mit 4 multipliziert wird, ist kaum einzusehen [85]).

Aufgabe 29:

Wie viele rechteckige Häuser (30 × 20 Fuß) passen in eine runde Stadt des Umfangs $U = 8000$ *pedes?*

In der Rezension I wird der Kreis durch ein umfanggleiches Rechteck ersetzt, dessen Seiten sich wie 3 : 2 verhalten (analog zu den Maßen der Häuser). Man erhält $8000 = 4800 + 3200$, also die Seitenlängen 2400 und 1600 Fuß. Da eine Hauslänge $2400 : 30 = 80$mal in das große Rechteck und die Hausbreite ebenfalls $1600 : 20 = 80$mal in die Breite hineinpassen, gibt es insgesamt $80 \cdot 80 = 6400$ Häuser.

In der Rezension II wird der Kreis nicht in ein Rechteck umgeformt, sondern man berechnet nach demselben ungewöhnlichen Verfahren wie in Aufgabe 25 die Fläche und dann durch Division und anschließende Multiplikation die Zahl der Häuser:
$8000 : 4 = 2000$, $8000 : 3 = 2666$; $2000 : 2 = 1000$, $2666 : 2 = 1333$; $1333 : 30 \langle = 44$, $1000 : 20 = 50$; $44 \cdot 50 \rangle = 2200$; $2200 \cdot 4 = 8800$ [86]). Dieses Verfahren stimmt im Gedanken mit der *Geometria incerti auctoris* überein (IV 37 = S. 355, 8—16 BUBNOV): Hier wird

---

[81]) On the value equivalent to π in ancient mathematical texts. A new interpretation, in: *Archive for history of exact sciences* 6 (1970) 249—270, vor allem 249—251. SMEUR weist nach, daß CANTORS Schluß, der Autor habe $\pi = 4$ angenommen, nicht zulässig ist.

[82]) Versehen für 3300.

[83]) O. NEUGEBAUER, *Vorlesungen über Geschichte der antiken mathematischen Wissenschaften*, Band 1, Berlin 1934, S. 126.167f.

[84]) VOGEL (4), S. 47.122f.

[85]) Man beachte, daß die Zwischenergebnisse 133 und 66 auch in der *Geometria incerti auctoris* vorkommen, obwohl der Ausgangswert des Umfangs verschieden ist.

[86]) Die in spitzen Klammern stehenden Rechnungen wurden ergänzt: Die einzige Handschrift M_1 weist hier ein Lücke auf.

nach der bei Aufgabe 25 besprochenen Methode zunächst d bestimmt, dann $F = \dfrac{d}{2} \cdot \dfrac{U}{2}$ und schließlich $F : (20 \cdot 30) =$ Zahl der Häuser.

Aufgabe 30:

Eine mit Ziegelsteinen gepflasterte Basilika hat die Maße 240 × 120 *pedes;* ein Stein mißt $1\frac{1}{2}$ × 1 *pedes* = 23 × 12 *unciae.* Wie viele Steine braucht man?

Auch hier divergieren die Verfahren: In Rezension I berechnet man, wie viele Steinlängen bzw. -breiten auf den Boden passen, findet $240 : 1\frac{1}{2} = 126$ bzw. 120 und erhält als Ergebnis 120 · 126 = 15120. In Rezension II dagegen werden Länge und Breite der Basilika in *unciae* verwandelt (12 · 240 = 2880, 12 · 120 = 1440); dann dividiert man durch Längen- und Breitenzahl der Steine, wobei allerdings andere Zahlen gewählt werden: 15 × 8 statt 23 × 12 *unciae.* Schließlich werden die Verhältniszahlen 180 und 192 multipliziert, und es ergibt sich die Zahl 34560. — Auch diese Aufgabe ist in der *Geometria incerti auctoris* vorhanden (IV 38 = S. 355, 17—27 BUBNOV). Hier wird zunächst die Fläche der Basilika (in *unciae*) bestimmt und dann durch die Fläche eines Steins (12 × 23) dividiert.

Alle drei Verfahren sind richtig und in sich schlüssig. Unklar bleibt, warum die Handschriften der Rezension II die Zahlen änderten[87]).

Aufgabe 31:

In einem rechteckigen Keller (100 × 64 *pedes*) sind Fässer aufgestellt, die einen Platz von 7 × 4 *pedes* einnehmen. Wie viele passen hinein, wenn ein Weg von 4 Fuß freibleiben soll?

In der Rezension I und auch in der *Geometria incerti auctoris* (IV 39 = S. 355, 28—356,8 BUBNOV) nimmt man offenbar an, daß nur ein Gang parallel zur Längsseite freibleibt, und rechnet daher 100 : 7 = 14, (64 — 4) : 4 = 15, 15 · 14 = 210 (Fässer). Ganz anders in der Rezension II. Hier gibt es offenbar zwischen zwei Breitreihen je einen Gang und zusätzlich zwei an der Breitwand, wobei die Breite des Ganges im Gegensatz zur gestellten Aufgabe mit 3 *pedes* angenommen wird. Außerdem liegt die längere Seite der Fässer (7 *pedes*) parallel zur Breitseite des Kellers. Dann ergeben sich wegen $64 \approx 63 = 6 \cdot 7 + 7 \cdot 3$ sechs Reihen und sieben Gänge. An der Längsseite stehen jeweils 100 : 4 = 25 Fässer. Insgesamt sind also 25 · 6 = 150 Fässer vorhanden. Diese eigenwillige Deutung ist grundsätzlich von der anderen Fassung verschieden.

Versucht man, die Eigenheiten der beiden Versionen aufgrund dieser Stellen herauszuarbeiten, so fällt auf, daß bei diesen Aufgaben das Verfahren der Rezension I überall der Rezension II vorzuziehen ist. Die Rezension II ist in sich uneinheitlich, die Methoden sind unnötig kompliziert (Aufgaben 22, 25), die Zahlen in den Lösungen weichen von den gegebenen ab (Aufgaben 30, 31). Demgegenüber sind die Lösungsverfahren in der Rezension I klarer dargestellt und einheitlicher. Sie stimmen im allgemeinen besser mit denjenigen der *Geometria incerti auctoris* überein, deren Methoden im übrigen die besten Ergebnisse liefern. Es hat den Anschein, als ob die abweichenden Beweise der Rezension II auf das Konto eines oder mehrerer Bearbeiter gehen, die ohne jegliche Systematik unterschiedliche Verfahren benutzten, die gegebenen Zahlen veränderten und kaum Wert auf große Genauigkeit legten.

Zum Schluß sollen noch die beiden unterschiedlichen Lösungen der Aufgaben 9 und 10 in der Handschrift R erwähnt werden. Sie dürfen nicht ohne weiteres der Rezension II zugerechnet werden, da sie weder in S noch im Anhang zu M_1, sondern nur in R überliefert werden[88]). Bei beiden Aufgaben soll ein rechteckiges Stoffstück in kleinere Rechtecke

---

[87]) Da in der Rezension II die Zahlenwerte in den Lösungen zu Aufgabe 30 und 31 von denjenigen im Text der Aufgaben abweichen, könnte man vermuten, daß auch der Aufgabentext in dieser Rezension ursprünglich vom *textus communis* abwich und später durch diesen ersetzt wurde. Daß diese Hypothese weitreichende Konsequenzen für die Entstehungsgeschichte der *Propositiones* hat, ist klar.

[88]) In R_1 fehlen die Aufgaben 9 und 10.

geteilt werden (Maße: Aufgabe 9: 100 × 80 in 5 × 4, Aufgabe 10: 60 × 40 in 6 × 4). Hier ist der Text in R der üblichen Fassung vorzuziehen[89]): In R wird, nachdem man wie in Aufgabe 22 die gleichen Rechteckseiten unnötig gemittelt hat, festgestellt, wie oft die kleineren Seiten in den entsprechenden größeren enthalten sind; das Produkt der Verhältniszahlen liefert das richtige Ergebnis. Ganz anders die Lösungen in den übrigen Handschriften: In Aufgabe 9 findet man die Rechnungen $400 : 80 = 5$, $400 : 100 = 4$, also (!) $80 \cdot 5 = 100 \cdot 4 = 400 = x$; in Aufgabe 10: $60 : 10 = 6$, $40 : 10 = 4$, also $10 \cdot 10 = 100 = x$. Beide Rechnungen sind kurz, aber trotz des richtigen Ergebnisses unklar. Aus den oben genannten Gründen hat es aber den Anschein, als ob die bessere Lösung in R nicht der Rezension II zuzuschreiben ist, so daß die negative Einschätzung dieser Fassung nicht revidiert werden muß.

6. MUTMASSUNGEN ÜBER DIE ENTSTEHUNG DER SCHRIFT; IHRE QUELLEN

Wie man aus der Beschreibung der Handschriften (Kapitel 2) ersieht, wird in keinem Codex ein Verfasser der *Propositiones* genannt. Wir sind also gezwungen, nach anderen Ansatzpunkten zu suchen, um den Autor oder wenigstens Entstehungsort und -zeit zu ermitteln.

Die meisten Handschriften lassen sich einigermaßen genau datieren und lokalisieren: R_1: s. 9 ex. St. Denis (Paris); O: um 1000 Westdeutschland/Ostfrankreich; A: um 1000 Reichenau; W: um 1010 Süddeutschland (Füssen?); M_2: um 1020 Chartres/St. Emmeram; V: um 1025 St. Martial (Limoges); B: s. 11[1] Westdeutschland/Ostfrankreich; M: s. 11[1] Ostfrankreich; R: s. 11 St. Mesmin (Orléans); M_1: s. 12 St. Emmeram; C: um 1250 St. Albans; S: s. 15 Bukfastleigh (Devon). Alle älteren Handschriften stammen also aus (Ost/Nord-) Frankreich bzw. (West/Süd-)Deutschland, und die Handschriften belegen, daß unsere Schrift im 10./11. Jahrhundert in diesem Raum recht verbreitet war. Eine inhaltliche Besonderheit stützt diese Feststellung noch: Man hat schon lange bemerkt[90]), daß in den Rechenrätseln das gallische Wegemaß, die *leuga (leuva)*, benutzt wird. Da dies in den Handschriften einheitlich an zwei Stellen geschieht (A 1, A 52), muß man an französische Einflüsse denken.

Es scheint also, als ob die Schrift in ihrer jetzigen Gestalt in Frankreich entstand. Als *terminus ante quem* ergibt sich das ausgehende 9. Jahrhundert, da die älteste Handschrift (R_1) damals geschrieben wurde. Der bewußt verkürzte Text in R_1 (siehe Seite 24) setzt zwar schon eine gewisse Tradition voraus, andrerseits wird man die Entstehung der Schrift nicht viel weiter zurückverlegen, da man nicht plausibel erklären könnte, warum die *Propositiones* zunächst kaum und dann im 10./11. Jahrhundert so oft abgeschrieben worden wären. So entstammt das Werk vermutlich dem 9. Jahrhundert.

Durch die Untersuchung der Handschriften werden die *Propositiones* also in die Zeit der karolingischen Renaissance gerückt. Ein weiteres Indiz weist in Richtung auf den karolingischen Kaiserhof: In zwei Aufgaben (A 39, A 52) werden Kamele erwähnt, die von einem orientalischen Händler gekauft werden. Man weiß, daß in dieser frühen Zeit Beziehungen zwischen den Höfen der Kalifen und Karolinger bestanden[91]). Somit gewinnt die aufgrund der handschriftlichen Überlieferung nicht zu belegende Vermutung[92]), ALKUIN sei der Verfasser dieser Schrift, wieder an Wahrscheinlichkeit.

Wir wollen einmal der Frage nachgehen, ob sich positive Gründe für die Autorschaft des ALKUIN finden lassen. ALKUIN liebte Rätselfragen, die er etwa in seinen Gedichten[93]) oder

---

[89]) Beide Texte sind im kritischen Apparat wiedergegeben.
[90]) THIELE (2), S. CLXIII.
[91]) VOGEL (5), S. 2f. (deutsch) = S. 250f. (russisch).
[92]) Keine der Handschriften nennt einen Autor.
[93]) MANITIUS[7]), S. 278.

in der *Disputatio Pippini cum Albino scholastico* [94]) stellte, und besaß großes Interesse für die Mathematik und ihre Anwendungen [95]). In einem Brief aus dem Jahre 800 an KARL teilt ALKUIN mit, er schicke ihm arithmetische Scherzfragen, und bittet ihn, diese dem EINHART mitzuteilen [96]). Es ist gut möglich, allerdings nicht sicher zu beweisen, daß mit diesen *figurae arithmeticae subtilitatis* unsere *Propositiones* gemeint sind.

Unter den erhaltenen und sicher ALKUIN zuzuschreibenden Texten gibt es keinen, der die Mathematik im engeren Sinne betrifft. Wenn bisweilen eine *Arithmetica Alcuini* genannt wird, so handelt es sich vermutlich um ein untergeschobenes Werk, das vielleicht im Zusammenhang steht mit einem Text desjenigen Anhangs, den in frühkarolingischer Zeit ein Gelehrter, kaum ALKUIN selbst, im Bestreben, verschiedene Materialien zu den Artes zu vereinigen, einer Handschrift der Alkuinschen Dialoge einfügte und die in den Drucken zwischen die Rhetorik und die Dialektik ALKUINs eingeschoben wurde [97]).

Wie steht es nun mit Erwähnungen in alten Katalogen? Aus mittelalterlichen Bibliothekskatalogen sind mir drei Stellen bekannt, an denen mathematische Schriften unter ALKUINS Namen genannt werden:

1. In einem Katalog der Bibliothèque de la Cathédrale du Puy aus dem 11. Jahrhundert wird unter den *Dialectice libri* auch folgender Codex genannt:

34. *Post, liber Augustini de magistro, cum quo Alcuinus de dialectica, rethorica, musica, arimetica, geometria, astronomia.* [98])

Schon DELISLE bemerkte, daß diese Handschrift mit Paris, BN lat. 2974 (s. 9) identisch ist. Dieser Pariser Codex enthält tatsächlich auf f. 50—70 unter der Überschrift *Incipit liber Albini dialecticus dialocus.* ALKUINs *Liber dialecticus* [99]), nicht aber Schriften zum Quadrivium. Wenn die Erwähnung im alten Katalog also genau und die Identifikation gesichert ist, so muß die Pariser Handschrift unvollständig sein [100]).

---

[94]) MANITIUS [7]), S. 284.

[95]) MANITIUS [7]), S. 285; CANTOR (3), S. 834f.

[96]) *Misi . . . et aliquas figuras arithmeticae subtilitatis laetitiae causa:* Ep. 172, ediert von E. DÜMMLER, *MGH Epp.* 4 (1895), S. 285, 8.

[97]) In der *Patrologia Latina* 101 steht *De rhetorica* auf Sp. 919—946, *De dialectica* auf Sp. 951—976. Über den Anhang, der eine Anzahl von Schemata mit Erläuterung enthält (Sp. 945—950), und seine Bedeutung für die Einteilung der Artes im Mittelalter siehe B. BISCHOFF, Eine verschollene Einteilung der Wissenschaften, in: *Mittelalterliche Studien*, Band 1, Stuttgart 1966, S. 273—288, vor allem S. 275f. und S. 286f. — Auch im Codex Wien 2269 befindet sich eine *Arithmetica Alcuini*. Dieser Codex enthält nach der Beschreibung in den *Tabulae . . .* [23]), Band 2, Wien 1868, S. 44f. folgende Schriften ALKUINS: f. 1r—3v *Dialectica*, 3v—6v *Rhetorica*, 7rv *Arithmetica*, 7v *Musica*, 8v *Astrologia*. Bei der Astrologie handelt es sich gemäß dem Incipit *(Duo sunt extremi vertices)* um ARATOS oder HYGINUS: THORNDIKE/KIBRE, *A catalogue of incipits . . .*, Sp. 473. Die Musikabhandlung beginnt mit *Octo tonos in musica consistere musicus scire debet* (Expl.: *superiorum*). Dieser Text, der ALKUIN zugeschrieben wird (siehe J. SMITS VAN WAESBERGHE, The theory of music from the Carolingian era up to 1400, vol. I = *Répertoire international des sources musicales* 3, 1, München/Duisburg 1961, S. 40f.), ist ediert bei M. GERBERT, *Scriptores ecclesiastici de musica*, Sankt Blasien 1783, Bd. 1, S. 26. Die auf vier Spalten befindliche *Arithmetica Albini M.* beginnt auf f. 7r mit *Mathematica dicitur latine doctrinalis scientia, quae abstractam considerat quantitatem* (Expl.: *omnes infiniti sunt*). Es handelt sich hierbei um ISIDOR, *Etymologiae* III 1—9. Am oberen Rand von fol. 7v steht rechts noch *Geometria Albini*, aber die folgenden fünf Spalten (bis f. 8v) sind leer geblieben. Diese Handschrift, die bisher allgemein ins 13. Jahrhundert datiert wurde, entstand schon in der 1. Hälfte des 11. Jahrhunderts in Frankreich (siehe B. BISCHOFF, Literarisches und künstlerisches Leben in St. Emmeram [Regensburg] während des frühen und hohen Mittelalters, in: *Mittelalterliche Studien*, Band 2, Stuttgart 1967, S. 81, Anm. 26).

[98]) L. DELISLE, *Le cabinet des manuscrits de la Bibliothèque Nationale*, Band 2, Paris 1874 = Amsterdam 1969, S. 444.

[99]) Ediert in PL 101, Sp. 951—976. Zu diesem Werk siehe MANITIUS [7]), S. 283f. Die Pariser Handschrift ist beschrieben in *Bibliothèque Nationale, Catalogue général des manuscrits latins*, Bd. 3, Paris 1952, S. 354f.

[100]) Man könnte vermuten, daß der oben erwähnte Anhang mit dem Text über die Artes oder eine Sammlung von Schriften ähnlich wie in Wien 2269 (siehe Anm. 97) fehlt.

V

2. Von den Buchbeständen des Benediktinerklosters der Christ Church, Canterbury, ließ der Prior HENRY OF EASTRY (1284—1331) einen Katalog anfertigen, der im British Museum, Cotton, Galba E IV (um 1300) erhalten ist[101]). Hier wird eine Sammelhandschrift[102]) folgendermaßen beschrieben:

Compilaciones Ieronimi. In hoc vol. cont.: Sententie Prosperi. Liber constructionum. Libellus Benedicti monachi de compoto. Libellus eiusdem de Augmento et decremento Lune. Albinus de arismetrica. Tractatus de Iohanne presbitero rege Indie. Tractatus de accentu. De Rethorica, libri ii. Huguncio de declinacionibus. Tractatus de Barbarismo et ceteris viciis artis gramatice. Tractatus de signis artis dialectice. Vita sancti Zozime monachi, versifice.

Der Codex ist offenbar nicht erhalten[103]), und die Angaben zur angeblichen ALKUIN-Schrift sind zu vage, um daraus weiterreichende Schlüsse zu ziehen.

3. Interessant ist eine Erwähnung in Fulda. Zwar ist die Reihe der *Mittelalterlichen Bibliothekskataloge Deutschlands und der Schweiz* noch nicht bei Fulda angelangt, aber durch die Monographie von KARL CHRIST[104]) sind die Bücherverzeichnisse des 16. Jahrhunderts erschlossen, und PAUL LEHMANN hat sich in mehreren Arbeiten um die Erforschung der Frühgeschichte verdient gemacht[105]). Durch zwei Bruchstücke eines Katalogs um 850 kennen wir offenbar den Fuldaer Gesamtbestand der Schriften von AUGUSTINUS, HIERONYMUS, ALKUIN und HRABAN. Dabei gab es unter den *Opuscula Alcuini* auch:

21. eiusdem quaestiones in genesim. eiusdem de formulis arithmeticae artis. eiusdem de grammatica in I. vol.[106])

Möglicherweise verbergen sich hinter dem zweiten Titel die *Propositiones*. Dieser Verdacht verhärtet sich noch, wenn man berücksichtigt, daß die arithmetische Schrift in Fulda auf ALKUINS *Quaestiones in genesin* folgte: Dieselbe Anordnung ist auch im Augiensis 205 anzutreffen[107]). Wenn die Zuschreibung der *formulae arithmeticae artis* im alten Katalog auf Tatsachen beruht, haben wir hier also ein wichtiges Indiz für die Autorschaft des ALKUIN. Da der Titel *formulae arithmeticae artis* an die *figurae arithmeticae subtilitatis* im ALKUIN-Brief[108]) erinnert, dürfte in beiden Fällen dieselbe Schrift gemeint sein.

Es gibt also keinen sicheren Beweis, aber eine ganze Reihe von Andeutungen, die dafür sprechen, daß die *Propositiones* von ALKUIN verfaßt sind; mit Sicherheit gehören sie etwa in seine Zeit und in den Bereich des fränkischen Hofes.

Die *Propositiones* können ihre Beziehungen zur klösterlichen Welt nicht verleugnen. So handelt die Aufgabe 30 von einer Basilika, deren Boden mit Steinen ausgelegt werden

---

[101]) Erstmals, allerdings schlecht, ediert von EDWARD EDWARDS, Memoirs of libraries: including a handbook of library economy, Band 1, London 1859; dann erneut herausgegeben von M. R. JAMES, The ancient libraries of Canterbury and Dover, Cambridge 1903, S. 13—142.

[102]) Nr. 98 bei JAMES (auf S. 27).

[103]) Jedenfalls wird er nicht unter den erhaltenen Handschriften genannt bei JAMES[101]), S. 505ff., und bei KER[48]), S. 29—40.

[104]) Die Bibliothek des Klosters Fulda im 16. Jahrhundert, Leipzig 1933.

[105]) Vor allem: Fuldaer Studien, in: Sitzungsberichte der philosophisch-philologischen und der historischen Klasse der Bayerischen Akademie der Wissenschaften zu München, Jahrgang 1925, 3. Abhandlung, München 1925, und: Quot et quorum libri fuerint in libraria Fuldensi, in: Bok- och Biblioteks-historiska studier tillägnade Isak Collijn på hans 50-årsdag, Uppsala 1925, S. 47—57. Danach gibt es aus der Fuldaer Frühzeit das Fragment eines Verzeichnisses von vor 800 (LEHMANN, 1925, S. 48, und Fuldaer Studien, Neue Folge, 1927, S. 52), ferner zwei Bruchstücke eines Katalogs, der kurz vor 850 entstand (abgedruckt bei G. BECKER, Catalogi bibliothecarum antiqui, Bonn 1885, Nr. 128, dort irrtümlich ins 12. Jahrhundert gesetzt, und Nr. 13), und kleine Teile eines weiteren Verzeichnisses vor allem theologischer Texte aus dem 9. Jahrhundert (LEHMANN, 1925, S. 10). Die Bruchstücke des 1. und 3. Katalogs nennen keine mathematische Schrift des ALKUIN.

[106]) BECKER[105]), S. 31. Diese Handschrift ist in den von CHRIST[104]) abgedruckten Verzeichnissen des 16. Jahrhunderts nicht mehr zu finden.

[107]) Siehe S. 17.

[108]) Siehe S. 31.

soll. In Aufgabe 47 verteilt ein Bischof Brote unter seinem *clerus*, und es wird nach der Zahl der *presbyteri*, *diaconi*, *lectores* gefragt. Ganz ähnlich in Aufgabe 53: Hier sind es eine Anzahl Eier, die der Abt seinen Mönchen zuweist. Auch die Einkleidungen der übrigen Aufgaben passen gut in den durch die vermutliche Entstehungszeit geprägten Rahmen: Handwerk und Handel spielen nur eine sekundäre Rolle[109]); Motive aus dem Landleben überwiegen[110]). Auch die genannten Gegenstände[111]) und Tiere[112]) gehören in diesen Bereich. Nur an einer Stelle (Aufgabe 13) deuten Soldaten auf kriegerische Auseinandersetzungen hin.

Natürlich sind in die *Propositiones* auch andersartige Elemente eingeflossen. Auf die Erwähnung der Kamele, die auf frühe Beziehungen zum islamischen Kulturkreis hindeuten könnten, haben wir schon hingewiesen[113]). Aufschlußreicher noch sind die Erbschaftsaufgaben, die immerhin dreimal vertreten sind (Nr. 12, 35, 51). Ebenso wie die vier Aufgaben über die Verwandtschaftsgrade (Nr. 11, 11a—c) entstammen sie der römischen Gedankenwelt und bestätigen, daß die *Propositiones* in der Tradition der Römer stehen. CANTOR hat richtig bemerkt, daß die Lösung der Aufgabe über die Zwillingserbschaft (Nr. 35) in den *Propositiones* den Willen des Erblassers völlig auf den Kopf stellt[114]): Der Sohn bekommt nicht, wie im Testament verfügt, das Dreifache des Erbes der Mutter, sondern nur unwesentlich mehr als diese. Die Aufgabe selbst stammt aus den römischen Rechtsschulen und wird mit der richtigen Verteilung in den Digesten bei SALVIANUS JULIANUS, CAECILIUS AFRICANUS und JULIUS PAULUS erwähnt. Die erstaunliche Auflösung in den *Propositiones* zeigt, daß ihr Autor von der römischen Rechtspraxis nichts mehr verstand. — Noch eine weitere Aufgabengruppe steht in römischer Tradition: die Flächenberechnungen und -verwandlungen. Über sie wird im folgenden Kapitel zu reden sein.

Möglicherweise sind die *Propositiones* auch von spätgriechisch-byzantinischen Aufgabensammlungen beeinflußt. Hier muß man vor allem an die arithmetischen Epigramme in der Anthologia Palatina, Buch XIV, denken, die gemäß der Überlieferung überwiegend von METRODOROS stammen[115]) (XIV 1—4, 6, 7, 11—13, 48—51, 116—146). In der griechischen Anthologie nehmen wie in den *Propositiones* die Hau-Rechnungen einen überaus breiten Raum ein; auch Brunnenaufgaben sind vertreten[116]). Es ist gut möglich, daß die METRODOROS-Aufgaben gerade im 8./9. Jahrhundert in Westeuropa bekannt wurden, da wir

---

[109]) Kaufleute werden in den Aufgaben 5, 6, 38, 39 erwähnt, Handwerker nur an zwei Stellen: Maurer (A37), Stellmacher (A49).

[110]) Wanderer: A2, 3; Pflügen des Ackers: A14, 15; Teilen von Äckern: A21—25; Stadtformen: A27—29; Weinkeller: A31; Verteilung der Ernte: A32—34, 33a; weidende Tiere: A4, 21, 40; Schule und Schüler: A48; Getreidetransport: A52.

[111]) Schüssel: A7; Tonne: A8, 31; Weinkrug: A50, 51; Ölflasche: A12; Leiter: A42; Schweinestall: A41; Wagenrad: A49; Tuch für Kleidung: A9—10; Geldbörse: A46; Brot: A47; Eier: A53.

[112]) Esel: A39; Hase: A26; Hund: A26; Igel: A20; Kamel: A39, 52; Pferd: A4, 38; Rind/Ochse: A14, 16, 38; Schaf: A21, 38—40; Schnecke: A1; Schwalbe: A1; Schwein: A5, 6, 41, 43; Storch: A3; Taube: A42, 45; Wolf: A18; Ziege: A18.

[113]) Natürlich sind aus der Erwähnung der Kamele allein keine zwingenden Schlüsse zu ziehen. Im Alten Testament beispielsweise werden *cameli* oft genannt, z. B. Gen. 12, 16 und 24, 35; 1. Reg. 15,3 und 27,9 (überall zusammen mit *oves* und *asini*); Gen. 30, 43; Exod. 9, 3; 1. Par. 12, 40; Isai. 21, 7 (zusammen mit *asini*); Gen. 32, 7; Judith 3, 3; Job 1, 3 (gemeinsam mit *oves*) und öfter.

[114]) CANTOR (2), S. 146—149; CANTOR (3), S. 562.838.

[115]) Die Aufgaben mit Scholien sind ediert von PAUL TANNERY, *Diophanti Alexandrini opera omnia*, Band 2, Leipzig 1895, S. 43—72. Ergänzend heranzuziehen ist TANNERYS kurzer, aber inhaltsreicher Aufsatz Sur les épigrammes arithmétiques de l'Anthologie Palatine, in: *Revue des études grecques* 7 (1894) 59—62; wiederabgedruckt in: *Mémoires scientifiques de Paul Tannery*, Band 2, Toulouse/Paris 1912, S. 442—446. Hier hat TANNERY anhand des Inhalts und der Überlieferung die ursprüngliche Reihenfolge und bestehende Lücken in den METRODOROS-Epigrammen ermittelt und die Probleme klassifiziert. METRODOROS lebte vielleicht um 500; ein genaues Datum ist nicht bekannt.

[116]) Zu diesen Aufgabentypen siehe Kapitel 7 (S. 35—38).

wissen, daß von einer byzantinischen Gesandtschaft an Karl den Großen im Jahre 781 der Grieche ELISSAIOS als Lehrer der Prinzessin ROTRUD zurückblieb[117]).

Die *Propositiones* beruhen also auf drei Wurzeln: In erster Linie stehen sie in der römischen Tradition; daneben sind griechisch-byzantinische und arabische Einflüsse anzunehmen. Diese zunächst recht grobe Feststellung läßt sich anhand der Geschichte der Probleme erhärten (siehe Kapitel 7). Dabei wird es sich auch zeigen, daß einige Aufgaben erstmals in den *Propositiones* auftreten.

Wir müssen noch der Frage nachgehen, ob die *Propositiones* von Anfang an in ihrer heutigen Gestalt selbständig vorlagen oder als Teil eines größeren Ganzen konzipiert wurden. THIELE hat als erster und wohl auch als einziger den Standpunkt vertreten, die Rechenrätsel seien seit alter Zeit mit den äsopischen Fabeln verbunden[118]). Zu dieser Annahme führte ihn die Tatsache, daß in den Handschriften *B* und *V* beide Texte aufeinander folgen, daß die erste Aufgabe in Form einer Tierfabel gekleidet ist und die Rechenaufgaben in ADEMARS Handschrift *V* wie die vorangehenden Fabeln illustriert sind. Aufbauend auf diesen Beobachtungen und auf dem Text des Augiensis *A*, konstruierte THIELE drei Entwicklungsstufen: Ursprünglich waren die Aufgaben integraler Bestandteil der Fabelsammlung des ROMULUS-AESOP (wie in *V*). Später wurden die Rätsel von den Fabeln gelöst, erhielten den Titel *Propositiones arithmeticae* (!) *ad acuendos iuvenes* und außerdem Einzelüberschriften (wie in *B*). Im letzten Stadium wurden die Aufgaben ganz von den Fabeln getrennt (wie in *A*). — Es bedarf keiner Phantasie, um diese Ansicht als nicht sehr wahrscheinliche Hypothese abzutun. Hätte THIELE nicht nur die Handschriften *ABV*, sondern alle 12 Codices gekannt, so wäre ihm bewußt geworden, daß die Verbindung der Rätsel mit den Fabeln eine Besonderheit der Handschriften *BV* ist und daß demgegenüber in den anderen zum Teil älteren und zuverlässigeren Handschriften die Rechenrätsel isoliert dastehen. Insbesondere dürfen natürlich aus dem relativ jungen Codex *V*, der auch sonst die individuellen Eigenheiten des Kompilators ADEMAR zeigt, keine weiterreichenden Schlüsse gezogen werden. Die Übereinstimmung in der Überlieferung bestätigt, daß die *Propositiones* als selbständiges Werk konzipiert wurden. Dies schließt nicht aus, daß in einem späteren Zeitpunkt die Rechenrätsel an Fabelsammlungen angehängt und zum Bestandteil einer Tradition in der AESOP-Überlieferung wurden. Tatsächlich legt die thematische Übereinstimmung einiger Fabeln eine solche Verbindung nahe, ähnlich wie in anderen Handschriften unter die *Propositiones* verschiedene andere mathematische Rätselaufgaben gemischt sind[119]). Wohl aus denselben Gründen begegnen uns in den Handschriften die *Propositiones* oft nach oder vor Pseudo-BEDAS *De arithmeticis propositionibus*[120]), ohne daß man daraus auf denselben Ursprung oder Verfasser schließen könnte[121]).

7. INHALT DER PROPOSITIONES; HISTORISCHE BEMERKUNGEN

Um den Stellenwert der *Propositiones* innerhalb der Darstellungen des angewandten Rechnens einschätzen zu können, ist zu untersuchen, welcher Art die Probleme sind, die uns in dieser Schrift begegnen. Hier genügt ein verhältnismäßig pauschaler Überblick, da die in den verschiedenen Kulturen auftauchenden Probleme des angewandten Rechnens in der

---

[117]) VOGEL (5), S. 2 (deutsch) = S. 250 (russisch).

[118]) THIELE (1), S. 22—25.

[119]) Etwa in *V* (nach Aufgabe 2); in *M*; in *S* (zwischen Aufgaben und Lösungen) und in den kontaminierten Fassungen in Cotton Cleop. B. IX sowie Oxford, Bodleian Library, Digby 98 (siehe Kapitel 2). In *A* folgen auf die *Propositiones* noch einige Scherzrätsel, die CANTOR erstmals ediert hat: CANTOR (2), S. 219f. (als Anmerkung 269).

[120]) In den Handschriften BMM_1OR_1.

[121]) Siehe FOLKERTS (2), S. 36.

voraussichtlich 1978 erscheinenden Neuauflage von J. Tropfkes *Geschichte der Elementar-mathematik* behandelt und klassifiziert werden[122]). Die im folgenden gewählte numerische Einteilung entspricht derjenigen bei Tropfke.

1. Probleme des täglichen Lebens

Hierzu zählen etwa Prozent-, Rabatt-, Zins-, Gesellschafts- und Mischungsrechnungen sowie Aufgaben, die mit Geldwechsel zu tun haben. Probleme dieser Art sind in Westeuropa insbesondere in Sammlungen des 14. und 15. Jahrhunderts stark vertreten. Es fällt auf, daß in den *Propositiones* keine Aufgabe dieser Gruppe zuzurechnen ist. Auch die Nummern 1, 46, 49, 50, 53, die sich mit Umrechnungen verschiedener Art bzw. einfachen Divisionen befassen, behandeln nur scheinbar Probleme des täglichen Lebens; in Wirklichkeit sind es Phantasieaufgaben[123]).

2. Probleme der Unterhaltungsmathematik

2. 1 *Lineare Probleme mit einer Unbekannten*

Der Hauptanteil der arithmetischen Aufgaben gehört dieser Gruppe an. Von den 13 Aufgaben dieser Abteilung sind 8 Hau-Rechnungen (2. 1. 1), eingekleidet in die Frage nach der Anzahl („Gott-Grüß-Euch-Aufgaben": 2. 1. 1. 1; Nr. 2—4, 40, 45, 48) oder nach dem Alter (2. 1. 1. 2; Nr. 36, 44). Diese Aufgaben führen stets auf die lineare Gleichung $nx + p = 100$, wobei n die Summe gewisser rationaler Zahlen und p eine natürliche Zahl oder Null ist[124]). Der Verfasser der *Propositiones* gibt in keinem Fall das Lösungsverfahren an, sondern beschränkt sich auf die Erwähnung der Lösungszahl und die Probe. — Hau-Rechnungen begegnen uns, angefangen von den ältesten ägyptischen Texten, in fast allen Sammlungen der Unterhaltungsmathematik, so auch in indischen und arabischen Texten und bei Anania von Schirak. Vielleicht beruhen die Aufgaben in den *Propositiones* auf der griechischen Anthologie, in der Hau-Rechnungen etwa $\frac{2}{3}$ aller Probleme ausmachen. Die Aufgabe 2 = 40 findet man mit denselben Zahlenwerten auch in dem byzantinischen Rechenbuch aus dem 15. Jahrhundert sowie bei Abraham ibn Ezra und Elia Misrachi[125]).

Eine Aufgabe (Nr. 26) gehört zu den Bewegungsproblemen (2. 1. 4): Ein Hund verfolgt einen Hasen, der 150 Fuß Vorsprung hat. In der Zeit, in der der Hund 9 Schritte läuft, legt der Hase 7 Schritte zurück. Da der Hund also in der Zeiteinheit zwei Schritte näherkommt, holt er den Hasen nach 150 : 2 = 75 Schritten ein. „Alkuin" gibt ohne Begründung lediglich an, daß man als Lösung die Hälfte des Abstandes nehmen müsse. Diese Aufgabe vom Hasen und Hund, die in späteren Sammlungen mit vielen Varianten behandelt wird, taucht hier in der einfachsten Fassung erstmals in der Literatur auf; eine kompliziertere Version liegt schon in den chinesischen 9 Büchern arithmetischer Technik vor[126]). Überhaupt werden in diesem chinesischen Rechenbuch viele Bewegungsaufgaben verschiedenster Art behandelt. Merkwürdigerweise kennen Inder und Araber diese Aufgabe nicht.

Die restlichen vier Aufgaben (Nr. 7, 8, 37, 52) sind untypische Vertreter dieser Gruppe 2. 1. Sie sollen hier kurz erwähnt werden.

---

[122]) Diese 4. Auflage soll die drei ersten Bände der 3. Auflage ersetzen.

[123]) Nr. 1: Umwandlung *leuva—uncia* sowie Tage in Jahre; Nr. 46: Talente in *solidi*; Nr. 49: aus der Zahl der Räder wird die Anzahl der Wagen ermittelt; Nr. 50: Verwandlung von *metra* in *sextarii* und *meri*; Nr. 53: Verteilung von Eiern unter Mönche.

[124]) Die Gleichungen lauten in moderner Schreibweise:
Nr. 2 und Nr. 40: $x + x + \frac{1}{2}x + \frac{1}{4}x + 1 = 100$; Nr. 3: $x + x + x + \frac{1}{2}x + 2 = 100$; Nr. 4: $x + x + \frac{1}{2}x = 100$; Nr. 36: $(4x) \cdot 3 + 1 = 100$; Nr. 44: $2x \cdot 3 + 1 = 100$; Nr. 45: $x + x + x + 1 = 100$; Nr. 48: $x \cdot \frac{2 \cdot 3}{4} + 1 = 100$.

[125]) Vogel (2), S. 95.

[126]) Buch 6, Nr. 14: siehe Vogel (4), S. 126f.

In Aufgabe 7 sind bei einer Wurfscheibe die Gewichtsanteile zu bestimmen, wenn die Scheibe aus Gold, Silber, Messing und Zinn besteht und jeder Bestandteil in dreifacher Menge wie der vorhergehende vorhanden ist. Diese Aufgabe könnte man auf den ersten Blick unter die Mischungsrechnung einreihen, jedoch fehlen wesentliche Kriterien dieses Typus, da es nicht auf den Feingehalt der Legierung ankommt, sondern nur auf die Gewichte. In Wirklichkeit erfordert die Aufgabe nur das Lösen einer linearen Gleichung. In den *Propositiones* sind auch hier nur Lösung und Probe angegeben.

In Nr. 8 kann eine Tonne von 300 *modia* Inhalt durch drei Röhren entleert werden, wobei durch die erste Röhre $\frac{1}{3} + \frac{1}{6}$ des Inhalts fließt, durch die zweite $\frac{1}{3}$ und durch die dritte $\frac{1}{6}$. Wie viele *sextarii* fließen aus jeder Röhre? Diese Einkleidung erinnert an die Brunnenaufgaben, einen Problemtypus, der ebenfalls in der griechischen Anthologie relativ zahlreich vertreten ist[127]). Vom Inhalt her unterscheidet sich das Problem jedoch von den üblichen Zisternenaufgaben: Hier ist nicht der Dreisatz anzuwenden, sondern es genügen simple Divisionen sowie elementare Umrechnungen. Der Autor gibt wiederum nur die Lösung an.

Nr. 37: Fünf Meister und ein Lehrling bauen gemeinsam ein Haus und erhalten dafür zusammen 25 *denarii* pro Tag, wobei der Lehrling nur die Hälfte eines Meisters verdient. Um die einzelnen Anteile zu bestimmen, werden zunächst 22 *denarii* in $5 \cdot 4 + 1 \cdot 2$ zerlegt. Die restlichen 3 *denarii* werden in je 11 Teile geteilt, von denen jeder Meister 6 und der Lehrling 3 erhält. Somit bekommen die Meister je $4\frac{6}{11}$ und der Lehrling $2\frac{3}{11}$ *denarii*. Hier wird also ausnahmsweise der Lösungsweg verraten, doch macht die Rechnung einen etwas unbeholfenen Eindruck. Auch diese Aufgabe gehört nur scheinbar in eine bekannte Problemgruppe (Leistungsprobleme: 2. 1. 2); letzten Endes handelt es sich um eine einfache Division durch 11.

Nr. 52: Ein Kamel transportiert in Fuhren zu je 30 *modia* insgesamt 90 *modia* Getreide über eine Entfernung von 30 *leuvae*, wobei es pro *leuva* 1 *modium* auffrißt. Wieviel Getreide kommt ans Ziel? Die Lösung besagt, daß in drei Fuhren je 30 *modia* zunächst nur über 20 *leuvae* transportiert werden. Hierbei verbleibt ein Rest von jeweils 10 *modia*. Zuletzt werden diese 30 *modia* über die restlichen 10 *leuvae* gebracht, so daß 20 *modia* übrigbleiben. Diese eigenwillige Aufgabe, bei der ebenfalls außer der Lösung auch der Lösungsweg angegeben wird, paßt in kein Aufgabenschema.

2. 2 *Lineare Probleme mit mehreren Unbekannten*

Insgesamt neun Aufgaben gehören in diese Abteilung: eine (Nr. 16) unter die Rubrik „Geben und Nehmen" (2. 2. 4), die übrigen acht (Nr. 5, 32, 33, 33a, 34, 38, 39, 47) zu den „Zechenaufgaben" (2. 2. 6).

In Nr. 16 verlangt A von B zwei Ochsen; dann hätten beide gleichviel Tiere. Wenn dagegen B von A zwei Ochsen erhält, besitzt B doppelt so viele wie A. Es handelt sich also um die einfachste Form des „Gebens und Nehmens", bei der nur zwei Personen beteiligt sind. Als Lösung wird nicht, wie zu erwarten, 10 bzw. 14 Ochsen angegeben, sondern 4 und 8. Somit hat man davon auszugehen, daß der zweite nicht von der ursprünglichen, sondern von der schon veränderten Anzahl etwas hergibt[128]). Eine ähnliche Variante liegt auch im *Algorismus Ratisbonensis*, Nr. 231, vor[129]). — Das Problem „Geben und Nehmen" ist ein typisches Stück griechischer Algebra[130]). In späterer Zeit begegnet es auch bei den Indern und Arabern und in anderen abendländischen Sammlungen.

---

127) VOGEL (2), S. 97.
128) Statt $x + 2 = y - 2$, $2(x - 2) = y + 2$ lautet das Gleichungssystem $x + 2 = y - 2$, $2(x + 2 - 2) = (y - 2) + 2$.
129) VOGEL (1), S. 171.
130) VOGEL (1), S. 218.

Zu den unbestimmten Aufgaben gehören die Zechenaufgaben, die in den *Propositiones* stark vertreten sind: Unter eine bestimmte Anzahl von Personen wird eine gleichgroße Anzahl von Gegenständen je nach Würdigkeit verteilt, oder 100 Tiere verschiedenen Wertes kosten 100 *solidi*. Es handelt sich stets um zwei Gleichungen mit drei Unbekannten. Auch hier verzichtet der Autor auf Angabe des Verfahrens; er greift nur unter den meist mehreren ganzzahligen Möglichkeiten eine Lösung heraus und zeigt ihre Richtigkeit[131]). — Das Problem der 100 Vögel, aus dem später die Zechenaufgaben hervorgingen, stammt ohne Zweifel aus China. Von den Chinesen gelangte die Aufgabe zu den Indern und Arabern[132]). Es liegt nahe, für die *Propositiones* hier eine arabische Vorlage anzunehmen, zumal die Aufgabe 39 von der Einkleidung her[133]) auf orientalische Quellen weist und genau dieselben Zahlenwerte uns bei ABŪ KĀMIL begegnen[134]).

2. 4 *Aufgaben mit Folgen und Reihen*

Aufgaben, in denen arithmetische und geometrische Folgen und Reihen eine Rolle spielen, sind seit der ältesten Zeit in fast allen Kulturen anzutreffen. Das wohl älteste Beispiel einer geometrischen Reihe mit dem Quotienten 7 steht im Papyrus Rhind (Nr. 79). Auch aus China sind schon früh Aufgaben dieser Art belegt[135]). In den *Propositiones* gehören drei Aufgaben zu dieser Gruppe: Nr. 13, 41, 42.

In Aufgabe 13 sammelt ein König ein Heer, indem er nacheinander in 30 Häuser geht und aus jedem so viele Soldaten mitnimmt, wie insgesamt das Haus betreten haben. Dies führt zur geometrischen Reihe $1 + 1 + 2 + 4 + \ldots + 2^{29} = 2^{30}$ [136]). Der Autor der *Propositiones* benutzt zur Bestimmung der Summe keine Formel, sondern zählt alle Zwischenergebnisse auf.

In ähnlicher Weise wird in Aufgabe 41 die geometrische Folge $8, 8^2, \ldots, 8^6 = 262144$ bestimmt. Als Einkleidung wählt man dabei eine Schweinefamilie: Eine Sau wirft in der Mitte des Schweinestalls 7 Junge; alle 8 Schweine und ihre Nachkommen werfen in jeder Ecke und schließlich noch einmal in der Stallmitte wieder je 7 Ferkel. Wie viele Tiere sind es insgesamt?

In Aufgabe 42 soll die Zahl der Tauben auf einer 100sprossigen Leiter bestimmt werden, wenn auf der 1. Sprosse eine Taube und auf jeder weiteren jeweils eine Taube mehr als auf der vorhergehenden sitzt. Hier wird ausnahmsweise die Lösung nicht einfach genannt, sondern ein Verfahren zur Bestimmung des Wertes dieser arithmetischen Reihe $1 + 2 + + 3 + \ldots + 99 + 100$ angegeben: Durch Zusammenfassen von $1 + 99$, $2 + 98$, ..., $49 + 51$ erhält man 49 Paare zu je 100 Tauben, denen noch der Wert des 50. und 100. Gliedes zuzurechnen ist. Dies ergibt insgesamt $49 \cdot 100 + 50 + 100 = 5050$. Eine entsprechende

---

[131]) Die einzelnen Aufgaben lauten:
Nr. 5: 100 Schweine kosten 100 *denarii*; Werteinstufung 10, 5, $\frac{1}{2}$; Lösung 1, 9, 90.
Nr. 32: 20 Maß Getreide an 20 Personen; Würdigkeit 3, 2, $\frac{1}{2}$; Lösung 1, 5, 14.
Nr. 33: 30 Maß Getreide an 30 Personen; Würdigkeit 3, 2, $\frac{1}{2}$; Lösung 3, 5, 22.
Nr. 33a: 90 Maß Getreide an 90 Personen; Würdigkeit 3, 2, $\frac{1}{2}$; Lösung 6, 20, 64.
Nr. 34: 100 Maß Getreide an 100 Personen; Würdigkeit 3, 2, $\frac{1}{2}$; Lösung 11, 15, 74.
Nr. 38: 100 Tiere kosten 100 *solidi*; Werteinstufung 3, 1, $\frac{1}{24}$; Lösung 23, 29, 48.
Nr. 39: 100 Tiere kosten 100 *solidi*; Werteinstufung 5, 1, $\frac{1}{20}$; Lösung 19, 1, 80.
Nr. 47: 12 Brote an 12 Geistliche; Würdigkeit 2, $\frac{1}{2}$, $\frac{1}{4}$; Lösung 5, 1, 6.
[132]) VOGEL (2), S. 98; VOGEL (1), S. 222.
[133]) Kamele, Orient: siehe Seite 33.
[134]) HEINRICH SUTER, Das Buch der Seltenheiten der Rechenkunst von Abū Kāmil el Miṣrī, in: *Bibliotheca Mathematica*, 3. Folge, Band 11, 1910/11, S. 100—120, hier: S. 102.
[135]) VOGEL (4), S. 127. Beispiele für ähnliche Aufgaben in Byzanz bei VOGEL (3), S. 148.152.
[136]) Die Annahme bei THIELE (1), S. 64f., in der Rechnung sei ein Fehler, da in der 3. Stadt die Zahl erst auf 6 anwachsen durfte, ist nicht zwingend: Offenbar muß der König mitgezählt werden, so daß $4 + 4$ Personen das 3. Haus verließen, da $1 + 3 = 4$ Personen es betreten hatten.

Aufgabe findet man auch in der byzantinischen Sammlung des frühen 14. Jahrhunderts[137]), wobei dort aber die Formel $s = \dfrac{100^2 + 100}{2}$ benutzt wird.

2. 7 Anordnungsprobleme; Sonstiges

Insgesamt 11 Aufgaben aus den *Propositiones* kann man den Anordnungsproblemen im weiteren Sinn zurechnen. Nr. 6 ist ein typischer Vertreter des Problems „Gewinn beim Verkauf um den Einkaufspreis" (2. 7. 2): Zwei Händler kaufen 250 Ferkel für 100 *solidi*. Obwohl sie sie scheinbar zum selben Preis (d. h. je 5 für 2 *solidi*) verkaufen, machen sie $4\frac{1}{6}$ *solidi* Gewinn: Der eine verkauft nämlich die schlechtere Hälfte zum Stückpreis von $\frac{1}{3}$ *solidus*, der andere die bessere für je $\frac{1}{2}$ *solidus*. — Dieses Problem findet sich hier wohl zum erstenmal. Später begegnet es uns etwa im *Liber augmenti et diminutionis* und bei LEONARDO VON PISA[138]).

Das Problem der Zwillingserbschaft (2. 7. 4) ist Thema von Aufgabe 35. Über dieses aus dem römischen Bereich stammende Problem und die ungerechte Lösung wurde schon gesprochen[139]).

Gleich viermal werden Transportprobleme (2. 7. 5) behandelt, und zwar in Nr. 17—20. Dabei sind in Nr. 17 drei Männer mit ihren Frauen über den Fluß zu setzen; in Nr. 18 handelt es sich um Ziege, Wolf und Kohlkopf, in Nr. 19 um Mann, Frau und zwei Kinder und in Nr. 20 um zwei Igel mit ihren Jungen. In allen Fällen wird sehr ausführlich angegeben, welche Lebewesen jeweils überzusetzen sind. Es hat den Anschein, als ob dieser Aufgabentyp, der vor allem in der Einkleidung mit Ziege, Wolf und Kohlkopf später sehr populär wurde, erstmals hier auftritt.

Zweimal sind „Umfüllaufgaben" (2. 7. 6) vertreten: In Nr. 12 werden 10 volle, 10 halbvolle und 10 leere Flaschen unter drei Söhne so verteilt, daß jeder gleichviel Öl und Flaschen bekommt. Ohne Begründung greift der Autor unter den fünf Möglichkeiten eine heraus. — In Nr. 51 fällt die triviale Lösung auf: Vier Söhne sollen vier Weingefäße mit den Inhalten 40, 30, 20, 10 *modia* gerecht teilen. Der Autor meint, die ersten beiden Söhne sollten die Gefäße mit 40 und 10 *modia* nehmen, die beiden anderen die mit 30 und 20; dann hätten je zwei Söhne zwei Gefäße mit insgesamt 50 *modia*. — Diese beiden trivialen Aufgaben erinnern an sinnvollere Verteilungsaufgaben in den *Annales Stadenses* und anderen abendländischen Sammlungen, bei denen neun Fässer mit den Inhalten 1, 2, . . ., 9 unter drei Personen gerecht verteilt werden sollen[140]).

Die restlichen drei Probleme, die man vielleicht diesem Komplex zurechnen könnte (Nr. 11, 11a, 11b), behandeln komplizierte Verwandtschaften. Diese nicht mathematischen Aufgaben stehen der römischen Gedankenwelt nahe. Bezeichnenderweise verzichtet man darauf, eine Lösung zu geben.

Drei weitere Aufgaben haben mit Mathematik wenig oder nichts zu tun. Sie lassen sich in keinen der gängigen Aufgabenkomplexe einordnen. Zwei von ihnen (Nr. 14, 43) sind Scherzaufgaben: In Nr. 14 wird nach der Zahl der Spuren gefragt, die ein pflügender Ochse erzeugt, und in der Lösung wird richtig bemerkt, daß der Pflug die Spuren löscht. In Aufgabe 43, deren Sinn ich nicht verstehe, bemerkt der Autor ausdrücklich: *Haec fabula est tantum ad pueros increpandos*. In Aufgabe 15 schließlich wird die triviale Frage gestellt, wie viele Furchen beim Pflügen entstehen, wenn der Pflug an beiden Seiten des Feldes je dreimal gewendet wird. Natürlich lautet die richtige Antwort: 7.

---

137) VOGEL (3), S. 126—129 = Nr. 111.
138) Zur Geschichte der Aufgabe siehe VOGEL (1), S. 229f.
139) Siehe Seite 33.
140) Siehe VOGEL (1), S. 229.

2. 3 *Aufgaben der rechnenden Geometrie*

Wie in vielen anderen Sammlungen der Unterhaltungsmathematik sind auch in den *Propositiones* zahlreiche Aufgaben der rechnenden Geometrie vertreten. Fast ein Viertel des Gesamtbestandes gehört in diesen Komplex (12 Aufgaben: Nr. 9, 10, 21—25, 27—31). Diese Aufgaben unterscheiden sich jedoch von denjenigen in anderen Sammlungen dadurch, daß jene im allgemeinen Anwendungen des pythagoreischen Lehrsatzes oder der Ähnlichkeitssätze bringen, während diese sich ausschließlich mit Flächenberechnungen bzw. -verwandlungen beschäftigen. Auch innerhalb der *Propositiones* nehmen die genannten Aufgaben eine Sonderstellung ein, da im Gegensatz zu den meisten anderen Problemen hier im allgemeinen der Lösungsweg angedeutet wird. Dies alles macht es erforderlich, auf diese Aufgabengruppe etwas ausführlicher einzugehen.

Die geometrischen Aufgaben sind keine typischen Vertreter der Unterhaltungsmathematik; vielmehr dienen sie dazu, geometrische Formeln an Beispielen zu verdeutlichen. Diese Probleme sind also der ernsthaften Mathematik zuzurechnen. Einkleidung und Lösungsverfahren zeigen, daß diese Aufgaben in der Tradition der römischen Feldmesser stehen; ähnliche Probleme begegnen uns schon bei HERON, im 5. Buch von COLUMELLAS *De re rustica* und in Schriften des *Corpus agrimensorum*. Besonders starke Ähnlichkeit besteht mit der anonymen Geometrie, die mit GERBERTS Geometrie in Verbindung steht (*Geometria incerti auctoris;* im folgenden abgekürzt mit G. i. a.): Abgesehen von den Aufgaben 9, 10 der *Propositiones*, werden die übrigen Probleme in ganz ähnlicher Weise auch in der G. i. a. behandelt, und zwar entsprechen die Nummern 21—25, 27—31 der *Propositiones* den Aufgaben IV 30—39 in der G. i. a. So ist es notwendig, auch die G. i. a. in die Betrachtung einzubeziehen.

Die G. i. a. galt lange Zeit als Teil der GERBERTschen Geometrie. Erst BUBNOV, der die Schrift erstmals kritisch edierte[141]), konnte durch seine Kenntnis der handschriftlichen Überlieferung überzeugend nachweisen, daß dieses Werk vor GERBERT entstand: Da die G. i. a. arabische Schriften über das Astrolab voraussetzt, sie andrerseits aber in der Agrimensoren-Handschrift vorhanden war, die GERBERT 983 benutzte, kann die Schrift nicht vor dem 9. und nicht nach dem 10. Jahrhundert entstanden sein[142]). Somit ist sie etwa zeitgleich mit den *Propositiones*. Da nicht sicher zu klären ist, welche von beiden Schriften die jüngere ist, muß ein Vergleich des Inhalts über eine mögliche Abhängigkeit entscheiden. Im folgenden werden die geometrischen Aufgaben in den *Propositiones* einzeln behandelt; dabei wird auf Ähnlichkeiten wie Unterschiede zur G. i. a. hingewiesen.

Die Aufgaben 9 und 10, die nicht in der G. i. a. vorhanden sind, behandeln die Frage, wie viele kleine Tücher aus einem großen rechteckigen Tuch hergestellt werden können. Die konfuse Lösungsmethode, die wohl nur zufällig zum richtigen Ergebnis führt, ist schon auf Seite 29f. erwähnt worden.

Nr. 21 = G. i. a. IV 30: Wie viele Schafe können auf einem rechteckigen Feld weiden (100 × 200 Fuß), wenn jedes Tier 4 × 5 Fuß benötigt? Das Ergebnis lautet: $(200 : 5) \cdot (100 : 4)$ $= 40 \cdot 25 = 1000$. Zahlenwerte und Lösungsverfahren stimmen in beiden Texten überein; sogar die Formulierung ist fast identisch. Die Übereinstimmung geht so weit, daß fast dieselbe verderbte Ausdrucksweise für „Teile 200 durch 5" benutzt wird: „ALKUIN" hat: *Duc bis quinquenos de CC,* G. i. a.: *Duc bis vicenos, vel quintam partem de ducentis.*

Die folgenden drei Aufgaben (22—24 = G. i. a. IV 31—33) beschäftigen sich mit der Berechnung der Fläche eines schrägen, viereckigen und dreieckigen Ackers. Bei „ALKUIN" wie in der G. i. a. benutzt man dafür die üblichen Feldmesserformeln, bei denen durch

---

141) BUBNOV, S. 310—365.
142) BUBNOVS Argumente sind zusammengefaßt auf S. 310, Anm. 1 (mit Ergänzungen auf S. 560), S. 400 und S. 471 f.

Mittelbildung die ungeraden Figuren in annähernd gleiche Rechtecke verwandelt werden[143]). Beide Texte weisen dieselben Zahlenwerte auf, verwenden dieselben[144]) Formeln und ähneln sich auch im lateinischen Text. In allen drei Fällen ist jedoch das Ergebnis in der G. i. a. genauer, weil hier bei der Verwandlung der *perticae* in *aripenni* auch der Rest berücksichtigt wird[145]).

Im Gegensatz zu den erwähnten Aufgaben wird die Fläche eines runden Ackers in beiden Schriften (*Propositiones*, Nr. 25; G. i. a. IV 34) auf verschiedene Arten bestimmt, wobei das Verfahren in der G. i. a. wesentlich besser ist[146]).

Die Flächenformeln für den viereckigen und dreieckigen Acker wiederholen sich in Nr. 27 und 28 (= G. i. a. IV 35 und 36). Dabei befinden sich in einer entsprechend geformten Stadt rechteckige Häuser. Um ihre Anzahl zu bestimmen, verwandelt man die Städte durch Mittelbildung in annähernd flächengleiche Rechtecke und prüft, wie oft die Längen bzw. Breiten der Häuser in ihnen enthalten sind[147]). Auch hier stimmen die Texte in den *Propositiones* und in der G. i. a. weitgehend überein. Es fällt auf, daß in Nr. 28 = G. i. a. IV 36 in beiden Fällen statt des genauen Wertes 45 die abgerundete Zahl 40 benutzt wird. In IV 35 erwähnt der Autor der G. i. a. abweichend von „ALKUIN" den Rest.

Bei der analogen Aufgabe, die Anzahl der Häuser in einer runden Stadt zu errechnen („ALKUIN" 29 = G. i. a. IV 37), verwenden beide Autoren grundsätzlich verschiedene Verfahren und Zahlenwerte. Wie schon erwähnt[148]), ist die Methode in der G. i. a. gegenüber dem eigenwilligen Vorgehen bei „ALKUIN" bei weitem vorzuziehen.

In Aufgabe 30 = G. i. a. IV 38 (Bestimmung der Ziegelsteine in einer Basilika) stimmen wenigstens die Zahlen überein; im Text der Aufgabe gibt es nur geringe Anklänge, und die Verfahren differieren. Auch hier ist das Ergebnis der G. i. a. genauer[149]).

Die letzte gemeinsame Aufgabe (Nr. 31 = G. i. a. IV 39) stellt die Frage nach der Anzahl der Fässer in einem Weinkeller. Hier ist das Verfahren in beiden Fällen identisch, und sogar der lateinische Text stimmt über weite Strecken überein.

Fassen wir zusammen: Der geometrische Aufgabenkomplex in den *Propositiones* (Nr. 21—25. 27—31) begegnet uns in derselben Reihenfolge geschlossen auch in der G. i. a. (IV 30—39). Die Zahlenwerte sind bis auf Ausnahmen identisch, die Verfahren meistens auch; oft stimmen sogar die Formulierungen überein. Wo Differenzen bestehen, ist die Methode in der G. i. a. durchweg besser und die Lösung genauer. Die Aufgaben in der G. i. a. machen einen einheitlichen Eindruck; man merkt, daß der Autor die geometrischen Grundformeln und das Rechnen mit Brüchen beherrscht. Demgegenüber wirken die Lösungen bei „ALKUIN" oft unbeholfen und sind fehlerhaft. Sicherlich hat der Autor der *Propositiones* die Texte in der G. i. a. nicht gekannt, da die G. i. a. vermutlich etwas jünger ist und sonst auch nicht einzusehen wäre, warum er sie nicht dem besseren Text nicht übernommen hätte. Daß umgekehrt der Autor der G. i. a. die geometrischen Aufgaben in den *Propositiones* kannte und mit Hilfe seines mathematischen Wissens verbesserte, ist möglich, aber nicht sehr wahrscheinlich. Eher sollte man eine gemeinsame Quelle in Form einer heute verlorenen Agrimensoren-Handschrift annehmen[150]): Die in diesen Aufgaben benutzten Verfahren stehen in uralter Feldmesser-Tradition. Jedenfalls entstammen die geometrischen Aufgaben

---

[143]) Siehe CANTOR (2), S. 144.

[144]) In Nr. 22 allerdings bildet „ALKUIN" das arithmetische Mittel aus den drei Breiten, während die G. i. a. nur die beiden verschiedenen Breitseiten mittelt.

[145]) Zu den ungenauen Umrechnungen in den *Propositiones*, Aufg. 22 und 24, siehe S. 27 bzw. 25.

[146]) Siehe S. 27 f.

[147]) Siehe CANTOR (2), S. 145.

[148]) Seite 28 f.

[149]) Siehe S. 29.

[150]) So schon BUBNOV, S. 352, Anm. 85: *Videntur incertus auctor et Alcuinus haec omnia ex communi fonte, id est ex Epaphroditi textus plenioris parte deperdita, sumpsisse.*

(Nr. 21—25. 27—31 und wohl auch 9. 10) einer anderen Wurzel als die restlichen Probleme; sie gehören nicht der Unterhaltungsmathematik, sondern der ernsthaft betriebenen Geometrie an.

Während also über die Quellen des geometrischen Aufgabenkomplexes relativ genaue Angaben gemacht werden können, wissen wir über die benutzten Vorlagen in den anderen Teilen der Schrift weniger, weil direkte Paralleltexte nicht bekannt sind. Wir müssen uns mit der Aussage begnügen, daß einige Probleme, die in den *Propositiones* behandelt werden, erstmals hier auftauchen; für andere gibt es ältere Belege vor allem aus dem griechisch-römischen Bereich, aber auch aus arabischen Quellen. Dies paßt gut zu den Ergebnissen des vorhergehenden Kapitels und bestätigt, daß der Autor in der Tradition der griechisch-römischen Antike steht, daneben aber auch schon Beziehungen zu den Arabern bestanden.

V

HÄUFIGER ZITIERTE LITERATUR

BUBNOV   N. BUBNOV, *Gerberti opera mathematica*, Berlin 1899

CANTOR (1)   M. CANTOR, *Mathematische Beiträge zum Kulturleben der Völker*, Halle 1863

CANTOR (2)   M. CANTOR, *Die römischen Agrimensoren und ihre Stellung in der Geschichte der Feldmeßku* Leipzig 1875

CANTOR (3)   M. CANTOR, *Vorlesungen über Geschichte der Mathematik*, 1. Band, ³Leipzig 1907

FOLKERTS (1) M. FOLKERTS, Mathematische Aufgabensammlungen aus dem ausgehenden Mittelalter, *Sudhoffs Archiv* 55 (1971) 58—75

FOLKERTS (2) M. FOLKERTS, Pseudo-Beda, De arithmeticis propositionibus. Eine mathematische Schrift aus Karolingerzeit, in: *Sudhoffs Archiv* 56 (1972) 22—43

THIELE (1)   G. THIELE, *Der illustrierte lateinische Aesop in der Handschrift des Ademar. Codex Vossianus Lat.* 15, fol. 195—205, Leiden 1905

THIELE (2)   G. THIELE, *Der lateinische Äsop des Romulus und die Prosa-Fassungen des Phädrus*, Heidelberg 1

VOGEL (1)   K. VOGEL, *Die Practica des Algorismus Ratisbonensis*, München 1954

VOGEL (2)   H. HUNGER/K. VOGEL, *Ein byzantinisches Rechenbuch des 15. Jahrhunderts*, Wien 1963

VOGEL (3)   K. VOGEL, *Ein byzantinisches Rechenbuch des frühen 14. Jahrhunderts*, Wien 1968

VOGEL (4)   K. VOGEL, *Chiu Chang Suan Shu. Neun Bücher arithmetischer Technik*, Braunschweig 1968

VOGEL (5)   K. VOGEL, Byzanz, ein Mittler — auch in der Mathematik — zwischen Ost und West, *XIII. Internationaler Kongreß für Geschichte der Wissenschaft, Moskau 1971: Colloquium: Wiss schaft im Mittelalter; Wechselbeziehungen zwischen dem Orient und Okzident;* Wiederabdr in russischer Sprache: ВИЗАНТИЯ КАК ПОСРЕДНИК МЕЖДУ ВОСТОКОМ ЗАПАДОМ В ОБЛАСТИ МАТЕМАТИКИ, in: *ИСТОРИКО-МАТЕМАТИЧЕСК ИССЛЕДОВАНИЯ* 18 (1973) 249—263

BEMERKUNGEN ZUR EDITION

Der abgedruckte Text beruht auf allen Handschriften und den beiden Drucken *a* und *b*. Er folgt im allgemeinen der Mehrzahl der besseren Textzeugen, so daß die Schreibweise einzelner Ausdrücke oder stereotyper Wendungen gemäß dem Zeugnis der Handschriften bisweilen schwankt (z. B. *Dicat, qui potest | valet | velit; modius | modium*; Schreibweise der Zahlen).

Im kritischen Apparat sind alle wesentlichen Varianten jeder Handschrift und der beiden Drucke verzeichnet. Es fehlen lediglich unbedeutende orthographische Differenzen wie *t | c*; Groß- und Kleinschreibung; Aspiration; Schreibart der Zahlen (z. B. *IIII | IV | quat(t)uor*). Alle Abkürzungen wurden aufgelöst, sofern dies eindeutig möglich war. Schwierigkeiten ergaben sich bei den Handschriften *S* und *W*, von denen mir nur Photokopien vorlagen: In *S* sind gegen Ende einige Wortgruppen sehr blaß geschrieben oder ganz unleserlich, so daß bisweilen nicht erkennbar ist, ob sie überhaupt vorhanden sind. In *W* wurden Zahlen und Überschriften (sofern vorhanden) mit einer hellen Tinte geschrieben; auf den Kopien sind sie oft nicht lesbar. Die Überschriften scheinen zu fehlen. Die Varianten der Handschrift *V* schließlich wurden nur bis zur Aufgabe 23 vermerkt und auch dort nur an Stellen, wo sie in Beziehung zu den Lesarten anderer Handschriften stehen. Der Umfang des kritischen Apparats hätte sich sehr stark vergrößert, wenn alle Sonderlesarten von *V* aufgenommen worden wären. Für die Textgestaltung ist *V* unerheblich, da der Schreiber oft ganze Aufgaben oder Lösungen frei bearbeitete.

V

A = Karlsruhe, Augiensis 205, f. 54r—70r (ca. 1000)
B = London, BM, Burney 59, f. 7v—11r (s. 11[1])
C = London, BM, Cotton Iul. D. VII, f. 132r (s. 13)
M = Montpellier 491, f. 94r—108r (s. 11[1])
M_1 = München, Clm 14689, f. 13v—20r (s. 12)
M_2 = München, Clm 14272, f. 181v (ca. 1020)
O = Vat. Ottob. lat. 1473, f. 28r—35v (ca. 1000)
R = Vat. Reg. lat. 208, f. 57v—61v (s. 11)
R_1 = Vat. Reg. lat. 309, f. 16rv. 3v—4r (s. 9ex.)
S = London, BM, Sloane 513, f. 43v—48r. 52r—56v (s. 15)
V = Leiden, Voss. lat. oct. 15, f. 203v—205v. 206v—210r (ca. 1025)
W = Wien, Ms. lat. 891, f. 4v—27v (ca. 1010)
a = ALKUIN-Ausgabe von FORSTER, St. Emmeram 1777, S. 440—448
b = BEDA-Ausgabe von HERWAGEN, Basel 1563, Sp. 135—146

add. = hinzugefügt
corr. = berichtigt
del. = getilgt
hab. = vorhanden
om. = ausgelassen
supr. = darübergeschrieben
⟨ ⟩ = in allen Texten fehlende, aber zu ergänzende Stücke
[ ] = in allen Texten vorhandene, aber zu tilgende Stücke
///// = Rasur
† = verderbte Stelle
M_1^1, M^2 usw.: Hochgestellte Zahlen bezeichnen Korrekturen oder Ergänzungen einer mit
dem Schreiber gleichzeitigen ([1]) oder einer späteren ([2]) Hand

INCIPIUNT PROPOSITIONES AD ACUENDOS IUVENES.

(1) PROPOSITIO DE LIMACE.

Limax fuit ab hirundine invitatus ad prandium infra leuvam unam. In die autem non potuit plus quam unam unciam pedis ambulare. Dicat, qui velit, in quot diebus ad idem
5 prandium ipse limax perambulaverit.

---

1 inscr. om. CM₂RSV, INCIPIVNT PROPOSITIONES AD EXERTITIVM ACVENDORVM IVVE-NVM *W*, PROPOSITIONES AD ACVENDOS IVVENES *R₁a*, Incipiunt aliae propositiones ad acuendos iuvenes *b*, DE CONIECTVRIS DILIGENTIBVS OPPOSITIS *M*
2—5: ABCMM₁M₂ORSVWab, om. R₁
2 inscr. om. MM₂RSVW, PROPOSITIO DE LYMACE *M₁*, DE LIMACE *B*, Proposicio *C*, Et primo de Limace *b* | *3—5*: Quaedam hirundo celebrans festivitatem dicitur ad convivium invitasse limacem. Vicini namque fuerant, et mutuo ut fit convivia sepe exercere solebant. Non enim maiori distabant spatio quam integro et semi miliario. Limax solito arripiens viam unam tantum in die conficiebat untiam, ex quibus XII pedem faciunt unum, pedes quinque coniuncti unum tantum passum, passus CXXV unum stadium. Stadia octo miliarium, quattuor adiuncta faciunt dimidium. Dicat, qui scit, per quot annos limax pervenit ad convivium. *M* | *3* fuit *post* hirundine *C*, *post* prandium *V* | HYRVNdine *W*, yrundine *C*, irundine *S*, /// arundine *A*, hierundine (MS. harundine) *a* | invitatus ab hirundine *R* | invitatus] vocatus *M₂* | intra *B* | leuvam *ARVW*] leugam *BM₁M₂OS*, leucam *Cab* | autem *om. CV* | *4* unciam unam *CS* | pedibus *ABOW* | velit] vult *BR*, potest *S* | quot] annis (annos *Sb*) vel *add. CM₁M₂Sab* | dies *BRSVWb* | *4sq.* ad idem prandium] *post* perambulaverit (*l. 5*) *M₁*, illuc *R* | idem] eundem *W*, illud *O* | *5* ipse *om. BV* | limax *om. OV* | perambulaverit *B*] perambulavit *CM₂(post* prandium)*ORSWb*, *corr. M₂¹*, perambulabat *Aa*, pervenisset *M₁*, pervenit *V*

---

6 SOLUTIO DE LIMACE.

In leuva una sunt mille quingenti passus, $\overline{\text{VIID}}$ pedes, $\overline{\text{XC}}$ unciae. Quot unciae, tot dies fuerunt, qui faciunt annos CCXLVI et dies CCX.

6—8: ABCMM₁M₂ORSVWab, om. R₁
6 inscr. om. MM₁M₂SV, PRIMA SOLUTIO DE LIMACE *B*, SEQUITUR SOLUTIO DE LIMACE *Aa*, SOLVTIO *CW*, R' *R*, De limace solutio *b* | *7—8*: Cuius questionis problema tali solvitur coniectura. Sume passus unius stadii, quorum pluralitas in CXXV dicitur aggregari. Quorum CXXV passuum pedes poteris invenire, is per V supradictam summam volueris multiplicare, quorum pedum multiplicationem in DCXXV videbis redundare. Si vero progrediens horum pedum untias studueris reperire, totam hanc summam per duodecim temptes coaugmentare, quos invicem coaugmentatos in septem milia et D aspicies exuberare. At si totius leugae untias secteris invenire, supradictas quae sunt unius stadii debes multiplicare per duodecim, quippe quia (*?*) tot stadia leuga dicitur continere, in quibus invenies $\overline{\text{XC}}$ untias excrescere, quas pro numero dierum debemus computare. De quibus si annos volueris facere, per dies anni, scilicet CCCLXV, debes dividere. Qui divisi videntur CCXLVI remanentibus CC^tis decem diebus incunctanter procreare. *M* | *7* leuva *AM₁OVW*] leuuua *R*, leuga *BM₂S*, leuca *Cab* | pedes 7 M. et quingenti *S* | uncie XC.M. *S* | $\overline{\text{XC}}$] 90 *b* | *8* qui — CCX] annos CC^tos XLVI et dies CCX (*ante* 7 In) *M₂* | CCXLVI] CC et XL. VII *S* | et *om. BV* | CCX] CC et X *C*

---

(2) PROPOSITIO DE VIRO AMBULANTE IN VIA.

10   Quidam vir ambulans per viam vidit sibi alios homines obviantes et dixit eis: Volebam, ut fuissetis alii tantum, quanti estis, et medietas medietatis, et rursus de medietate medietas; tunc una mecum C fuissetis. Dicat, qui vult, quot fuerint, qui in primis ab illo visi sunt.

9—12: ABCM₁M₂ORSVWab, om. MR₁
9 inscr. om. M₂RSVW, ITEM ALIA PROPOSITIO *BO*, PROPOSITIO (PROPOSITIO *om. b*) DE HOMINE ALIIS HOMINIBVS (alios homines *b*) IN VIA SIBI OBVIANTIBVS *M₁b*, Proposicio *C* | *10* sibi

post homines M_1 / homines *om.* B / 10*sq.* Volebam, ut] Utinam CM_1M_2Sb, O V / 11 fuissetis] essetis B / tantum] tanti W, tot B / quanti] quantum M_1M_2S, quot B, *corr.* M_1^1 / estis] est S / medietatis] et huius numeri medietas *add. Aa,* etiam huius numeri medietatis *add.* W / et (2.) — medietas (2.) CM_1M_2Sab] *om.* ABORVW / rursum C / 12 fuissetis] essetis B / vult] velit M_1M_2a, potest S / quot] quanti AORWa / fuerint, qui *om.* CM_1M_2SVb / fuerunt AB^1ORWa / in primis] primum BM_1(*post* illo)M_2, primo O, primi W, primis b, *om.* S / ab illo *om.* C / sunt] fuerint M_1M_2, fuerunt CSb

SOLUTIO.

Qui imprimis ab illo visi sunt, fuerunt XXXVI. Alii tantum fiunt LXXII, medietas
15 medietatis sunt XVIII, et huius numeri medietas sunt VIIII. Dic ergo sic: LXXII et XVIII
fiunt XC. Adde VIIII, fiunt XCVIIII. Adde loquentem, et habebis C.

    13—16: $ABCM_1M_2ORR_1SVWab$, *om.* M
    13 *inscr. om.* M_2SV, SOLVTIO DE ILLIS QVI OBVIAVERVNT CVIDAM HOMINI B, SOLVTIO DE EADEM PROPOSITIONE *Aa,* ALIA SOLVTIO M_1, R' R, INCIPIT SOLVTIO PRIMA R_1, De ciconijs b / 14 inprimis AWa] primis CM_1Rb, primi B, primum M_2, primitus S, prius OR_1 / fuerunt *om.* C / fuerunt XXXVI] XXXVI sunt S / XXXV C / Alii tantum] Si alii XXXVI addantur B, *om.* b / tanti W / fiunt *om.* $ASab$ / LXXII] *om.* b, et huius *add.* CM_1M_2Sb / 15 medietatis *om.* S / sunt (1.) *om.* AVa / et (1.) — VIIII *om.* CW / sunt (2.) *om.* O / Dic — 16 XCVIIII *om.* B / 16 VIIII] VIII A / habes M_2

(3) PROPOSITIO DE DUOBUS PROFICISCENTIBUS VISIS CICONIIS.

Duo viri ambulantes per viam videntes ciconias dixerunt inter se: Quot sunt? Qui conferentes numerum dixerunt: Si essent aliae tantae et ter tantae et medietas tertii, adiectis
20 duabus C essent. Dicat, qui potest, quantae fuerunt, quae imprimis ab illis visae sunt.

    17—20: $ABM_1M_2ORSVWab$, *om.* CMR_1
    17 *inscr. om.* M_2RSVW, DE DVOBVS PROFICISCENTIBVS A, Propositio. De duobus proficiscentibus a, De duobus proficiscentibus visis ciconijs b, ITEM ALIA PROPOSITIO DE DUOBUS PROFICISCENTIBVS UISIS CYCONIIS O / 18 viri] homines *Aa* / ambulantes] ambulaverunt S / videntes] videntesque *Aa,* cum vidissent B / cyconias M_2OW, cicunias R / inter se] intra se AM_2S, ad invicem B / 19 numerum] intra *add.* S / essent aliae tantae] bis tantae essent quantum nunc sunt M_1 / tantae (1.)] tot BV / ter] tercie S, etiam b / tantae (2.)] tot BV / tertii] tercie S, tanta R / 20 duobus *Aa* / essent] fuissent M_1Sb / quantae] quot BOV / fuerunt] fuerint M_2 / imprimis] primis RWb, primum BM_2, primo (*post* illis) M_1, primitus V, prius O / illo B

21 SOLUTIO DE CICONIIS.

XXVIII et XXVIII et tertio sic fiunt LXXXIIII, et medietas tertii fiunt XIIII. Sunt in totum XCVIII. Adiectis duabus C apparent.

    21—23: $ABM_1M_2ORSVWab$, *om.* CMR_1
    21 *inscr. om.* M_2SV, SOLVTIO OW(?), ALIA M_1b, R' R / 22 XXVIII (1.) — fiunt (1.)] Ter viginti octo faciunt B / XXVIII (1.)] fuerunt XXVIII M_2 / XXVIII (2.)] XVIII a / LXXXIII S / et medietas] medietas vero B / tertii] terci(a)e M_1ORSWb / tertii fiunt] viginti octo sunt B / Sunt] ergo *add.* B / in] per M_2 / 23 duobus *Aa* / apparent centum BM_2

(4) PROPOSITIO DE HOMINE ET EQUIS IN CAMPO PASCENTIBUS.

25   Quidam homo videns equos pascentes in campo optavit dicens: Utinam fuissetis mei, et essetis alii tantum, et medietas medietatis: certe gloriarer super equos C. Discernat, qui vult, quot equos imprimis vidit ille homo pascentes.

    24—27: $ABCM_1M_2ORR_1SVWab$, *om.* M
    24 *inscr. om.* M_2RR_1SVW, PROPOSITIO DE HOMINE ET EQUIS *Aa,* Proposicio C / PROPOSITIO *om.* b / CAMPIS O / 25 homo] ut *add.* B^1 / videns CSV] vidit $ABM_1M_2ORR_1Wab$ / in campo pascentes RR_1V / campo] et *add.* M_1O / optavit dicens] dixit M_2 / fuissetis] essetis Ra, *corr.* R^1 / 26 tantum] tanti R_1W, tot B / medietatis medietas M_2 / C equos M_2 / equos] illos R_1 / Discernat] Dicat B / 27 vult] potest S / viderit M_2 / homo ille C / pascentes *om.* C

SOLUTIO DE EQUIS.

XL equi erant, qui pascebant. Alii tantum fiunt LXXX. Medietas huius medietatis, id
30 est XX, si addatur, fiunt C.

28—30: ABCM₁M₂ORSVWab, om. MR₁
28 inscr. om. M₂SV, SOLVTIO *COW,* AL' *M₁,* R' *R,* De equis *b* | *29* errant *A* | pascebantur *M₁M₂Ob*
| tanti *W* | fiunt] faciunt *B,* id est *S* | LXXX] 800 *b* | huius medietatis] medietatis huius *Aa* | id — *30* ad-
datur] unum XX^ti sic *C,* si addatur *S* | *30* addantur *B*

(5) PROPOSITIO DE EMPTORE IN C DENARIIS.

Dixit quidam emptor: Volo de centum denariis C porcos emere; sic tamen, ut verres
X denariis ematur, scrofa autem V denariis, duo vero porcelli denario uno. Dicat, qui intelligit, quot verres, quot scrofae, quotve porcelli esse debeant, ut in neutris nec superabundet
35 numerus nec minuatur.

31—35: ABCM₁ORR₁SVWab, om. MM₂
31 inscr. om. RR₁SVW, PROPOSITIO DE EMPTORE DENARIORUM *Aa,* PROPOSITIO DE
EMPTORE IN DENARIIS *B,* ITEM PROPOSITIO DE EMPTORE IN DENARIIS C *O,* De emptore in
denarijs centum *b,* Propositio *C* | *32* quidem *W* | emptor] mercator *M₁b* | ut *om. C* | verris *AM₁OR₁W,*
verrus *BR* | *33* X] vel C *supr.* M₁¹ | scropha *S* | autem *om. VW* | denariis *(2.) om.* CM₁RR₁SVb | vero]
autem *S* | uno denario *CM₁RR₁SVb* | intellegit *ABRR₁W,* vult *C* | *34* quot verres *om. V* | verri *B* |
scrophe *S* | quotve] quot vero *R,* quot *SV* | esse debeant] essent *O* | in *om. R* | *34 sq.* numerus nec super-
abundet *Aa* | *35* minuetur *A*

36 SOLUTIO DE EMPTORE.

Fac VIIII scrofas et unum verrem in quinquaginta quinque denariis, et LXXX porcellos
in XL. Ecce porci XC. In quinque residuis denariis fac porcellos X, et habebis centenarium
in utrisque numerum.

36—39: ABCM₁ORR₁SVWab, om. MM₂
36 inscr. om. R₁SV, SOLVTIO *COW,* AL' *M₁,* R' *R,* De centum denarijs *b* | *37* scrofas VIIII *R₁* |
et unum verrem] emi *B* | quinquaginta *B,* quadraginta *B²* | denariis] et unum verrem precio X denariorum
add. B² | *38* XL] LX *M₁,* 60 *b,* vel XL *supr.* M₁¹, denariis *add. C* | Ecce porci XC *post* numerum *(39) hab.*
M₁ | residuis quinque *a* | denariis residuis *M₁* | fac porcellos] faporcellos *R* | *39* numerum *ante* in *hab. Ca*

40 (6) PROPOSITIO DE DUOBUS NEGOTIATORIBUS C SOLIDOS COMMUNES HABENTIBUS.

Fuerunt duo negotiatores habentes C solidos communes, quibus emerent porcos. Eme-
runt autem in solidis duobus porcos V volentes eos saginare atque iterum venundare et in
solidis lucrum facere. Cumque vidissent tempus non esse ad saginandos porcos et ipsi eos
45 non valuissent tempore hiemali pascere, temptaverunt venundando, si potuissent, lucrum
facere, sed non potuerunt, quia non valebant eos amplius venundare, nisi ut empti fuerant,
id est, ut de V porcis duos solidos acciperent. Cum hoc conspexissent, dixerunt ad invicem:
Dividamus eos. Dividentes autem et vendentes, sicut emerant, fecerunt lucrum. Dicat, qui
valet, imprimis quot porci fuerunt, et dividat et vendat ac lucrum faciat, quod facere de
50 simul venditis non valuit.

40—50: ABM₁ORSVWab, om. CMM₂R₁
40 sq. inscr. om. RSVW | PROPOSITIO *om. Ab* | ITEM PROPOSITIO *O* | negotiationem *b* | COM-
MUNES *om. Aa* | HABENTIBUS *post* duobus *b,* *ante* C *BM₁,* *om. O* | *42* communes *om. O* | emerent]
emerant *O,* emerunt *S* | *43* autem *om. S* | in *(1.) del.* M₁¹ | solidos duos *R* | *44* Cumque] Cum *S* | *44 sq.*
non valuissent eos *M₁O* | *45* hiemali tempore *M₁* | tentavere *ab* | *46* valebant] valuerunt *M₁* | empti] temp-
tati *R* | *47* ut *om. S* | Cum] Cumque *O* | *48* emerunt *S* | *49* valet] velit *M₁* | et *(2.) BORW]* ac *M₁Sab* |
et vendat *om. A* | ac] et *ab* | *50* venditis] vendentes *R* | valuit] potuit *V,* poterat *O,* libras XXX *add.* M₁b

SOLUTIO DE PORCIS.

Imprimis CCL porci erant, qui C solidis sunt comparati, sicut supra dictum est, duobus solidis V porcos: quia sive quinquagies quinos sive quinquies L duxeris, CCL numerabis. Quibus divisis unus tulit CXXV, alter similiter. Unus vendidit deteriores tres
55 semper in solido, alter vero meliores duos in solido. Sic evenit, ut is, qui deteriores vendidit, de CXX porcis XL solidos est consecutus, qui verọ meliores, LX solidos est consecutus, quia de inferioribus XXX semper in X solidis, de melioribus autem XX in X solidis sunt venundati. Et remanserunt utrisque V porci, ex quibus ad lucrum IIII solidos et duos denarios facere potuerunt.

51—59: *ABM₁ORSVWab*, *om. CMM₂R₁*
51 inscr. om. SV, SOLVTIO *OW*, ALIA *M₁*, R' *R*, De communi negocio *b* | *52* CCL] CL *R* | solidis sunt] solidīs *O*, solidis *b* | comparati sunt *M₁* | duobus *ABOW*] in duobus *M₁RSab*, id est duobus *V* | *53* sive *(1.) om. A* | quinquagies] quinquages *W*, quiquagies *O* | quinquies] quinquagies *BO* | duxeris *BOV*] dixeris *AM₁SWab*, *om. R* | numerabis] numeris *A* | *54* tulit *om. O* | similiter] CXXV *add. M₁S*, 125 *add. b* | vendidit *post* tres *O* | *55* in *(1.)*] an *R* | vero *om. Aa* | duos] semper *add. B* | is] his *B*, hijs *S* | *56* CXX] CXXX *O* | solidos *(1.)*] solidum *A* | consecutus *(1.)*] adeptus *M₁Rb* | vero — *57* quia *om. S* | LX] quadraginta similiter *B*, XL *O*, vel XL *supr. M₁*¹ | *57* solidis *(1.)*] solidos *BOVW* | autem XX *BM₁RSVb*] viginti autem *AWa*, vero *O* | solidis *(2.)*] solidos *OW* | *58* V *om. S* | solidorum *b* | *59* facere potuerunt] potuerunt facere *B*, fecerunt *W*

---

60 (7) PROPOSITIO DE DISCO PENSANTE LIBRAS XXX.

Est discus qui pensat libras XXX sive solidos DC habens in se aurum, argentum, auricalcum et stagnum. Quantum habet auri, ter tantum argenti; quantum argenti, ter tantum auricalci; quantum auricalci, ter tantum stagni. Dicat, qui potest, quantum unaquaeque species penset.

60—64: *ABMM₁ORSVWab*, *om. CM₂R₁*
60 inscr. om. MRSVW, De disco *b* | ITEM PROPOSITIO *O* | *61* auricalchum *Aab*, aurichalchum *W*, aurichalcum *MR* | *62* et *om. MV* | stannum *ab* | aurum *RV* | tantum *(1.)*] habet *add. RSab*, habet et *add. M₁* | argenti *(1.)*] argentum *RV* | quantum argenti *om. OW* | quantum *(2.)*] habet *add. M₁RSb* | argenti *(2.)*] argentum *RV* | ter *(2.) om. OV* | tantum *(2.)*] habet et *add. M₁* | *63* auricalci *(1.) M₁OS*] auricalchi *ABM₁*¹ *ab*, aurichalci *M*, aurichalchi *W*, auricalcum *R*, auricalci *(2.) OS*] auricalchi *ABM₁ab*, aurichalci *M*, aurichalchi *W*, auricalcum *R*, auricalcum *V* | stanni *ab*, stagnum *RV* | quantum *(2.)*] in *add. Aa* | *63 sq.* unaquaeque species] unaquaque specie *a*, unaquaque *A* | *64* penset *BORSb*] pensat *AMM₁VWa*

---

65 SOLUTIO DE DISCO.

Aurum pensat uncias novem. Argentum pensat ter VIIII uncias, id est libras duas et tres uncias. Auricalcum pensat ter libras duas et ter III uncias, id est libras VI et uncias VIIII. Stagnum pensat ter libras VI et ter VIIII uncias, hoc est libras XX et III uncias. VIIII unciae et II librae cum III unciis et VI librae cum VIIII unciis et XX librae cum III unciis adunatae
70 XXX libras efficiunt.

Item aliter ad solidos. Aurum pensat solidos [argenteos] XV. Argentum ter XV, id est XLV. Auricalcum ter XLV, id est CXXXV. Stagnum ter CXXXV, hoc est CCCCV. Iunge CCCCV et CXXXV et XLV et XV, et invenies solidos DC, qui sunt librae XXX.

65—73: *ABMM₁ORSVWab*, *om. CM₂R₁*
65 inscr. om. MSV, SOLVTIO *AWa*, De disco *b*, AL' *M₁*, R' *R* | *66* pensat *(1.) om. O* | pensat *(2.) om. Aa* | VIIII uncias] untias VIIII *Aa* | id] hoc *R* | libras *om. M₁* | duas] duo *A* | *67* tres] ter *A* | tres uncias] uncias III *M₁S*, uncias 3 *b* | auricalcum *OSV*] Auricalchum *AM₁a*, Aurichalchum *W*, aurichalcum *BMRb* | ter *(1.) om. S* | libras duas] duas libras *M* | uncias *(2.)*] uncius *W* | et uncias VIIII] et VIIII uncias *Ma*, *om. A* | *68* Stannum *ab* | VIIII uncias] uncias VIIII *a* | III uncias] uncias III *M₁b*, III unciae *W*, Et *add. b* | III] in *O* | *69* unciis *(3.) om. S* | *71* solidos *(1.)*] solidum *Aab* | solidos *(2.) om. O* | argenteos

om. *MSV* | id] idem *S* | 72 auricalcum *OSV*] Auricalchum *AM₁Wa*, Aurichalcum *BMRb* | CXXXV *(1.)*]
CXXV *MM₁a*, corr. *M₁¹* | stannum *ab*, pensabat *add. M* | CXXXV *(2.)*] centum viginti quinque *M* |
hoc] id *MRS* | 72sq. Iunge CCCCV om. *A* | Iunge] ergo *add. S* | 73 et *(1.)* om. *b* | CXXXV] CXXV
M | et XV om. *M₁*, supr. *M₁¹* | solidos om. *a* | DC] DCtos *AM₁OVa*, 100 *b*, vel CC *supr. M₁¹* | qui] quod
AM₁RWb, quot *O* | sunt librae] fiunt librae *V*, faciunt libras *BMS* | XXX om. *O*

(8) PROPOSITIO DE CUPA.

75   Est cupa una, quae C metretis impletur capientibus singulis modia tria habens fistulas
III. Ex numero modiorum tertia pars et sexta per unam fistulam currit, per alteram tertia
pars sola, per tertiam sexta tantum. Dicat, qui vult, quot sextarii per unamquamque fistulam
cucurrissent.

74—78: *ABMM₁M₂ORR₁SVWab*, om. *C*
74 inscr. om. *MM₂RR₁SVW*, ITEM PROPOSITIO DE CUPA *O*, PROPOSITIO DE CAVPA. VEL
CUPA *M₁*, De cupa *b* | 75 cuppa *R*, culpa *W*, copa *M*, caupa *M₁M₂*, cuba *V* | quae] qui *R₁*, namque
A | metretas *b* | tria modia *B*, modios tres *MM₁M₂R*, ternos modios *V* | 76 III] tres (*ante* habens, 75) *W* |
ex modiorum numero. Tercia *M₁b* | unam] una *W* | tertia *(2.)* om. *O* | 77 sola] solo *R* | Dicat] nunc
add. Aa, ergo *add. RR₁S* | vult] velit *M₂* | quot] quod *B* | sextaria *R*

SOLUTIO.

80   Per primam fistulam $\overline{\text{III}}$ DC sextarii cucurrerunt, per secundam $\overline{\text{II}}$ CCCC, per tertiam
$\overline{\text{I}}$ CC.

79—81: *ABMM₁ORR₁SVWab*, om. *CM₂*
79 inscr. om. *MR₁SV*, SOLVTIO DE CVPA *B*, R' *R*, AL' *M₁*, De cuppa *b* | 80 fistula *W* | secun-
dum *a*, secunda *O* | $\overline{\text{II}}$ CCCC] 1400 *b* | 81 $\overline{\text{I}}$ CC] $\overline{\text{II}}$ CC *W*, hoc sunt sextarii $\overline{\text{VII}}$ CC ^{ti} *add. R*

(9) PROPOSITIO DE SAGO.

Habeo sagum habentem in longitudine cubitos C et in latitudine LXXX. Volo exinde
per portiones sagulos facere, ita ut unaquaeque portio habeat in longitudine cubitos V et in
85 latitudine cubitos IIII. Dic, rogo, sapiens, quot saguli exinde fieri possint.

82—85: *ABMM₁ORSVWab*, om. *CM₂R₁*
82 inscr. om. *MRSVW*, ITEM PROPOSITIO DE SAGO *O*, De sago *b* | 83 Habeo — habentem]
Est mihi sagus habens *M* | cubitus *S* | latitudine] longitudine *b* | 84 per om. *M* | sagulorum *B*, singulos
A, om. *V* | ita om. *W* | habeat] habeo *S* | in longitudine om. *M* | V] senos *S* | 85 cubitos om. *BS* | IIII]
III *A*, ut sufficiat ad tunicam consuendam *add. S* | Dic — saguli] Dicat qui potest quot invicem *S* | rogo]
ergo *M₁Ob* | quot] quod *B* | exinde] inde *M* | possunt *R*

86 SOLUTIO.

De quadringentis octogesima pars V sunt et centesima IIII. Sive ergo octogies V sive
centies IIII duxeris, semper CCCC invenies. Tot sagi erunt.

86—88: *ABMM₁ORSVWab*, om. *CM₂R₁*
86 inscr. om. *MSV*, SOLVTIO DE SAGO *B*, De sago *b*, R' *R*, AL' *M₁* | 87sq.: Iunge duas longi-
tudines sagi istius, et fiunt CC. Iunge duas latitudines, et fiunt CLX. Duc mediam de ducentis, et fiunt C. Similiter et de CLX duc mediam, et fiunt LXXX. Et quia unumquodque sagum debet habere in longitudine cubitos
V et in latitudine IIII, duc quintam partem de C et fiunt XX, et quartam de LX *(!)*, fiunt similiter XX. Duc
ergo vicies vicenos et fiunt CCCC ^{ti}. Tot siquidem sagi exinde cubitorum longitudinis V et latitudinis IIII
fieri possunt. *R (cf. p. 30)* | 87 quadragintis *M₁W*, vel dringentis *supr. M₁¹* | V sunt] sunt V *M* | et] C *add.*
S, de *add. M₁b*, de *del. M₁¹* | IIII] quarta *W*, III *O* | octogies *Wb*] octoagies *BM₁OSV*, octuagies *AMa* |
88 IIII] quater *BW* | duxeris] dixeris *M₁Sb* | invenies. Tot om. *M* | saga fuerunt *M₁* | erant *O*

(10) PROPOSITIO DE LINTEO.

90 Habeo linteamen unum longum cubitorum LX, latum cubitorum XL. Volo ex eo portiones facere, ita ut unaquaeque portio habeat in longitudine cubitos senos et in latitudine quaternos, ut sufficiat ad tunicam consuendam. Dicat, qui vult, quot tunicae exinde fieri possint.

> *89—93*: $ABMM_1ORVWab$, om. CM_2R_1S
> *89 inscr. om. MRVW,* PROPOSITIO DE LINTEAMINE M_1, De linteamine *b* | *90* Habeo] Est mihi *M* | linteum *B* | cubitorum *(bis)*] cubitis *M* | XL] LX *OW*, 60 *b* | ex eo] autem exinde M_1, exinde *b*, om. *V* | *91* ita *om. MV* | habat *R* | et *om. M* | *92* tonicam *A* | consuetam *O* | Dicat — vult] Dic *M* | tonicae *A* | exinde *om. MRV* | *92sq.* fieri possint] possunt fieri *R*, facere possint *W*, erunt *M*, sunt *V*

SOLUTIO.

95 Decima pars sexagenarii VI sunt, decima vero quadragenarii IIII sunt. Sive ergo decimam sexagenarii sive decimam quadragenarii decies miseris, centum portiones VI cubitorum longas et IIII cubitorum latas invenies.

> *94—97*: $ABMM_1ORSVWab$, om. CM_2R_1
> *94 inscr. om. MSVb,* SOLVTIO DE LINTEO *B,* R' *R,* AL' M_1 | *95—97*: Iunge duas longitudines huius linteaminis, et fiunt CXX. Similiter iunge duas latitudines, et fiunt LXXX. Dic *(!)* mediam de CXX, et fiunt LX, similiter et de LXXX, et XL. Quia igitur unaquaeque tunica debet habere in longitudine cubitos senos et in latitudine quaternos, duc sextam de LX partem, et fiunt XX *(!)*, et de XL quartam similiter assume partem, et repperies X. Decies ergo denos si duxeris, centum tunicas sex cubitorum longas et IIII cubitorum latas repperies. *R (cf. p. 30)* | *95* sunt *(2.) om. MS* | *96* sexagenarii — decimam *om. S* | sive] seu *M* | miseris] duxeris *M* | centum] hae *M* | *97* longas] longe *M* | latas] lates *A,* late *M* | invenies] incunctanter. et absque dubio provenient *M*

(11) PROPOSITIO DE DUOBUS HOMINIBUS SINGULAS SORORES ACCIPIENTIBUS.

100 Si duo homines ad invicem alter alterius sororem in coniugium sumpserit, dic, rogo, qua propinquitate filii eorum sibi pertineant.

> *98—101*: $ABMM_1OR_1SVab$, om. CM_2RW
> *98 inscr. om.* MR_1SV | PROPOSITIO] ITEM PROPOSITIO *O, om. Ab* | SINGULAS *om. Aa* | SINGULAS SORORES] ALTER ALTERIVS SOROREM M_1b | *100* sumpserit in coniugium *O* | sumpserint *S* | rogo] ergo *O, om. M* | *101* eorum *om. M* | sibi *om. B* | pertineant] vel coniungantur *add.* M_1 | *solutionem add.* AR_1a: SOLUTIO EIUSDEM. Verbi gratia. Si ego accipiam sororem socii mei et ille meam et ex nobis procreentur (filii *add. a*), ego denique sum patruus filii sororis meae et illa amita filii mei, et ea propinquitate sibi invicem pertinent. *Aa,* Filius igitur meus et filius sororis meae oquolibet *(!)* generat consobrinerite vocantur. R_1

(11a) PROPOSITIO DE DUOBUS HOMINIBUS SINGULAS MATRES ACCIPIENTIBUS.

Si duo homines alter alterius matrem similiter in coniugium sumpserit, quali cognatione
105 filii eorum sibi coniungantur.

> *102—105*: BMM_1OR_1SVb, om. ACM_2RWa
> *102 inscr. om.* MR_1SV, De duobus hominibus alter alterius matrem accipientibus *b* | PROPOSITIO] ITEM PROPOSITIO *O* | *104* Si *om.* R_1 | similiter *om.* MR_1V | in *om. b* | sumpserint M^1S, sumpserunt R_1, dic ergo *add. O* | quali] qua *MV* | *105* sibi] si *O, om. M* | coniunguntur M_1V, pertineant *M* | *solutionem add.* R_1: Filius igitur meus et filius matris mee avunculi et nepotes sunt.

V

51

106 (11b) PROPOSITIO DE PATRE ET FILIO ET VIDUA EIUSQUE FILIA.

Si relictam vel viduam et filiam illius in coniugium ducant pater et filius, sic tamen, ut filius accipiat matrem et pater filiam, filii, qui ex his fuerint procreati, dic, quaeso, quali cognatione sibi iungantur.

106—109: BMM_1OR_1SVb, om. ACM_2RWa
106 inscr. om. MR_1SV, De patre et filio matrem et eius filiam accipientibus b / PROPOSITIO] ITEM PROPOSITIO O / ET FILIO om. B / 107 vel om. R_1S / et (1.)] vel M_1, corr. $M_1{}^1$ / illius] ipsius R_1 / ducant in coniugium M / ducant om. S / sic — 108 filius om. O / 108 his] eis OSV / procreati fuerint S / dic om. MR_1V / quaeso om. MV / quali — 109 iungantur] quomodo sunt propinqui M / 109 sibi iungantur] subiungantur M_1b, corr. $M_1{}^1$ / iunguntur S / solutionem add. R_1: Filius igitur meus et filius patris mei avunculus et nepos est unus alteri.

110 (12) PROPOSITIO DE QUODAM PATREFAMILIAS ET TRIBUS FILIIS EIUS.

Quidam paterfamilias moriens dimisit in hereditate tribus filiis suis XXX ampullas vitreas, quarum decem fuerunt plenae oleo, aliae decem dimidiae, tertiae decem vacuae. Dividat, qui potest, oleum et ampullas, ut unicuique eorum de tribus filiis aequaliter obveniat tam de vitro quam de oleo.

110—114: $ABMM_1ORSVWab$, om. CM_2R_1
110 inscr. om. $MRSVW$ / PROPOSITIO] ITEM PROPOSITIO M_1O, om. b / QUODAM om. b / EIUS] SUIS O, corr. O^1 / 111 divisit b / in hereditate] in hereditatem AV, haereditatem a, hereditate R, corr. R^1, om. M / 112 plenae fuerunt MM_1W / aliae decem] X aliae O, decem V / tertiae] et aliae W, Aliae V / 113 oleum post Dividat S / de — 114 tam] equa veniat portio W / filiis] suis add. S / obtineat R, corr. R^1 / 114 tam ... quam] quam ... tam S / vitreo M_1b, corr. $M_1{}^1$ / quam] et add. BRS

115 SOLUTIO.

Tres igitur sunt filii et XXX ampullae. Ampullarum autem quaedam X sunt plenae et X mediae et X vacuae. Duc ter decies, fiunt XXX. Unicuique filio veniunt X ampullae in portionem. Divide autem per tertiam partem, hoc est, da primo filio X semi⟨plena⟩s ampullas, ac deinde da secundo V plenas et V vacuas, similiterque dabis tertio, et erit trium aequa
120 germanorum divisio tam in oleo quam in vitro.

115—120: $ABMM_1ORSVWab$, om. CM_2R_1
115 inscr. om. MSV, SOLVTIO DE PATRE ET FILIIS EIVS TRIBVS B, De ampullis relictis b, R' R, AL' M_1 / 116 ampullae] pullae A / quaedam] id est add. O / plenae] mediae O, vacuae M / et (2.) om. O / 117 mediae] plenae O, sunt mediae M_1 / vacuae] plenae M / Duc — 118 portionem om. M / veniunt] eveniunt RV / 118 portione R / autem om. BV / hoc est om. M / semiplenas M] semiplenos S, S V, semis ABM_1ORWab / 119 ac — secundo] Secundo autem M / plenas] pleno O / V (2.) om. A / vacuas] et add. M / similiter MVa / dabis] dabit W, om. M / 119sq. germanorum aequa MR / 120 geminorum BV / divisio germanorum W / quam] et add. ABM_1SWb / vitro] XVI add. M_1

(13) PROPOSITIO DE REGE ET DE EIUS EXERCITU.

Quidam rex iussit famulo suo colligere de XXX villis exercitum eo modo, ut ex unaquaque villa tot homines sumeret quotquot illuc adduxisset. Ipse tamen ad primam villam solus venit, ad secundam cum altero; iam ad tertiam tres venerunt. Dicat, qui potest, quot
125 homines fuissent collecti de his XXX villis.

121—125: $ABMM_1ORSVWab$, om. CM_2R_1
121 inscr. om. $MRSVW$, PROPOSITIO DE REGE Aa, PROPOSITIO DE REGE ET DE (DE om. B) EIVS EXERCITV IN XXX VILLIS COLLECTO BM_1, De rege et eius exercitu in triginta villis collecto b, ITEM PROPOSITIO DE REGE QUODAM ET DE EIUS EXERCITUS IN XXX UILLIS COLLECTIO O / 122 colligere post villis M_1Vb / 123 villa om. M_1O / quotquot] quot S / tamen] tunc W, autem V / villam primam Aa / 124 venit post primam (123) B / iam om. MS / iam — venerunt om. W / tres] 4 S / potest] velit M_1 / 125 homines om. M / collecti post villis $BMORSV$ / collecti — villis om. W / his om. ASa / villis] postremo add. M

SOLUTIO.

In prima igitur mansione duo fuerunt, in secunda IIII, in tertia VIII, in quarta XVI, in quinta XXXII, in sexta LXIIII, in septima CXXVIII, in octava CCLVI, in nona DXII, in decima $\overline{\text{I}}$ XXIIII, in undecima $\overline{\text{II}}$ XLVIII, in duodecima $\overline{\text{IIII}}$ XCVI, in tertia decima $\overline{\text{VIII}}$

130 CXCII, in quarta decima $\overline{\text{XVI}}$ CCCLXXXIIII, in quinta decima $\overline{\text{XXXII}}$ DCCLXVIII, in sexta decima $\overline{\text{LXV}}$ DXXXVI, in septima decima $\overline{\text{CXXXI}}$ LXXII, in octava decima $\overline{\text{CCLXII}}$ CXLIIII, in nona decima $\overline{\text{DXXIIII}}$ CCLXXXVIII, in vicesima mille milia $\overline{\text{XLVIII}}$ DLXXVI, in vicesima prima bis mille milia $\overline{\text{XCVII}}$ CLII, in vicesima secunda quater mille milia $\overline{\text{CXCIIII}}$ CCCIIII, in vicesima tertia octies mille milia $\overline{\text{CCCLXXXVIII}}$ DCVIII, in vicesima

135 quarta XVI mille milia $\overline{\text{DCCLXXVII}}$ CCXVI, in vicesima quinta XXXIII mille milia $\overline{\text{DLIIII}}$ CCCCXXXII, in vicesima sexta LXVII mille milia $\overline{\text{CVIII}}$ DCCCLXIIII, in vicesima septima $\overline{\text{CXXXIIII}}$ mille milia $\overline{\text{CCXVII}}$ DCCXXVIII, in vicesima octava CCLXVIII mille milia $\overline{\text{CCCCXXXV}}$ CCCCLVI, in vicesima nona DXXXVI mille milia $\overline{\text{DCCCLXX}}$ DCCCCXII, in tricesima villa milies LXXIII mille milia $\overline{\text{DCCXLI}}$ DCCCXXIIII.

126—139: ABMM₁ORSVWab, om. CM₂R₁
126 inscr. om. MSV, SOLVTIO DE TRIGINTA VILLIS B, SOLVTIO SVPRADICTE A, De exercitu in 30 villis collecto b, R' R, AL' M₁ / 127—139: b praebet tabulam numerorum verbis In villa ... fuerunt collecti milites ... additis / 127 primi M / igitur om. MRV / duo fuerunt] fuerunt duo. ipse et quem ibi accepit M / in (3.) — 139: quod si summam omnium vis invenire, numerum villarum in unum dispone, et alternatim dispositos per duplicem numerum debes multiplicare, et sic multiplicatis videbis et miraberis summam rite coaugmentatam in unum redundare. M / XVI] XII O / 128 LXIIII] XLIIII B / 129 IIII XCVI] IIII CXV W / in (3.) — VIII CXCII om. a / 130 CXCII] CXII B, CCCII O / CCCLXXXIIII] CCCLXXXII M₁¹O, CCCCLXXXIIII R / XVI — 131 DXXXVI] ///// In sexta decima XVI CCCLXXXIIII W / in (3.) — 139 DCCCXXIIII] &c. a / 131 LXV DXXXVI, in septima decima om. O / CXXXI LXXII] vel II XXII supr. M₁¹ / 132 CXLIIII] CLIIII W, om. A / XLVIII] XVIII O, om. S / DLXXVI] DLXXV W, CCCLXXVI S, om. A / 133 XCVII] XCVII AM₁, XCVIII R, XCIII B, om. S / quater — 134 CCCIIII] 4 194 214 b / quater] quatuor M₁ / mille (2.) om. B / 134 CXCIIII om. S / CCCIIII] CCCCIIII R / octies — DCVIII] 8 388 428 b / CCCLXXXVIII DCVIII] CCCLVIII.DCVIII W, et DC et VIII S / 135 XVI — CCXVI] 16 776 856 b, XVI milia mille DCCLXXVII CCXXVI A, XVI mille milia DCCLXXVII.CCXVII R, XVI mille milia DCCLXXVII CCVI B, XVII mille millia et C et XVI S / mille (1.)] milia O / vicesima quinta] viginti quinque B, corr. B¹ / XXXIII — 136 CCCCXXXII] 33 553 712 b, XXXIII DLIII CCCCXXXII B, XXX mille milia DCIIII.CCCCXXXII R, XXXIII mille milia DCIIII (vel DLIIII supr. M₁¹) CCCCXXXII M₁, XXXIIII mille millia et CC et XXXII S / 136 CCCCXXXII] CCCCXXXIII W / LXVII — DCCCLXIIII] 67 107 424 b / LXVII] LXVI BO, LXIIII A, LXVI M₁, vel VII supr. M₁¹, LXVIII S / CVIII] CVIII A, CVIII BM₁, VIII supr. M₁¹, om. S / DCCCLXIIII] DCCCCLXIIII R, et CCCC et LXIIII S / 137 CXXXIIII — DCCXXVIII] 134 214 848 b / CXXXIIII mille milia] CXXXIII M W, C mille millia et XXXVII mille millia S / CCXVII] CCXIII B, om. S / DCCXXVIII] DCCCXXVIII R, et XXVIII S / CCLXVIII — 138 CCCCLVI] 268 429 696 b / 137sq. CCLXVIII mille milia] CLXVIII mille milia A, CC mille millia et LXXIIII mille millia S / 138 CCCCXXXV] CCCXXXV A, CCCC V O, om. S / CCCCLVI] CCCLVI A, et LVI S, vel CCC supr. M₁¹ / DXXXVI — 139 DCCCXII] 536 859 392 b / DXXXVI mille milia] D mille milia et XLVIII mille milia S / DCCCLXX] DCCCC LXX O, DCCCCᵉˢLXX (vel DCCCC supr. M₁¹) DCCCCLXX M₁, om. S / 139 DCCCCXII] et CXII S / villa om. M₁RS, supr. M₁¹ / milies] miles A / milies — milia] milies LXXII M. milia O, milies mille milia et XCVI mille milia S / milies — DCCCXXIIII] 1 073 718 184 b / DCCXLI] DCCXLII M₁, om. S / DCCCXXIIII] DCCCCXXIIII OR, DCCC.XXXIII M₁, et CC et XXIIII S

140 **(14) PROPOSITIO DE BOVE.**

Bos qui tota die arat, quot vestigia faciat in ultima riga?

140sq.: ABM₁ORSVWab, om. CMM₂R₁
140 inscr. om. RSVW, De bove b / 141 aratur M₁Rb / facit OV

SOLUTIO.

Nullum omnino vestigium bos in ultima riga facit, eo quod ipse praecedit aratrum et hunc aratrum sequitur. Quotquot enim hic praecedendo inexculta terra vestigia figit, tot
145 illud subsequens excolendo resolvit. Propterea illius omnino nullum reperitur in ultima riga vestigium.

142—146: $ABM_1ORSVWab$, om. CMM_2R_1
142 inscr. om. SV, SOLVTIO DE BOVE B, De vestigijs boum b, R' R, AL' M_1 | 143 omnino] enim A | bos vestigium M_1b | facit ante bos Aa | 144 enim] et add. R, om. W | praebendo O, om. M_1 | inexculta terra] in exultra terra B, in cultura S | figit] facit BM_1b | 145 illud] ille $ABM_1ORVWab$, id S | resolvitur b | illius] id S | omnino om. a | reperitur nullum A | repperitur BORW, invenitur M_1, id est add. b | 146 vestigium ante in (145) Aa

147 (15) PROPOSITIO DE HOMINE.

Quaero a te, ut dicas mihi, quot rigas factas habeat homo in agro suo, quando de utroque capite campi tres versuras factas habuerit.

147—149: $ABMM_1ORSVWab$, om. CM_2R_1
147 inscr. om. BMRSVW, ITEM PROPOSITIO DE HOMINE M_1, De homine b | 148 Quaero — mihi om. SV | Queso R | a] ad M | facta b | habeat om. O | homo] hoc modo O, post rigas M_1b | suo om. R | 149 campi om. BS | factas] ex omni circuicione add. M

150 SOLUTIO.

Ex uno capite campi III, ex altero III, quae faciunt rigas versuras VII.

150sq.: $ABM_1ORSVWab$, om. CMM_2R_1
150 inscr. om. SV, ITEM SOLVTIO B, De rigis quas facit homo b, R' R, AL' M_1 | 151 capite om. OS | campi III] capita tria B | campo O | ex (2.)] et ex M_1Sb | III (2.) om. b | quae] qui R, corr. R^1 | rigas om. R, atque add. O | versuras] versuras/suras B, sunt add. S | VII] VI ARa, om. W

(16) PROPOSITIO DE DUOBUS HOMINIBUS BOVES DUCENTIBUS.

Duo homines ducebant boves per viam, e quibus unus alteri dixit: Da mihi boves duos, et habebo tot boves, quot et tu habes. At ille ait: Da mihi, inquit, et tu duos boves, et habebo
155 duplum quam tu habes. Dicat, qui velit, quot boves fuerunt, quot unusquisque habuit.

152—155: $ABCMM_1ORSVWab$, om. M_2R_1
152 inscr. om. MRSVW, Proposito C | PROPOSITIO om. b | 153 e] ex CM_1OS | unus om. W | dixit alteri CSb, dicebat alteri M_1 | 154 habebo (1.) B^1CMM_1Wb] habeo ABORSVa | habes om. M_1 | At] Et W | ille ait] alius respondit C | inquit om. CMVa | tu (2.)] mihi add. C | boves duos M_1b | habebo (2.)] habeo AORSVa | 155 quam tu] quantu B, corr. B^1, quantum tu CS | velit] vult ACMa | fuerunt] et W, om. M | quot (2.)] quod b, quos B, om. M | unusquisque] inprimis quisque M

SOLUTIO.

Prior, qui dari sibi duos rogavit, boves habebat IIII. At vero, qui rogabatur, habebat VIII. Dedit quippe rogatus postulanti duos, et habuerunt uterque sex. Qui enim prior acceperat, reddidit duos danti priori, qui habebat sex, et habuit VIII, quod est duplum a quattuor,
160 et illi remanserunt IIII, quod est simplum ab VIII.

156—160: $ABCMM_1ORSVWab$, om. M_2R_1
156 inscr. om. MSV, SOLVTIO DE HOMINE BOVES DVCENTE B, De bove ducto b, R' R, AL' M_1 | 157 sibi dari M | duos] dari sibi C | boves rogavit M | quatuor habebat boves C | vero om. MV | qui rogabatur] alius C | 158 VIII] VII O | Dedit quippe] deditque prope O, Deditque C | Dedit — rogatus] rogatur a R | habuerunt] habebat S | utrique M | enim] autem M | prior] prius Aa, propior b | acceperat] qui add. O | 159 reddit C | danti duos M | dandi b, dati B | priori om. M | habuit] habebat C | duplum post quattuor M | 160 et — IIII om. C | illi] illae M_1, corr. M_1^1, sibi R | IIII om. O | est om. R | simplum ab VIII] ab octo indubitanter simplum M

(17) PROPOSITIO DE TRIBUS FRATRIBUS SINGULAS HABENTIBUS SORORES.

Tres fratres erant, qui singulas sorores habebant et fluvium transire debebant. Erat enim unicuique illorum concupiscentia in sorore proximi sui. Qui venientes ad fluvium non in-
165 venerunt nisi parvam naviculam, in qua non poterant amplius nisi duo ex illis transire. Dicat, qui potest, qualiter fluvium transierunt, ut ne una quidem earum ex ipsis maculata sit.

> 161–166: $ABMM_1ORSVWab$, om. CM_2R_1
> 161 inscr. om. $MRSVW$, De duobus fratribus singulas sorores habentibus b / PROPOSITIO om. A / TRIBUS] DVOBVS M_1b, corr. M_1^1 / SORORES] CORONAM A / SORORES HABENTIBVS BM_1b / 163 Tres] igitur add. M_1Rb / erant] fuerunt MOV / debebant] volebant M / Erat enim] et erat M / enim] autem M_1 / 164 sororem S / 165 poterant] potuerunt a / nisi (2.)] quam S / transire] simul intrare M / 166 transierunt] transirent O / ut] et (?) O, om. a / una – ipsis] una ex illis quidem earum M_1, corr. M_1^1 / quidem om. S / earum] illarum MW, ipsarum S / ex ipsis post sit S / ex – sit] corrumpatur M / ipsis] illis M_1 / muculata O

SOLUTIO.

Primo omnium ego et soror mea introissemus in navem et transfretassemus ultra, transfretatoque fluvio dimisissem sororem meam de navi et reduxissem navem ad ripam. Tunc
170 vero introissent sorores duorum virorum, illorum videlicet, qui ad litus remanserant. Illis itaque feminis navi egressis soror mea, quae prima transierat, intraret ad me navemque reduceret. Illa egrediente foras duo in navem fratres intrassent ultraque venissent. Tunc unus ex illis una cum sorore sua navem ingressus ad nos transfretasset. Ego autem et ille, qui navigaverat, sorore mea remanente foris ultra venissemus. Nobisque ad litora vectis una ex
175 illis duabus quaelibet mulieribus ultra navem reduceret, sororeque mea secum recepta pariter ad nos ultra venissent. Et ille, cuius soror ultra remanserat, navem ingressus eam secum ultra reduceret. Et fieret expleta transvectio nullo maculante contagio.

> 167–177: $ABMM_1ORSVWab$, om. CM_2R_1
> 167 inscr. om. MSV, SOLVTIO DE TRIBVS FRATRIBVS SINGVLIS SOROREM HABENTIBVS B, De sororibus b, R' R, AL' M_1 / 168 Primum B / mea] cum add. M_1 / navim MM_1Vb / in – transfretassemus om. O / transfretavissemus M, transfretamus A / transfretatoque] transfretoque A / 169 meam om. b / navi] nave Aa / reduxi M_1V, corr. M_1^1 / navem] navim MM_1Ob / 170 vero – 172 Tunc om. O / virorum duorum M / videlicet] scilicet R / qui] quae B / remanserunt M / 170sq. remanserant illis. Itaque B / 171 itaque] igitur a / feminis] de supr. M_1^1 / quae – transierat om. A / ad me om. ASa, post navemque M / navemque] in navem quae W, in navem et S, navimque b / reduceret] ad nos add. Aa / 172 foras om. S / navem] navim M_1, navi $BMRb$, nave W / fratres] post duo S, cum supr. M_1^1 / ultraque] utraque W, ultra quae B / 173 sua om. S / navim MM_1OWb, navi BR / ingressi Aa / ad] ut ad O / transfretassent Aa, transfretaret O / autem om. b / 174 sorore] soror O, cum sorore S / mea] cum sua add. S / foras Aa / nobisque . . . vectis SW] Nosque . . . vectos $ABMM_1ORab$ / litora] littera a / 175 illis] ex add. M_1 / duabus quaelibet] quibuslibet S / ultra navem] navim ultra M_1 / navem] navim M_1b, navigare O, om. S / reduceret – 176 venissent] venissent M_1, reduceret M_1^1 / sororeque] soror aeque b / 176 venisset M / Et – 177 reduceret om. R / navem] navim MM_1Ob, navi BVW, corr. B^1 / eam] eamque W / ultra (3.)] ultro O, om. $ASVa$, post eam B / 177 Et – contagio] Tali (talis S) igitur sicque (sicque sicque S) sollicitante studio facta est navigatio nullo fuscante inquinationis (nec inquinante S) contagio M_1RSb / expleta om. M / transvectio] navigatio B, corr. B^1 / nullam M / contagio] XX add. M_1

(18) PROPOSITIO DE LUPO ET CAPRA ET FASCICULO CAULI.

Homo quidam debebat ultra fluvium transferre lupum et capram et fasciculum cauli,
180 et non potuit aliam navem invenire, nisi quae duos tantum ex ipsis ferre valebat. Praeceptum itaque ei fuerat, ut omnia haec ultra omnino illaesa transferret. Dicat, qui potest, quomodo eos illaesos ultra transferre potuit.

> 178–182: $ABCMM_1ORSVWab$, om. M_2R_1
> 178 inscr. om. $MRSVW$, PROPOSITIO DE HOMINE ET CAPRA ET LUPO Aa, Proposito C / PROPOSITIO om. b / FASCULO M_1 / 179 debebat] habebat W / ultra] quidem add. O / transferre

$CM_1{}^1a$] transire $ABMM_1ORSVWb$ / et *(1.) om. AMa* / capram — cauli] fasciolum (fasciculum *a*) cauli et capram M_1a / fasciculam *C,* fassiculum *W* / caulis *M,* secum ducens *add.* S / 180 navim M_1O / invenire *post* valebat M_1, *corr.* $M_1{}^1$ / duos] duo *O, om.* A / ex ipsis] ex illis *M,* ex eis *O, om.* S / 180*sq.* Praeceptum itaque] Preceptumque *B* / 181 ultra *om.* M / MV / inl(a)esa omnino *Aa,* inl(a)esa *RSV* / omnino *post* transiret *M* / transferret AM_1Sa] transiret *MORVWb,* transirent *C,* transponeret *B* / potest] vult *C* / 182 eis inl(a)esis *Aa* / eos *om.* C / ultra *om. Aa,* fluvium *add. MS,* flumen *add.* C / transferre *BCS*] transire *AMORWa* / transferre potuit] transiret *b,* transferret M_1 / potuerit *R*

SOLUTIO.

Simili namque tenore ducerem prius capram et dimitterem foris lupum et caulum. Tum
185 deinde venirem lupumque ultra transferrem, lupoque foras misso rursus capram navi re-
ceptam ultra reducerem, capraque foras missa caulum transveherem ultra, atque iterum
remigassem, capramque assumptam ultra duxissem. Sicque faciente facta erit remigatio
salubris absque voragine lacerationis.

183—188: $ABCMM_1ORSVWab$, *om.* M_2R_1
183 *inscr. om.* MM_1SV, SOLVTIO DE LVPO ET CAPRA ET FASCICVLO CAVLI *B,* De lupo et
capra *b,* R' *R* / 184 Simili — prius] Primo ducam *C* / namque *om.* M / ducere *W* / dimittam *C* / foris
om. CV / caulem *OV,* caules *M* / Tum] Tunc M_1RSb, *om.* CV / 185 veniam *C* / ultra] inde *O, om.*
ACa / transferrem *ABa*] transferam *C,* transirem $MM_1ORSVWb$ / lupoque] lupo quoque *BORW* / foras
BCRSW] foris AMM_1Oab / misso] dimisso *C* / rursus CM_1] rursum *S,* rursusque *BMWb,* rursumque *OR,*
om. Aa / capra . . . recepta *BV* / navi] in navem *C,* in navim *S* / receptum *C* / 186 reducam *C* / capraque
CMS] capra *V,* capramque ABM_1ORWab / foras *BCSWb*] foris AMM_1ORa / missa *CMSVW*] missam
ABM_1ORab / caulem *O,* caules *M* / transveherem] transferam *C,* transferentem *S* / atque] et *V,* ut ultra
C / 187 remigrassem *M,* remigarem *B,* remigavi *V,* remigam *C* / capramque — duxissem *om.* OV / Capram *R* / ducerem *B,* ducam *C* / faciente AM_1OWb] facienti $M_1{}^1S$, facientem *R,* faciendo *BMa, om.* C /
facta *post* remigatio *C* / facta erit] esset sane *R* / erit] est *S* / 188 voragione *O*

(19) PROPOSITIO DE VIRO ET MULIERE PONDERANTIBUS PLAUSTRUM.

190 De viro et muliere, quorum uterque pondus habebat plaustri onusti, duos habentes
infantes inter utrosque plaustrali pondere pensantes flumen transire debuerunt. Navem invenerunt, quae non poterat ferre plus nisi unum pondus plaustri. Transfretari faciat, qui se
putat posse, ne navis mergatur.

189—193: $ABMM_1ORSVWab$, *om.* CM_2R_1
189 *inscr. om.* MRSVW / PROPOSITIO *om.* Ab / PONDERANTIBUS PLAUSTRUM *om.* B /
PLAUSTRUM O] ONVS PLAVSTRI ONVSTI M_1, plaustri pondus onusti *ab, om.* A / 190 De — muliere]
Vir et mulier *M* / pondus habebat] pensabat pondus *M* / habebat] pondus *add.* O, *om.* A / honusti M_1S /
duos — 191 infantes] habentes pueros duos *M* / 191 inter] duos *R* / plaustrali] plaustri *OSW* / plaustrali —
pensantes] pensantes plaustrale pondus *M* / flumen] frumenti *O,* fluvium *ab* / debuerunt] voluerunt. et *M* /
invenerunt navem *M* / Navim *O* / 192 poterant *A,* potuit *S* / plus *om.* BMM_1ORSWb / Transfretari *Aa*]
Transfretare BMM_1OSWb, Tranfretare *R* / faciat] faç *A* / 193 posse putat *M* / posse *om.* B / navis] in
flumine ullo modo *add.* M

SOLUTIO.

195 Eodem quoque ordine, ut superius: Prius intrassent duo infantes et transissent, unusque
ex illis reduceret navem. Tunc mater navem ingressa transisset. Deinde filius eius reduceret
navem. Qua transvecta frater illius navem ingressus ambo ultra transissent, rursusque unus
ex illis ad patrem reduceret navem. Qua reducta filio foris stante pater transiret, rursusque
filius, qui ante transierat, ingressus navem eamque ad fratrem reduceret, iamque reductam
200 ingrediantur ambo et transeant. Tali subremigante ingenio erit expleta navigatio forsitan
sine naufragio.

194—201: $ABMM_1ORSVWab$, *om.* CM_2R_1
194 *inscr. om.* MS, SOLVTIO DE VIRO ET MVLIERE *B,* De pondere plaustri *b,* R' *R,* AL' M_1 /
195 Eodem quoque] Eodemque *S* / ut — Prius *om.* S / et *om.* W / unusquisque *M* / 196 illis] eis *O* / navem

56

(1.)] navim M_1Ob / matre B / ingressa navim W / navem *(2.)*] navim MM_1OWb, navi R, *om.* B / eius *om.* W / *196sq.* navim reduxisset M / *197* navem *(1.)*] navim MM_1Ob / frater — *198* reducta *om.* R / frater illius] illius frater M, filius eius W / navem *(2.)* SW] navim BMM_1Oab, navi A / ambo — transissent] ad matrem redissent M / ambo] est ambo / ambo S / transirent M_1b / rursumque M_1Ob / unus — *198* rursusque *om.* b / *198* navim reduxisset M / navim MM_1OW / filio] eius *add.* W / pater] foris R / pater transiret] pertransiret B / transiret] transisset M / rursusque] rursumque BO, Rursus M / *199* transierat] transiret A / navem M_1S] navim $BMM_1{}^1OWab$, navi AR / eamque] eam B, *om.* M / reduceret] reducat $ABOW$ / reductam *om.* M / *200* Tali] et tali R / subremigante] supra migrante S, remigante M / erit] fiet R / erit — *201* naufragio] transibunt cuncti absque naufragio M / forsan $BOSW$ / forsitan — *201* naufragio *om.* R

202 **(20) PROPOSITIO DE ERICIIS.**

De ericiis masculo et femina habentibus duos natos libram ponderantibus flumen transire volentibus.

202—204: $ABM_1ORSVWab$, om. CMM_2R_1
202 inscr. om. $BRSW$, ITEM ALIA DE HYRICIIS M_1, De hiricis b / ERICIIS] HERICIIS O, HYRICIIS M_1, hiricis b, HIRTITIIS Aa / *203* De] Item de B / ericiis] hericiis BM_1O, hiriciis R, iriciis W, hiritiis A, hiricis b, hirtitiis a, *om.* S / feminae W / habentes M_1 / duos — ponderantibus *bis habet* O / fluvium Rb, et flumen S / *204* volentibus] cupientibus R

205 **SOLUTIO.**

Similiter, ut superius, transissent prius duo infantes, et unus ex illis navem reduceret. In quam pater ingressus ultra transisset, et ille infans unus, qui prius cum fratre transierat, navem ad ripam reduceret. In quam frater illius rursus ingressus ambo ultra venissent, unusque ex illis foras egressus, et alter ad matrem reduceret navem, in quam mater ingressa ultra 210 venisset. Qua egrediente foras filius eius, qui ante cum fratre transierat, navem rursus ingressus eam ad fratrem ultra reduceret. In quam ambo ingressi ultra venissent, et fieret expleta transvectio nullo formidante naufragio.

205—212: $ABM_1ORSVWab$, om. CMM_2R_1
205 inscr. om. S, SOLVTIO DE HERICIIS B, R' R, AL' M_1, Aliud b / *206* transsissent AW / prius *om.* O / navim M_1OWb / *207* quam] qua M_1ORWb / transisset] transiret BOS / unus *om.* Aa / prius] primus W / fratre M_1, patre *supr.* $M_1{}^1$ / *208* navem BM_1RS] navim $AM_1{}^1OWab$ / quam] qua BM_1ORWb / rursum S / venissent] transissent O / unusque] unus W, unusquisque AO, propterea *add. a*, prope *add.* A / *209* foras *om.* S / egressus] egredens *(?)* B^1, est *add.* S / et *om.* R, *del.* B^1 / reduceret] et O / navem $B(?)$ M_1RS] navim $AOWab$ / in quam mater] illa S / quam] qua videlicet navi $BORWb$, quam videlicet navim M_1 / ingressa] cum filio *add.* S / *210* venissent M_1Sb, veniret B / foras *om.* S / fratre] patre Aa / navem S] navi AR, navim BM_1OWab / *211* quam] qua BM_1ORWb / venirent B / expleta *om.* M_1, salubris *add.* M_1RSb / *212* formidante] formidine W, mortis *add.* M_1RSb / naufragium BO

(21) PROPOSITIO DE CAMPO ET OVIBUS IN EO LOCANDIS.

Est campus, qui habet in longitudine pedes CC et in latitudine pedes C. Volo ibidem 215 mittere oves, sic tamen, ut unaquaeque ovis habeat in longitudine pedes V et in latitudine pedes IV. Dicat, rogo, qui valet, quot oves ibidem locari possunt.

213—216: $ABMM_1ORSVWab$, om. CM_2R_1
*213 inscr. om. MRSVW / PROPOSITIO om. b / ET] DE add. A / *214* CC] C V / et om. MV / pedes *(2.)* om. M / *215* ut om. W / unaquaque A / longitudine M_1ORVb] longo $ABMSWa$ / latitudine M_1ORVW] lato $ABMSab$ / *216* pedes om. M_1Ob / rogo] quaeso R, ergo M_1, om. MS / valet] vult M / ibidem] illic M_1, omnino *add.* M / possunt $BORSVW$] possint AMM_1ab

SOLUTIO.

Ipse campus habet in longitudine pedes CC et in latitudine pedes C. †Duc bis quinquenos de CC†, fiunt XL, ac deinde C divide per IIII. Quarta pars centenarii XXV sunt.
220 Sive ergo XL vicies quinquies, sive XXV quadragies ducti, millenarium implent numerum.
Tot ergo oves ibidem collocari possunt.

217—221: *ABMM₁ORSVWab*, *om. CM₂R₁*
217 *inscr. om. MSV*, SOLVTIO DE CAMPO ET OVIBVS IN EO LOCANDIS *B*, De campo et ovibus
b, AL' *M₁*, R' *R* | 218 *et om. R* | pedes *(2.) om. R* | Duc — 219 CC: *textus corruptus. R habet* Divide per
quintam CC^tos partem, et / Duc bis] ducat binos *S* | quinquenos] quinquennos *M₁b*, quindenos *W*, quin-
quagies *M*, quinos *S* | 219 ac *AMM₁ORSVWb*] at *a*, *om. B* | divide *om. R* | sunt *om. ABOWa* | 220 qua-
dragies] quinquagies *O* / ducti] duxeris *M₁RSb* / implent] complent *M*, impletum *S*, *om. A* / numerum
om. W / 221 ergo *om. MR* / ibidem oves *Aa* / oves] boves *M₁* / ibidem] in eo campo *M* / collocari]
locari *BMOR* / possunt] XXVII. (XXVIII. *M₁¹*) *add. M₁*

(22) PROPOSITIO DE CAMPO FASTIGIOSO.

Est campus fastigiosus, qui habet in uno latere perticas C et in alio latere perticas C et
in fronte perticas L et in medio perticas LX et in altera fronte perticas L. Dicat, qui potest,
225 quot aripennos claudere debet.

222—225: *ABMM₁ORR₁SVab*, *om. CM₂W*
222 *inscr. om. MRR₁SV* / PROPOSITIO *om. M₁b* / FASTIGIOSO] FASTIDIOSO *ABM₁O* / 223
fastigiosus *ab*] fastidiosus *ABMM₁ORR₁SV* / et *(1.)* — C *(2.) om. OR₁V*, *supr. R₁¹* / alio] altero *a* / 224 LX]
XL *O*, 40 *S* / altera] altero *ABO*, tertio *R₁* / fronte *(2.)*] parte *R* / qui potest] rogo qui valet *RR₁*, quis
M / 225 aripennos *MM₁RSb*] arripennos *O*, arpennos *V*, arpennas *B*, aripennas *Aa*, arp *R₁* / cludere *A*,
concludere *V*

SOLUTIO.

Longitudo huius campi C perticis et utrius-que frontis latitudo L, medietas vero LX in-cluditur. Iunge utriusque frontis numerum 230 cum medietate, et fiunt CLX. Ex ipsis assu-me tertiam partem, id est LIII, et multiplica centies, fiunt V̄CCC. Divide in XII aequas partes, et inveniuntur CCC⟨C⟩XLI. Item eosdem divide in XII partes, et reperiuntur 235 XXXVII. Tot sunt in hoc campo aripenni numero.	Iunge duas longitudines, fiunt CC. Duc mediam de ducentis, fiunt C. Et iunge L et LX et L, fiunt CLX. Duc vero tertiam partem de CLX, fiunt LIII. Et duc centies LIII, fiunt V̄ CCC. Divide per duodecimam partem V̄ CCC, hoc est, fac ex eo bis XII. Verbi gratia: de V̄ CCC duc XII. partem, fiunt CCCCXLI. Rursusque de CCCCXLI duc XII., fiunt XXXVII. Tot sunt in hoc aripenni numero.

226—236: *recensio I* (Longitudo — numero): *ABMM₁OSVab*, *recensio II* (Iunge — numero): *M₁RR₁*,
om. CM₂W
recensio I: 226 *inscr. om. MM₁SV*, SOLVTIO DE CAMPO *A*, SOLVTIO DE CAMPO FASTIDIOSO
B, De campo fastigioso *b* / 227 perticis C *M₁* / utrisque *O* / utriusque frontis] ex utrisque frontibus *M* /
228 frontis] numerum *add. A* / latitudo] longitudo *S* / L *om. A* / LX] XL *AOS* / 229 utrisque *O* / 230 cum
medietate] tum medietates *M₁* / et *om. MM₁OV* / fient *B* / 231 LIIII *A* / multiplicata *A* / 232 fiunt] fient
B, *om. O* / V̄CCC] V̄CCCCti *O*, CCC *S*, V̄.CCCC *M₁* (vel CCC *supr. M₁¹*) / Divide] deinde *b* / 232*sq.*
partes aequas *M₁* / 233 invenientur *B* / CCCCXLI] CCCXLI *ABMM₁OS*, 351 *b* / 234 eos *O* / repperiun-
tur *A*, repperientur *B* / 235 XXXVI *MS* / hoc campo] huius *M₁b*, hoc *ABMOS* / aripenni *AM₁*, arpenni
MV / 235*sq.* numero arripenni *A* / 236 numero *om. a*
recensio II: 226 *inscr. om. R₁*, AL' *M₁*, R' *R* / 227 fiunt] faciunt *R₁* / 228 medium *R₁* / de *om. R₁* /
iunge *om. R₁* / 228*sq.* L et *om. R* / 229 Duc] dic *R₁* / 230 duc centies] ducenties *M₁R₁* / 231 LIII] LIIII
M₁ / Divide — 232 V̄ CCC *om. M₁* / duodecimam] XII *R₁* / 233 XII.] X. *M₁* / partem] et *add. R₁* /
234 CCCCXLI *(1.)*] CCCXLI *M₁* / Rursusque de CCCCXLI *om. M₁* / 235 XII.] partem *add. R₁* / 235*sq.*
aripenni in hoc *R* / 236 arripenni *M₁* / numero] Sunt in huius arripenni XCVI. *add. M₁* (*cf. v. 269*)

V

58

(23) PROPOSITIO DE CAMPO QUADRANGULO.

Est campus quadrangulus, qui habet in uno latere perticas XXX et in alio perticas XXXII et in fronte perticas XXXIIII et in altera fronte perticas XXXII. Dicat, qui potest,
240 quot aripenni in eo concludi debent.

237—240: ABMM₁ORR₁SVab, om. CM₂W
237 inscr. om. MRR₁SV | PROPOSITIO om. M₁b | 238 quadrangulus] quadratus M | XXX — perticas (2.) om. R | XXX — 239 perticas (1.) om. B | alio] latere add. O | 239 et (1.) — XXXII (2.) om. O | perticas (1.) om. S | altera] altero AR₁, alio RV | fronte (2.) om. Aa | perticas (2.) om. R, del. (?) R₁¹ | 240 arripenni AM₁O, aripennii Mb, arpenni BV | in eo] eo O, om. BS | claudi MS | debent M₁, debeant M₁¹

SOLUTIO.

Duae eiusdem campi longitudines faciunt LXII. Duc dimidiam de LXII, fiunt XXXI. Atque duae eiusdem campi latitudines iunctae fiunt LXVI. Duc vero mediam de LXVI, fiunt XXXIII. Duc namque tricies semel XXXIII, fiunt $\bar{\text{I}}$ XX. Divide per duodecimam
245 partem bis sicut superius, hoc est, de mille viginti duc duodecimam, fiunt LXXXV, rursusque LXXXV divide per XII, fiunt VII. Sunt ergo in hoc aripenni numero septem.

241—246: ABMM₁ORR₁SVab, om. CM₂W
241 inscr. om. MM₁R₁SV, SOLVTIO DE CAMPO QVADRANGVLO B, Aliud b, R' R | 242 LXII (1.)] XLII A | Duc] dic MRR₁ | dimidiam] mediam M₁RR₁Sb | de om. AR₁a, supr. R₁¹ | LXII (2.)] XLII A | fiunt] et fiunt MR₁, faciunt S | XXXI] XXI A, XXXII O | 243 Atque] Ac Aa | duae] duc S, post campi M | eiusdem] ipsius RR₁ | eiusdem campi] campi ipsius R | longitudines O | fiunt] faciunt M | Duc] Dic MRR₁ | 244 fiunt (1.)] et fiunt R₁, faciunt M | namque] vero Aa | tricies] trecies M₁Sb, terties Aa | XXXIII (2.) om. a | $\bar{\text{I}}$ XX] $\overline{\text{M}}$ III S, $\bar{\text{I}}$ XXI V | Divide — 245 duodecimam om. O | 245 bis post Divide (244) RR₁ | sicut] ut RR₁ | hoc] id M | de om. M | mille viginti] $\overline{\text{M}}$ III S, 120 b | duc] dic MRR₁, per add. Aa, om. S | duodecimam] X iunge simul M, et add. R₁ | LXXXV] LXXV O, LXXXIII S | rursusque] Rursus MV | 245sq. rursusque LXXXV om. O | 246 LXXXV] LXXXIII S | VII] $\overline{\text{VII}}$ B, VI MS | Sunt — septem om. S | ergo om. M | numero arpenni M | arpenni MV, aripa (?) R, arripenni AM₁O

247 ## (24) PROPOSITIO DE CAMPO TRIANGULO.

Est campus triangulus, qui habet in uno latere perticas XXX et in alio perticas XXX et in fronte perticas XVIII. Dicat, qui potest, quot aripennos concludere debet.

247—249: ABMM₁ORR₁SVab, om. CM₂W
247 inscr. om. MRR₁S | PROPOSITIO om. AM₁b | 249 perticas om. S | XVIII] X et VIII M₁, novendecim b | arripennos AM₁O, arpennos B | concludere] omnino add. M | debeat M

250 ## SOLUTIO.

Iunge duas longitudines istius campi, et fiunt LX. Duc mediam de LX, fiunt XXX, et quia in fronte perticas XVIII habet, duc mediam de XVIII, fiunt VIIII. Duc vero novies triginta, fiunt CCLXX. Fac exinde bis XII, id est, divide CCLXX per duodecimam, fiunt XXII et semis. Atque iterum XXII et semis per duodecimam divide partem: fit aripennus
255 unus et perticae X ac dimidia.

250—255: ABMM₁ORR₁SVab, om. CM₂W
250 inscr. om. MM₁R₁S, SOLVTIO DE CAMPO TRIANGVLO B, R' R, Aliud b | 251 latitudines b | et (1.) om. MM₁RR₁Sb | LX (1.)] et add. MM₁ORR₁Sb | Duc] dic MRR₁ | media R, dimidiam O | XXX] Atque duae eiusdem campi latitudines iunctae fiunt LXVI. Duc vero mediam de LXVI fiunt XXXIII add. M₁¹ in margine | 252 perticas] pedes ABMM₁OR₁b, om. R | duc (1.)] dic MRR₁ | media R₁ | XVIII (2.)] XVIIII A, X et VIII S | triginta] et add. R₁ | 253 CCLXX (1.)] 260 b | Fac — CCLXX (2.) om. O | duodecimam] partem add. MM₁RR₁Sb | fiunt (2.) — 254 partem om. M | 254 semis (1.)] S RR₁, ꝩ add.

M_1 / semis *(2.)*] S R_1, semis .S. M_1R / divide *om.* R / partem divide M_1b / partem] et fiunt II et remanent IIII, quae est tertia pars (de *add.* R) XII. Sunt (Sint S) ergo aripenni (arripenni M_1) in hoc numero II et tertia pars de aripenno (arripenno M_1) tertio (tertio *om.* *(?)* R). *add.* M_1RR_1Sb (*cf. p. 25*) / fit — 255 dimidia *om.* RR_1, vel fit — dimidia *supr.* $M_1{}^1$ / aripennus $M_1{}^1S$] aripennis *Bab*, arripennis *AO*, arripenis M_1, arpennus *M* / 255 et] ex O / perticae — dimidia] per terciam XII S / pertica O / ac] et AM_1ab

256 (25) PROPOSITIO DE CAMPO ROTUNDO.

Est campus rotundus, qui habet in gyro perticas CCCC. Dic, quot aripennos capere debet.

256—258: $ABMM_1ORR_1Vab$, *om.* CM_2SW
256 *inscr. om.* MRR_1 / PROPOSITIO *om.* M_1b / 257 giro $ABMM_1a$, rigo O / CCCC] CCC B / Dic] dicat qui potest R_1, dicat aliquis *M*, quaeso *add.* R / arripennos M_1O, arripenos *A*, arpennos *B* / capere] claudere *B*

SOLUTIO.

260 Quarta quidem pars huius campi, qui CCCC includitur perticis, in C consistit. Hos si per semetipsos multiplicaveris, id est, si centies duxeris, fiunt $\overline{X}$. Hos in XII partes dividere debes. Etenim de decem milibus 265 duodecima est DCCCXXXIII, quam cum item in XII partitus fueris, invenies LXVIIII. Tot enim aripennos huiusmodi campus includit.	Duc ergo quartam partem de CCCC, fiunt C. Et iterum de CCCC duc tertiam partem, fiunt CXXXIII. Duc quoque mediam de C, fiunt L. Rursusque duc mediam de CXXXIII, fiunt LXVI. Duc vero quinquagies LXVI, fiunt $\overline{III}$ CLI. Divide hos per XII^{mam} partem, fiunt CCLXXX. Rursusque CCLXXX divide per XII^{mam} partem, fiunt XXIIII. Duc vero quater XXIIII, fiunt XCVI. Sunt in totum aripenni XCVI.

259—269: *recensio I* (Quarta — includit): $ABMM_1OSVab$, *recensio II* (Duc — XCVI): M_1RR_1, *om.* CM_2W

recensio I: 259 *inscr. om.* MM_1S, SOLVTIO DE CAMPO ROTVNDO *B*, Aliud *b* / 261 CCCC] trecentis *B* / in C consistit] ince *A*, est C *a* / C] X *M* / 262 Hos] has S / semetipsas S, seipsos *M* / multiplicari O / id] hoc *M* / 263 X milia (millia) fiunt *Aa* / Hos] has S / 264 de *om.* S / 265 duodecima] pars *add. M* / DCCCXXXIII] DCCCCI.XXII *A*, DCCC^{ti} XXXII *B*, DCCti et XXXIII *M* / quam cum] Quacum *B* / 266 in XII *om. M* / fueris partitus O / pertitis S / LXVIII *A* / 267 enim — includit] sunt ar$\overline{p}$ *M* / enim *om.* S / aripennos BSb] arripennos M_1O, aripennis *a*, arripennis *A* / huiusmodi] hic S / includitur ABa
recensio II: 259 *inscr. om.* M_1R_1, R' R / 260 Duc] Dic RR_1 / 261 duc] dic RR_1 / 262 partem *om.* RR_1 / CXXXIII R_1 / Duc] Dic RR_1 / 263 Rursusque] rursus R / duc] dic RR_1 / 264 Duc] Dic R / quinquagies] quinquies R_1, L. M_1 / 267 CCLXXX *om.* R_1 / XII^{mam}] XII R_1 / 269 Sunt — XCVI *(2.) om.* M_1 (*cf. v. 236*) / agripenni R, ar$\overline{p}$ R_1

270 (26) PROPOSITIO DE CAMPO ET CURSU CANIS AC FUGA LEPORIS.

Est campus, qui habet in longitudine pedes CL. In uno capite stabat canis, et in alio stabat lepus. Promovit namque canis ille post ipsum, scilicet leporem, currere. Ast ubi ille canis faciebat in uno saltu pedes VIIII, lepus transmittebat VII. Dicat, qui velit, quot pedes quotve saltus canis persequendo vel lepus fugiendo, quoadusque comprehensus est, con- 275 fecerint.

270—275: $ABMM_1OR_1SVab$, *om.* CM_2RW
270 *inscr. om.* MR_1S, PROPOSITIO DE CVRSV CBNKS BC FXGB (FVGB *a*) LFPPRKS *Aa*, ITEM PROPOSITIO DE CAMPO ROTVNDO *B*, DE CAMPO E CURSV CANIS ET FVGA LEPORIS M_1, De campo et cane ac fuga leporis *b* / 271 CL] et *add. M* / stabat] stabit R_1 / alio] altero *b* / 272 stabat] stabit R_1, *om.* MM_1b / lepus stabat S / namque — ille *(1.)*] ipse R_1 / post — currere] ut curreret post leporem *M* / ipsum BOR_1S] illum AM_1a, *om. b* / scilicet *om. b, del.* $M_1{}^1$ / Ast] At MO / ubi *om.* S / ille *(2.) om. b* / 272sq. canis ille R_1 / 273 faciebat *om. M* / pedes *(2.)*] pedus *A* / 274 quotve] quotque *a*, vel quot O / saltus] $\overline{p}$saltus M_1, vel *add.* M_1R_1Sb / insequendo *M* / vel] et *a*, *om. M* / quoadusque] quousque *ABO*, quo ad *M* / confecerint BM_1R_1b] confecerunt *OS*, fecerint *A*, fecerunt *Ma*

V

SOLUTIO.

Longitudo huius videlicet campi habet pedes CL. Duc mediam de CL, fiunt LXXV. Canis vero faciebat in uno saltu pedes VIIII. Quippe LXXV novies ducti fiunt DCLXXV, tot pedes leporem persequendo canis cucurrit, quoadusque eum comprehendit dente tenaci.
280 At vero, quia lepus VII pedes in uno saltu faciebat, duc ipsos LXXV septies: fiunt DXXV. Tot pedes lepus fugiendo peregit, donec consecutus fuit.

276−281: ABMM₁OR₁SVab, om. CM₂RW
276 inscr. om. MR₁S, SOLVTIO DE CAMPO ET CVRSV CANIS ET FVGA LEPORIS B, De campo, cane et lepore b, AL' M₁ / 277 Longitudo] igitur add. S / huius videlicet] scilicet istius O, huius Mb / campi videlicet S / Duc] dic M, dic vero R₁ / medium M / LXXV] LXXX M₁, 80 b, vel LXXV supr. M₁¹ / 278 VIIII] VIII M / 279 consequendo Aa / quod adusque B / eum] eam M, om. A / comprehendet R₁ / dente tenaci om. M / 280 At] Aut R₁ / pedes VII Aa / faciebat post uno M₁R₁Sb, post lepus Aa / fiunt DXXV om. Aa / DXXV] de XV O, DXXVI M₁, 526 b, vel V supr. M₁¹ / 281 Tot] vero add. Aa / consecutus fuit] consequebatur M, persecutus fuit S / fuit] est Aa

(27) PROPOSITIO DE CIVITATE QUADRANGULA.

Est civitas quadrangula, quae habet in uno latere pedes mille centum, et in alio latere pedes mille, et in fronte pedes DC, et in altera pedes DC. Volo ibidem tecta domorum
285 ponere sic, ut habeat unaquaeque casa in longitudine pedes XL et in latitudine pedes XXX. Dicat, qui valet, quot casas capere debet.

282−286: ABMM₁OSVWab, om. CM₂RR₁
282 inscr. om. MSW / PROPOSITIO om. M₁b / CIVITATE QUADRANGULA] CAMPO QVADRANGVLO A / 283 pedes om. S / centum om. M / 283 sq. latere pedes mille] totidem M / 284 pedes (1.) om. O / mille] .I.C S / pedes (2.) om. M / et (2.) − DC (2.) om. ABS / altera] altera fronte M, altero fronte W, alio fronte O / pedes DC (2.)] totidem M / 285 ponere] et add. B / sic] volo add. ABM₁SW, sic tamen M, ita tamen O / casa] cara A / pedes (1.)] pede S / 286 qui om. W / valet BMOSW] velit AM₁ab / capere] ponere ex toto M / debet] valet M₁b

SOLUTIO.

Si fuerint duae huius civitatis longitudines iunctae, faciunt I̅I̅ C. Similiter duae, si fuerint latitudines iunctae, fiunt I̅ CC. Ergo duc mediam de I̅ CC, fiunt DC, rursusque duc mediam
290 de I̅I̅ C, fiunt I̅ L. Et quia unaquaeque domus habet in longo pedes XL et in lato pedes XXX, deduc quadragesimam partem de mille L, fiunt XXVI. Atque iterum assume tricesimam de DC, fiunt XX. Vicies ergo XXVI ducti fiunt DXX. Tot domus capiendae sunt.

287−292: ABMM₁OSVWab, om. CM₂RR₁
287 inscr. om. MS, SOLVTIO DE CIVITATE QVADRANGVLA B, SOLVTIO DF CKXKTBTF QXADRBNGVLB A, De civitate quadrangula b, AL' M₁ / 288 Si fuerint (1.) om. M / fuerint (1.)] fuerunt M₁Ob, corr. M₁¹ / faciunt] facient Aa / fuerint (2.)] fuerunt Ob / 289 latitudines om. M / fiunt (1.)] faciunt AMa / I̅ CC (1.)] mille I̅ CC O / duc (1.)] dic M / I̅ CC (2.) − de (290) om. BS / fiunt (2.)] faciunt Aa / Rursumque O / duc (2.)] dic M, duo A / 290 longo] longitudine Aa / lato] latitudine M / pedes (2.) om. AMa / XXX] viginti M / 291 deduc] duc MM₁Sb / de] id est ABOW, om. M / iterum om. ABMOW / assume] adsumme W / tricesimam] trigintam O, trigesimam partem M / 292 DC] quingentis M / ducti om. B / DXX] DXXII O / domos ABOS / capiendae sunt] sunt M, huius capienda est civitas S

(28) PROPOSITIO DE CIVITATE TRIANGULA.

Est civitas triangula, quae habet in uno latere pedes C, et in alio latere pedes C, et in
295 fronte pedes XC. Volo enim ibidem aedificia domorum construere, sic tamen, ut unaquaeque domus habeat in longitudine pedes XX et in latitudine pedes X. Dicat, qui potest, quot domus capi debent.

293−297: ABMM₁OSVWab, om. CM₂RR₁
293 inscr. om. MSW / PROPOSITIO om. AM₁b / 294 in uno habet Aa / et (1.) − C (2.) om. B / latere pedes (2.) om. MM₁Sb / C (2.)] totidem M / et (2.) − 295 XC om. M / 295 enim om. M / enim

— construere] ut fiat ibi (ibi fiat *S*) domorum constructio (constructio domorum *M*₁) *M*₁*Sb* / aedificia — construere] aedificare domos *O* / constituere *M* / sic] sit *O* / una quaque *A* / *296* pedes *(1.)*] pedis *O* / XX — pedes *(2.) om.* W / *post* latitudine *repetit* B ut *(295)* — latitudine / qui potest *om.* b / potest] velit *M*₁ / *297* domos *ABOSW* / capere debet *O*

SOLUTIO.

Duo igitur huius civitatis latera iuncta fiunt CC, atque duc mediam de CC, fiunt C.
300 Sed quia in fronte habet pedes XC, duc mediam de XC, fiunt XLV. Et quia longitudo uniuscuiusque domus habet pedes XX et latitudo ipsarum habet pedes X, itaque in C quinquies XX et in XL quater X sunt. Duc igitur quinquies IIII, fiunt XX. Tot domos huiusmodi captura est civitas.

*298—303: ABMM*₁*OSVWab, om. CM*₂*RR*₁
*298 inscr. om. MM*₁*Sb,* SOLVTIO DE CIVITATE TRIANGVLA *B,* SOLVTIO DE CIVITATE *A* / *299* Duae *O* / igitur *om. M* / civitatis *om. S* / latera] latitudines *O* / iunctae *O* / fiunt *(1.)*] faciunt *M* / duc] dic *M* / CC *(2.)*] trecentis *B* / *300* Sed] se *B* / duc] dic *M* / de] deis *O, om. A* / quia *(2.) om. M* / *300 sq.* uniuscuiusque longitudo *S* / *301* domi *BW* / pedes XX] XX pedes *S* / habet *(2.) om. Aa* / itaque — *302* sunt] duc XX (vigesimam *S*) partem de (in *a*) C, fiunt V. Et pars decima (decima *om. S*) quadragenarii IIII sunt *M*₁*Sab*; itaque — sunt *in margine add. M*₁¹ / *302* XX *(1.)* — quinquies *om. A* / in *om. B* / Duc] Duo *M*₁ / igitur] itaque *a, om. M* / fiunt] sunt *M* / huiusmodo *MO* / *303* captura est] capienda est *ABMM*₁*SWb,* capit *O*

(29) PROPOSITIO DE CIVITATE ROTUNDA.

305   Est civitas rotunda, quae habet in circuitu pedum V̄III. Dicat, qui potest, quot domos capere debet, ita ut unaquaeque domus habeat in longitudine pedes XXX et in latitudine pedes XX.

*304—307: ABMM*₁*OSVab, om. CM*₂*RR*₁*W*
304 inscr. om. MS / PROPOSITIO *om. AM*₁*b* / *305* pedum] pedes *M*₁ (es *in rasura*) *S,* ped' *A* / V̄III] IIII *A,* VIII *M*₁, octo *S* / domus *MM*₁ / *306* debet] poterit *M* / ita] tamen *add. M,* tamen *supr. M*₁¹ / domus *om. Aa* / *307* pedes *om. M* / XX] XXII *B*

SOLUTIO.

In huius civitatis ambitu V̄III pedes nume-
310 rantur, qui sesqualtera proportione dividuntur in ĪĪĪĪ DCCC et in ĪĪĪ CC. In illis autem longitudo domorum, in istis latitudo versatur. Subtrahe itaque de utraque summa medietatem, et remanent de maiore ĪĪ
315 CCCC, de minore vero Ī DC. Hos igitur Ī DC divide in vicenos et invenies octoagies viginti, rursumque maior summa, id est ĪĪ CCCC, in XXX partiti octoagies triginta dinumerantur. Duc octoagies LXXX, et
320 fiunt V̄I CCCC. Tot in huiusmodi civitate domus secundum propositionem supra scriptam construi possunt.

Ambitus huius civitatis V̄III complectitur pedum. Duc ergo quartam de V̄III partem, fiunt ĪĪ. Rursusque duc tertiam de V̄III partem, fiunt ĪĪDCLXVI. Duc vero mediam de duobus milibus, fiunt Ī, atque iterum de duobus milibus DCLXVI mediam assume partem, fiunt Ī CCCXXXIII. Deinde duc partem tricesimam de Ī CCCXXXIII, ⟨fiunt XXXXIIII, rursusque duc partem vigesimam de Ī, fiunt L. Duc vero quinquagies XXXXIIII⟩, fiunt ĪĪ CC. Deinde duc simul bina milia CC quater, fiunt V̄III DCCC. Hoc est summa domorum.

308—322: recensio I (In — possunt) : *ABMM*₁*OSVab, recensio II* (Ambitus — domorum): *M*₁, *om. CM*₂*RR*₁*W*
recensio I: 308 inscr. om. MS, SOLVTIO DE CIVITATE ROTVNDA *B,* AL' *M*₁, Aliud *b* / *309* V̄III] novem milia *B,* VIII *A* / pedes *BMM*₁*OS*] pedum *ab,* vel um *supr. M*₁¹, *om. A* / numeratur *B* / *310* sex-

q̄altera *S*, sesquialtera *ab* / portione M_1, *om. S* / *311* in *(2.) om. ABM* / illis] illa *B* / *312* in istis latitudo *om.* O̲ / istis] autem *add.* M_1 / *313* Subtrahe] Sumpta *S* / itaque] utique *M*, *post* summa *S* / *314* de maiore *post* II CCCC *M* / maiore M_1] maiori *ABMOSab* / *315* minori *S* / I̅ DC *(2.)*] mille quingentos *B*, vel D *supr.* M_1^1 / *316* DC] Dtos *A* / vicenos] vigenos *S*, XX O / octogies *M*, octuagies M_1^1 / *317 sq.* I̅I̅I̅ CCCC ᵗᵗ O / *318* partita *A*, partitio *B* / octuagies *AM* / triginta — *319* octoagies *om. B* / *319* dinumerant *MO* / octuagies AMM_1^1 / et *om. S* / *321* domus *post* scriptam *M* / domos ABM_1OSb / proportionem O / supra-dictam MM_1 / *322* construi] constitui M_1b, constitui construique *M* / possunt] XXXII *add.* M_1

recensio II: 308 inscr.] AL' M_1 / *316* fiunt — *319* XXXXIIII *supplevi*; *om.* M_1 / *320* VIII.DCCC M_1 / *321* domorum] XXXII. *add.* M_1

(30) PROPOSITIO DE BASILICA.

Est basilica, quae habet in longitudine pedes CCXL et in latitudine pedes CXX. Later-
325 culi vero stratae eiusdem unus laterculus habet in longitudine uncias XXIII, hoc est, pedem unum et XI uncias, et in latitudine uncias XII, hoc est, pedem I. Dicat, qui velit, quot laterculi eam debent implere.

323—327: $ABMM_1OSVWab$, om. CM_2RR_1
323 inscr. om. MSW / PROPOSITIO *om.* M_1b / *324* latitudine M_1OS] lato *ABMWab* / Laterculus O / *325* eiusdem stratae M_1 / strate *B*, strati *B¹* / laterculus] vero *add.* O / *326* unum et *spatio relicto om. S* / et *(1.)* — I *om. B* / XI uncias] uncias XI M_1SWb / XI] XII *A*, vel XII *supr.* M_1^1 / XII] viginti tres *W* / qui velit] aliquis sapiens *M* / *327* eam] eandem *Aa* / implere debent *M*

SOLUTIO.

CCXL pedes longitudinis implent CXXVI
330 laterculi et CXX pedes latitudinis CXX laterculi, quia unusquisque laterculus in la-titudine pedis mensuram habet. Multiplica itaque centum vicies CXXVI, in X̅V̅ CXX summa concrescit. Tot igitur in huiusmodi
335 basilica laterculi pavimentum contegere possunt.

Si duxeris duodecies CCXL, fiunt I̅I̅ DCCCLXXX. Et quia uncias XV habet unus laterculus in longitudine, duc XV ᵐᵃᵐ partem de I̅I̅ DCCCLXXX, fiunt CXCII. Iterum duc duodecies CXX, fiunt I̅ CCCCXL. Et quia VIII uncias habet unus-quisque laterculus in latitudine, duc partem octavam de I̅ CCCCXL, fiunt CLXXX. Duc quippe centies octoagies CXCII, fiunt X̅X̅X̅I̅I̅I̅I̅ DLX. Tot laterculi implebunt.

328—338: recensio I (CCXL — possunt): $ABMM_1OSVWab$, recensio II (Si — implebunt): M_1. om. CM_2RR_1
recensio I: 328 inscr. om. MM_1S, SOLVTIO DE BASILICA *AB*, De basilica *b* / *329* CCXL] CXL AM_1a, Centum XL O, Centum quadraginta *BW*, 140 *b* / longitudines O / implent *om.* O / CXXVI — *330* lati-tudinis *om. S* / *330* latitudinis pedes *M* / *331* quia] et quia *S* / latitudines *Mb*, *corr.* M¹ / *332* pedes MM_1, *corr.* M_1^1 / *333* itaque] ergo *S* / CXXVI] CXXV *S* / X̅V̅ CXX] III.X̅V̅.CXX *W*, X̅V̅ *S*, vel XV *supra* CXX *supr.* M_1^1 / *335* basilicam *S* / pavimentum *om.* M_1, *supr.* M_1^1 / contingere *b*, constitui M_1
recensio II: 328 inscr. om. M_1 / *329 sq.* I̅I̅.CC.LXXX M_1 / *332* DCCCLXXX] CC.LXXX. M_1 / *333 sq.* I̅.CCC.XL M_1 / *337* Duc] Duae M_1 / *338* XXX.IIII.DLX M_1

(31) PROPOSITIO DE CANAVA.

340 Est canava, quae habet in longitudine pedes C et in latitudine pedes LXIIII. Dicat, qui potest, quot cupas capere debet, ita tamen, ut unaquaeque cupa habeat in longitudine pedes VII et in lato, hoc est in medio, pedes IIII, et pervius unus habeat pedes IIII.

339—342: $ABMM_1OSVWab$, om. CM_2RR_1
339 inscr. om. MSW, De cavana *b* / PROPOSITIO *om.* M_1 / CANEUA O / *340* caneva O, cavana *b*, canna *M* / LXIIII] quadraginta quatuor *W* / *341* potest] velit M_1b / cuppas *M* / unaquaque *A* / cuppa *M* / *342* lato — medio] latitudine M_1, lato *b* / est] media *add. ABSW*, mediam *add.* O / in medio] medio *M* / habeat *om. M* / IIII *(2.)*] et unaqu(a)eque cupa habeat (habeat *om. S*) pedes VII *add.* M_1Sb

SOLUTIO.

In centum autem quaterdecies VII nume-
345 rantur, in LXIIII vero sedecies quaterni continentur, ex quibus IIII ad pervium depu-
tantur, quod in longitudinem ipsius canavae
ducitur. Quia ergo in LX quindecies quaterni sunt et in centum quaterdecies septeni,
350 duc quindecies XIIII, fiunt CCX. Tot cupae
iuxta suprascriptam magnitudinem in huiusmodi canava contineri possunt.

Si duxeris sexies VII, fiunt XLII, hoc sunt
VI ordines cuparum. Et ut ad pervium
venies, qui habet pedes III, duc septies III,
fiunt XXI. Igitur iunge simul XLII et XXI,
fiunt LXIII. Ecce pervius sex ordines cuparum. Deinde quartam assume partem de C,
fiunt XXV, hoc est, in uno semper ordine
sunt cupae XXV. Et quia VI ordines sunt
cuparum, sexies XXV ducti fiunt CL. Ipse
est totus numerus cuparum.

343—353: recensio I (In — possunt): ABMM₁OSVWab, recensio II (Si — cuparum): M₁, om. CM₂RR₁
recensio I: 343 inscr. om. MS, SOLVTIO DE CANAVA *AB,* De cavana *b,* AL' *M₁ | 344* quaterdecies]
quatuor Xes *W,* quater sex *O | VII] VI A | 345* LXIIII vero *om. M |* LXIIII] XLIIII *B,* LXVI *A,* vel XL
supr. M₁¹ | quaterni] quater III *M | 346* ad *om. O |* reputantur *Aa | 347* longitudine *M₁OSW |* canev(a)e
BW, cavanae *M₁b,* vel canave *supr. M₁¹,* cavene *S,* cannae *M | 348* deducitur *M |* ergo] igitur *A |* LX]
XLX *O,* hoc *(?) S | 350* CCX] CCXX *S |* cuppae *M | 351* supradictam *MW |* magnitudinem supra-
dictam *M |* magnitudinem] longitudinem *M₁ |* in *om. B | 352* canavi *B,* cavana *M₁b,* canna *M |* conti-
neri] continere *B,* inveniri absque dubitatione ab astuto *M*
recensio II: 343 inscr. om. M₁ | 345 VI] VII *M₁ | 348* LXIII] LXVII *M₁ |* cuparum *M₁ | 351* cubae
M₁ | 352, 353 cubarum *M₁*

(32) PROPOSITIO DE QUODAM PATREFAMILIAS DISTRIBUENTE ANNONAM.

355 Quidam paterfamilias habuit familias XX, et iussit eis dare de annona modios XX: sic
iussit, ut viri acciperent modios ternos et mulieres binos et infantes singula semodia. Dicat,
qui potest, quot viri aut quot mulieres vel quot infantes esse debent.

354—357: ABCMM₁OR₁SVWab, om. M₂R
354 inscr. om. MR₁SW, Propositio *C,* PROPOSITIO DE QUODAM PATREFAMILIAS *a,* De quodam
patrefamilias *b,* DE PATRE FAMILIAS DISTRIBVENTE ANNONA *M₁ |* PROPOSITIO *om. Ab |* DIS-
TRIBUENTE ANNONAM] DISTRIBVENTI *A | 355* familias] familiam *CMS,* familia *OR₁ |* eis] ei
M | dari *BM |* XX mod' de annona *R₁ |* sic] et sic *B | 356* iussit] tamen *M |* accipiant *M₁b |* et infantes]
infantesque *S,* Infantes quoque *C |* singula semodia] singulos semodios *C,* singulos modios *S,* duo singulos
modios *B |* Dicat — *357* potest] Dic *M | 357* potest] vult *C |* aut] et *R₁, om. CMM₁b |* quot *(2.) om.
S |* vel] et *C, om. M |* vel quot] quotve *M₁R₁Sb |* infantes — debent] fuerint infantes sic et de XXX *R₁ |*
esse debent] sint *M,* fuerunt *M₁Sb |* debeant *C*

SOLUTIO.

Duc semel ternos, fiunt III, hoc est, unus vir III modios accepit. Similiter et quinquies
360 bini, fiunt X, hoc est, quinque mulieres acceperunt modios X. Duc vero septies binos, fiunt
XIIII, hoc est, XIIII infantes acceperunt modios VII. Iunge ergo I et V et XIIII, fiunt XX. Hae
sunt familiae XX. Ac deinde iunge III et VII et X, fiunt XX, haec sunt modia XX. Sunt
ergo simul familiae XX et modia XX.

358—363: ABCMM₁OR₁SVWab, om. M₂R
358 inscr. om. MM₁R₁S, DE SVPRADICTA QVESTIONE *A,* ITEM SOLVTIO *B,* De patrefamilias
b | 359 Duc] Dic *R₁ |* et *om. O | 360* bini] binis *b |* X *(1.) —* fiunt *(2.) om. M₁, in margine add. M₁¹ |*
hoc — X *(2.) om. C |* hoc est] vel hae sunt *add. B¹ |* est] sunt *AMW |* modios *BMM₁¹Ob]* modia *Aa,* mod'
SW, md' *R₁ |* Duc] Dic *MR₁,* Due *C,* Duo *S |* vero] ergo *O |* fiunt *(2.)]* fient *B | 361* ergo *om. BCS |*
et *(1.) om. S |* V] 2 *b |* et *(2.)]* ad *B |* Hae — *362* XX *(2.) om. O |* haec *CMR₁SWb,* Heae *B | 362*
familia *M,* famil' *R₁ |* VII] VI *R₁ |* et *(2.)]* ad *B |* haec — XX *(3.) om. CS |* haec] hoc *ABMM₁OR₁W
|* modii *MO |* Sunt — *363* XX *om. R₁ | 362sq.* Sunt ergo] sic et *M | 363* simul *om. BM |* familia *MOS |*
modii *MOb,* modios *S*

(33) ALIA PROPOSITIO.

365 Quidam paterfamilias habuit familias XXX, quibus iussit dare de annona modios XXX. Sic vero iussit, ut viri acciperent modios ternos et mulieres binos et infantes singula semodia. Solvat, qui potest, quot viri aut quot mulieres quotve infantes fuerunt.

364—367: $ACMM_1OSVWab$, om. BM_2RR_1
364 inscr. om. MSW, PROPOSITIO DE ALIO PATRIFAMIL' (patrefamilias *a*) EROGANTI (erogante *a*) SVAE FAMILIAE ANNONAM *Aa,* Propositio *C,* ALIA M_1b | *365* familias] familiam *CMS,* familia *O* | quibus — XXX *(2.)* om. *b* | quibus om. *M* | dari iussit *M* | dari *Ma* | modia *CS* | *366* Sic vero] Sicque *C* | vero iussit om. *M* | et *(2.)* om. *M* | singulos *CS* | *367* Solvat] Dicat *C,* igitur add. M_1 | potest] vult *CM* | aut om. CMM_1b | quotve] quot *CM* | infantes fuerunt] fuerunt infantes *O,* infantes *A,* infantes esse debeant *M,* fuere infantes similiter et de $\overline{C}$ faciendum est R_1

SOLUTIO.

Si duxeris ternos ter, fiunt VIIII. Et si duxeris quinquies binos, fiunt X. Ac deinde duc
370 vicies bis semis, fiunt XI: hoc est, tres viri acceperunt modios VIIII, et quinque mulieres acceperunt X, et XXII infantes acceperunt XI modios. Qui simul iuncti III et V et XXII faciunt familias XXX. Rursusque VIIII et XI et X simul iuncti faciunt modios XXX. Quod sunt simul familiae XXX et modii XXX.

368—373: $ABCMM_1OSVWab$, om. M_2RR_1
368 inscr. om. MM_1S, ITEM SOLVTIO B, Aliud *b* | *369* duxeris *(bis)*] dixeris *M* | ternos duxeris III (ter M_1^1) M_1 | ter ternos *M* | ternos] tres *O* | Ac — *370* XI om. *C* | deinde om. *A* | duc] dic *M* | duc — *370* XI] bis sex fiunt XII *C* | *370* semis] seni *M* | fiunt] sunt *W* | XI] XXI *S* | est om. *A* | tres] vel VI *supr.* M_1^1, om. *W* | viri III *Aa* | modia *Aa* | VIIII] VIII *B* | mulieres V *CS* | *371* acceperunt *(1.)* om. *CO* | acceperunt *(2.)* om. *C* | XI] III *O* | modios XI CMM_1S, modios 11 *b* | modios] modia *Aa,* om. *B* | Qui] Quos *C,* quod *AWb,* Quot M_1 *(del.* M_1^1*),* om. *a* | iuncti] id est add. *O* | III] VI *W* | XXII *(2.)*] XX *C* | *372* faciunt familias] familiam II faciunt *C* | familias] familiam *CS,* familia *MW,* familiae *b* | VIIII] VIII *B* | X et XI *CS* | XI] XII *O* | modia *Aa* | Quod — *373* simul] suntque sic *M* | Quod] quot *ABCOb,* igitur add. *C* | *373* sunt om. *O* | famili(a)e simul *BC* | familia *MOW* | et modii] mod' *M* | XXX] tantum add. BCM_1S, tantum 36 add. *b*

(33a) ITEM ALIA PROPOSITIO.

375 Quidam paterfamilias habuit familias XC, et iussit eis dare de annona modios XC. Sic quoque iussit, ut viri acciperent modios ternos et mulieres binos et infantes singula semodia. Dicat, qui se arbitratur scire, quot viri aut quot mulieres quotve fuere infantes.

374—377: BMM_1OR_1SVb, om. ACM_2RWa
374 inscr. om. MR_1S, ALIA PROPOSITIO B, Alia *b* | *375* paterfamilias] pater familia *O* | familias] familios *S,* familia R_1, defamilia *M,* om. *O* | eis] ei *M* | dari *BM* | de annona dare *O* | modia *S* | *376* iussit om. *M* | et *(2.)* om. *M* | infantes] autem add. *M* | singula semodia] singulos modios *BO* | singula] singulos *BOS,* singuli *M,* corr. M^2 | *377* Dicat] Solvat R_1 | se — scire] potest M_1R_1Sb, vult *M* | aut] et R_1, om. MM_1b | quotve] quodve R_1, quotqueve *O,* aut quot *b,* quot MM_1 | infantes fuerunt MM_1Sb, fuerunt infantes *O* | infantes] similiter et de C faciendum est add. R_1

SOLUTIO.

Duc sexies ter, fiunt XVIII, et duc vicies binos, fiunt XL. Duc vero sexagies quaternos
380 semis, fiunt XXXII. Id est, sex viri acceperunt modios XVIII, et XX mulieres acceperunt modios XL, et LXIIII infantes acceperunt modios XXXII. Qui simul iuncti, hoc est VI et XX et LXIIII, familias XC efficiunt. Iterumque iunge XVIII et XL et XXXII, fiunt XC, qui faciunt modios XC. Qui simul iuncti faciunt familias XC et modios XC.

378—383: BMM_1OR_1SVb, om. ACM_2RWa
378 inscr. om. MR_1S, ITEM SOLVTIO B, AL' M_1, Aliud *b* | *379* Duc *(1.)*] Dic *M,* Due *O* | XVIII] X et VIII MM_1S | duc *(2.)*] dic *M,* om. *S* | vices *O* | *380* XXXII] XXII *O,* vel XXII *supr.* M_1^1 | XVIII]

X et VIII MM_1 / mulieres XX R_1 / *381* LXIIII] XLIIII *B*, vel XLIII *supr.* $M_1{}^1$ / modios *(2.)*] semodios O / Qui] Quod M_1R_1Sb, quot O, *corr.* $M_1{}^1$ / iuncti] coniuncti *S* / hoc] id MM_1R_1Sb / et *(2.) om. M* / *382* et *(ter) om. M* / et *(1.) om.* MO / LXIIII] XLIIII *B* / familia *MO*, familiam *S* / efficiunt] sunt *M* / Iterumque iunge] iterum *(?)* R_1 / XVIII] X et VIII M_1, vel VIIII *supr.* $M_1{}^1$ / XL] LX M_1, 60 *b*, vel XL *supr.* $M_1{}^1$ / *382 sq.* qui faciunt] quod sunt R_1 / *383* XC modios *S* / Qui] Quod M_1Sb, quot O, *corr.* $M_1{}^1$ / familiam *S*, familia O / modia O / XC *(3.)*] ducentos *B*, tantum *add.* BM_1OS, tantum 37 *add. b*, et fiunt aequa sibi invicem comparata *add. M*

(34) ITEM ALIA PROPOSITIO.

385   Quidam paterfamilias habuit familias C, quibus praecepit dari de annona modios C, eo vero tenore, ut viri acciperent modios ternos, mulieres binos, et infantes singula semodia. Dicat ergo, qui valet, quot viri, quot mulieres aut quot infantes fuerunt.

384—387: $ABM_1OSVWab$, *om.* CMM_2RR_1
384 inscr. om. SW, ITEM ALIA PROPOSITIO DE PATRE FAMIL' PARTIENTI FAMIL' SUAE ANN *A*, PROPOSITIO ALIA De Patrefamilias partiente familiae suae annonam *a*, Alia *b* / *385* familiam *S*, familia O / de annona dari *B* / dare AM_1OSWab / modia *S* / *386* ut *om. S* / ternos] et *add.* M_1Sb / singula semodia] singulos semodios *S*, duo singulos modios *B* / *387* valet] vivalet O, velit *B* / viri] aut *add. S* / aut quot] quotve *W* / fuere infantes *W* / fuerint *B*

SOLUTIO.

    Undecies terni fiunt XXXIII, et XV bis ducti fiunt XXX. Duc vero septuagies quater
390 semis, fiunt XXXVII: id est, XI viri acceperunt XXXIII modios, et XV mulieres acceperunt XXX, et LXXIIII infantes acceperunt XXXVII. Qui simul iuncti, id est XI et XV et LXXIIII, fiunt C, quae sunt familiae C. Similiter iunge XXXIII et XXX et XXXVII, faciunt C, qui sunt modii C. His ergo simul iunctis habes familias C et modios C.

388—393: $ABMM_1OSVWab$, *om.* CM_2RR_1
388 inscr. om. MM_1S, ITEM SOLVTIO *B*, Aliud *b* / *389* Undecies *S*] Undecim $ABMM_1OWab$ / Duc] Due *S* / Duc — *390* XXXVII *om. a* / septuagies] octoagies M_1, octuagies $M_1{}^1$, vel septu *supr.* $M_1{}^1$, octogies *b* / *390* senis O / XXXVII] XXXVI M_1 / XXXIII — *391* acceperunt *om. b* / XXXIII] XXXVII M_1, vel III *supr.* $M_1{}^1$ / modios — *391* XXXVII *om.* M_1, *supr.* $M_1{}^1$ / et *om. B* / XV] XII *A* / acceperunt *(2.)*] modios *add.* S, *om.* $M_1{}^1$ / *391* XXX] mod' *add.* $M_1{}^1$ / XXXVII] XXXVI *M* / Qui] Quod AM_1SWb, quot O / simul] semel M_1, *corr.* $M_1{}^1$ / et *(2., 3.) om. M* / *392* fiunt] faciunt *BM* / quae] qui *B* / sunt familiae] faciunt familias *B* / sunt] simul *add. S* / familia *MO* / XXXIII] XXXII *B* / et *(1.)*] ad *B*, *om. M* / XXX et *om. A* / et *(2.) om. M* / XXXVII] XXXVI *M* / *392 sq.* faciunt C *om. M* / faciunt] fiunt M_1Sb / *393* qui] quae *M* / modii] modia O, modios *S* / ergo *om. M* / similiter O / iunctis] coniunctis *M* / habes — C *(4.)*] fit familia cum mod' *M* / familia O / C *(4.) om. W*

(35) PROPOSITIO DE OBITU CUIUSDAM PATRISFAMILIAS.

395   Quidam paterfamilias moriens reliquit infantes et in facultate sua solidos DCCCCLX et uxorem praegnantem. Qui iussit, ut, si ei masculus nasceretur, acciperet de omni massa dodrantem, hoc est, uncias VIIII, et mater ipsius acciperet quadrantem, hoc est, uncias III. Si autem filia nasceretur, acciperet septuncem, hoc est, VII uncias, et mater ipsius acciperet quincuncem, hoc est, V uncias. Contigit autem, ut geminos parturiret, id est, puerum et
400 puellam. Solvat, qui potest, quantum accepit mater vel quantum filius quantumve filia.

394—400: $ABMM_1OSVWab$, *om.* CM_2RR_1
394 inscr. om. MSW / PROPOSITIO *om. b* / PATRISFAMILIAS] PATRIS *B* / *395* et *(1.) om.* M_1 / in *om. B*, de *supr.* B^1 / facultatibus suis *B* / solidos MM_1] solidorum *ABOWab* / DCCCCLX] DCCCC et

V

66

LX *M*, 860 *b* | *396* Qui] cui *M* | ut] et *A* | ei *om.* O | *397* dodrantem].ᴖ dodrantem *B*, 𝔰𝔰 dodrans (tem *supr.* W¹) *W*, 𝔖 dodrans *O*, 𝔠 dodrans *S*, dodrantis *M*,ᴇ𝔰 dograns (tem *supr.* $M_1$¹) M_1, dodrans *Aa*, *om. b* | hoc est *(1.)*] id est *M*, *om. b* | ipsius] eius *S* | acciperet *om. M* | quadrantem] quadrantem *B*, •𝔥·quadrans *A*, ᵧ quadrans *O*, 𝔠. quadrans *S*, 𝔷 quadrans (tem *supr.* W¹) *W*, ᴇ𝔰 quadrans (tem *supr.* $M_1$¹) M_1, quadrantes *M*, quadrans *a*, *om. b* | hoc est *(2.)*] id est *M*, *om. b* | *398* nasceretur $MM_1 Sb$] nata esset *ABOWa* | septuncem — acciperet *(2.) om. B* | septuncem] ᴇ𝔰 septuncem M_1, 𝔏 septunx *O*, ᴖ.septunx *S*, 𝔶 id est semptunx (cem *supr.* W¹) *W*, septuncem *M*, septunx *Aa*, *om. b* | hoc est] id est *M*, hoc *W*, *om. b* | VII] V M_1, quinque *b*, *corr.* $M_1$¹ | untias VII M_1O | ipsius *om. M* | *399* quincuncem] •𝔥·quincunx *A*, 𝔷 quincunx *O*, 𝔲 quincunx *S*, ᴋ quincunx *B*, 𝔶·id est quincunx *W*, ᴇ𝔰 quincunx (cem *supr.* $M_1$¹) M_1,˙ quincunces *M*, quincunx *a*, *om. b* | hoc est] id est *M*, etiam *b* | unci(a)e quinque $M_1 W$ | ·unti(a)e $AM_1 W$ | Contingit *b* | parturiet *S*, pareret *BM* | id est *om. M* | *400* Solvat] igitur *add.* $M_1 S$, ergo *add. b* | accipit *b* | vel] et *a*, id est *W*, *om. M* | quantumve] vel quantum *B*, quantum *M*

SOLUTIO.

Iunge ergo VIIII et III, fiunt XII. XII namque unciae libram faciunt. Prorsusque iunge similiter VII et V, faciunt iterum XII. Ideoque bis XII faciunt XXIIII. XXIIII autem unciae faciunt duas libras, id est, solidos XL. Divide ergo per vicesimam quartam partem DCCCCLX
405 solidos: vicesima quarta pars eorum fiunt XL. Deinde duc, quia facit dodrans, XL in nonam partem. Ideo novies XL accepit filius, hoc est, XVIII libras, quae faciunt solidos CCCLX. Et quia mater tertiam partem contra filium accepit et quintam contra filiam, III et V fiunt VIII. Itaque duc, quia legitur, quod faciat [bis sive] bisse, XL in parte octava. Octies ergo XL accepit mater, hoc est, libras XVI, quae faciunt solidos CCCXX. Deinde duc, quia legitur, quod faciat septunx [sive septus], XL in VII partibus. Postea duc septies XL, fiunt
410 XIIII librae, quae faciunt solidos CCLXXX. Hoc filia accepit. Iunge ergo CCCLX et CCCXX et CCLXXX, fiunt DCCCCLX solidi et XLVIII librae.

401–412: ABMM₁OSVWab, om. CM₂RR₁
401 inscr. om. MM₁S, SOLVTIO DE OBITV CVISDAM PATRIS FAMILIAS *B*, SOLVTIO SVPRADICTE QVESTIONE *A*, De animalibus emptis *b* | *402* XII *(2.) om.* MO | namque] simul *M* | libram — *404* libras] faciunt et libram XXIIII untiae faciunt duas *M* | faciunt libram *O* | Prorsusque] Rursusque *a*, prorsus *Sb* | *403* simul $BM_1 SWb$ | faciunt *(1.)*] fiunt *Sa* | XXIIII *(1.)*] et *add. W* | unciae *om. a* | *404* Divide — *405* XL *(1.) om.* $M_1 b$, *in margine add.* $M_1$¹ | Divide] Deinde *a* | ergo] duc *add. a* | vigesimam *M* | quartam] quartem *A* | DCCCXL *B*, nongentos XL *M* | *405* solidos] et *add. a* | vigesima *M* | fiunt] fac̄ $M_1$¹ | duc] dic *M* | quia] qui *O* | *post* facit *desinit b verbis* Reliquae solutiones desiderantur *additis* | dodrans] dodrans sive dodras *AMa*, ᴖ dodrans *B*, 𝔥 dodrans sive dodras M_1,𝔖 dodrans sive dodras *O*, dodrans sive doras *S*, 𝔰 id est dodrans sive dodras *W* | XL *(2.)*] quadragesimam *B* | in *om. B* | *406* Ideo] idest M_1 | XVIII] X et VIII MM_1 | libras *om.* M_1, *supr.* $M_1$¹ | faciunt *post* CCCLX *O* | *407* Et *om.* MS | III *(?) S* | III et *om. S* | *408* Itaque] Ideoque *O* | duc] dic *ABMOW* | legetur *O* | quod] quot *O*, quid *M* | bis sive bisse] bes vel bisse *M*, bes sive bisse *S*, bis seu bisse *a*, •𝔥·*add. A*, ᴖ *add. B*, ᵗⁿ *add.* M_1, 𝔖 *add. O*, 𝔰𝔰 *add. W* | XL] id est XL *W* | *409* XL *om.* M_1, *supr.* $M_1$¹ | XVI] vel XV *supr.* $M_1$¹ | duc] dic *M*, *om. O* | quia *om. O* | *410* quot *BW*, quid *M* | septunx sive septus *M*] septunx *Aa*,𝔰 septunx sive septus *B*, 𝔏 septunx sive septus M_1, ᴗ septunx sive septux *O*, •𝔠·septum sive ///// *S*, 𝔶 *supr.* W¹ | VII] septima BM_1, II *S*, *om. O*, *corr.* O¹ | parte $BM_1 OS$, partes *MW* | Postea] Pot ea *A*, post *W* | Duc postea M_1 | duc] dic *M* | fiunt] faciunt M_1, fiunt *supr.* $M_1$¹ | *411* XIIII] XXIIII *B* | CCLXXX] CCLXX *O* | Hoc] est *add. W* | CCCLX] CCCXL *M* | *412* CCXX] CCCXXX *W* | XLVIII] XLVII *A*, LXVIII *S*, XLVIIII M_1, vel VIII *supr.* $M_1$¹

(36) PROPOSITIO DE SALUTATIONE CUIUSDAM SENIS AD PUERUM.

Quidam senior salutavit puerum, cui et dixit: Vivas, fili, vivas, inquit, quantum vixisti,
415 et aliud tantum, et ter tantum, addatque tibi deus unum de annis meis, et impleas annos centum. Solvat, qui potest, quot annorum tunc ipse puer erat.

413–416: ABCMM₁OR₁SVWab, om. M₂R
413 inscr. om. MR₁SW, Propositio *C* | PROPOSITIO *om.* $M_1 b$ | *414* et *om. Bb* | inquit] vivas *add. M*, *om. O* | quantum] tu *add. C* | *415* deus tibi *C* | deus] dominus *M* | impleas] adimpleas R_1 | *416* centum] C̄ M_1 | Solvat — potest] Dic *M* | Soluit *A* | ipse *B*] tempore ipse $M_1 OSb$, temporis ipse *CW*, tp̄ripse R_1, tempore *Aa*, temporis *M* | puer erat] erat puer ille *M*

SOLUTIO.

In eo vero, quod dixit: vivas, quantum vixisti, vixerat ante annos VIII et menses tres. Et aliud tantum fiunt anni XVI et menses VI, et alterum tantum fiunt anni XXXIII, qui ter
420 multiplicati fiunt anni XCVIIII. Uno cum ipsis addito fiunt C.

417—420: ABCMM₁OR₁SVWa, om. M₂Rb
417 inscr. om. MM₁R₁S, SOLVTIO DE SALVTATIONE CVIVSDAM SENIS AD PVERVM B /
418 vero om. M / quod] quot A / dixit] senex ad puerum. fili add. B / vivas] in add. S / vixerat] vix erat O / ante om. MS / VIII] XVIII S / tres — 419 menses om. C / 419 aliud] alii W / XVI — anni (2.) om. B / alterum] aliud M / tantum (2.) om. C / XXXIII] XXIII M₁ / 420 anni om. B / XCVIII S / unum AM₁OR₁a, unus M / cum om. a / additum AM₁OR₁a, additus M¹, additis MS / fiunt C om. S

(37) PROPOSITIO DE QUODAM HOMINE VOLENTI AEDIFICARE DOMUM.

Homo quidam volens aedificare domum locavit artifices VI, ex quibus V magistri et unus discipulus erat. Et convenit inter eum, qui aedificare volebat, et artifices, ut per singulos dies XXV denarii eis in mercede darentur, sic tamen, ut discipulus medietatem de eo, quod
425 unus ex magistris accipiebat, acciperet. Dicat, qui potest, quantum unusquisque de illis per unumquemque diem accepit.

421—426: ABMM₁OSVWab, om. CM₂RR₁
421 inscr. om. MSW, De quodam aedificante domum b / PROPOSITIO] ALIA PROPOSITIO O, om. A / VOLENTE M₁a / 422 Quidam homo M / VI] V S / 423 erat discipulus M / 424 eis om. O / in om. M₁ / de] ex M / eo] hoc M₁Sb / 425 accipiebat] accipiebant A / Dicat — potest] Dic M / qui om. A / unusquisque de illis] quisque M / 426 unumquemque] unumquotque b, unamquamque Aa / acciperet MS

SOLUTIO.

Tolle primum XXII denarios et divide eos in VI partes. Sic da unicuique de magistris, qui quinque sunt, IIII denarios. Nam quinquies quattuor XX sunt. Duos, qui remanserunt,
430 quae est medietas de IIII, tolle et da discipulo. Et sunt adhuc III denarii residui, quos sic distribues: Fac de unoquoque denario partes XI. Ter undecim fiunt XXXIII. Tolle illas triginta partes, divide eas inter magistros V. Quinquies seni fiunt XXX. Accidunt ergo unicuique magistro partes VI. Tolle tres partes, quae super XXX remanserunt, quod est medietas senarii, ac dabis discipulo.

427—434: ABMM₁OSVWa, om. CM₂RR₁b
427 inscr. om. MM₁S, SOLVTIO DE QVODAM HOMINE VOLENTI AEDIFICARE DOMVM B / 428 XXII] XII S / VI] VII S / da] tamen ut M₁₁, om. Aa / 429 IIII] III M₁, vel IIII supr. M₁¹ / denariis W / quinquies quattuor] quater VI M, VIIIIᵉˢ O / quattuor] quater B, III M₁, IIII supr. M₁¹ / sunt (2.)] fiunt M / remanserunt] remanserant O, remanent S, tolle add. W / 430 quae est] qui sunt M₁¹ / IIII] uno a / et (1.) om. M₁ / 431 de om. M₁, supr. M₁¹ / undecim] de unoquoque add. O (del. O¹) / fiunt om. M₁, supr. M₁¹ / 431 sq. triginta partes] partes XXX O / 432 partes] et add. M₁ / eas om. O / Accidunt] Accipiunt O / 433 VI — partes (2.) om. B / VI] VII S / tres] VI M / supra MM₁ / remanserant M / quod] quae M / est om. B / 434 denarii O / ac] hac B, et MOa, id (?) S / dabis] da Ma

435 ## (38) PROPOSITIO DE QUODAM EMPTORE IN ANIMALIBUS CENTUM.

Voluit quidam homo emere animalia promiscua C de solidis C, ita, ut equus tribus solidis emeretur, bos vero in solido I, et XXIIII oves in solido I. Dicat, qui valet, quot caballi, vel quot boves, quotve fuerunt oves.

435—438: ABMM₁OSVWab, om. CM₂RR₁
435 inscr. om. MSW, De animalibus emptis b / 436 Voluit post homo M / homo om. M₁Sb, supr. M₁¹ / emere post promiscua S, om. AO / C (1.)] scilicet C S / de solidis C om. O / solidis C] C solidis S / equos S / 437 in (bis) om. M / solido (1.)] solidum I / oves XXIIII O / XXXIII M₁, 33 b / Dicat — valet] compararet. qui vult dicat S / qui valet om. O, scilicet qui valet supr. O¹ / valet] velit abacista M / caballi — 438 oves] cameli vel asini sive oves in negocio C solidorum fuerunt S / 438 vel om. M / quotve] quotque O, quot M / fuerint B / fuerunt oves] oves ad ultimum emerentur M

68

SOLUTIO.

440   Duc ter vicies tria, fiunt LXVIIII. Et duc bis vicies quattuor, fiunt XLVIII. Sunt ergo caballi XXIII et solidi LXVIIII, et oves XLVIII et solidi II, et boves XXVIIII in solidis XXVIIII. Iunge ergo XXIII et XLVIII et XXVIIII, fiunt animalia C. Ac deinde iunge LXVIIII et II et XXVIIII, fiunt solidi C. Sunt ergo simul iuncta animalia C et solidi C.

> *439–443: ABMM₁OSVWa, om. CM₂RR₁b*
> *439 inscr. om. MM₁S,* SOLVTIO DE QVODAM EMPTORE IN ANIMALIBVS CENTVM *B* |
> *440* Duc ter] Ductae *M₁* | vicies *(1.)*] vigies *B, om. M* | tria] vicies *add. O* | LXVIIII] XLVIII *A,* VIIII *M* | Et *om. M* | duc *(2.)*] II *M₁, om. S* | bis vicies quattuor] duobus XXIIII *S* | vicies *(2.)*] vigies *B* | XLVIII] XLIIII *B* | ergo *om. M* | *441* XXIII] XXIIII *AO* | LXVIIII] XLVIIII *O,* LXVIII *M* | boves] oves *M* | in] et *S* | *442* XXIII] XXIIII *A* | XLVIII] XLVII *O,* XLVIIII et duo *M₁,* vel VIII *supr. M₁¹* | *443* et *(2.) om. AW* | Sunt — C *(3.)*] Qui simul adiuncti resonant aequalia cuncti *M* | iuncta] iuncti *M₁W, om. BS*

(39) PROPOSITIO DE QUODAM EMPTORE IN ORIENTE.

445   Quidam homo voluit de C solidis animalia promiscua emere C in oriente. Qui iussit famulo suo, ut camelum V solidis acciperet, asinum solido uno, XX oves in solido uno compararet. Dicat, qui vult, quot cameli vel asini sive oves in negotio C solidorum fuerunt.

> *444–447: ABMM₁OSVWab, om. CM₂RR₁*
> *444 inscr. om. MSW,* ALIA PROPOSITIO DE EADEM RE *M₁,* ALIA PROPOSITIO *BO,* Alia *b* |
> *445* promiscua animalia *M* | C in oriente] scilicet C *S* | *446* camelum] in *add. M₁Sb* | solidis] solidos *O¹b* | asinum] in *add. M₁Sb* | uno *(1.)*] et *add. M₁Sb* | XX] XXX *M₁* | in *om. M* | uno *(2.) om. Aa* | *447* Dicat — vult] Dic *M* | vel] quot *MO* | sive] quot *M* | in — solidorum *om. M*

SOLUTIO.

    Si duxeris X novies quinos, fiunt XCV, hoc est, XVIIII cameli sunt empti in solidis XCV
450 per X novies quinos. Adde cum ipsis unum, hoc est, in solido uno asinum unum, fiunt XCVI. Ac deinde duc vicies quater, fiunt LXXX, hoc est, in quattuor solidis oves LXXX. Iunge ergo XVIIII et I et LXXX, fiunt C. Haec sunt animalia C. Ac deinde iunge XCV et I et IIII, fiunt solidi C. Simul ergo iuncti faciunt pecora C et solidos C.

> *448–453: ABMM₁OSVWa, om. CM₂RR₁b*
> *448 inscr. om. MM₁S,* ITEM SOLVTIO *B* | *449* dixeris *M* | novies] et *add. a* | quinos *W*] V *AMM₁OSa,* quinque *B* | XCV *(1.)*] XCVI *M,* XXV *S* | est *om. W* | cameli XVIIII *a* | XVIIII] X et VIIII *MM₁,* XVIII *S* | cameli *om. O* | XCV *(2.)*] XC *B* | *450* per — quinos *om. a* | quinos] quinquenos *B,* V *W,* quinque solido *S* | in] cum *O* | uno] id est *M₁* | *451* duc] ducti *M* | vigies *B* | quattuor solidis] solidis III *B,* solidis IIII°ʳ *M₁* | *452* XVIIII] X et VIIII *M₁* | Haec — C *(2.) om. M* | C *(2.) om. a* | XCV] LXXXV *M₁,* vel XC *supr. M₁¹* | IIII] III *M₁,* vel IIII *supr. M₁¹* | *453* ergo *om. M* | faciunt] fiunt *W* | pecora — solidos] solidos C et pecora *M₁* | peccora *BMO* | et *om. S* | C *(3.) om. (?) M*

(40) PROPOSITIO DE HOMINE ET OVIBUS IN MONTE PASCENTIBUS.

455   Quidam homo vidit de monte oves pascentes, et dixit: Utinam haberem tantum et aliud tantum et medietatem de medietate et de hac medietate aliam medietatem, atque ego centesimus una cum ipsis meam ingrederer domum. Solvat, qui potest, quot oves vidit ibidem pascentes.

> *454–458: ABMM₁OR₁SVWab, om. CM₂R*
> *454 inscr. om. MR₁SW,* De homine et ovibus pascentibus *b* | IN MONTE *om. M₁b* | *455* de monte vidit *M* | pascentes oves *M* | dixit] infit *M* | haberem *om. M* | *456* aliud] alium *A,* alii *S* | aliam] idcirco *add. M₁R₁b,* idcirco *del. M₁¹* | medietatem *(2.) om. S* | atque] Tunc *R₁S, om. M₁b, corr. M₁¹* | *457* ipsis] illis *R₁,* ovibus *M,* unam cum ipsis *add. S* | ingrederer meam *a,* ingrederem meam *A,* ingrederer *R₁* | Solvat — potest] dicat aliquis *M* | *457sq.* ibidem vidit *R₁S* | vidit] viderit ille homo *M* | *458* ibidem] ibi *BM₁Ob,* inprimis *M*

SOLUTIO.

460 In hoc ergo, quod dixit: haberem tantum, XXXVI oves primum ab illo visae sunt. Et aliud tantum fiunt LXXII, atque medietas de hac videlicet medietate, hoc est de XXXVI, fiunt XVIII. Rursusque de hac secunda scilicet medietate assumpta medietas, id est XVIII, fiunt VIIII. Iunge ergo XXXVI et XXXVI, fiunt LXXII. Adde cum ipsis XVIII, fiunt XC. Adde vero VIIII cum XC, fiunt XCVIIII. Ipse vero homo cum ipsis additus erit centesimus.

459—464: ABMM₁OSVWa, om. CM₂RR₁b
459 inscr. om. MM₁S, ITEM SOLVTIO B | 460 ergo, quod om. S | dicit M, dixerat W | habere W | XXXVI oves post sunt M | ab illo primum M₁ | primum om. S | 461 LXXII] LXII S | videlicet om. M | 462 XVIII (1.)] X et VIII AM₁a, vel VIIII supr. M₁¹ | secunda scilicet] scilicet M₁, secunda MM₁¹ | est] de add. Wa | XVIII (2.)] VIIII O, X et VIII M₁S | 463 VIIII] vel VIII supr. M₁¹ | ergo om. M₁S, supr. M₁¹ | et XXXVI om. B | cum del. M₁¹ | XVIII] X et VIII M₁, decem et octo M | XC — 464 fiunt om. S | XC] XCVIII O, XX M₁, corr. M₁¹ | 464 VIIII] VIII A, post XC M | Ipse] Iste O | vero (2.) om. M | erat S | centesimus] et tali modulo solvitur questio add. M

465 **(41) PROPOSITIO DE SODE ET SCROFA.**

Quidam paterfamilias stabilivit curtem novam quadrangulam, in qua posuit scrofam, quae peperit porcellos VII in media sode, qui una cum matre, quae octava est, pepererunt igitur unusquisque in omni angulo VII. Et ipsa iterum in media sode cum omnibus generatis peperit VII. Dicat, qui vult, una cum matribus quot porci fuerunt.

465—469: ABMM₁OSVWab, om. CM₂RR₁
465 inscr. om. MSW | PROPOSITIO om. b | 466 Quidam] homo add. M | stabilivit] statuit M | curtim O, curiam iogam (?) S | quadrangulam om. ABMOW | 467 medio A | qui] quia b | peperunt A, peperit S | 468 igitur om. MM₁S | omni] suo W | generantibus B, generantur O, generaliter MM₁SWb | 469 una om. b | matribus] ex omni numero add. M | porcis A | fuerunt] videntur esse M

470 **SOLUTIO.**

In prima igitur parturitione, quae fuit facta in media sode, fuerunt porcelli VII et mater eorum octava. Octies ergo octo ducti fiunt LXIIII. Tot porcelli una cum matribus suis fuerunt in primo angulo. Ac deinde sexagies quater octo ducti fiunt DXII. Tot cum matribus suis porcelli in angulo fuerunt secundo. Rursumque DXII octies ducti fiunt $\overline{IIII}$ XCVI. Tot
475 in tertio angulo cum matribus suis fuerunt. Qui si octies multiplicentur, fiunt $\overline{XXXII}$ DCCLXVIII. Tot cum matribus suis in quarto fuerunt angulo. Multiplica quoque octies $\overline{XXXII}$ DCCLXVIII, fiunt $\overline{CCLXII}$ et CXLIIII. Tot enim creverunt, cum in media sode novissime partum fecerunt.

470—478: ABMM₁OSVWa, om. CM₂RR₁b
470 inscr. om. MS, SOLVTIO DE SODE ET SCROFA B, AL' M₁ | 471 partitione M | 472 ergo] igitur Aa | ducti om. M | suis om. a | 473 sexagies] vel sexies supr. M₁¹ | ducti] ductu O | 474 suis porcelli om. O | porcelli post Tot (473) B | fuerunt] fuere BM₁W, om. AOa | Rursusque Sa | fiunt om. S | IIII XCIII A | 475 angulo] una add. M | suis] porcelli add. B | fuerunt] fucre porcelli M | Qui — 478 fecerunt] Si vero eodem modo per VIII. summam multiplices, numerum quem queris absque dubietate invenies. M | Qui si] Quasi S | octies] vicies B, septies AOW, septies supr. M₁¹ | 476 DCCLXVIII] DCCLXV B, DCCLXXVIII A, DCCLXXXVIII a, vel LXX supr. M₁¹ | Tot] In tot B | suis om. a | fuerunt post suis BO | angulo fuerunt M₁ | octies] nonies S | 477 DCCLXVIII] DCCLXXVIII AM₁, DCCLXXXVIII a | fiunt om. S | CCLXII W | CXLIIII] CXLIII A, C et XLIIII S, CCCIIII a | cum in media] milia cum S | in om. O

(42) PROPOSITIO DE SCALA HABENTE GRADUS CENTUM.

480   Est scala una habens gradus C. In primo gradu sedebat columba una, in secundo duae, in tertio tres, in quarto IIII, in quinto V. Sic in omni gradu usque ad centesimum. Dicat, qui potest, quot columbae in totum fuerunt.

<hr>

    479—482: ABMM₁OR,SVWab, om. CM₂R

    479 inscr. om. MR₁SW, De scala cum centum gradibus *b* / HABENTE GRADUS] CVM GRADIBVS *M₁* / HABENTI *O* / *480* Est scala una] Quaedam scala est *M* / sedet *M* / una columba *M* / columbam *b* / secunda *O* / duae] 2° *S* / *481* IIII, in quinto *om. O* / in quinto V *om. S* / Dicat — *482* potest] Dic *MR₁* / *482* columbae — totum *om. MR₁* / fuerunt] sunt *M*

SOLUTIO.

    Numerabis autem sic: A primo gradu, in quo una sedet, tolle illam, et iunge ad illas 485 XCVIIII, quae in nonagesimo nono gradu consistunt, et erunt C. Sic secundum ad nonagesimum octavum, et invenies similiter C. Sic per singulos gradus unum de superioribus gradibus et alium de inferioribus hoc ordine coniunge, et reperies semper in binis gradibus C. Quinquagesimus autem gradus solus et absolutus est non habens parem. Similiter et centesimus solus remanebit. Iunge ergo omnes simul, et invenies columbas $\overline{V}$ L.

<hr>

    483—489: ABMM₁OSVWa, om. CM₂RR₁b

    483 inscr. om. MS, SOLVTIO DE SCALA HABENTE GRADVS C *B,* AL' *M₁* / *484* Numerabitur *a* / una] prima *M₁* / *485* quae] sedent *add. M* / in *om. AM₁a* / nonagesimo] nonagentesimo *S* / nono *om. AB* / consistunt *om. M* / secundam *S* / *486* et invenies *om. W* / invenies] fiunt *M* / superibus *W* / *487* gradibus *(1.)*] gradum *BM₁S, om. M* / alium] unum *MO* / ordine] modo *M* / repperies *OW,* invenies *M* / semper in binis] in binis ubique *M* / *488* C *om. W* / autem *om. O* / et *(1.) om. BM₁* / *489* Iunge — $\overline{V}$ L] simul iuncti, faciunt $\overline{V}$ et quinquaginta. Est et alia regula ex coacervatione omnium numerorum secundum ordinem naturalem positorum. Si vis scire quanta summa concrescat, per ipsius medietatem qui ultimus aggregatur, si par fuerit, subsequens multiplicetur. Verbi gracia: Si I, II, III, IIII, V, VI velis scire quot sint, per senarii medietatem qui est ultimus aggregatus sequens numerus, id est septenarius, multiplicetur, et fiunt XXI, quam summam reddet supradicta coadunatio. Si autem impar numerus aggregatur, per maiorem sui partem ipse multiplicetur. *M* / simul *om. Aa* / $\overline{V}$ L] quinque milia et L *M₁*

490 ## (43) PROPOSITIO DE PORCIS.

    Homo quidam habuit CCC porcos et iussit, ut tot porci numero impari in III dies occidi deberent. Similis est et de XXX sententia. Dicat modo, qui potest, quot porci impares sive de CCC sive de XXX in tres dies occidendi sunt. [Haec ratio indissolubilis ad increpandum composita est.]

<hr>

    490—494: ABMM₁OSVab, om. CM₂RR₁W

    490 inscr. om. MS, De porcis *b* / PORCIS] CCC VEL XXX *add. B,* CCC SIVE DE XXX *add. M₁O* / *491* Quidam homo *M* / CCC] ducentos *B, corr. B¹* / impari] et *add. S* / in III dies *om. M* / *492* occidi deberent] occiderentur *MM₁¹S,* III occiderentur *M₁,* tres occiderentur *b* / deberentur *AB* / sententia *ante* est *M* / Dicat — potest] Dic tu girbertista *M* / modo *om. Aa* / *493* de *(1.) om. M₁* / CCC … XXX] XXX … CCC *M* / in] inter *Aa,* infra *M₁S* / dies] ter *add. a* / Haec — *494* est *seclusi; habent mss. ab* / *493 sq.* increpandum composita] acuendos sensus iuvenum ingeniosissime reperta *M* / *494* compositum *S*

495 ## SOLUTIO.

    Ecce fabula, quae a nemine solvi potest, ut CCC porci sive triginta in tribus diebus impari numero occidantur. Haec fabula est tantum ad pueros increpandos.

<hr>

    495—497: ABMM₁OSVa, om. CM₂RR₁Wb

    495 inscr. om. MM₁S, SOLVTIO DE PORCIS CCC SIVE XXX *B* / *496* Ecce] Hec *S* / porcos *M₁* / diebus *om. A* / *497* occidamus *M₁S* / Haec — est] quae est inventa *M* / increpandos pueros *M*

(44) PROPOSITIO DE SALUTATIONE PUERI AD PATREM.

Quidam puer salutavit patrem: Ave, inquit, pater. Cui pater: Valeas, fili. Vivas, quan-
500 tum vixisti. Quos annos geminatos triplicabis, et sume unum de annis meis, et habebis annos
C. Dicat, qui potest, quot annorum tunc tempore puer erat.

498—501: ABCMM₁OSVWab, om. M₂RR₁
498 inscr. om. MSW, Propositio C | PROPOSITIO om. b | AD SALUT̄ O | 499 salutavit puer O |
Have O | Valeas — Vivas] vivas inquit fili M | Valeas] Benevaleas W | fili] et add. S, om. C | Vivas]
vive O, pater add. C | 500 triplicabis] triplicatos Aa | 501 qui potest] quis M | potest] valet M₁b | tempore]
tpī O, temporis BMW, ipse M₁b, tempore ipse S

502 SOLUTIO.

Erat enim puer annorum XVI et mensium VI. Qui geminati cum mensibus fiunt anni
XXXIII. Qui triplicati fiunt XCVIIII. Addito uno patris anno C apparent.

502—504: ABCMM₁OSVWa, om. M₂RR₁b
502 inscr. om. MS, SOLVTIO DE SALVTATIONE PVERI AD PATREM B, AL' M₁ | 503 enim
om. BC(?)M | puer] tunc C, om. M₁OSW | mensium] mensum A | anni om. B | 504 XXXIII — appa-
rent] etc. S (reliquae solutiones desunt) | fiunt] anni add. C | Addito] Anno C | anno om. O | C om. B |
C apparent] sunt C M, Ecce alia ratio de naturali numero. Sive par sive impar sit, ultimus numerus qui aggre-
gatur per suam denominationem in se ipsum multiplicatur, et si sibi suum proprium latus adiungitur, et si tota
illa summa in duo dividitur, ablata medietate divisionis remanet summa coacervationis. add. M

505 ## (45) PROPOSITIO DE COLUMBA.

Columba sedens in arbore vidit alias volantes et dixit eis: Utinam fuissetis aliae tantum
et tertiae tantum. Tunc una mecum fuissetis C. Dicat, qui potest, quot columbae erant in
primis volantes.

505—508: ABCMM₁OR₁SVWab, om. M₂R
505 inscr. om. MR₁SW, Propositio Ca, De columba b | 506 Quaedam columba M | alias] columbas
add. R₁ | et om. AR₁Sab | tantum] tantae CM₁R₁Sb | 507 tertiae OW] tertio B, ter CMM₁R₁S, ternae
Aa, adhuc b | tantum] tant(a)e CM₁R₁Sb | fuissetis C AMOWa] centum fuissetis BCM₁Sb | C om. R₁ |
Dicat — potest] dic MR₁ | potest] potes O, vult C | columbae — 508 primis] fuerunt prius R₁

SOLUTIO.

510   Triginta III erant columbae, quas prius conspexit volantes. Item aliae tantae fiunt LXVI,
et tertiae tantae fiunt XCVIIII. Adde sedentem, et erunt C.

509—511: ABCMM₁OVWa, om. M₂RR₁Sb
509 inscr. om. M, SOLVTIO DE COLVMBA B, AL' M₁ | 510 Trigintae M₁ | Erant XXXIII M |
prius] primum M, quidam add. B | conspexit] vidit M | tantae] tantum C | 511 tantae] tantum AMa |
et erunt C] cum aliis. et fiunt centum in ultimis M | C] XLVIIII add. M₁

(46) PROPOSITIO DE SACCULO AB HOMINE INVENTO.

Quidam homo ambulans per viam invenit sacculum cum talentis duobus. Hoc quoque
alii videntes dixerunt ei: Frater, da nobis portionem inventionis tuae. Qui renuens noluit
515 eis dare. Ipsi vero irruentes diripuerunt sacculum, et tulit sibi quisque solidos quinquaginta.
Et ipse postquam vidit se resistere non posse, misit manum et rapuit solidos quinquaginta.
Dicat, qui vult, quot homines fuerunt.

512—517: ABMM₁OSVWab, om. CM₂RR₁
512 inscr. om. MSW, De sacculo b | AB HOMINE om. M₁ | 513 cum om. AO | duobus talentis M |
514 Frater om. W | da] dona B | tuae] tantum Aa | rennuens AOW | 515 dare eis O | eis om. MM₁b |
vero om. W | quisque sibi M | quisque om. O | 516 Et—quinquaginta om. M₁Sb, in margine add. M₁¹ | Et
— vidit] Ille vero videns M | Et] At B | postquam] post W | se — posse] quia non valebat repugnare M |
resistere post posse M₁¹O, illis add. BO(post se)W, eis add. M₁¹ | 517 Dicat — vult] Dic M | vult] potest M₁

SOLUTIO.

Apud quosdam talentum LXXV pondo vel libras habet. Libra vero habet solidos
520 aureos LXXII. Septuagies quinquies LXXII ducti fiunt $\overline{V}$ CCCC, qui numerus duplicatus
facit $\overline{X}$ DCCC. In X milibus et octingentis sunt quinquagenarii CCXVI. Tot homines idcirco fuerunt.

518—522: *ABMM₁OVWa, om. CM₂RR₁Sb*
518 *inscr. om. M*, SOLVTIO DE SACCVLO INVENTO AB HOMINE *B*, SOL' AL' *M₁* / 519 LXXV]
LXXII *Aa*, vel *add. Aa* / pondos *M₁*, pondera *W* / habet libras *Aa* / solidos habet *M* / habet *(2.)*] inter
O / 520 aureos *om. M* / Septuagies] Sexagies *a* / quinquies] vigies *B* / LXXII *(2.)*] LXX *W* / fiunt *om.*
M₁ / VCCC *A* / 521 facit] fiunt *Aa* / $\overline{X}$ DCCC] X DCCC *W*, decies DCCC *Aa*, $\overline{X}$ DCCCC *MO*, vel
DCCCC *supr. M₁¹* / et *om. M* / octingentis sunt] octingentis *O* / quinquegenari *O* / idcirco *om. M*

(47) PROPOSITIO DE EPISCOPO QUI IUSSIT XII PANES IN CLERO DIVIDI.

Quidam episcopus iussit XII panes dividi in clero. Praecepit enim sic, ut singuli presby-
525 teri binos acciperent panes, diaconi dimidium, lector quartam partem. Ita tamen fiat, ut
clericorum et panum unus sit numerus. Dicat, qui valet, quot presbyteri vel quot diaconi
aut quot lectores esse debent.

523—527: *ABCMM₁OR₁SVWab, om. M₂R*
523 *inscr. om. MR₁SVW*, De episcopo *b*, Propositio *C* / QUI—DIVIDI *om. M₁* / IN CLERO *om. Aa* /
DIVIDI *om. A* / 524 iussit *om. O* / panes XII *M₁b* / in clero dividi *CM₁Sb* / Praecepit — sic] sic tamen
iussit *M* / enim] autem *M₁* / sic, ut] sicut *R₁* / presbiteri *AS* / 525 binos *del. M₁¹* / diacones *OW*, diaconos *R₁*, diaconus *AMa* / lector — partem *om. M₁*, *supr. M₁¹* / lectores *B* / Ita] Item *W* / tamen] inquit
add. B / fiet *S* / 526 Dicat — valet] Dic *M* / valet] vult *ABCWa*, potest *O* / presbiteri *S* / vel *om. CMR₁*
/ quot *(2.)* *om. S* / diaconos *b*, diacones *ACM₁OWa* / 527 aut quot] quotve *b*, vel quot *R₁*, quot *CM* /
esse debent] sint *M* / debeant *M₁R₁Sb*

SOLUTIO.

Quinquies bini fiunt X, id est, V presbyteri decem panes receperunt, et diaconus unus
530 dimidium panem, et inter lectores VI habuerunt panem et dimidium. Iunge V et I et VI in
simul, et fiunt XII. Rursusque iunge X et semis et unum et semis, fiunt XII. Hi sunt XII
panes, qui simul iuncti faciunt homines XII et panes XII. Unus est ergo numerus clericorum
et panum.

528—533: *ABCMM₁OVWa, om. M₁RR₁Sb*
528 *inscr. om. M*, SOLVTIO DE EPISCOPO QVI IVSSIT XII PANES IN CLERO DIVIDI *B*, AL'
M₁ / 529 acceperunt *MM₁O* / 530 inter *del. M₁¹* / habuerunt VI *M* / panem *(2.)*] panes *MO* / 530 sq. in
simul] simul *O*, *om. CM*, in unum *M₁* / 531 et *(1.)* *om. BMOW, add. B¹* / Rursusque — XII *(2.) om. B*
/ Rursumque *O*, Rursus *AM* / fiunt *(2.)*] panes *add. C* / Hi] Hii *BC*, Et illi *a* / 531 sq. XII panes] panes
duodecim *C* / 532 faciunt] fiunt *CO* / panes XII. Unus] sic *C* / Unus — 533 panum] sic fit aequalis numerus
M / est *om. O* / ergo *om. C* / numerus] unus *add. C*, et *add. W*

(48) PROPOSITIO DE HOMINE QUI OBVIAVIT SCOLARIIS.

535 Quidam homo obviavit scolariis et dixit eis: Quanti estis in scola? Unus ex eis respondit
dicens: Nolo hoc tibi dicere. Tu numera nos bis, multiplica ter. Tunc divide in quattuor
partes. Quarta pars numeri, si me addis cum ipsis, centenarium explet numerum. Dicat, qui
potest, quanti fuerunt, qui pridem obviaverunt ambulanti per viam.

534—538: *ABMM₁OR₁SVWab, om. CM₂R*
534 *inscr. om. MR₁SW*, De homine scholasticis obviante *b* / PROPOSITIO *om. A* / SCOLARIBVS *A*,
scholaribus *a* / 535 scholarii *b*, scolarios *R₁*, scolaribus *AM*, scholaribus *a*, in via *add. R₁* / dixit eis] inquit
M / schola *ab* / Unus] Uno *b*, autem *add. M₁R₁Sb* / 536 hoc — dicere *om. M* / tibi hoc *O* / Tu] autem

add. M | bis] et *B* | multiplica ter] multiplicatur *AR₁* | ter *om. M* | Tunc] et *R₁* | in *om. A* | *537* numeri] numerum *MM₁R₁*, vel numeri *supr. M₁¹*, nostrum *b* | si me] sume *W* | addas *O* | exples *A*, complet *M* | Dicat — *538* potest] dic *R₁* | *538* potest] vult *M₁b* | fuerunt — viam] obviaverunt pridem homini illi *M* | ambulanti] ambulaverunt *B*

SOLUTIO.

540   Tricies ter bini fiunt LXVI. Tanti erant, qui pridem obviaverunt ambulanti. Qui numerus bis ductus CXXXII reddit. Hos multiplica ter, fiunt CCCXCVI. Horum quarta pars XCVIIII sunt. Adde puerum respondentem, et reperies C.

> *539—542: ABMM₁OVWa, om. CM₂RR₁Sb*
> *539 inscr. om.* M, SOLVTIO DE HOMINE OBVIANTI SCOLARIIS *B*, AL' *M₁* | *540* Tricies] Trigies *B*, Terties *Aa* | ter bini] terni *O*, *corr.* O¹ | ter *om. W* | fiunt] sunt *W* | LXVI] LXIIII *A*, CXVI *W* | pridem] primum *M* | *541* ductus] datus *O* | CXXXII] CXXII *BW* | multiplica ter] multiplicatos *A*, simul triplicat *W* | *542* repperies *AW*, invenies *BMO* | C] sine impedimento probationis *add. M*

543 **(49) PROPOSITIO DE CARPENTARIIS.**

    Septem carpentarii septenas rotas fecerunt. Dicat, qui vult, quot carra erexerunt.

> *543 sq.: ABM₁OR₁SVWab, om. CMM₂R*
> *543 inscr. om.* R₁SW | PROPOSITIO *om. b* | CARPENTARIIS] ROTIS *BO* | *544* carpentari *A* | vult] vul *R₁*, potest *Aa* | carras *S*, carrae *Aa* | erexerunt] fecerunt *M₁b*, rexerunt *AOa*

545 **SOLUTIO.**

    Duc septies VII, fiunt XLVIIII. Tot rotas fecerunt. XII vero quater ducti XL et VIII reddunt. Super XL et VIIII rotas XII carra sunt erecta, et una superfuit rota.

> *545—547: ABMM₁OVWa, om. CM₂RR₁Sb*
> *545 inscr. om.* M, SOLVTIO DE SEPTEM CARPENTARIIS *B*, AL' *M₁* | *546* VII] VI *A* | XII — *547* rota] carpentarii. Qui per quatuor divisi, tot enim rotae faciunt carrum, reddunt duodecim carra remanente rota *M* | quater] equaliter *W* | XL et VIII] XL et VIIII *O*, XLVIII *Ba* | *547* et *(1.) om. B* | VIIII] VIII *M₁*, octo *W*, vel IX *supr. M₁¹* | carre *O* | aerectae *O*

(50) ALIA PROPOSITIO.

    Centum metra vini, rogo, ut dicat, qui valet, quot sextarios capiunt, vel ipsa etiam 550 centum metra quot meros habent.

> *548—550: ABM₁OSVab, om. CMM₂RR₁W*
> *548 inscr. om.* AS, ITEM ALIA PROPOSITIO *BM₁*, De vino in vasculis *b*, Propositio de vino in vasculis *a* | *549* qui velit ut dicat *S* | valet *B*] velit *M₁OSb*, vult *Aa* | etiam ipsa *M₁S* | ipsa *om. O* | *550* meros] numeros *B*

551 **SOLUTIO.**

    Unum metrum capit sextarios XL et VIII. Duc centies XLVIII, fiunt IIII DCCC. Tot sextarii sunt. Similiter et unum metrum habet meros CCLXXXVIII. Duc centies CCLXXXVIII, fiunt XXVIII DCCC. Tot meri sunt.

> *551—554: ABM₁OVa, om. CMM₂RR₁SWb*
> *551 inscr.*] ITEM SOLVTIO *B*, AL' *M₁* | *552* sectarios *a* | XL et VIII] XLVIII *B*, XLVIIII *O* | IIII] III *M₁*, vel IIII *supr. M₁¹*, quatuor *A* | DCCC] DCCCC *B* | *553* meros] metros *B* | CCLXXXVIII *OM₁*] CCLXXVIII *B*, CCLXXXVIIII *Aa*, vel IX *supr. M₁¹* | *554* CCLXXXVIII *BM₁O*] CCLXXXVIIII *Aa*, vel IX *supr. M₁¹* | XXXVIII DCCC *O*, XXVIIII milia DCCCC *A*, XXVIII DCCCC *a* | metri *B* | sunt meri *a*

555 (51) PROPOSITIO DE VINO IN VASCULIS A QUODAM PATRE DISTRIBUTO.
Quidam paterfamilias moriens dimisit IIII filiis suis IIII vascula vini. In primo vase erant modia XL, in secundo XXX, in tertio XX, et in quarto X. Qui vocans dispensatorem domus suae ait: Haec quattuor vascula cum vino intrinsecus manente divide inter quattuor filios meos, sic tamen, ut unicuique eorum aequalis sit portio tam in vino quam et in vasis.
560 Dicat, qui intelligit, quomodo dividendum est, ut omnes aequaliter ex hoc accipere possint.

555—560: ABMM₁OSVWab, om. CM₂RR₁
555 inscr. om. MSW, De patrefamilias distribuente *b | VINO] VINI Aa | PATRE] PATRE FAMILIAS B, om. M₁ | DISTRIBUTO] divisione a, om. AO | 556 divisit M₁b, vel m supr. M₁¹ | suis om. Aa | 557 modii M | et om. MM₁S | 558 ait] inquit M | vasa S | manente om. M₁b, supr. M₁¹ | divide] dividite S, dividantur O | inter] in S, om. O | 559 filiis meis OS | aequalis] aequa O, una Aa | in (1.)] et in Ob, de M₁ | et om. BSa | 560 intelligit ABM₁W | ut] quot M₁, corr. M₁¹ | accipere possint] accipiant M*

SOLUTIO.
In primo siquidem vasculo fuerunt modia XL, in secundo XXX, in tertio XX, in quarto X. Iunge igitur XL et XXX et XX et X, fiunt C. Tunc deinde centenarium idcirco numerum per quartam divide partem. Quarta namque pars centenarii XXV reperitur, qui numerus bis
565 ductus quinquagenarium de se reddit numerum. Eveniunt ergo unicuique filio in portione sua XXV modia, et inter duos L. In primo XL et in quarto sunt modii X. Hi iuncti faciunt L. Hoc dabis inter duos filios. Similiter iunge XXX et XX modios, qui fuerunt in secundo et tertio vasculo, et fiunt L, et hoc quoque similiter ut superius dabis inter duos, et habebunt singuli XXV modia, eritque id faciendo singulorum aequa filiorum divisio tam in vino
570 quam et in vasis.

561—570: ABMM₁OVWa, om. CM₂RR₁Sb
561 inscr. om. M, ITEM SOLVTIO *B,* AL' *M₁ | 562 quidem W | modii BM | XX] et add. W | 563 Igitur iunge ABM₁ | igitur — X (2.)] simul M | fiunt] faciunt M | idcirco om. B, del. M₁¹ | numerum idcirco M | 564 Quarta] Quartam B | namque om. O | repperitur AOW, reperiuntur M₁, repperiuntur B | 565 Evenit O | 565sq. portione sua] portionem vasa M | 566 modia] modii O, modios W | et (2.) om. M | sunt modii om. | modii] modia B, alii O | Hii BW | faciunt] fiunt M₁ | 567 filios om. Aa | modios MO] modia ABa | qui] quae a | 568 et (1.)] in add. B | vascula a | et (2.) om. OW | similiter] simul M₁ | ut superius om. O | duos] filios add. O, fratres add. M | 569 modios MM₁O | id faciendo] infaciente B, id faciente AOW, ad facientes M₁ | filiorum aequa M | divisio filiorum O | tam — 570 vasis] vini et vasorum M*

(52) PROPOSITIO DE HOMINE PATREFAMILIAS.

Quidam paterfamilias iussit XC modia frumenti de una domo sua ad alteram deportari, quae distabat leuvas XXX, ea vero
575 ratione, ut uno camelo totum illud frumentum deportaretur in tribus subvectionibus et in unaquaque subvectione XXX modia portarentur, camelus vero in unaquaque leuva comedat modium unum. Dicat, qui
580 velit, quot modii residui fuissent.

Quidam paterfamilias habebat de una domo sua ad alteram domum leugas XXX et habens camelum, qui debebat in tribus subvectionibus ex una domo sua ad alteram de annona ferre modia XC, et in unaquaque leuga isdem camelus comedebat semper modium I. Dicat, qui valet, quot modia residua fuerunt.

571—580: recensio I (Quidam — fuissent): ABMM₁¹OVWa, recensio II (Quidam — fuerunt): M₁R₁Sb, om. CM₂R
recensio I: 571 inscr. om. MW | PROPOSITIO om. A | PATREFAMILIA A, PATRI FAMILIAS O, PATRIS FAMILIAS B | 572 modios BM | 573 alteram] domum suam add. M | 574 leugas BM, leucas a | vero om. O | 576 subiectionibus M₁¹ | 577 et] ut M | subvectione — 578 unaquaque om. O | subiectione M₁¹, subvectionibus B, corr. B¹ | modii BM | 578 deportarentur BM | vero] quoque Aa | 579 leuga

M, leuca *a* / comedit *ABMOW*, comedebat (*ante 578* in) M_1^1 / Dicat — *580* fuisssent *om.* M_1^1 / Dicat — *580* velit] Dic *M* / *580* fuerunt *W*, sunt *M*

  recensio II: *571 inscr. om.* R_1S, De camelo *b*, PROPOSITIO DE CAMELO CVIVSDAM PATRIS FAMILIAS. QVI IN III^{bus} SVBVECTIONIBVS FEREB̄ MOD̄ M_1 / *572* habuit *S* / *573* leuvas M_1, leugas *b* / *574* debebat] habebat *S* / subiectionibus *b* / *576* ferre] fere *Sb* / unaquaeque *Sb* / *577* leuva M_1, leuca *b* / camelis $R_1(?)S$ / *579* fuerint M_1b

SOLUTIO.

In prima subvectione portavit camelus modios XXX super leuvas XX et comedit in unaquaque leuva modium unum, id est, modios XX comedit, et remanserunt X. In secunda subvectione similiter deportavit modios XXX, et ex his comedit XX, et remanserunt X.
585 In tertia vero subvectione fecit similiter: Deportavit modios XXX, et ex his comedit XX, et remanserunt decem. Sunt vero de his, qui remanserunt, modii XXX et de itinere leuvae X. Quos XXX in quarta subvectione domum detulit, et ex his X in itinere comedit, et remanserunt de tota illa summa modia tantum XX.

  581—588: *ABMM₁OVWa, om. CM₂RR₁Sb*
  581 inscr. om. MM₁, SOLVTIO DE CAMELO *B* / *582* vectione *W* / portavit — *588* XX *om.* O *in exemplari foliis quibusdam mutatis* (*cf. p. 17 notam 17*) / portabat M_1 / XXX modios *M* / leuvas] leugas *BM*, leucas *a* / XX] X *a* / *583* unaquaeque *W* / leuva] leuga *BM*, leuca *a* / unum modium *W* / unum — comedit] qui simul sunt viginti *M* / unum *om. BM* / modios XX] XX modios *BW*, XII mod' M_1, vel XX *supr.* M_1^1 / comedebat M_1 / *584* vectione *W* / *585* In — *586* decem *om. W* / vero *om. BMM₁* / subportatione *M* / fecit *om. M* / similiter] et *add. BM₁* / *586* modia *Aa* / et — *587* XXX *om. B* / itinere] sunt *add. M* / leugae *M*, leucae *a* / *587* detulit domum *M* / domum] donum *B* / *588* illa — tantum] summa modii *M* / modii *BMM₁W* / XX tantum *BM₁W*

(53) PROPOSITIO DE HOMINE PATREFAMILIAS MONASTERII XII MONA-
590 CHORUM.

Quidam pater monasterii habuit XII monachos. Qui vocans dispensatorem domus suae dedit illi ova CCIIII iussitque, ut singulis aequalem daret ex eis omnibus portionem. Sic tamen iussit, ut inter V presbyteros daret ova LXXXV et inter quattuor diaconos LXVIII et inter tres lectores LI. Dicat, rogo, qui valet, quot ova unicuique ipsorum in por-
595 tione evenerunt, ita ut in nullo nec superabundet numerus nec minuatur, sed omnes, ut supra diximus, aequalem in omnibus accipiant portionem.

  589—596: *ABMM₁OSVWab, om. CM₂RR₁*
  589 inscr. om. MSW, De dispensatore in monasterio *b* / PROPOSITiO *om. A* / PATREFAMILIAS — MONACHORUM] PATRIS FAMIL' MONAST̄ *B* / XII MONACHORUM *om.* M₁O / *591* habuit] et *add. S* / vocans] convocans M₁*b* / *592* illi] illis *a* / iussitque] iussit *S* / daret *om.* O / eis] his *SW* / omnibus *om. a* / *593* iussit] ei *add. B* / presbiteros *S* / daret ova *om.* O / et — *594* LXVIII *om. A* / diacones *OW* / *594* lectores] subdiaconos M₁, clericos *S*, *om. b* / rogo] ergo *BW* / valet] et *add. b* / portionem *Aa* / *595* venerunt *ABMa* / nec (1.) *om. M* / omnes — *596* portionem] omnibus aequaliter *M* / omnes] et omnes *B*, omnis *a* / *596* equale *S* / omnibus] omni *a* / accipiat *Aa*

SOLUTIO.

Ducentos igitur quattuor per XII^{am} divide partem. Horum quippe pars XII^{a} in septima decima resolvitur parte, quia sive duodecies XVII sive decies septies XII miseris, CCIIII
600 reperies. Sicut enim octogenarius quintus numerus septimum decimum quinquefarie de se reddit numerum, ita sexagenarius octavus quadrifarie et quinquagesimus primus trifarie.

Iunge ergo V et IIII et III, fiunt XII. Isti sunt homines XII. Rursusque iunge LXXXV et LXVIII et LI, fiunt CCIIII. Haec sunt ova CCIIII. Veniunt ergo singulorum ex his in parte ova XVII per duodecimam partem septimum decimum numerum aequa lance divisum.

597–604: ABMM₁OVWa, om. CM₂RR₁Sb

597 inscr. om. M, AL' M₁, ITEM B | 598 partem divide a, divide partes W | Horum] orum O | 598 sq. septima decima] XVI M₁, vel XVII supr. M₁ⁱ, septima O | 599 parte] partes M₁ | duodecies] XIIes. XIIes et VII M₁, XIIes (2.) del. M₁¹ | XVII] X et VI A, XVI W, decem et septem M | CIIII BMW, C et IIII M₁, CVII A | 600 repperies BMOW | quinquefarie] quinqueferiae O, quinari(a)e ABMM₁W, quinarium a | de se post numerum (601) Aa, post decimum O | 601 ita] et add. a, de add. A | sexagenarius O | quadrifacie M | trifacie M | 602 ergo om. Aa | Isti – XII (2.) om. O | LXXXV] LXXV A | et (3.) om. M | 603 LXVIII] LXVIIII W, LXXIII M₁, om. M | LI] L W | Venient O | singulorum] singulis B | ex] et M₁ | parte] partes Aa, partem M₁ | XVII] XIIII M₁ | 604 numerum om. AM₁a | aequa] aequo iure BM₁, (a)equa iure AW | divisum] dividi fiunt Aa, sine ulla falsitate regulariter divisum M

1. NAMENVERZEICHNIS

2. HANDSCHRIFTENVERZEICHNIS

VI

The Names and Forms of the Numerals on the Abacus in the Gerbert Tradition[*]

An interesting chapter of the history of mathematics is formed by the transmission of Hindu-Arabic numerals[1]. In Western Europe two periods of this transmission can be distinguished: one beginning with the 12th century, when the well-known translations from Arabic to Latin were made; and before this the time of Gerbert of Aurillac. We are here concerned with the second, but I may perhaps summarize the first. In al-Khwārizmī's arithmetical treatise the Hindu-Arabic numerals and the rules of arithmetic manipulation were presented. This gave rise to several Latin treatises on the subject, the so-called *algorismus* treatises. The most famous of these are the *Algorismus* of Johannes de Sacrobosco and the *Carmen de algorismo* of Alexander de Villa Dei, both written in the first half of the 13th century. They were both studied in the universities, but we should note that the "Rechenschulen" throughout Europe also taught the new numerals and how to calculate with them and were involved in the further development of their use.

1 Gerbert's abacus and the forms of the numerals on it

Let us now turn to the first appearance of the numerals in the West at the end of the 10th century. It is well known that Gerbert here played a special role. On a journey to the Spanish March, which he undertook in 967 as a companion to Borel of Barcelona, he studied mathematics, as we know from the remarks of his biographer Richer[2]. One has assumed that Gerbert came to know the Hindu-Arabic numerals on this journey.

---

[*]This is a slightly modified reprint of the article published in F. G. Nuvolone (ed.), *Gerberto d'Aurillac da Abate di Bobbio a Papa dell'Anno 1000. Atti del Congresso internazionale. Bobbio, Auditorium di S. Chiara, 28–30 settembre 2000* (= *Archivum Bobiense, Studia IV*), Bobbio / Pesaro 2001, pp. 245–265.

[1]See, for example, Tropfke, 'Elementarmathematik', pp. 61–70, and the literature cited there.

[2]Richer writes that Gerbert studied in Vich with Bishop Hatto: "apud quem etiam in mathesi plurimum et efficaciter studuit" (*Historiae* III,43; see Bubnov, 'Gerberti Opera', p. 376).

To support this assumption, we may note the Hindu-Arabic numerals in the so-called "Codex Vigilanus", which was written in the monastery Albelda in Asturia in 976.

Gerbert was active from 971 to 996 – with breaks – in Rheims and taught there in the cathedral school. In his teaching he used a counting-board, or abacus. Richer informs us precisely about the form of this instrument. He writes:

> In teaching geometry he gave himself no less trouble (than in the other disciplines). As a preparation he introduced an abacus, i. e. a table of an appropriate size, with the help of a shield-maker. He divided its length in 27 parts and he placed marks, nine in number, to represent any number. In the like-ness of which he had a thousand characters made out of horn, which he distributed through the 27 parts of the abacus, to designate the multiplication or division of each number: (the characters) divide or multiply the multitude of the numbers with such a shortening (of effort) that with the excessive mul-titude it is easier to understand than to describe in words. Whoever wishes to know fully the science of these (charac-ters), may read his book which he wrote to Constantinus the grammarian. There he will find this enough and abundantly treated[3].

From Richer's description it follows that Gerbert's abacus was divided into 27 columns and that Gerbert used nine *notae* for the representation of numbers. These *notae* were special pieces of horn on which the nu-merical value was marked. Unfortunately Richer does not say how these *notae* looked.

To answer this question, it is no use to refer to Gerbert's book men-tioned by Richer. It is extant, but it tells us nothing about the appearance and division of the abacus, but only gives – in tedious fashion – the rules

---

[3] "In geometria vero non minor in docendo labor expensus est. Cujus introductioni abacum, id est tabulam dimensionibus aptam, opere scutarii effecit. Cujus longitudini in XXVII partibus diductae novem numero notas omnem numerum significantes dis-posuit. Ad quarum etiam similitudinem mille corneos effecit caracteres, qui per XXVII abaci partes mutuati, cujusque numeri multiplicationem sive divisionem designarent: tanto compendio numerorum multitudinem dividentes vel multiplicantes, ut pre nimia numerositate potius intelligi, quam verbis valerent ostendi. Quorum scientiam qui ad plenum scire desiderat, legat ejus librum, quem scribit ad Constantinum gram-maticum. Ibi enim hæc satis habundanterque tractata inveniet." (*Historiae* III,54, cited from Bubnov, 'Gerberti Opera', pp. 380f.).

for place value in multiplication and division[4]. These rules are very similar to multiplication rules given in the "Codex Vigilanus", just before the presentation of the Hindu-Arabic numerals; and there are good reasons to assume that Gerbert knew this text, which might have been present in a manuscript he saw during his visit to the Spanish March[5].

According to our present knowledge there are in principle only two possibilities for the form of the *notae*: the Greek alphabetic numerals or the Hindu-Arabic numerals. We know that in the 10th century, shortly before Gerbert's time, a counting-board was in use on which the value of a counter was indicated by the corresponding Greek letter. On the other hand Gerbert had very probably become acquainted with the Hindu-Arabic numerals on his journey in the Spanish March, so that we may well assume that he used these signs to indicate the value of the counter. This supposition is supported by texts on calculation on the abacus and by pictorial representations of counting-boards which are extant in manuscripts of the eleventh century. I should now like to discuss these sources more closely.

2 Pictorial representations of the counters: Bernelinus, Ps.-Boethius

Gerbert's treatise on calculation on the abacus was well known in the Middle Ages. This is shown by the 35-odd manuscripts which are extant today; and there are several commentaries. With Gerbert began a tradition which led in the next 150 years to numerous new treatises. More than 15 abacus treatises are extant, some of them very compendious. I should here like to limit myself to the oldest of these texts, namely the six treatises written before 1050.

Two of them were written in Gerbert's lifetime. Their authors are Heriger of Lobbes (ca. 950 – after 1007) and Abbo of Fleury (ca. 945/50 – 1004). Both authors treat the rules for calculation on the abacus, but say nothing about the form of the counting-board or whether the operations

---

[4] *Regulae de numerorum abaci rationibus*, edited in Bubnov, 'Gerberti Opera', pp. 1–22. The treatise was probably written about 980. For the dating see Bergmann, 'Innovationen', p. 180, note 48.

[5] The text of these rules in the "Codex Vigilanus" is printed in Lattin, 'Origin', p. 191. John Burnham has evolved the theory that the prototype of the "Codex Vigilanus" may have exist in Ripoll and that Gerbert saw it when he was there (Lattin, 'Origin', p. 190).

were carried out with marked counters or not[6]. The same goes for two texts which were written in the second quarter of the 11th century, by Laurentius of Amalfi[7] and by Hermannus Contractus[8].

There remain only two texts which are important for our investigation, because they do describe the form of the numerical symbols written on the counters.

The oldest treatise on calculation on the abacus which gives the form of the counting-board and of the counters was written by Bernelinus. His treatise cannot have been written before 999, but was perhaps composed in Gerbert's lifetime[9]. In the introduction Bernelinus says that Gerbert precisely taught and made known the rules about calculation on the abacus, but that these were in his own time almost forgotten[10]. According to Bernelinus the counting-board is drawn on a polished table. It is divided into 30 columns. Of these the first three are reserved for fractions and the remaining 27 for the units, tens, hundreds, etc. Bernelinus expressly states that every three columns are closed off by a semi-circle and within each group of three there are two smaller semi-circles, one for the units and one for the tens and hundreds together[11]. He justifies this grouping by pointing to the special meaning of the unity. Over the three columns of a group there are written (from left to right) the letters *C*, *D*, *M*, which characterize the "hundreds" (*centenum*), "tens" (*decenum*),

---

[6]For Heriger's life and his treatise on the abacus see Lindgren, 'Gerbert', p. 55, and Bergmann, 'Innovationen', pp. 195f. His work was edited in Bubnov, 'Gerberti Opera', pp. 205–221. The treatise was probably written between 997 and 1007; see Bubnov, 'Gerberti Opera', p. 205. For Abbo's life and work see Lindgren, 'Gerbert', pp. 49–55, and van de Vijver, 'Abbon'. The relevant parts on the abacus in Abbo's commentary to Victorius' *Calculus* were edited by Bubnov, 'Gerberti Opera', pp. 197–203; further parts in Christ, 'Victorius'. Bergmann dates the treatise to "about 975/985" ('Innovationen', p. 180).

[7]Edited in Newton, 'Laurentius', from the only known manuscript.

[8]Edited in Treutlein, 'Intorno'. Three manuscripts are known.

[9]Little is known about Bernelinus; see Bergmann, 'Innovationen', pp. 199f. The text is edited in Olleris, 'Oeuvres', pp. 357–400, and in Bakhouche, 'Bernelin', pp. 16–104. For the dating see Bergmann, 'Innovationen', p. 199, note 167.

[10]"... multiformes abaci rationes ... negligentia quidem apud nos iam paene demersas, sed a domino papa Gerberto quasi quaedam seminaria breviter et subtilissime seminatas" (Bakhouche, 'Bernelin', p. 16).

[11]"Tabula ... diligenter undique prius polita, per XXX dividatur lineas, quarum tres primas unciarum minutiarumque dispositioni reservamus, reliquarum vero XXVII, per ternas et ternas, haec certa mensurandi proveniat regula: primam de tribus lineam circinus in hemispherii modum teneat amplexam; maior autem circinus duas reliquas amplectatur; sed easdem tres maximus complectatur pariter" (Bakhouche, 'Bernelin', p. 21).

"units" or "thousands" (*unitas, monos, mille*); in addition there is the letter S (for *singulare*) for the first column of the whole numbers[12]. For the symbols on the counters, which Bernelinus, like Gerbert, calls *caracteres*, he suggests two possibilities (Fig. 1): either the (new) numerals, i. e. the numerals coming from the Arabic[13], or the Greek alphabetical numerals, Alpha to Theta. Compared with his predecessors, Bernelinus describes the calculations on the counting-board (multiplication, the different kinds of division, fractions) very clearly. In none of the numerous extant manuscripts[14], however, is a picture of the abacus to be found.

The oldest pictorial representation of a counting-board connected with a treatise is in the so-called *Geometry II* of Pseudo-Boethius. It was compiled from various sources in the first half of the 11th century by an unknown scholar in Lorraine. For the section on the abacus he used Gerbert's text[15]. Calculation on the abacus is done with "variously formed characters" (*diverse formatos apices vel caracteres*). These are the forms of the Western Arabic numerals used in Spain. These numerals appear not only in the text of *Geometry II* (Fig. 2), but (with the addition of a circle) also on the illustration of the counting-board (Fig. 3). In the oldest manuscripts it consists of 12 columns. The numerals 1, 2, ..., 9, 0 are written in these columns; over them the words *igin, andras, ormis, arbas, quimas, calctis, zenis, temenias, celentis, sipos* are also inscribed (see Fig. 3)[16]. In later abacus treatises these words are used as names for the ten numerals[17]. I shall later come back to these names.

3   Isolated pictorial representations of the counters

As I have mentioned, in the *Geometry II* we find an illustration of the abacus, which is in the Gerbert tradition. Similar pictures also exist, isolated, in some manuscripts. I should like to consider now the pictorial representations of this kind in three manuscripts datable to the end of

---

[12] "... per omnem abaci tabulam omnes eaedem praescribantur litterae, id est C, D, M, S, hoc modo: C super centenum, D super decenum, M super unitatem, sive ut monos designet, sive ut mille significet; cui supponatur S singulare significans" (Bakhouche, 'Bernelin', p. 22).

[13] Bernelinus gives the forms without any hypothesis about their origins.

[14] Bakhouche mentions 11 manuscripts ('Bernelin', p. 11). I know of 17 manuscripts, 15 of these of the 11th or 12th century.

[15] *Geometry II* is edited from the 22 known manuscripts in Folkerts, 'Geometrie II'. The question of the sources of the part on the abacus is treated on pp. 83–94.

[16] The illustrations of the abacus and of the forms of the numerals in the text of all manuscripts are reproduced in Folkerts, 'Geometrie II', plates 1–21.

[17] For instance in the texts of Gerland of Besançon and Radulph of Laon (see below).

the 10th or to the 11th century. These show close connections with the writings of Bernelinus and Gerbert. On two of these we find the numerals in the Western Arabic forms and on one the names associated with them; further, a verse occurs twice which connects the symbols on the abacus with Gerbert. The manuscripts are: Bern, Burgerbibliothek 250; Paris, BnF, lat. 8663; Vatican, lat. 644. In the following I shall consider the contents and origin of these manuscripts and then turn to the illustrations of the abacus which they contain[18].

Codex Bern 250 (*B*) begins with the picture of the counting-board (f. 1r; see Fig. 4). This is followed by the *Calculus* of Victorius, i. e. a series of tables for multiplying Roman numerals and fractions, and by numerous appendices – such appendices were added until the time of Odo of Cluny[19]. There follow *computus* texts by Abbo of Fleury and others. The codex is variously dated to the 10th or even the 9th century[20]; the picture of the counting-board could well stem from the end of the 10th century.

Manuscript Paris, Bibliothèque nationale de France, lat. 8663 (*P*) was probably written at the beginning of the 11th century, apparently in Fleury[21]. It is in three parts. The last, which begins on folio 47, contains, first, some abacus texts by Gerbert and Herigerus[22] and then, on fol. 49v, the representation of a counting-board (Fig. 5). There follow further texts on astronomy, music theory and metrology[23].

---

[18]For details see Folkerts, 'Darstellungen'.

[19]For the content of these additions see Friedlein, 'Calculus', pp. 49–53. Some of these additions are edited in Arrighi, 'Sulla matematica', pp. 262–268.

[20]*B* is described in Hagen, 'Catalogus', pp. 285f., and in Cordoliani, 'Manuscrits'. Cordoliani (p. 135) dates the manuscript to the 10th century and connects it closely to Fleury. As to Victorius' *Calculus* in manuscript *B* and to the parallel manuscript Basel, University Library, O.II.3, which appears to have been written in Fulda, see Bischoff, 'Studien', pp. 55–72 (here: p. 64). Bischoff assumes that the part that contains Victorius' work was written as early as the second quarter of the 9th century.

[21]See the excellent description in Vernet, 'Notes', pp. 41–44.

[22]f. 47r–v: Gerbert, *Regulae de numerorum abaci rationibus* (ed. in Bubnov, 'Gerberti Opera', pp. 6,3–22,13); f. 47v–48r: Heriger of Lobbes, *Regulae de numerorum abaci rationibus* (ed. in Bubnov, 'Gerberti Opera', pp. 208,4–209,11); f. 48r–49r: commentary of Gerbert's *Regulae*, fragment of part III (ed. in Bubnov, 'Gerberti Opera', pp. 262,5–268,6); f. 49r: *Gerberti regulae de numerorum abaci rationibus*, interpolated version, chapters 3a, 9a, 5a, 6a, 9bis a (ed. in Bubnov, 'Gerberti Opera', pp. 13f., 19f., 15–17, 20f.).

[23]f. 50r: Gerbert's letter to Brother Adam on the change of the length of day in the course of a year, edited by Bubnov, 'Gerberti Opera', pp. 39–41; f. 50r–57r: shorter texts on the theory of music; f. 57v: lists of measures of weights and volumes and of fractions.

Codex Vaticanus latinus 644 (*V*), of unknown provenance, was written in the 10th century; only the last three folios could come from the 11th century[24]. The biggest part of the codex includes writings by Bede[25]. There follow short texts on *computus* and astrology[26]. On the last three folios are the drawing of an abacus (f. 77v–78r; see Fig. 6–7) and some short texts on metrology and on calculation on the abacus[27].

The drawings of the abacus in the three manuscripts *B*, *P* and *V* have both similarities and differences. I confine myself here to the forms and names of the numerals on them.

In *B* the forms of the numerals are not explicitly exhibited. There is a connection to Gerbert through a one-line hexameter at the very top of the drawing: "Gerbertus Latio numeri abacique figuras" (Gerbert gave Latium [i. e. the Latin world] the numbers and the figures of the abacus). The representation of the abacus in *P* is, in comparison to *B*, much simplified. It is clear that the man who drew the counting-board did not understand it. The hexameter found in *B* is not in *P*. On the other hand, the forms of the Hindu-Arabic numerals from 1 to 9 are present in *P*; they are written in the first column of every group of three. As in *P*, the picture of the abacus in *V* was clearly drawn by someone who did not understand it, though the artist evidently took a great deal of trouble with it. Near the top of the page the Hindu-Arabic numerals are displayed; over each is its name (*igin* etc.), and over this is the Roman numeral indicating the denomination of the column in which they stand. In the same square of the table as the Hindu-Arabic "9" there is also

---

[24] The manuscript is described in *Codices* 1902, pp. 495f.

[25] *De natura rerum, De temporibus, De temporum ratione.*

[26] On f. 75v–77r. They concern, *inter alia*, the limits of the date of Easter and for other movable feasts and also the so-called "Egyptian days", which were supposed to bring bad luck.

[27] On f. 78v: a verse in 20 hexameters on the pound and its parts (*De libra et partibus eius*, ed. in Riese, 'Anthologia latina', pp. 206f., no. 741); four lines from the work *De minutiis* of an unknown author (III,3 end – III,5; ed. in Bubnov, 'Gerberti Opera', pp. 242,1–244,8); a reworking of Heriger's treatise on the abacus (*Ratio numerorum abaci secundum Herigerum*, chapter 1; ed. in Bubnov, 'Gerberti Opera', pp. 221,20–222,3); a text on simple and compound division beginning with "Simplex divisio cum differencia taliter fit" (edited in Arrighi, 'Sulla matematica', p. 273; there is an earlier edition based upon manuscript Munich, Clm 14689, f. 120v, in Treutlein, 'Intorno', p. 629). On f. 78r–79r: another text on division by fractions (*Racio dividencium minuciarum. De simplici et singulari minucia*). The text begins with "Si quelibet minucia assis, unciae aut scripuli in divisionibus dividens sola et simpliciter apponatur". It is edited in Arrighi, 'Sulla matematica', pp. 273f.; there is an earlier edition based on manuscript Munich, Clm 14689, f. 121r–v, in Treutlein, 'Intorno', pp. 629f.

a circle to which the name *sipos* is attached. We find in *V* the same hexameter as in *B*: "Gerbertus latio numeros abacique figuras".

On the basis of this description we can make some deductions. The drawings of the counting-board in manuscripts *B* and *P* show their common origin: common to them are the subdivision in groups of three by different-sized semicircles and the labels of the individual columns of a group through the letters *M*, *D* and *C*. The counting-board in *V* also appears to belong to this tradition: the hexameter about Gerbert connects it with *B*, the Hindu-Arabic numerals connects it with *P*.

There are several good reasons to believe that these pictorial representations describe the same type of counting-board as the texts of Gerbert, Abbo and Bernelinus:

- The description of the counting-board by Bernelinus agrees in every detail with the board in *B* and in most of the characteristics also to that in *P*. Therefore it is very probable that the drawing in *B* represents the counting-board described by Bernelinus.
- On the board in *B*, just as in *V*, there is the verse describing Gerbert as the bringer of numbers and figures to the Latin world. It is very likely that these *figure abaci* referred to the Western Arabic numerals, because they are present in the pictures in *P* and *V*.
- Since the counting-board of *P* consists of 27 columns and we know from Richer that Gerbert's abacus was also divided in this way, it seems likely that the drawing in *P* is closely related to Gerbert's abacus.
- We know that Gerbert dedicated his treatise on calculation on the abacus to Constantinus of Fleury. Therefore it was known in Fleury. Certainly Abbo knew it. How far these drawings may be connected with Abbo it is difficult to say[28]. We may mention that manuscript *P* was most probably, and *B* possibly, written in Fleury and that in *B* there are two writings by Abbo: on the *computus* and his commentary to Victorius' *Calculus*.

Even if some questions remain open, there is no doubt that the manuscripts mentioned contain a pictorial representation of the abacus as it was used about the year 1000. They broaden our knowledge of the calculation on the counting-board in the time of Gerbert, Abbo and Bernelinus. The fact that there appear in these illustrations not Greek, but the Western Arabic, numerals makes it very probable that Gerbert

---

[28]The fact that Gerbert and Abbo do not mention each other in their writings is mostly to be explained by their bad relations; see Lindgren, 'Gerbert', pp. 49f.

himself had used Arabic and not Greek numerals as markings of the counters.

There are at least three further illustrations of a counting-board on which Western Arabic numerals are marked, all in manuscripts from the 11th or early 12th century: Rouen, Bibliothèque Municipale, Ms. 489; Oxford, St. John's College, Ms. 17; and Luxembourg, Bibliothèque Nationale, Ms. 770. The picture of the abacus in codex Rouen 489 (Fig. 8) appears within a number of texts on calculation on the counting-board[29]. In the Oxford manuscript the abacus is drawn immediately after the table for multiplying Roman fractions to be ascribed to Hermannus Contractus[30]. In the picture (Fig. 9), which has considerable aesthetic qualities, the names of the numerals and both Greek and Western Arabic forms appear. The Rouen manuscript also has the names and both forms. The abacus table in the manuscript now in Luxembourg is on a single sheet which was used as a pastedown in the binding of a large Echternach bible and thus seems to originate from Echternach. The sheet must have been written before 1081. It is considerably larger than any of the other tables (603 × 420 mm) and is the only one which was drawn on a detached sheet and not included in a manuscript; thus, it seems to have been used for demonstration, for teaching. There are neither the names of the numerals nor the Greek letters representing them, but we have the Arabic numerals in their Western form[31].

4   The names of the numerals in the Gerbert tradition

I have said that on some illustrations of the abacus not only are the new Hindu-Arabic numerals to be found, but also their names. I should now like to say something about these names, what they mean and where they are found. The usual names are (the numbers in brackets indicate the numerical value):

---

[29] On f. 59r–67r are two anonymous texts on division on the counting-board. f. 67r–68r is empty. The picture of the abacus is on f. 68v–69r. On f. 72v–73v there are parts of Gerbert's treatise on calculation with the abacus and of a commentary on it; see Bubnov, 'Gerberti Opera', p. LXXXV.

[30] On f. 48v–49r; see Yeldham, 'Fraction tables'. For a reproduction of the illustration of the abacus see Evans, 'Difficillima', pp. 22f., with explanations on p. 24. Evans (1979, p. 82) thinks it possible that this illustration is connected with the treatise on the abacus by Radulph of Laon.

[31] On this table see the forthcoming article by Ch. Burnett in *SCIAMUS*.

igin (1)	*calc(t)is* (6)
andras (2)	*zenis* (7)
ormis (3)	*temenias* (8)
arbas (4)	*celentis* (9)
quimas (5)	*sipos.*

The word *sipos* and the corresponding form of the numeral requires special mention. It looks like a modern zero, but it has another function: on the counting-board one does not need a zero, for where we would write a zero, one can simply leave the column empty. The word *sipos* and the circular sign which it denotes stand for the marker which may be put during the calculation anywhere on the abacus to mark the place. The name *sipos* probably comes from the Greek word for counter, *psēphos*[32].

There have been several attempts, some quite curious, to derive the etymology of these words. L. P. Sédillot and other French scholars of the 19th century were of the opinion that these names were scrambled Arabic words; A. J. H. Vincent believed he could show that there were Greek and Hebrew influences and claimed the system was Alexandrian[33]. M. Cantor maintained that the names *arbas*, *quimas* and *temenias* were "recognized by all scholars as Semitic"; for the remaining words he suggested a number of possible hypotheses[34]. Bubnov thought that of the names of the numerals "more than half" were "Semitic, but not necessary Arabic", and that it was likely "that the origin of the names of the numerals is to be found in the land of the Euphrates and Tigris, and thus in ancient Chaldaea"[35].

Another position was taken by J. Ruska[36]. He thought he could prove that the names for the numerals 1 to 9 *must* be derived from Arabic, and tried to show that all names *could* be explained as Arabic, if one admits some changes of place or switches in value. But a glance at the words seems to contradict this theory: only the words for 4 and 8 show evident similarity (4: *arbas*, Arabic *arba'a*; 8: *temenias*, Arabic *tamāniya*).

---

[32]For the special function of this counter see also the remarks by Evans, 'Difficillima', pp. 36f; she also discusses its possible meaning as zero.

[33]Vincent, 'Premier supplément'. For the two opinions see the explanation in Olleris, 'Oeuvres', pp. 579–581.

[34]Cantor, 'Vorlesungen', pp. 894–897.

[35]Bubnow, 'Selbstständigkeit', pp. 63–68. On the occurence and possible origins of this name see the numerous books and articles by Bubnov in Russian on the origin and history of the Western numerals (see the bibliography in Lattin, 'Origin', pp. 181f.).

[36]"Die Namen der arabischen Ziffern", in Ruska, 'Algebra', pp. 82–92.

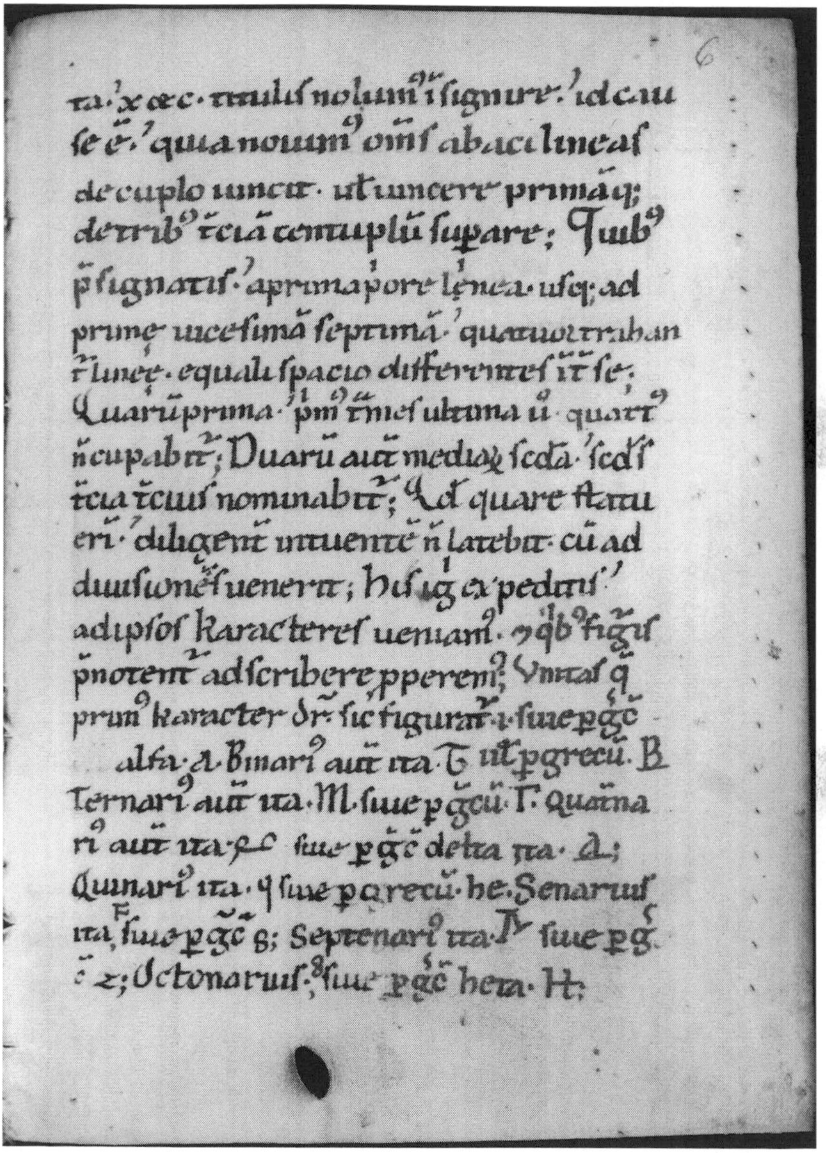

ta. x c̄c̄. titulis nolum̄ ī signur̄. id c̄au
se ē. quia nouum oīs abaci lineas
decuplo uincit. ut uincere prima q;
de trib̄ tcia centuplu supare; Tub9
p̄signatis. a prima pore l̄inea. usq; ad
prime uicesimā septimā. quatuor trahan
ī linee. equali spacio differentes īt se;
Quaru prima. p̄m ēmes ultima ū quar̄
n̄cupab̄r; Duaru aut mediaq secda. secd̄s
tcia tcius nominab̄r; Ad quare statu
eri. diligent intuente n̄ latebit. cu ad
diuisiones uenerit; his iq expeditis
ad ipsos karacteres ueniam. 9 qb9 figu̅s
p̄notem̄ ad scribere p̄perem; Vnitas q̄
prim9 karacter d̄r sic figura ·1· siue p̄g̅c̅ ·
   alfa. A. Binari aut ita. б ut p̄grecū. B.
Ternari aut ita. M. siue p̄ग̅cū Γ. Quatna
ri aut ita p̄ siue p̄g̅c̅ delta ita. A;
Quinari ita. q siue p̄ crecū. he. Senarius
ita. siue p̄g̅c̅ ʒ; Septenari ita. Jʸ siue p̄g̅.
ē ʒ; Octonarius. siue p̄g̅c̅ heta. H;

Fig. 1: Hindu-Arabic and Greek numerals in Bernelinus' treatise. (Berlin, Staatsbibliothek, lat. oct. 162, f. 6r.)

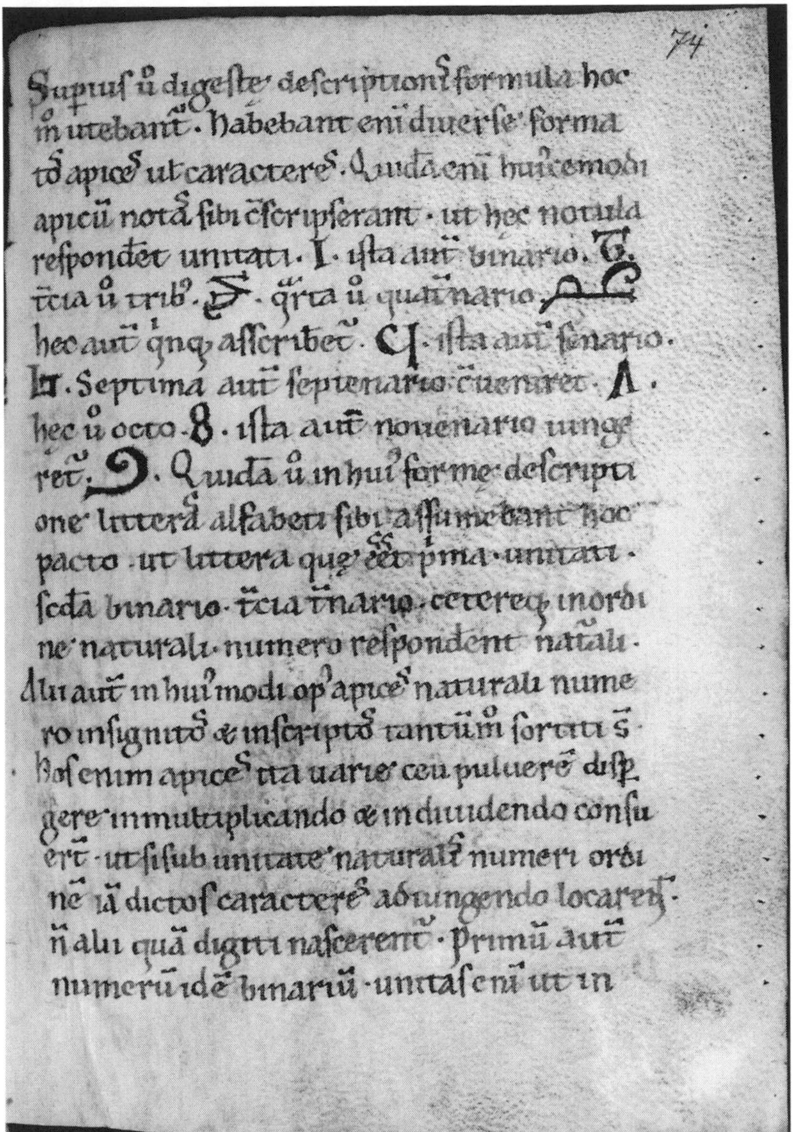

Fig. 2: Hindu-Arabic numerals in the text of Ps.-Boethius, *Geometry II*. (Berlin, Staatsbibliothek, lat. oct. 162, f. 74r.)

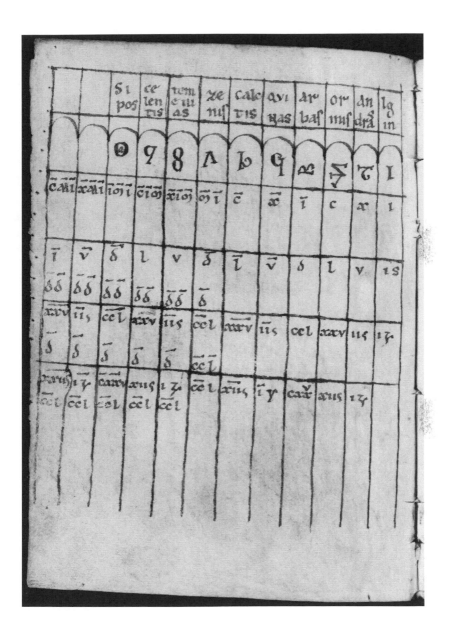

Fig. 3: Hindu-Arabic numerals and their names in the abacus table of Ps.-Boethius, *Geometry II*. (Berlin, Staatsbibliothek, lat. oct. 162, f. 73v.)

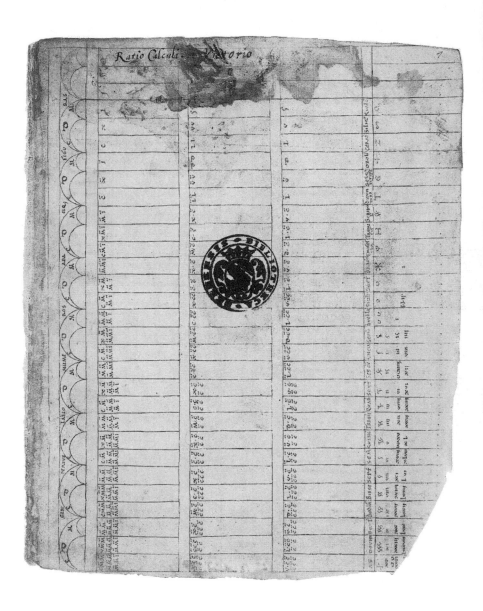

Fig. 4: Illustration of a Gerbertian abacus. On the top is the hexameter: "Gerbertus Latio numeros abacique figuras". (B = Bern, Burgerbibliothek, Ms. 250, f. 1r.)

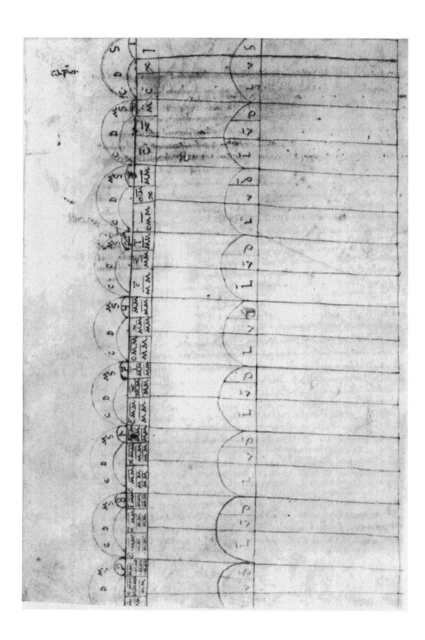

Fig. 5: Illustration of an Gerbertian abacus, with Hindu-Arabic numerals. (*P* = Paris, BnF, lat. 8663, f. 49v.)

Fig. 6 + 7: Illustration of a Gerbertian abacus, with Hindu-Arabic numerals, their names and the hexameter: "Gerbertus Latio numeros abacique figuras". (*V* = Vatican, lat. 644, f. 77v–78r.)

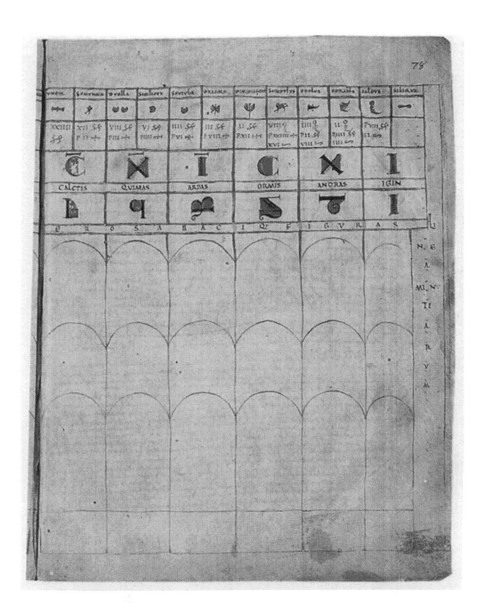

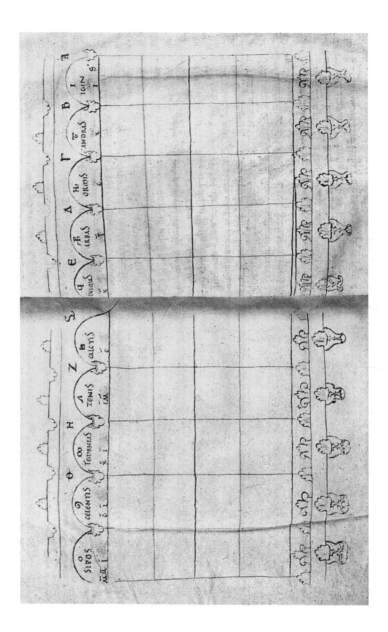

Fig. 8: Illustration of a Gerbertian abacus, with Greek and Hindu-Arabic numerals and their names. (Rouen, Bibliothèque Municipale, Ms. 489, f. 68v–69r.)

Fig. 9: Illustration of a Gerbertian abacus, with Greek and Hindu-Arabic numerals and their names. (Oxford, St. John's College 17, f. 48v–49r.)

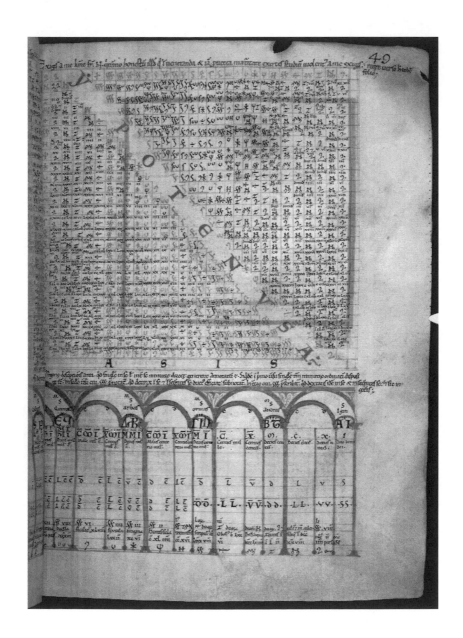

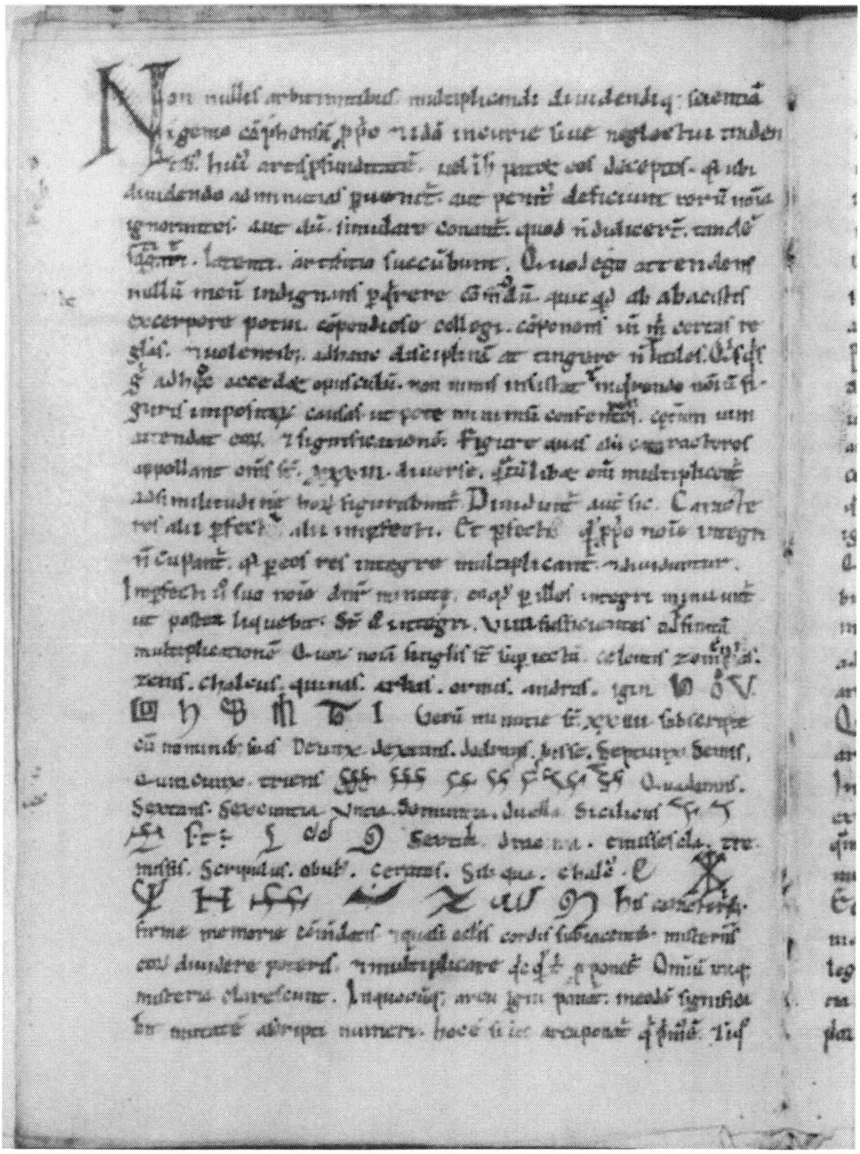

Fig. 10: Hindu-Arabic numerals and their names in the text of Gerland's abacus treatise. (Leiden, University Library, Voss. lat. oct. 95, f. 6v.)

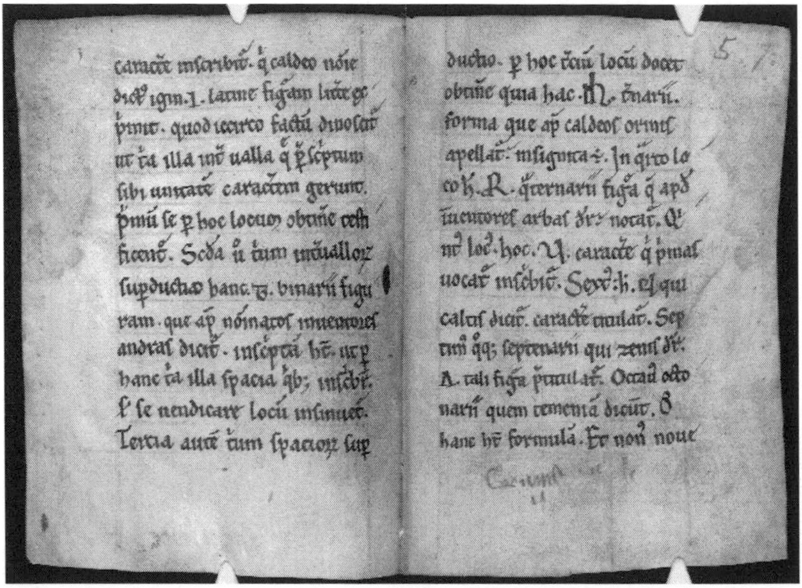

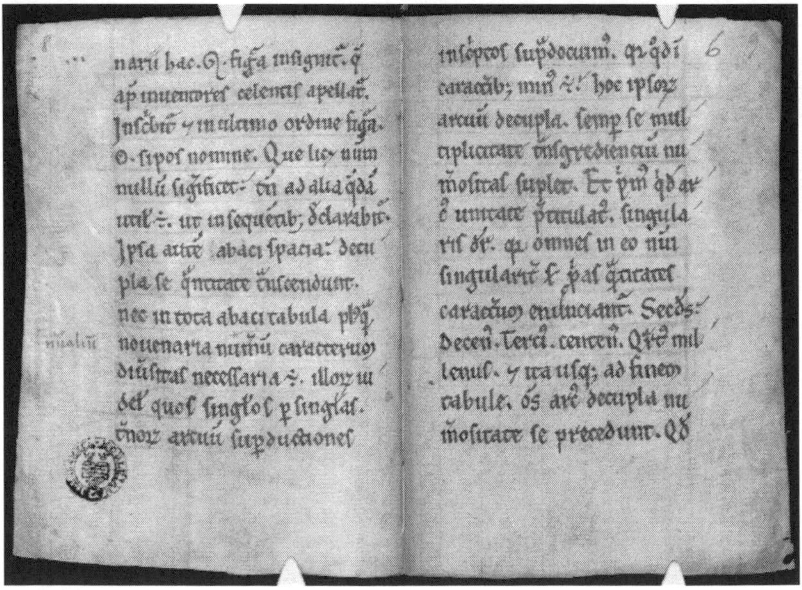

Fig. 11 + 12: Hindu-Arabic numerals and their names in the text of Radulph's abacus treatise. (Oxford, Bodleian Library, Hatton 108, f. 4v–6r.)

The word for 5 (*quimas*) can also be derived from the Arabic equivalent (*khamsa*), if *quimas* was pronounced as *kimas*.

P. Kunitzsch also thinks that the derivation of the names of the numerals 4, 5 and 8 from the Arabic is probable[37]. He points out that the word *igin* for 1 is strongly reminiscent of the Berberish word for "one"[38]. He thinks that there is no certain etymology of the remaining names[39].

Despite the lack of certainty we can agree with Ruska's statement[40]:

> There seems to remain the assumption that a manuscript transmission, in itself unreliable, was passed on early to a place where there was no understanding of Arabic – where the words were not understood as Arabic words for numerals, but for foreign names of the symbols. From that original text the literary curiosity must have spread: the introduction of *Arabic* names for numerical symbols which could be, and were, just as well named in Latin remains a literary curiosity without any meaning of substance.

5 Texts in which the names of the numerals were used

Now to the use of these names. They do not occur in the earliest writings, about 1000 A.D., on calculation on the abacus, i. e. in the treatises by Gerbert, Heriger and Bernelinus. Even in the writings of Laurentius of Amalfi and Hermannus Contractus, which were a few decades later, we still do not find these names. As I have mentioned, we do find them on certain pictorial representations of the abacus in some manuscripts of the 11th century: manuscripts Vatican, lat. 644; Rouen 489; and Oxford, St. John's College 17. The names are also to be found in the picture of an abacus in the so-called *Geometry II* of Pseudo-Boethius. Accordingly the names were already in use before 1050. We should notice that in

---

[37] Private communication.

[38] In Berberish there is no single language of the educated, but only dialects. Other pronunciations of *igin* are *yun*, *yèin* and *yau* (private communication of P. Kunitzsch).

[39] Kunitzsch also makes the following remark (private communication): it is noticeable that several of these names end in -*as* (2, 4, 5, 8); of these the names for 4 and 8 (possibly also of 5) go back to Arabic number words ending in -*a*. It is not impossible that Byzantine-Greek influence, genuine or simulated, is responsible for this. Words similarly derived from Arabic words ending in -*a* but carrying the ending -*as* are to be found in the literature of chronology of the 9th century and later: for example, *Mauhias* and *Abdelas*.

[40] Ruska, 'Algebra', pp. 90f.

Geometry II the names of the numerals are found only on the illustration, but not in the text, where the Latin names and the corresponding Hindu-Arabic symbols are given[41]. We may suppose that these names were taken over from a picture of the abacus – such as those mentioned above.

The earliest known writings on the abacus in which the names of the Hindu-Arabic numerals appear in the text itself are those of Gerland of Besançon and Radulph of Laon. Gerland's work was written between 1081 and 1084 and was much copied[42]. He explained in his clear and detailed treatise exactly how the abacus looked and how it was used. Right at the beginning he introduced the nine Hindu-Arabic numerals with the names belonging to them (Fig. 10) and used these names freely for the numerals when describing the calculations.

Radulph of Laon was a younger contemporary of Gerland; he probably died in 1133[43]. In his treatise on the abacus[44] Radulph goes into great detail in describing the form of the counting-board, the representation of numbers and calculation on the counting-board. His abacus is – like Gerbert's – divided into 27 columns. Radulph used the Hindu-Arabic numerals to mark the counters and denoted them by the names listed above (Fig. 11–12). He said that they were derived from the Chaldaeans[45]. In contrast to Gerland, Radulph did not use the above-listed names in his description of calculation, though he did use the Hindu-Arabic numerals. In multiplication Radulph also used the tenth symbol, which he represented as a small circle with a dot in the middle and which he called *sipos* or *rotula*; he described in detail its use as a marker in complex multiplications in order to avoid mistakes[46].

In the early 12th century there are two other treatises on calculation with the abacus which show the knowledge of the names of the numerals. Both were written by Englishmen. Thurkil (Turchillus), a member of the

---

[41]Folkerts, 'Geometrie II', p. 139, lines 456–459.

[42]For the life and work of Gerland see e.g. Borst, 'Zahlenkampfspiel', p. 111, and the literature cited there. Gerland's text was edited by Treutlein (1877, pp. 595–607). Besides the manuscript used by Treutlein there are 15 others which transmit the whole or part of the text.

[43]For Radulph see e.g. Evans, 'Schools', pp. 71f., 81–85.

[44]Edited in Nagl, 'Radulph', pp. 96–133.

[45]Nagl, 'Radulph', p. 97: "qui chaldeo nomine dictus igin" and "quae apud chaldeos ormis appellatur".

[46]Nagl, 'Radulph', p. 97: "quae, licet numerum nullum signitet, tantum ad alia quaedam utilis, ut in sequentibus declarabitur". The application during multiplication is explained on pp. 107–110.

royal court, wrote the first, before 1117[47]. Like Gerland and Radulph, Thurkil used the names of the Hindu-Arabic numerals in the text to denote the corresponding values of the numbers; he said that they came from the Arabic[48]. Much better known than Thurkil is his contemporary Adelard of Bath, who likewise wrote a treatise on calculation on the counting-board[49]. Since Adelard was at some time in France, and possibly in Laon, it is probable that he knew Radulph's treatise. In contrast to Gerland, Radulph and Thurkil, Adelard does not mention the names of the Hindu-Arabic numerals at the beginning of his work. But he must have known them because he used two of them: *zenis*[50] and *siposcelentis*[51]. *Siposcelentis* here has the same meaning as *sipos* in the writings of other authors; for instance, Adelard used the corresponding counter to mark the zeros in representing the number 90707 on the abacus[52]. The portmanteau word made from the names for 9 and the marker-counter can be easily explained: on the pictures of the abacus which contain these names the word *sipos* always stands to the left of the word *celentis*, and the distance between the two is sometimes very small.

In the 11th and early 12th centuries the names for the new numerals occur not only in the large works already mentioned, but also in various other places. I will not here list the numerous scattered small texts in question that are known to me[53]. A curiosity is a poem in ten hexameters which is exclusively devoted to these names. It is transmitted in 13 manuscripts and was apparently written in France or England before 1100[54]. Probably these verses were written to help in the process of learning the strange names.

To sum up: The names for the nine units and for the marker-counter first appear in isolated pictorial representations of the abacus from the 11th century. The oldest document where these names are in an illus-

---

[47]Edited in Narducci, 'Intorno', pp. 135–154. For Turchillus see Haskins, 'Studies', pp. 327–331.

[48]"Has autem figuras ... a pytagoricis habemus, nomina vero ab arabibus": Narducci, 'Intorno', p. 136.

[49]Edited in Boncompagni, 'Intorno'. See also Burnett, 'Diligentia', pp. 227–229.

[50]"... zenis vero caracterem": Boncompagni, 'Intorno', p. 99,29. In Boncompagni's edition the text is wrongly emended into "Eius ... caracterem".

[51]E. g. Boncompagni, 'Intorno', pp. 97,13; 97,31; 99,6f.; 99,28; 130,32.

[52]"Adde ... deceno et milleno vacantibus siposcelentis caracterem": Boncompagni, 'Intorno', p. 99,6f.

[53]These are mentioned in Folkerts, 'Benennungen', pp. 228–230.

[54]Critical edition and comments about the supposed origin in Folkerts, 'Benennungen', pp. 223–228.

tration which is integrated into a text is the so-called *Geometry II* of Pseudo-Boethius. In both cases the names are immediately associated with the corresponding Hindu-Arabic numerals. A little later the names appear in sundry texts, sometimes in connection with tables. In 1080 at the latest the names were written in texts, for instance in Gerland's. They were also mentioned by later authors like Radulph and Turchillus, though not by all writers on calculation on the abacus. The names were the subject of a ten-hexameter verse, apparently written shortly before 1100. With the disappearance of this form of the abacus in the first half of the 12th century the names of the numbers disappeared too.

Bibliography

G. Arrighi, "Sulla matematica attorno all'anno Mille", *Rendiconti Accademia Nazionale delle Scienze detta dei LX, Memorie di Matematica* 102, vol. 8 (1984), 261–274.

B. Bakhouche, *Bernelin, élève de Gerbert d'Aurillac. Livre d'abaque. Texte latin établi, traduit et annoté. Notes complémentaires de J. Cassinet*, Pau 1999.

W. Bergmann, *Innovationen im Quadrivium des 10. und 11. Jahrhunderts. Studien zur Einführung von Astrolab und Abakus im lateinischen Mittelalter*, Stuttgart 1985.

B. Bischoff, *Mittelalterliche Studien*, vol. 3, Stuttgart 1981.

B. Boncompagni, "Intorno ad uno scritto inedito di Adelardo di Bath intitolato 'Regule abaci'", *Bullettino di bibliografia e di storia delle scienze matematiche e fisiche* 14 (1881), 1–90.

A. Borst, *Das mittelalterliche Zahlenkampfspiel*, Heidelberg 1986.

N. Bubnov, *Gerberti postea Silvestri II papae Opera Mathematica (972–1003)*, Berlin 1899; reprint Hildesheim 1963.

N. Bubnow, *Arithmetische Selbstständigkeit der europäischen Kultur. Ein Beitrag zur Kulturgeschichte*, Berlin 1914.

Ch. Burnett, "Algorismi vel helcep decentior est diligentia: the Arithmetic of Adelard of Bath and his Circle", in M. Folkerts (ed.), *Mathematische Probleme im Mittelalter: der lateinische und arabische Sprachbereich*, Wiesbaden 1996, pp. 221–331.

M. Cantor, *Vorlesungen über Geschichte der Mathematik. Erster Band: Von den ältesten Zeiten bis zum Jahre 1200 n. Chr.*, 3rd edition, Leipzig 1907.

W. von Christ, "Über das argumentum calculandi des Victorius und dessen Commentar", *Sitzungsberichte der königl. bayer. Akademie der Wissenschaften zu München, Jahrgang 1863*, vol. I, pp. 100–152.

Codices Vaticani Latini, vol. 1, Città di Vaticano 1902.

A. Cordoliani, "Les manuscrits de la bibliothèque de Berne provenant de l'abbaye de Fleury au XIe siècle. Le comput d'Abbon", *Zeitschrift für Schweizerische Kirchengeschichte* 52 (1958), 135–150.

G. R. Evans, "Difficillima et Ardua: theory and practice in treatises on the abacus, 950–1150", *Journal of Medieval History* 3 (1977), 21–38.

G. R. Evans, "Schools and scholars: the study of the abacus in English schools c. 980 – c. 1150", *The English Historical Review* 94 (1979), 71–89.

M. Folkerts, *"Boethius" Geometrie II, ein mathematisches Lehrbuch des Mittelalters*, Wiesbaden 1970.

M. Folkerts, "Frühe Darstellungen des Gerbertschen Abakus", in R. Franci, P. Pagli, L. Toti Rigatelli (ed.), *Itinera mathematica. Studi in onore di Gino Arrighi per il suo 90° compleanno*, Siena 1996, pp. 23–43.

M. Folkerts, "Frühe westliche Benennungen der indisch-arabischen Ziffern und ihr Vorkommen", in M. Folkerts, R. Lorch (ed.): *Sic itur ad astra. Studien zur Geschichte der Mathematik und Naturwissenschaften. Festschrift für den Arabisten Paul Kunitzsch zum 70. Geburtstag*, Wiesbaden 2000, pp. 216–233.

G. Friedlein, "Der Calculus des Victorius", *Zeitschrift für Mathematik und Physik* 16 (1871), 42–79.

H. Hagen, *Catalogus codicum Bernensium*, Bern 1875.

Ch. H. Haskins, *Studies in the History of Mediaeval Science*, Cambridge / Mass. 1924.

H. P. Lattin, "The origin of our present system of notation according to the theories of Nicholas Bubnov", *Isis* 19 (1933), 181–194.

U. Lindgren, *Gerbert von Aurillac und das Quadrivium*, Wiesbaden 1976.

A. Nagl, "Der arithmetische Tractat des Radulph von Laon", *Abhandlungen zur Geschichte der Mathematik* 5 (1890), 85–133.

E. Narducci, "Intorno a due trattati inediti d'abaco contenuti in due codici Vaticani del secolo XII", *Bullettino di bibliografia e di storia delle scienze matematiche e fisiche* 15 (1882), 111–162.

F. Newton (ed.), *Laurentius Monachus Casinensis, Archiepiscopus Amalfitanus, Opera*, Weimar 1973.

A. Olleris, *Oeuvres de Gerbert*, Clermont-Ferrand / Paris 1867.

A. Riese, *Anthologia latina sive poesis latinae supplementum*, Part I.2, Leipzig 1870.

J. Ruska, *Zur ältesten arabischen Algebra und Rechenkunst*, Heidelberg 1917. (Sitzungsberichte der Heidelberger Akademie der Wissenschaften, Philosophisch-historische Klasse, Jahrgang 1917, 2. Abhandlung.)

VI

P. Treutlein, "Intorno ad alcuni scritti inediti relativi al calcolo dell'abaco", *Bullettino di bibliografia e di storia delle scienze matematiche e fisiche* 10 (1877), 589–656.

J. Tropfke, *Geschichte der Elementarmathematik. 4. Auflage. Band 1: Arithmetik und Algebra. Vollständig neu bearbeitet von K. Vogel, K. Reich, H. Gericke*, Berlin / New York 1980.

M.-Th. Vernet, "Notes de Dom André Wilmart † sur quelques manuscrits latins anciens de la Bibliothèque Nationale de Paris (fin)", *Bulletin d'Information de l'Institut de Recherche et d'Histoire des Textes* 8 (1959), 7–45.

A. van de Vijver, "Les oeuvres inédites d'Abbon de Fleury", *Revue Bénédictine* 47 (1935), 125–169.

A. J. H. Vincent, "Premier supplément à la note G. sur l'origine de nos chiffres", *Notices et extraits des manuscrits de la Bibliothèque du Roi et autres bibliothèques* 16/2 (1847), 143–150.

F. A. Yeldham, "Fraction Tables of Hermann Contractus", *Speculum* 3 (1928), 240–245.

Illustrations:

Fig. 1: Hindu-Arabic and Greek numerals in Bernelinus' treatise. (Berlin, Staatsbibliothek, lat. oct. 162, f. 6r.)

Fig. 2: Hindu-Arabic numerals in the text of Ps.-Boethius, *Geometry II.* (Berlin, Staatsbibliothek, lat. oct. 162, f. 74r.)

Fig. 3: Hindu-Arabic numerals and their names in the abacus table of Ps.-Boethius, *Geometry II.* (Berlin, Staatsbibliothek, lat. oct. 162, f. 73v.)

Fig. 4: Illustration of a Gerbertian abacus. On the top is the hexameter: "Gerbertus Latio numeros abacique figuras". (B = Bern, Burgerbibliothek, Ms. 250, f. 1r.)

Fig. 5: Illustration of an Gerbertian abacus, with Hindu-Arabic numerals. (P = Paris, BnF, lat. 8663, f. 49v.)

Fig. 6 + 7: Illustration of a Gerbertian abacus, with Hindu-Arabic numerals, their names and the hexameter: "Gerbertus Latio numeros abacique figuras". (V = Vatican, lat. 644, f. 77v–78r.)

Fig. 8: Illustration of a Gerbertian abacus, with Greek and Hindu-Arabic numerals and their names. (Rouen, Bibliothèque Municipale, Ms. 489, f. 68v–69r.)

Fig. 9: Illustration of a Gerbertian abacus, with Greek and Hindu-Arabic numerals and their names. (Oxford, St. John's College 17, f. 48v–49r.)

Fig. 10: Hindu-Arabic numerals and their names in the text of Gerland's abacus treatise. (Leiden, University Library, Voss. lat. oct. 95, f. 6v.)

Fig. 11 + 12: Hindu-Arabic numerals and their names in the text of Radulph's abacus treatise. (Oxford, Bodleian Library, Hatton 108, f. 4v–6r.)

Note:

For the purposes of this reprinting, figures 1–12 are reproduced between pages 10–11 of this chapter.

VII

THE IMPORTANCE OF THE PSEUDO-BOETHIAN *GEOMETRIA*
DURING THE MIDDLE AGES

Compared to the other writings of Boethius on the *Trivium* and *Quadrivium*, his *Geometria* takes a special place. We do not have this work in its original form, but only in two later adaptations: both contain only part of the original but on the other hand they are enlarged through a variety of insertions. Therefore any study of the importance of Boethius' *Geometria* in the Middle Ages should not only try to show the influence of geometrical writings which were transmitted under the name of Boethius, but also should try to understand the origin of such compilations. From these two demands, then, arises the organization of my essay: the first section will be concerned with the scanty evidence known about the authentic *Geometria* as well as with the contents and origins of both extant compilations. In the second part of my essay I will attempt to show the dissemination and impact of both writings during the Middle Ages. For this purpose I will make use of entries of the *Geometria* in medieval library catalogues and of allusions to it in other medieval writings. A broader understanding of the importance of these two writings and of their comprehension in this period can be gained from an analysis of the scholia to one of these compilations which have been neglected up to now. Such a study should make it possible to indicate the value of these treatises in comparison with the other two, authentic, works of the Boethian *quadrivium*–the *Arithmetica* and *Musica*.

I. The authentic Geometria and the origins of the first two basic texts.

1.1 The original translation; earlier fragments of uncertain origin.

Cassiodorus testifies in two places that Boethius had written a geometrical work. In a letter from Theodoric to Boethius, he says that through Boethius' translation, Euclid's work on geometry is now known in the Latin language.[1] Cassiodorus makes a similar reference to Boethius in the *Institutiones*.[2] Therefore there is no doubt that Boethius made a Latin translation of Euclid. To be sure, these testimonies give no information as to whether Boethius translated all of the 13 or 15 books of the *Elements* or only parts of it. But beyond doubt parts of the first five books belonged to that translation as surviving excerpts show (see ahead 1.2 and 1.3). Because of Boethius' approximate date of birth (c. 480) and the date of Cassiodorus' testimony, this translation must have been made in about the year 500. With this, our knowledge concerning the original is exhausted. This original must not have been in use very much after Boethius' time because existing manuscripts and the allusions of other authors are based on the two later compilations of the Boethius *Geometria*, but do not refer to the original work.

---

1   Cassiodorus, *Variae* I, 45.4: *Translationibus enim tuis Pythagoras inusicus, Ptolemaeus astronomus leguntur Itali. Nicomachus arithmeticus, geometricus Euclides audiuntur Ausonii.*
2   II, 6.3: *...ex quibus Euclidem translatum Romanae linguae idem vir magnificus Boethius edidit.*

There has been much discussion on the issue of whether two anonymous Latin Euclid fragments from the 5th and 9th centuries respectively have transmitted parts of the original Boethian translation or at least some material originating from the circle around Boethius. The older of these two texts is the famous geometry fragment now extant in the Biblioteca Capitolare of Verona, Ms. XL (38). This manuscript is a palimpsest in which the former text was replaced, in the beginning of the 8th century, by the *Moralia* of Gregory the Great. The original text, now removed, contained portions of Vergil and Livy along with a Latin translation of Euclid from which three double folios are yet decipherable.[3] The extant text contains sections of Euclid's *Elements*: parts of the propositions XI, 24-25; XII, 2-3.8; XIII, 2-3.7, though in the manuscript the books XII and XIII are marked as XIIII and XV. According to M. Geymonat who edited these texts,[4] the writing can be dated to the last years of the 5th century, that is, it is almost contemporary with the original Boethian translation. This fact more than anything else prompted M. Geymonat to conjecture that here we have part of that original translation. This attribution, however, appears to me somewhat risky because the existing manuscript contains only a few fragments–too small to draw solid, safe conclusions about the author. To attribute this translation to Boethius would mean that he prepared it before his 20th birthday.[5]

The second of these two fragments was written at the beginning of the 9th century in a north-east French scriptorium and is now found in the University Library at Munich (Univ. 2⁰ 757). Only two pages are extant, which concern Euclid's propositions I, 37 to 38 and II, 8 to 9. This fragmentary text has been edited by M. Geymonat, as well, based on earlier attempts of M. Curtze and A.A. Björnbo.[6] This text reveals that the translator obviously did not have control of the mathematical contents nor did he master the Latin grammar. Without understanding, the technical mathematical terms are entirely transliterated as Greek expressions having similar letters.[7] Greek letters attached to figures or used to denote the endpoints of segments are taken as number symbols and are accordingly transferred.[8] Already this astounding ignorance forbids us to attribute the text to Boethius. Such a conclusion receives further support from the fact that the only portions which are contained in the Munich fragments as well as in the texts Mc and Md (the last originating without doubt from Boethius),[9] in the Munich text deviate sharply from Mc and Md. Nevertheless Geymonat considers the Munich fragment to be connected with the Boethius translation. He is of the opinion that it goes back to a text of the end of the 5th century which was used as a basis for the true translation. This text was done by a scholar not well versed in the Greek language who came from the circle around Boethius. Perhaps this text descends from an original interlinear translation of a Greek Euclid manuscript.[10] Geymonat's theory is an interesting conjecture on the classification of this Euclid fragment. However, his opinion leaves us with no sure proof. In any case, at the very most the Munich fragment is of interest because of its possible connection with the authentic geometry, but not because of its actual influence in the Middle Ages.

---

3   ff. 331/326, 341/338, 336/343.
4   *Euclidis Latine facti fragmenta Veronensia* (Milano, 1964).
5   In this case, the Verona manuscript would be the autograph or a very early copy of it, if the date (end of the 5th century) is exact.
6   „Nuovi frammenti della geometria ‚boeziana‘ in un codice del IX secolo?" *Scriptorium*, 21 (1967), 3-16.
7   For example ἥμισυ = *nos quidem sic* (ἡμεῖς), γνώμων = *scito*.
8   For example ΔΕΖ = *quarto quinto et septimo.*
9   For Mc and Md, see the sections 1.2 and 1.3. In the Munich text there are the enunciations to I, 38 and II, 9 which also exist in Mc/Md.
10 Geymonat (note 6), pp. 7-9.

1.2 The Boethian Geometry in the 8th and 9th century. The origin of Geometria I (in 5 books).

Sure traces of the geometrical work of Boethius can first be found in the 8th and 9th centuries. From this time there are to be discovered excerpts from Euclid in three distinct writings. Parts of these conform with each other and almost certainly go back to Boethius' translation. As B.L. Ullmann has demonstrated convincingly,[11] the fate of Boethius' treatise was closely bound to Corbie, „the gromatic and geometric capital of the medieval world."[12] In the 9th century these three texts were present in that Cloister; in part they were assembled at Corbie in the form in which they came down to us. We are speaking of manuscripts of the *Corpus Agrimensorum* (Mb),[13] a version (the so-called „third" recension) of Cassiodorus' *Institutiones* (Ma)[13] and the Geometria I (*Ars Geometriae*) which has been attributed to Boethius in the manuscripts (Mc).[13]

Roman treatises that dealt with land measurements had, at an early date, already been put into a corpus. In this *Corpus Agrimensorum*, the archetype of which has been dated about 450,[14] geometrical works had a prominent part.[15] Not before the year 550, and probably much later, the gromatic texts in the *Corpus Agrimensorum* were rearranged to serve as geometrical material for the *quadrivium*.[16] This very probably happened at Corbie.[17] From this new recension of the *Corpus Agrimensorum*, two old manuscripts still exist: P (Vat. Palat. Lat. 1564) and G (Codex Guelf. 105 Gud.); both originate from the 9th century. P was probably written in France, very likely at Corbie;[18] G stems from the St. Bertin Cloister in neighboring St. Omer. Both P and G contain, among other things, fragments from Book I of Euclid[19] which, because of their similarity with the Ma and Mc text, may be said to go back to Boethius' translation of Euclid.

Euclidian excerpts from that same Boethian translation were also taken over into the text of the third recension of Cassiodorus' *Institutiones* (Ma) presumably also at Corbie.[20] This recension probably originated in the 8th century–in any case, certainly before Hrabanus Maurus.[21] In Ma, the definitions of Book I and V are present, but the enunciations are missing.[22]

Excerpts from the Boethian Euclid translation form an essential part of a treatise put

---

11   B.L. Ullman, „Geometry in the Mediaeval Quadrivium," *Studi di bibliografia e di storia in onore di Tammaro de Marinis*, 4 (Verona, 1964), 263-285.

12   Ullman, p. 283.

13   The *sigla* Ma, Mb and Mc for the Euclidian excerpts in this group go back to Bubnov. See N. Bubnov, *Gerberti Opera Mathematica* (Berlin, 1899).

14   G. Thulin, „Die Handschriften des *Corpus Agrimensorum Romanorum*," *Abhandlungen der Königlichen Preussischen Akademie der Wissenschaften*, Phil.-Hist. Classe (Berlin, 1911).

15   For example, Balbus, *Expositio et ratio omnium formarum.*

16   Ullman, p. 266.

17   Ullman, pp. 273-276.

18   Ullman, p. 274. In contrast to him, Professor B. Bischoff places P in the same scriptorium with the somewhat later agrimensorial manuscript N, Naples VA 13.

19   I def. 1-12.14.13.15-23, post. 1-5, ax. 1.3.2.7, prop. 1-3 with proof. The Mb excerpts are edited by M. Folkerts, „*Boethius*" Geometria II, ein mathematisches Lehrbuch des Mittelalters (Wiesbaden, 1970), pp. 173-217.

20   From the Ma manuscripts, Augiensis 106 indicates an origin from northeastern France; Paris, BN Lat. 12963 and 13048 also stem from Corbie. See *Cassiodori Senatoris institutiones*, edited from the manuscripts by R.A.B. Mynors (Oxford, 1937; 1961), introduction, pp. 32-33.

21   Mynors, introduction, p. 39.

22   Extant are: I def. 1-12.14.13.15-23, post. 1-5, ax. 1.3.2.7; II def. 2; V def. 1-8.11.9.10.13.12.14-16.18. 17. The text edited by Mynors was also made use of in the Folkerts edition, pp. 173-217.

together from various sources, presumably in the 8th century, also from Corbie: the so-called Geometria I, consisting of five books. This work, whose Euclidian parts will henceforth be designated as Mc, is, in the manuscripts, overwhelmingly accredited to Boethius and generally carries the title *Ars Geometriae et Arithmeticae*. The task of explaining the origins of Geometria I was essentially completed in the works of N. Bubnov,[23] C. Thulin,[24] and B.L. Ullman;[25] their results have been summarized by Menso Folkerts.[26]

Geometria I incorporates much material from the *Corpus Agrimensorum*; it was not, however, intended for use as a textbook for the student of land surveying, but rather as a mathematical schoolbook in the study of geometry and arithmetic within the *quadrivium*. This work is made up of three writings: arithmetic, gromatic, and geometric. The geometrical part consists of excerpts from Euclid's *Elements*, I-IV (without proofs); it is relatively intact.[27] This geometrical part includes books three and four in the Geometria I, but parts that originally belonged to Book IV ended up in a section marked Book V, and this because of page misplacement. These excerpts, which go back to Boethius' translation, make up the essential value of the collection. The portion on arithmetic takes up the greatest part of book two. It is arbitrarily composed of various portions of Boethius' arithmetic. The gromatic excerpts constitute Book I and part of Book V while the dialogue at the beginning of Book II also depends to some extent on the *Corpus Agrimensorum*. A second longer dialogue is found under the title *Altercatio Duorum Geometricorum de Figuris, Numeris et Mensuris* in the fifth book of Geometria I. Only recently has it been learned that this section does not go back to the original compilor of the collection but is based, instead, on Augustine's *De Quantitate Animae* and his *Soliloquia*.[28]

The Geometria I is a relatively bungling piece of work in which practically every sentence is adapted from some other known work. The compilor's main sources were: Boethius' Euclid translation, from which at least excerpts of Book I-IV were known in the 8th century; Boethius, *De Arithmetica*; a gromatic manuscript of the second class (P, G); some writings of St. Augustine. The compilor also used Isidore's *Etymologies*, Cassiodorus' *Institutiones* and Columella's work on farming.[29] One notes clearly that the compilor did not understand much of what he excerpted, and parts of the text are hopelessly corrupt. One wonders therefore all the more that the Geometria I was considered one of the most famous geometrical works in the Middle Ages up to the dissemination of the Arabic translations, to be compared only with the *Geometria* of Gerbert and its anonymous continuations, with the so-called Geometria II of the Pseudo-Boethius, and with some writings from the *Corpus Agrimensorum*.[30]

23   Bubnov, pp. 161-196.
24   C. Thulin, „Zur Überlieferungsgeschichte des Corpus Agrimensorum. Exzerptenhandschriften und Kompendien," *Göteborgs Kungl. Vetenskaps och Vitterhets-Samhälles Handlingar*, fjärde följden, XIV.1 (Göteborg, 1911).
25   See note 11.
26   M. Folkerts, „Die Altercatio in der Geometrie I des Pseudo-Boethius. Ein Beitrag zur Geometrie im mittelalterlichen Quadrivium," to appear in a volume on medieval „Fachprosa" (Berlin, 1980).
27   Extant are: I def. 1-12.14.13.15-23, post. 1-5, ax. 1.3.2.7; II def. 1.2, prop. 1; III def. 1-6.8-11; IV def. 1.2, prop. 1; III def. 6.8; I prop. 2-4. 6-8. (9). 10-18.21.23.26-28. 31-37.39-41.43.42.44-48; II prop. 1.3-6.9-12.14; III prop. 3.7 beginning. 22 end. 27.30-33; IV prop. 1-4.6.8.12.11; III prop. 7 end. 9.12.10.13. 14.16.18.19.24.22 beginning (all propositions without proof). On these excerpts see Folkerts, pp. 69-82 and the edition pp. 173-214.
28   Folkerts (see note 26).
29   Of this work only two manuscripts are known today which date before 1400; one of these originated in the 9th century at Corbie (now in Leningrad).
30   Especially Balbus as well as Epaphroditus and Vitruvius Rufus.

1.3 The Origin of the Geometria II (in 2 books) by the Pseudo-Boethius.

Besides the Geometria I, there is a second treatise which contains extensive parts of a Euclid translation and which agrees broadly with Mc and therefore must go back to the original Boethian Geometria. This is the so-called Geometria II, in two books, which in manuscript copies is generally attributed to Boethius. Since, in another publication,[31] the sources, origin, and provenience of this treatise have undergone clarification, it should suffice here merely to give a summary. This Geometria is made up of a section on geometry and a section on arithmetic. The geometrical section outweighs the rest; only at the end do both books deal with arithmetic questions, namely, with the arithmetic rules for the abacus. The author of this compilation has used two or three sources: an agrimensorial text from which he possibly took the Euclid excerpts too, and Gerbert's writing about the abacus. The first part of the Geometria II is composed of excerpts from Euclid I-IV (which are designated Md).[32] These excerpts very much resemble both the Geometria I (Mc) and the excerpts transmitted by the agrimensorial writings (Mb): Md has the same interpolations and transpositions as Mc, but since Md contains certain propositions which are absent in Mc, we may presume that the author of the Geometria II used a copy which was more complete than the Pseudo-Boethian Geometria I. Or, alternatively, the author could have used a copy of the Geometria I to which he added propositions lacking in that text but are taken from some source unknown to us and possibly, in accordance with this source, made some changes. There is also the possibility that the author used a manuscript now in Naples (N)[33] or one related to it; this manuscript contains the Geometria I and the demonstrations of Euclid I, 1-3 corresponding to Mb. It belongs to the group X[I] of the agrimensorial manuscripts, on which the other geometrical pieces of the Geometria II are based.[34] The rather unskillfully expressed mathematical rules on the abacus[35] show that for this section the compilor used the version „B" of Gerbert's treatise on the abacus, which did not originate before Gerbert's pontificate (999-1003).[36] This portion of the Geometria II is especially interesting for the history of mathematics since here for the first time a Latin mathematical text exists with early forms of our Indio-Arabic ciphers. Should these early ciphers have originated with Boethius, they would have already been known in the West around the year 500. No wonder then that the mathematical historians found interest in the Geometria II as early as the middle of the 19th century.[37]

Because of the sources used and the nature of the extant manuscripts,[38] the time of the origin of the Geometria II must be limited to the first half of the 11th century. Moreover, the work lends itself to easy localization–the B version of Gerbert's writing on the abacus was later included in the collection of the oldest abacus works, which originated about 995 in Lorraine.[39] Liège at this time was the „Lotharingian Athens" and the center for abacus studies.[40]

---

31 See Folkerts (note 19).
32 Extant are: I def. 1-12.14.13.15-23, post. 1-5, ax. 1.3.2.7; II def. 1.2; III def. 1-6.8-11; IV def. 1.2; I prop. 1-8. (9.) 10-41.43.42. 44-48; II prop. 1.3-6. 9-12.14; III prop. 3.7 beginning. 22.27.30-33; IV prop. 1-4.6.8.12.11 (all propositions without proof); further, I prop. 1-3 with proofs. See the Folkerts edition, pp. 109-135.
33 For this see p.
34 Folkerts, p. 104. Here the author deals chiefly with the calculations of areas.
35 These are at the end of Book I and II of Geometria II.
36 See Folkerts, pp. 83-94.
37 For preparatory works by other scholars, see Folkerts, p. 83.
38 The oldest stem from the middle of the 11th century. See section 2.1.
39 Bubnov, pp. 1-24; 294-296.
40 Heriger, Adelbold, Wazzo, Radulf, and Franco come from Liège. See also Bernelinus, *Liber Abaci* (ed. Bubnov, p. 383, 18-19): *Lotharienses... quos in his cum expertus sum florere.*

The above-mentioned N manuscript, or a closely related manuscript, was also located in Liège in the 11th century. Gerbert used it in 983 at Bobbio and later sent it to Adelbold of Liège.[41] Its author is, therefore, almost of necessity, a Lotharingian. This assumption is strengthened by the fact that the oldest existing manuscripts of this work stem exclusively from what are now Eastern France and Western Germany.[42]

The Geometria II in both theme and presentation fits well into the nature of the work of Lotharingian scholars of the 11th century, of whose work we can delineate a fairly clear picture for ourselves. Aside from the already mentioned writings on computations for the abacus, we are particularly well informed about their geometrical works. First of all, there is the correspondence between Regimbold and Radulph of Liège, of about 1025,[43] which concerned the sum of angles in a triangle. The solution to this problem is found in a somewhat later anonymous text.[44] Another geometrical writing from this setting is a treatise on the squaring of the circle by Franco of Liège written shortly before 1050.[45] As will be clearly demonstrated in section 2.3 below, this treatise is connected with the above mentioned correspondence and with the Geometria II. While the subject matter in the correspondence does not show familiarity with Geometria II, this Pseudo-Boethian treatise does reveal some striking similarities with the writing of Franco. We may, therefore, conclude that the Geometria II originated between 1025 and 1050.

We can now summarize what we know about the fate of Boethius' Euclid translation up to the 11th century. In two different places scholars were engaged, at different times, in this work: in the 8th century at Corbie and in the second quarter of the 11th century in Lorraine. In Corbie in a manuscript M (now lost), there existed at least the fragments of Md as well as the additional enunciations from Book III[46] and the definitions of Book V. Parts of M were then taken over into the third recension of Cassiodorus' *Institutiones* (Ma). A second copy (M_1), which possibly also originated at Corbie—which is now lost—contained at least the definitions and most of the propositions, without proof, from Books I-IV as well as the proofs to I, prop. 1-3. From this source M_1 came the Euclidian excerpts in the later recension of the *Agrimensores* collection (Mb) and a text (M_2) with the mistakes, glosses, and transpositions which are typical for the Geometria I (Mc). Both these (Mb and Mc) were put together at Corbie. Then in the 11th century an unknown Lotharingian scholar took from the original M_2 or from Mc most of the excerpts from Euclid I-IV and combined them as part of the Geometria II (Md). Indeed, he was trying to improve his original source, though with little success.

41   See Folkerts, p. 35.

42   See section 2.1.

43   Edited with commentary by P. Tannery and M. Clerval, ,,Une correspondance d'écolâtres du XIe siècle,'' *Notices et extraits des manuscrits de la Bibliothèque Nationale*, 36, no. 2 (Paris, 1901), 487-543; reprinted in P. Tannery, *Mémoires scientifiques*, 5 (Toulouse, Paris, 1922), 229-303.

44   This text was edited and classified, scientifically and historically, by J.E. Hofmann, according to manuscript 190 of Kues, ff. 1v-3r, ,,Zum Winkelstreit der rheinischen Scholastiker in der ersten Hälfte des 11. Jahrhunderts,'' *Abhandlungen der Preussischen Akademie der Wissenschaften*, math.-naturwiss. Klasse (1942), no. 8.

45   Edited by A.J.E.M. Smeur, ,,De verhandeling over de cirkelkwadratuur van Franco van Luik van omstreeks 1050,'' *Mededelingen van de Koninklijke Vlaamse Academie voor Wetenschappen, Letteren en Schone Kunsten van België*, 30.11 (Brussels, 1968). A new edition of this treatise with an English commentary by A.J.E.M. Smeur and M. Folkerts has appeared in *Archives internationales d'histoire des sciences*, 26 (1976), 59-105.225-253.

46   See note 27.

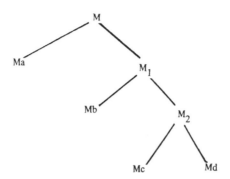

Boethius

2. The dissemination and importance of both Geometries.

2.1 The preserved and reconstructable manuscripts.

The existing manuscripts give us the first information about the dissemination of both geometries in the Middle Ages. Today there are still about 25 codices extant of each of the two writings. This is a small number when compared with the Arithmetic or Music of which more than a hundred manuscripts are at hand, yet it is significantly more than of some other important texts of late antiquity and the Middle Ages. This already shows that both Geometries were not of little importance in the Middle Ages. When one classifies the existing manuscripts according to their place of origin and time, it allows one to come to à conclusion as to when and where the manuscripts of both were especially read and copied. Fortunately today there is considerable clarity about the dating and placing of the important manuscripts, so that some secure conclusions can be made.

The so called „Geometria I" evidently blossomed rather soon after its birth, whereas it was hardly copied at all in the late Middle Ages: of the 26 existing manuscripts, 17 stem from the 9th-11th century and of those 10 were probably written before the year 1000. Three of the oldest manuscripts originated in Corbie, the home of Geometria I (Paris, BN lat. 13020, 13955, 14080). These manuscripts were written there in the 2nd or 3rd quarter of the 9th century, not around the year 1000, as it was earlier thought.[47] A codex of the 10th century found today in Prague (Universitní Knihovna, lat. 1717 = IX. C. 6) is related to these manuscripts. I know nothing of the origin of this manuscript. At Reichenau the geometrical text in the Munich codex CLM 560 was written during the 9th century.[48] This manuscript is closely

---

47 The early date has been proved by B. Bischoff in „Hadoard und die Klassikerhandschriften aus Corbie," *Mittelalterliche Studien*, Band 1 (Stuttgart, 1966), p. 60.
48 B. Bischoff, *Die südostdeutschen Schreibschulen und Bibliotheken in der Karolingerzeit*, Teil 1 (Wiesbaden, 2nd edition, 1960), p. 262, note 3.

related to a now lost codex, which was mentioned in the Reichenau catalogue of the year 822.[49] One of the oldest manuscripts is the famous codex Naples V A 13 from the beginning of the 10th century. This manuscript was possibly used by Gerbert in 983 at Bobbio and he sent it, or a copy of it, to Liège around 997.[50] The corrections in this manuscript were made by Gerbert or one of his students. Yet a second manuscript could have been associated with Gerbert: the codex Bamberg, Ms. class. 55. This manuscript, also originating from the beginning of the 10th century, perhaps came through Johannes Scottus (+ about 875) and Gerbert shortly after 1000 to the newly founded library of the Bamberg cathedral.[51] Further codices from the 10th century are Cambridge, Trinity College 939 (originally at Canterbury) and the Einsiedeln manuscripts 298 and 358. Both of the last two manuscripts, which were probably written towards the end of the century in Reims or Trier, were most probably conveyed with St. Wolfgang from Trier to Einsiedeln.

Seven manuscripts stem from the 11th century: the only precisely dated copy of Geometria I (Berne 87) was written in June of 1004 by a clergyman Constantius in Luxeuil. Bishop Werinharius (1002-1027) bestowed this codex to the Church of St. Mary in Strasbourg. A second manuscript from Berne (Ms. lat. 299) probably originated from some French source. Codex 830 from the Stiftsbibliothek in St. Gall, which was written there in the 1st half of the 11th century, shows handwritten notes of Ekkehart IV (+ about 1060). Likewise in the first half of the 11th century, not from the 13th century,[52] the manuscript Vienna 2269 originated in France and this contains a *Bibliotheca Septem Artium*. In England, perhaps at Canterbury, the Oxford manuscript Douce 125 was written and about the same time another copy originated in South Germany (Vienna 55). Vat. Barb. lat. 92 is more recent (Western Germany, about 1100).

From the following two centuries only six manuscripts of Geometria I are known. The oldest of them is Chartres 498. This codex, along with Chartres 497, makes up the *Heptateuchon* by Thierry of Chartres, an encyclopedia of the seven liberal arts which originated in the 2nd quarter of the 12th century.[53] There are two remaining manuscripts of the 12th century. One of these, the agrimensorial manuscript London, BM Add. 47679, is compiled from several sources.[54] The second, Vat. Ottob. Lat. 1862, is from the 2nd half of the 12th century and from eastern France. From the 13th century stem Munich, CLM 4024a (from France); Rostock, Ms. phil. 18; Florence, Laur. Plut. 29.19. The last manuscript was copied around 1250, perhaps from a Corbie codex, for Richard de Fournival from Amiens. It was brought to the Sorbonne in 1271 by Gerard de Abbeville.[55]

Only three manuscripts which are more recent are known, all in a humanistic hand of the 15th century: Breslau, Rehdig. 55 (copied from the Rostock manuscript); New York, Columbia University, Plimpton 164 (copied from Naples V A 14); Cesena, Plut. sin. XXVI.1 (written about 1450-1465 for the Malatesta Library).

The existing manuscripts testify therefore that in the 9th to 11th centuries the Geometria I was well circulated, most notably in the area of eastern France–western Germany. This probably relates to the presumed place of origin, Corbie. The work appears also relatively early

---

49   For this, see chapter 2.2 (p. 198).
50   Bubnov, *Gerberti Opera*, pp. 398.475. Thulin, „Zur Überlieferungsgeschichte", p. 5 f.
51   L. Traube, „Paläographische Forschungen," 4. Teil, *Abhandlungen der Königlich Bayerischen Akademie der Wissenschaften*, III. Klasse, 24 (1904), 10.
52   Bischoff, *Mittelalterliche Studien*, Band 2 (Stuttgart, 1967), p. 81, n. 26.
53   See chapter 2.3 (p. 201).
54   For this, see M. Folkerts, *Rheinisches Museum für Philologie*, NF 112 (1969), 53-70.
55   Ullman, „Geometry in the Medieval Quadrivium," pp. 279-282.

isolated in southern Germany, northern Italy, and England. Then from the 12th century the number of manuscripts diminished regularly, until the 15th century when a few humanists discovered and copied this writing anew.

The dissemination of the Geometria II is in many ways similar to that of the Geometria I. The existing manuscripts indicate that like the older Geometria I this treatise was also copied frequently very soon after its origin while in the subsequent centuries the manuscripts were less widely circulated.[56] Of the 23 extant manuscripts (eight of which contain only extracts of this work), 13 stem from the 11th or 12th century: the oldest codex (Erlangen 379) was probably written in southern Germany around 1050 – only a little later than the original. Closely related to this manuscript is the somewhat more recent incomplete codex Berlin, Ms. lat. oct. 162 (beginning of the 12th century) from the Benedictine monastery of St. Eucharius-Matthias in Trier. Likewise from the middle of the 11th century stems a London manuscript, BM Harley 3595. This codex is of west German origin. Towards the end of the 11th century in eastern France, perhaps in Reims, a manuscript was written which is presently found in Paris (BN lat. 7377C). Just a little more recent is Vat. Barb. lat. 92 (around 1100, from western Germany or Belgium).[57] From the 12th century come the following manuscripts: Chartres 498[58]; Munich, CLM 13021 (from Prüfening near Regensburg, after 1165); Vat. Lat. 3123 (western Germany or eastern France); Vat. Ottob. Lat. 1862 (eastern France);[59] Paris, BN lat. 7185 (end of the century, Normandy or England). Only parts of the treatise are found in Brussels, Bibliothèque Royale, Ms. Lat. 4499-4503 (beginning of the 12th century; this manuscript contains, among other things, mathematical texts of the agrimensores and writings from the Gerbert circle); Vat. Reg. lat. 1071 (France); London, BM Add. 47679.[60]

To the 13th century belong the following four manuscripts: Munich, CLM 23511 (southern Germany: Tegernsee?); London, BM Arundel 339 (Southern Germany: Kastl near Regensburg?); and the already mentioned Geometria I codices Munich, CLM 4024a (France) and Rostock, Ms. phil. 18, both of which contain only a small portion of Geometria II. The remaining six manuscripts are humanistic copies from the 15th century: London, BM Lansdowne 842 B (Italy, from the papal curia?); Cesena, Bibl. Malatestiana, Plut. sin. XXVI.1 (Cesena 1450-1465);[61] New York, Columbia University Library, Ms Plimpton 250 (copied from Vat. Lat. 3123, around 1500, Italy); Groningen 103 (monastery Thabor/West Friesland, about 1500) and the two incomplete manuscripts Breslau, Rehdigeranus 55 (Italy),[62] and Berne 87 (the beginning of the 16th century).[63]

Even more clearly than with the Geometria I, the extant manuscripts indicate that the Geometria II in the two centuries after its origin (1050-1250) was copied relatively often, then, however, it fell into oblivion. This treatise was also rediscovered first in the Renaissance by Italian humanists. All the older manuscripts (with the exception of Paris 7185) stem, in accordance to the character of the script, from western France or south and west Germany,

---

56 Only conjectural times and places of origin will be named below. A closer account on the contents of these manuscripts, their characteristics, and early ownership can be found in Folkerts, „Boethius" Geometrie II, pp. 3-33.
57 This manuscript also contains the Geometria I.
58 For this manuscript, which also contains the Geometria I, see pp. 194 and 201.
59 This manuscript also contains the Geometria I.
60 For this manuscript, which also contains the Geometria I, see p. 194.
61 Also contains the Geometria I.
62 Also contains the Geometria I.
63 Only the first lines of the Geometria II are added in this later hand. The greater part of the codex was written in the year 1004 at Luxeuil. See p.

respectively. This supports the likelihood[64] that the author of this treatise might have been a Lotharingian. This fact furthermore means that the Geometria II was well known in the Latin-speaking centers in Medieval Europe but was barely copied in the peripheral regions (England, Spain, Italy).

If one explores and compares the texts of the extant manuscripts, then one is able to see *a posteriori* the existence of manuscripts no longer surviving: common defects which are found in different manuscripts simultaneously demonstrate that these codices go back to a manuscript which today is lost. In this way one can find out the interdependence of these manuscripts and clarify it with a stemma. For the two geometries, the following stemmata are presented (taking the necessary precautions):[65]

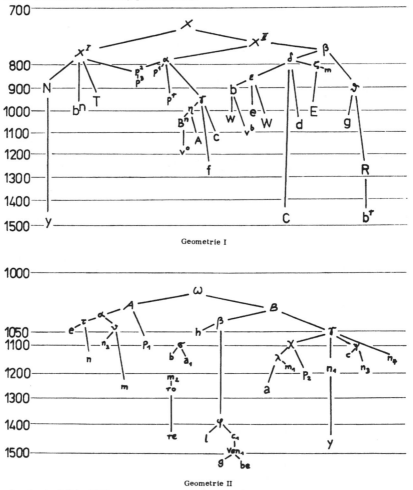

Geometrie I

Geometrie II

64   See chapter 1.3 (p. 192).
65   The stemma of the Geometria I results from the researches of Folkerts, „Die Altercatio,“ and the stemma of Geometria II from Folkerts, *Geometrie II*, pp. 51-61.

In these stemmata the lost codices are designated with Greek or cursively written letters, the extant manuscripts are designated with Latin characters. We know that there were at least twice as many manuscripts as are now extant.[66] Moreover, one sees how often the two Geometries were copied in the two centuries after their composition. So, for example, the autograph of Geometry II was written after 1000,[67] and the oldest extant manuscript (e = Erlangen 379) stems from about 1050. Between the autograph ω and e there are posited no less than three lost manuscripts, whose existence can be discovered from common flaws (A, a, τ). The four codices A, a, τ, e must have been written in a space of less than 50 years from each other. Geometria I has a similar situation: here there are as examples between the old manuscript b (Bamberg, Ms. class. 55) or m (CLM 560) and the autograph X four intermediate levels reconstructable ($\epsilon/\zeta, \delta, \beta$, X^{II}).

2.2 The two Geometries in medieval manuscript catalogues.

The number of existing manuscripts points out that the two Geometries of Boethius were not so rarely read from the 11th to 15th centuries, but were significantly less well known than the other extant writings of Boethius on the *Quadrivium*. Further indications have shown that both writings were copied in a relatively limited geographic region (Germany, eastern France, England) and that from about 1250 until the beginning of the 15th century no copy was made.

This will all be confirmed through the citations in medieval catalogues. In this we are able to draw only tentative conclusions, because on the one hand merely parts of the medieval library catalogues are published and evaluated in a scholarly way; on the other hand, the accounts in these catalogues are often so generalized that the writings are not clearly identifiable.[68]

At least 64 libraries contained the Boethius arithmetic, 52 his music, and many of them contained two or more manuscripts of these works.[69] In contrast, only 18 places mention the Geometry of Boethius. Four mentionings are drawn from surely identifiable manuscripts: Prague, University, 1370 = Prague, Lat. 1717; St. Gall, 1461 = Stiftsbibliothek 830; Canterbury, St. Augustine, at the end of the 15th century = Cambridge, Trinity College 939; Paris, Sorbonne, 1338, no. 49[70] = Florence, Laur. XXIX, 19.[71]

With some of the other mentionings, the probability is great that some already known manuscripts are designated: in three Corbie manuscript catalogues from the 10th and 12th centuries[72] geometries of Boethius are cited, and these may be identified with the existing Paris manuscripts 13020, 13955, and 14080. At Corbie four manuscripts of the Geometry

---

66 The total number of manuscripts must have been higher. The representation in the stemma is very much a simplification. It is practically impossible that in the Geometria I there were four direct manuscript copies from reconstructed codex a (p^2, p^1, p^r, γ): one must presume that a much greater number of other manuscripts came between.

67 Chapter 1.3 (p. 192).

68 The following account is based mainly on G. Becker, *Catalogi bibliothecarum antiqui* (Bonn, 1885); P. Lehmann and others, *Mittelalterliche Bibliothekskataloge Deutschlands und der Schweiz* (Munich, 1918 ff.); M. Manitius, *Handschriften antiker Autoren in mittelalterlichen Bibliothekskatalogen* (Leipzig, 1935). A more detailed discussion is found in Folkerts, *Geometrie II*, pp. 33-39.

69 Manitius, *Handschriften antiker Autoren*, pp. 275-300.

70 L. Delisle, *Le cabinet des manuscrits*, III (Paris, 1881), p. 68.

71 This manuscript is identical with number 44 in the catalogue of Richard de Fournival, about 1250; see Ullman, „Geometry in the Medieval Quadrivium," p. 281. To Fournival's books cf. R.H. Rouse, in *Revue d'Histoire des Textes* 3 (1973) 253 ff.

72 Becker, nos. 55, 79, 136.

were present around 1200.[73] Related to one of these manuscripts was very possibly a codex entitled *Boetius de geometria et astronomia* which was found in the 12th century in St. Bertin,[74] because between Corbie and the neighboring St. Bertin there existed close contacts. Gerbert saw in 983, at Bobbio, *VIII volumina Boetii de astrologia praeclarissima quoque figurarum geometriae, aliaque non minus admiranda.*[75] Bubnov,[76] Thulin,[77] and Ullman[78] have demonstrated that this manuscript is identical with Naples V A 15 or with its predecessor. Gerbert sent this manuscript or a copy of it around 997 to Adelbold of Liège.[79] Perhaps a manuscript which was found at Liège[80] in the 11th century[81] is identical with the one sent from Gerbert to Adelbold.[82]

We will now cite those libraries in which a copy of the Geometry was found during the Middle Ages but is now lost. Already by 821-822 there was at Reichenau a codex which among other things contained the Geometria I.[83] A second manuscript, enlarged by other texts, which is closely related to this codex is mentioned by the Reichenau monk Reginbert in his book inventory, about 840.[84] The Reichenau manuscript(s) are related to the Munich CLM 560 codex mentioned above,[85] and probably also to a codex which was present at Murbach in the 9th century and was designated as *Geometrica Liber I* in that library's catalogue.[86] Murbach was founded by Reichenau. Between these two monasteries there was a lively exchange of books: two thirds of the Reichenau books also appear in the Murbach catalogue, written about 840. A further foundation from Reichenau was the Benedictine house of Pfävers. Here also there is a library catalogue (which was written in 1155) where a *Liber Geometriae* is mentioned and which is probably ascribed to Boethius.[87] Because between Pfävers and Reichenau on the one hand, as well as St. Gall on the other, there was a lively exchange of books, it is possible that the Pfävers manuscript is related to the Reichenau codex from the 9th century or the Geometria I manuscript of St. Gall.[88] At Fulda there were in the 16th century three manuscripts entitled *Geometria Boetii.*[89] At least one of these was a codex of Geometria I, as the incipit of the Fulda catalogue shows.[90] The remaining citations are too vague to permit any certain conclusions which geometry is meant. In some of these citations it is even doubtful whether one is dealing with the writings of Boethius: for example, Weihenstephan (11th

73   *Ibid.*, no. 136, mss. 272, 274, 275, 281.
74   *Ibid.*, no. 77, ms. 61.
75   Bubnov, p. 99f.
76   *Ibid.*, p. 475.
77   Thulin, p. 5 f.
78   Ullman, p. 278.
79   Bubnov, p. 398.
80   It is not altogether sure that this catalogue draws on St. Laurent in Liège. See H. Silvestre, *Le Chronicon S. Laurentii Leodiensis* (Louvain, 1952), p. 40.
81   Becker, no. 60, ms. 24: *geometrica Boetii*. The library catalogue which is transmitted in Brussels 9668 was newly edited by J. Gessler (Tongres, 1927).
82   Ullman, p. 278, note 4.
83   Bubnov points this out persuasively on p. 181.
84   On this see Ullman, p. 277 f.
85   See p. 194.
86   W. Milde, *Der Murbacher Bibliothekskatalog des 9. Jahrhunderts* (Euphorion, Beiheft, 1967).
87   Lehmann, Band 1, p. 486, 11-14.
88   On this see p. 194.
89   No. IX 2, 14; X 2, 32 and X 4, 26 of catalogue F which includes the entire Fulda collection. See K. Christ, ,,Die Bibliothek des Klosters Fulda im 16. Jahrhundert,'' *Zentralblatt für das Bibliothekswesen*, Beiheft 64 (Leipzig, 1933).
90   V, no. 503; see Christ, p. 159.

century),[91] Toul (before 1084),[92] Durham (12th century)[93] and an unknown Italian or French library (13th century).[94]

The medieval library catalogues allow us to conjecture that Reichenau also played an important role in the dissemination of the Geometria I in the 9th-10th centuries. The influence that Fulda had in this early period is uncertain, because a Geometria of Boethius is only mentioned in relatively recent Fulda catalogues. We have seen that the accounts of the medieval catalogues are very generally stated so that one can only seldom know whether one is dealing with a codex of the Geometria I or II. None of these citations can lead to a certain conclusion that the codex was Geometria II: some of the citations must have been the Geometria I.

2.3 Citations by other authors.

To be able to answer the question to what extent the Boethian writings on Geometry were known and used in the Middle Ages, it would be helpful to know not only of references to these manuscripts in medieval catalogues, but also to have all references to these texts, whether in the form of direct quotation or as citations of the name of the author or of the title of the writings. That type of all-inclusive list of citations is presently not possible for many reasons, especially since only parts of the medieval texts are edited. Therefore a much more modest goal should be worked for: because the two geometries do contain much concrete mathematical materials, these writings must have been interesting particularly for the authors of medieval textbooks. I have examined the most important mathematical writings of the Latin Middle Ages in the hope of finding traces which would indicate that the writer was familiar with one of the two geometries.

According to the findings of the previous sections, it should be no surprise that in the mathematical writings up to the 12th century some knowledge of the Geometries is apparent. The name of the author was seldom mentioned, yet similarities in the Latin texts testify that one of the two geometries was known.

As one of the seven *artes liberales*, geometry was studied from late antiquity as part of the medieval framework of studies. It seems, however, as if geometry lapsed behind the other disciplines of the quadrivium—no wonder, from the scanty materials which the scholars up to the 12th century had at their disposal! In addition to the general geometrical concepts from the *encyclopaediae* of Cassiodorus, Martianus Capella and Isidore of Seville, it seems that the mathematicians of this time took their geometrical knowledge primarily from the writings of the *Corpus Agrimensorum*, from the two Geometries attributed to Boethius, and from the Geometry of Gerbert with expansions by some unknown author. Among these writings there is a complex interdependence in which the Geometries of Boethius are interwoven.

The anonymous geometry, which is connected with the Geometry of Gerbert (the *Geometria incerti auctoris*, hereafter shortened als *GIA*) passed for a long time as a part of Gerbert's geometry. It was Bubnov who made the first critical edition of this writing[95] and he was able to show convincingly that this work originated before Gerbert: The GIA is based on Arabic writings on the astrolabe. However, it was also present in the agrimensorial manuscript

---

91   Becker, no. 73, mss. 40.41; it is uncertain whether this refers to Boethius.
92   Becker, no. 68, ms. 251: *musica Boetii cum Euclide de geometria vol. I.* Because this catalogue was put together before the Arabic-Latin Euclid translations, it must concern itself with a manuscript of the Boethian tradition.
93   Becker, no. 117, ms. 186; it is uncertain whether this refers to Boethius.
94   No. 65: *Geometria Boecii et Hugonis.* See Manitius, p. 291.
95   Bubnov, pp. 310-365.

which Gerbert used in 983. The GIA therefore did not originate before the 9th or after the 10th century. The author of the GIA also utilized the Geometria I of Boethius, which he excerpted without mentioning either the work or its author.[96]

The most important geometrical work which was at the disposal of the West before the time of the translations from the Arabic was the *Geometria* of Gerbert which did not originate before 983.[97] The phraseology in this work often brings to mind the Geometria I, especially the Euclidian section. So it is very possible already, by reason of its contents, that Gerbert made use of the Geometria I in writing his geometry. This conjecture becomes more certain when one considers that Gerbert was acquainted with an agrimensorial manuscript at Bobbio which also contained the Geometria I.[98] Furthermore, Gerbert often cites the name of Boethius, though never in connection with geometry;[99] perhaps he perceived that the Geometria I was not genuine.

The two Geometries of Boethius, particularly the Geometria I, contain much gromatic material. So it is not surprising that the Geometria I was very quickly taken up by the monks (at the latest in the 9th century) who—obviously for the study of geometry within the quadrivium and less for surveyor—work[100]— would put together new geometrical and gromatic works from the transmitted texts. At least two of this type of compendia exist: a *Geometrica ars anonymi* (designated as Y by Thulin),[101] and a *Geometria Gisemundi* (designated as Z by Thulin).[102]

The *Geometrica ars*, divided into 34 chapters, contains few gromatic texts in the strict sense of the word. First of all, we find geometrical excerpts. The author presents the explanation given by Cassiodorus on geometry,[103] filled out and enlarged by excerpts from Censorinus, Balbus and the Geometria I of Pseudo-Boethius. Moreover, the Geometria I was particularly drawn upon and partly excerpted word for word.[104] The text is contained in five manuscripts: the oldest and best is CLM 13084, ff. 48v-69v, from the 10th century (Regensburg); from this was copied Munich, CLM 6406 (11th century, Freising) and CLM 14836 (11th century). A manuscript related to CLM 13084 is Vienna 51 (12th century). The fifth manuscript, Sélestat 1153bis (10th century, Worms)[105] is likewise akin to CLM 13084. It appears as if this version originated east of the Rhine and later it was particularly well known in that area.

The *Geometria Gisemundi* is a gromatic-geometric compendium of Spanish origin which is found in two manuscripts (Barcelona, Ripoll 106, ff. 76-89, 10th century; Paris, BN lat. 8812, about 900). The main source for the compilor who, according to his own account made use of more than one manuscript, was the Geometria I of Boethius from which he excerpted

---

96 Bubnov, p. 336, note 24; p. 400, 471.
97 The critical edition is by Bubnov, pp. 46-97. Hofmann (see note 44), p. 4.9.11, conjectures that the Geometry in its present form does not stem from Gerbert but rather is a version from the early 11th century.
98 See p. 198.
99 Gerbert cites Boethius' arithmetic and his commentary on Aristotle's *Categories*. That Gerbert also knew the Geometria I Bubnov has shown (p. 48, n. 3 with corrigenda on p. 556; p. 51, n. 1 with corrigenda on p. 556f.; p. 181, n. 3; p. 397f.).
100 Ullman, pp. 263-269, has shown convincingly that in the Middle Ages gromatical texts were used for the instruction of geometry but not to educate surveyors.
101 Analyzed by Thulin, pp. 44-54.
102 Analyzed by Thulin, pp. 55-68.
103 Chapter 6 from the *De Artibus ac Disciplinis Liberalium Artium*.
104 The contents are given by Thulin, pp. 44-48.
105 A. Giry, „Notes sur un ms. de la Bibliothèque de Schlestadt," *Revue de Philologie* (1879).

nearly all that concerned itself with land surveying.[106] He also used a gromatic manuscript, from the so called „Mischklasse."[107] The *Geometria Gisemundi* contains nothing of the arithmetic and geometric extracts which are present in books 2-4 of the Geometria I and nothing from the first dialogue. The excerpted Boethian text is also not a continuous work, but rather is broken down into about 30 chapters which are rarely in the same order as in the Pseudo-Boethius, and often they are separated by other excerpts. This Spanish handbook, which must have originated not later than in the 9th century, shows in any case that at this early time the Geometria I was also known in northern Spain (Ripoll).

Traces of the Geometria I also appear occasionally in the 12th century in the works of other authors. Thierry of Chartres, an important scholar of the 12th century, composed, probably before 1141, a book entitled *Heptateuchon*[108] or the *Bibliotheca septem artium liberalium*.[109] In this work Thierry presents not only a solid grounding for the encyclopedia of learning, but also arranged the learning which had evolved before his time into a self-contained unity and put it into order. In his *Heptateuchon* Thierry took up those writings which he valued most for the study of the *artes liberales*. For the quadrivium there were cited, naturally, Boethius' works on arithmetic and music. These geometrical writings are given: Adelard's Euclid translation (version II); the gromatical excerpts from Boethius' Geometria I, Book 5; other gromatical texts, among these the writings of Epaphroditus and Vitruvius Rufus and the *Liber Podismi*; excerpts from Gerbert's geometry; the Pseudo-Boethian Geometria II; Gerland's writing on the abacus.[110] These works can surely be taken as representative of the geometrical teachings in the early and high Middle Ages. Noticeably Thierry took into consideration both Geometries of Boethius. Yet he did not take all of the Geometria I–rather, typically, merely the *altercatio* of Book 5.[111] The excerpts from Boethius' Arithmetic and the *Elementa* of Euclid which are present in the Geometria I are notably missing. Both works were unnecessary since Thierry copied them in another part of his manuscript. Through what channel did Thierry know the writings of Boethius? Ullman has conjectured[112] that it took place through Fulbert who in the century before Thierry had been a teacher at Chartres and was probably a student of Gerbert. It is likely that Fulbert read the Corbie manuscript (Paris 13955) at Chartres which also contains the Geometria I.[113]

---

106 Contents by Thulin, pp. 55-58.
107 Thulin, pp. 58-61.
108 The name is openly based on the Pentateuch. In medieval catalogues manuscripts and commentaries on the Bible sometimes are entitled *Eptateuchus* or *Eptaticus*.
109 Chartres, Ms. 497-498. The manuscripts were burned in W W II but there has survived a microfilm. The text of the prologue to the *Heptateuchon* is published by E. Jeauneau, „Le Prologus in Eptatheucon de Thierry de Chartres," *Mediaeval Studies*, 16 (1954), 171-175 (reprinted in: *Lectio Philosophorum, Recherches sur l'Ecole de Chartres*, Amsterdam 1973, pp. 87-91). From the extensive literature on Thierry and the school of Chartres one may cite: A. Clerval, *L'enseignement des arts libéraux à Chartres et à Paris dans la première moitié du XIIe siècle d'après l'Heptateuchon de Thierry de Chartres* (Paris, 1889); A. Clerval, *Les écoles de Chartres au moyen âge* (Paris, 1895); Charles Haskins, *Studies in the History of Mediaeval Science* (Cambridge, 1924), especially p. 91; G. Paré, A. Brunet, and P. Tremblay, *La renaissance du XIIe siècle: Les écoles et l'enseignement* (Paris, Ottawa, 1933), pp. 94 ff.; R. Klibansky, „The School of Chartres," *Twelfth-century Europe and the Foundations of Modern Society*, ed. M. Clagett, Gaines Post, and R. Reynolds (Madison, 1961), pp. 3-14.
110 According to the order of texts in the manuscript, the Gerland writing belongs to Geometry. It was a wide-spread notion in the Middle Ages that the reckoning on the abacus is part of geometry.
111 Thulin's assumption (p. 18) that the end of the work is in the part of the manuscript now lost is wrong.
112 Ullman, p. 278.
113 A conjecture by B. Bischoff. Ullman's doubt concerning the connection between Gerbert and Thierry (see note 11, p. 279) seems now to be without reasons: a close analysis of the Geometria I text in Thier-

As we have seen, the Geometria I of Boethius was linked at least with Chartres by the 12th century in the established writings on the *quadrivium*. It may have been used in other places as well: there is a list of textbooks for the seven liberal arts probably made by Alexander Neckam at the end of the 12th century while he was teaching in Paris.[114] Neckam stated: *Institutis arsmetice informandus arismeticam Boecii et Euclidis legat.* It seems that this reference draws from the Goemetria I of the Pseudo-Boethius.[115]

So much for the citations of the Geometria I. The Geometria II originated, as we have seen, in the 11th century in Lorraine and was in the following hundred years spread mainly throughout western Germany and eastern France. This geometry must be seen as related to the works of a group of Liège mathematicians whose influence extended beyond Lorraine in the first half of the 11th century. These scholars wrote works on the abacus and also geometrical treatises. The most important of these texts–aside from the Geometria II– is a lengthy correspondence between the Cologne scholastic Regimbold and Radulph of Liège about the year 1025 and the treatise of Franco of Liège on the squaring of the circle (shortly before 1050). It is not surprising that in both cases there are connections to the Geometries of Boethius.

In the 11th century in geometry beside skimpy accounts in the *Corpus Agrimensorum* and in the encyclopedic writings of the Romans, only faulty excerpts from the first books of Euclid (without the proofs) were known. What was needed, above all, was a clarification on geometrical principles. In the course of the correspondence between Regimbold and Radulph,[116] the question comes up about what is an inner and an outer angle. In spite of various statements, no conclusive definition is found. Only a little later will the question concerning the sum of angles in a triangle be resolved experimentally, as it is known according to an anonymous text at Kues.[117] In the correspondence of these Rhineland scholars, there are no signs that they knew the Geometria II; probably this writing originated after 1125.[118] In two places, indeed, are Euclidian passages cited, both of which show that Regimbold knew the Geometria I of Pseudo-Boethius. The two statements of Regimbold read thus:[119]

I.16[120]   *Omnium triangulorum exterior angulus utrisque interioribus et ex adverso constitutis angulis maior existit.*

I.32[121]   *Omnium triangulorum et exterior angulus duobus angulis interius et ex adverso constitutis est aequalis, interiores tres anguli duobus rectis angulis sunt aequales.*

If one neglects insignificant changes, both phraseologies correspond with the Euclidian text which was carried over into the Geometria I and II. A small item, moreover, emerges to esta-

ry's manuscript Chartres 498 indicates that this codex (like Paris 13955) belongs to the subclass X[II], but that Thierry used another codex which, as Gerbert's manuscript, belonged to class X[I]. Using this manuscript, Thierry expertly corrected essential mistakes of his X[II] source. It is noteworthy that in the manuscript Berne 299 (on this see p. 194) there exist many corrections which correspond with the corrected text of Thierry's manuscript.

114   Charles H. Haskins, *Harvard Studies in Classical Philology*, 20 (1909), 75.
115   The citation of Euclid is not surprising in this connection because in the rubrics of the Geometria I Euclid's name occurs often.
116   Edited by Tannery and Clerval (see n. 43). Below, the page numbers will be cited from the *Mémoires Scientifiques*.
117   Edited by Hofmann (see. n. 44).
118   Tannery rightly remarks on p. 248 that at least for the metrological questions both correspondents would have drawn on the Geometria II if it had been known to them.
119   Regimbold cites here the name of Boethius but not his work: *testimonio ipsius Boetii ita scribentis* (Tannery, p. 280).
120   Tannery, p. 281.
121   *Ibid.*

blish that Regimbold used a manuscript copy of Geometria I that belongs to class X^{II}: only in this group of manuscripts does one find in I.32 the word *et* after *triangulorum*. In the manuscript group X^I and in the Geometria II the *et* does not occur. It is improbable that Regimbold inserted this word by himself. Probably Regimbold here followed the Geometria I manuscript Paris 13955.[122]

Also, a very striking resemblance exists between the Geometria II and the possibly somewhat more recent writing on the squaring of the circle which Franco of Liège wrote shortly before 1050.[123] The similar rhetorical style in both works is obvious, particularly in the prefaces of Franco's six books. The reckoning of the circle in Franco[124] recalls the formulae and substituted values in the corresponding place of the Geometria II.[125] In addition, the mention of *porticus, miliaria, stadia,* and *fluvii* in Franco find their correspondences in the Pseudo-Boethius.[126] The praise of Pythagoras and of Patricius Symmachus in relation with Boethius[127] could draw upon the arithmetic,[128] but also is in common with the Geometria II.[129] Many striking formulations in Franco bear such a strong resemblance to places in the Geometria II that one simply must consider it an influence.[130]

The Geometria II is the oldest extant writing known in the west in which an abacus of the Gerbert school and the Indic-Arabic figures (including the zero) are presented. For this reason the work is of special interest for the historians of science and it is conceivable that those authors who in the 11th and 12th century wrote discussions on reckoning with the abacus drew upon the Geometria II of the Pseudo-Boethius. The most important masters of the abacus whose works we know are: Odo,[131] Gerbert (+ 1003),[132] Abbo of Fleury (+ 1004),[133] Heriger of Lobbes (+ 1007),[134] Bernelinus (possibly a student of Gerbert),[135] Hermannus Contractus (1013-1054),[136] Turchillus (England, before 1117),[137] Gerland of Besançon (1st

---

122 This codex draws on class X^{II}; Fulbert (+ 1028) may have used it at Chartres (see p. 201). Regimbold cites in another letter of his correspondence (Tannery, p. 288) a discussion which he had at Chartres with Fulbert on the angles of a triangle. Therefore it is not improbable that Regimbold knew of the Boethius manuscript through Fulbert.

123 See note 45. The Franco text will be cited below in the edition by M. Folkerts and A.J.E.M. Smeur, „A Treatise on the Squaring of the Circle by Franco of Liège of about 1050," Part I, *Archives internationales d'histoire des sciences*, 26 (1976), 59-105.

124 *Ibid.*, p. 70, lines 111-116.

125 Folkerts, *Geometrie II*, p. 166, lines 885-888.

126 Franco, p. 71, l. 144f. = Boethius, p. 147, l. 560 f.

127 Franco, p. 78, l. 30-34.

128 In Friedlein's edition, p. 3, 1.1 11. 13; p. 7, 1.21-26.

129 See Folkerts, *Geometrie II*, p. 113, l.2; p. 138, lines 439-441; p. 139, l. 447-453; p. 169, l. 927-930.

130 E.g., *geometricae disciplinae peritissimi* (Franco, p. 76, 1. 148f.) = Pseudo-Boethius, p. 150, l. 607; *dubitationis obscuritate ... exempli luce* (Franco, p. 87, l. 83f.) = Pseudo-Boethius, p. 132, l. 324f.; *Pythagorica subtilitas* (Franco, p. 90, l. 10) = Pseudo-Boethius, p. 139, l.447 f.; *Patricius* (= Symmachus) (Franco, p. 90, l. 21) = Pseudo-Boethius, p. 113, l. 2.

131 It is questionable whether he is to be identified with Odo of Cluny (878-942); he could be better placed early in the 12th century. His *Regulae super abacum* are edited by M. Gerbert, *Scriptores Ecclesiastici de Musica*, I (Sankt Blasien, 1784), pp. 296-302 and later by Migne, *P.L.*, 133, col. 807-814.

132 His work on the abacus has been critically edited by Bubnov, pp. 6-22.

133 Edited by Bubnov, pp. 197-204.

134 Edited by Bubnov, pp. 205-225.

135 His *Liber abaci* has been edited by A. Olleris, *Oeuvres de Gerbert* (Clermont-Ferrand, Paris, 1867), pp. 357-400.

136 Edited by Treutlein in *Bullettino Boncompagni*, 10 (1877), 643-647.

137 Edited by E. Narducci in *Bullettino Boncompagni*, 15 (1882), 111-162. See also Haskins (note 109), pp. 327-335.

half of the 12th century),[138] Adelard of Bath (1st half of the 12th century),[139] Radulph of Laon (+ 1131).[140] In addition there are some commentaries on Gerbert's writing about the abacus[141] and anonymous writings, mostly from the 12th century.[142] In their works, many of these authors make comments about Gerbert as the inventor or the disseminator of the abacus.[143] At times Herman of Reichenau is also cited.[144] In addition, one often finds Pythagoras mentioned.[145] Boethius' name is cited only by Adelard, but only in connection with his arithmetic[146] and music[147]. In general, these authors make use of the works of Gerbert or of his students but not of the abacus portion of the Geometria II. On the other hand, a singular item emerges from the Geometria II: in this treatise it is stated for the first time that the abacus was invented by Pythagoras, and for that reason it was called the *mensa Pythagorea*.[148] One also finds this statement in Adelard[149] and Odo who, very clumsily in this connection, mentions the ,,translation activity" of Boethius.[150] The amazing similarity in the phraseology between the Geometria II and Adelard apparently indicates that Adelard knew of this work.[151] It is uncertain if Odo also knew of it.[152] Among all the other masters of the abacus there is no sure sign that they had read the Geometria II.[153]

2.4 The two Geometries and the Euclid translations from the Arabic.

As we have seen in chapter I,[154] the Geometria I as well as the Geometria II contain extensive excerpts from the *Elements* of Euclid, Book 1-4 (without proofs). The transmitted

138   Edited by Treutlein in *Bull. Bonc.* 10 (1877), 595-607.
139   Edited by B. Boncompagni in *Bull. Bonc.* 14 (1881), 91-134.
140   Edited by A. Nagl in *Zeitschrift für Mathematik und Physik*, Supplement, 34 (1889), 85-133.
141   Edited by Bubnov, pp. 245-284.
142   Edited by Treutlein in *Bull. Bonc.* 10 (1877), 607-629, 630-639, 639f.; Bubnov, pp. 225-244 (from the 10th century) and pp. 291-293.
143   For example, see Bernelinus (note 135), p. 357; Radulph (n. 140), pp. 100, 102, 103; Adelard (n. 139), pp. 91, 99, 100.
144   For example, see Radulph (n. 140), p. 100.
145   See notes 148, 149, 150.
146   Adelard (n. 139), pp. 108, 111.
147   Adelard, p. 111.
148   Folkerts, *Geometrie II*, p. 139, lines 447-453: *Pytagorici vero, ne in multiplicationibus et partitionibus et in podismis aliquando fallerentur, ut in omnibus erant ingeniosissimi et subtilissimi, descripserunt sibi quandam formulam, quam ob honorem sui praeceptoris mensam Pytagoream nominabant, quia hoc, quod depinxerant, magistro praemonstrante cognoverant--a posterioribus appellabatur abacus--ut, quod alta mente conceperant, melius, si quasi videndo ostenderent, in notitiam omnium transfundere possent, eamque subterius habita sat mira descriptione formabant.* On the origin of this legend, see Folkerts, *Ibid.*, p. 89.
149   Adelard (n. 139), p. 91, lines 7-10: *Pythagorici hoc opus* (i.e. *abacum) composuerunt, ut ea, quae magistro suo Pythagora docente audierant, oculis subiecta retinerent et firmius custodirent. Quod ipsi quidem mensam Pythagoream ob magistri sui reverentiam vocaverunt; sed posteri tamen abacum dixerunt.*
150   Odo (n. 131), p. 296: *Haec ars non a modernis, sed ab antiquis inventa, ideo a multis negligitur, quia numerorum perplexione valde implicatur, ut maiorum relatione didicimus. Huius artis inventorem Pythagoram habemus. ... Hanc* (i.e. *artem) antiquitus graece conscriptam a Boethio credimus in Latinum translatam.* Also in Radulph (n. 140), p. 90, the *mensa philosophorum* could have meant the *mensa Pythagorica.* On the identification of the abacus with the *mensa Pythagorica* see Bubnov, p. 157, note 17.
151   The other parts of Adelard's writing on the abacus yield no further arguments which support this assumption. Adelard mentions such names as Pythagoras (see n. 149), Boethius (note 146, 147) and Gerbert (n. 143) and then only Guichardus (p. 100), and a certain H. (p. 91).
152   It is very likely that Odo knew the assertions in the Geometria II only indirectly.
153   In a commentary on Gerbert's *Regulae* which originated about 1000, a passage was cited word by word from the Geometria I of Pseudo-Boethius: Bubnov, p. 250, note 6.
154   Pages pp. 187-188.

text is corrupt and often obscured. Scholars in western Europe drew their knowledge of Euclid up to the 12th century almost exclusively from these two works.

The situation advanced haphazardly in the 12th century when in the process of translations from the Arabic, the masterwork of Euclid was also translated into Latin more than once. Intensive studies of historians of science in the last decades have gone far to shed light on the various versions and treatises; even though the manuscript material is still not completely elucidated and especially have not all the Arabic sources for each translation been yet identified.[155] For my research the following condensed formulation may be made: one of the earliest translations from Arabic, the so called Version II by Adelard of Bath (originating about 1120) was disseminated rapidly already in the 12th century and supplanted the other Arabic to Latin translations of Euclid. Of Adelard II there exist over 50 manuscripts. More widespread in the Middle Ages was only the treatise which Johannes Campanus of Novara composed before 1260. His work, which is extant in more than 100 manuscripts, relies essentially on the Adelard II text. Both the Adelard II and the Campanus could be considered as *the* Euclid texts of the high and late Middle Ages, so that the Adelard II text especially in the 12th and 13th century, and the Campanus text in the 14th and 15th centuries became particularly important.

The question as to how Adelard produced his Euclid translation (or translations) cannot yet be conclusively answered despite the work of Clagett, Murdoch, and Busard.[156] Inasmuch as Adelard probably knew the Geometria II of Boethius,[157] the possibility cannot be excluded that he also took the Euclidian excerpts from this treatise for his translation; but this is not altogether clear.

With the Euclid translation of Adelard there was presented for the first time since late antiquity a complete Euclidian text in the Latin language which, in contrast to the previous texts of the Boethian tradition, also contained the proofs and encorporated hardly any translation or transmission errors. One would expect that these preferences for the Adelard text would have led to the disappearance of the insufficient Boethius tradition within a short time. But this does not seem to have been the case; rather, in the 12th century, some scholars attempted in various places to put the Euclidian excerpts in the Geometria I and II of the Pseudo-Boethius and the text of Adelard together into a new work and fit these with each other. To the modern observor it seems amazing that despite the widely read and good translation of Adelard, scholars held fast to the Boethian text full of flaws. The explanation for this could be found in the name „Boethius" which throughout the entire Middle Ages was highly honored. In the material to follow I would like to discuss this combination of Boethius and Adelard. Because this has been studied in another place with more detailed research,[158] it should suffice to bring together the results here and present some new supporting material.[159]

---

155 Good information is given by F. Sezgin, *Geschichte des arabischen Schrifttums, Band 5. Mathematik* (Leiden, 1974), pp. 83-120. The basic work on the Arabic-Latin Euclid translations is Marshall Clagett, „The Medieval Latin Translations from the Arabic of the Elements of Euclid, with Special Emphasis on the Versions of Adelard of Bath," *Isis*, 44 (1953), 16-42. See as well J.E. Murdoch, „The Medieval Character of the Medieval Euclid: Salient Aspects of the Translations of the Elements by Adelard of Bath and Campanus of Novara," *XIIe Congrès International d'Histoire des Sciences, Colloques (= Revue de Synthese*, vol. 89, 1968), pp. 67-94. H.L.L. Busard is preparing critical editions of all the significant Arabic-Latin Euclid translations.

156 See note 155.

157 See p. 201.

158 M. Folkerts, „Anonyme lateinische Euklidbearbeitungen aus dem 12. Jahrhundert," *Österreichische Akademie der Wissenschaften, Math.-nat. Klasse, Denkschriften*, 116. Band , 1. Abhandlung (Vienna, 1971).

159 Since the above work, I have found two further manuscripts: Leiden, Voss. lat. qu. 92, and San Juan Capistrano, Honeyman Ms. 50 (see below).

At two places in western Europe scholars undertook in the 12th century, independently from each other, an attempt to combine the Boethius work and Adelard's Euclidian translation. From one of these mélanges there is only a single manuscript now known (Luneburg, Ratsbibliothek, Ms. misc. D 4⁰ 48, ff. 13r-17v). This manuscript, which was written in northern Germany about 1200 and which an otherwise unknown *magister Helmoldus* presented to the Michaelis monastery in Hildesheim, contains most of the definitions, postulates, axioms, and propositions of the first four books of Euclid which also exist in the Geometria II. The author of the Luneburg manuscript used a codex of the Adelard translation and one of the Geometria II in such a manner that he preferred in his text of the definitions and postulates the Boethian formulation; the axioms and propositions, for the most part, follow Adelard's phraseology. The compilor carefully put his text together from the two sources; often he improved the defective text of the Boethian tradition. In many places the author writes his own text which deviates from both Boethius and Adelard, but is clearly intelligible and meaningful. The contaminated versions of the theorems and definitions are mathematically sound and unobjectionable. They indicate that the author had a great mathematical ability for his time. This means that the Luneburg excerpts are an advance over the very corrupt portions of the Euclidian excerpts in the Geometria II of the Pseudo-Boethius.[160]

Of the second compilation there are extant the following six manuscripts all of which come from the 12th century:

M_1 = Munich, CLM 13021, ff. 164r-186v[161]
M_2 = Munich, CLM 23511, ff. 1r-27r[162]
O = Oxford, Digby 98, ff. 78r-85v[163]
P = Paris, BN lat. 10257, ff. 1-88 (from Chartres)
V = Leiden, Voss. lat. qu. 92, ff. 2rv.1rv[164]
H = formerly San Juan Capistrano, Honeyman Ms. 50[165].

On the one hand, these six manuscripts show similarities; on the other, they represent distinctly different levels of treatment: the noteworthy but awkward text of the axioms indicates that all the known manuscripts were taken from a common source X. This source X used manuscripts of the Adelard and Boethius tradition such that it drew upon the Boethian text and completed it with the Adelard text. It appears that the author of X had a Boethius text of Geometria I and another one of the Geometria II at his disposal, but preferred the Geometria II. The Adelard text first is used to any great extent in book 3 where not all the propositions are found in the Boethius text. The manuscripts O and V are „cleanly" copied from source X. The mistakes in O are mainly the fault of the copyist. The author of the version O has drawn on no other manuscript. The short fragment in V[166] departs greatly from the text of manu-

---

160 A more complete account of the manuscript and the working method of the compilator may be seen in Folkerts, „Lateinische Euklidbearbeitungen," pp. 12-19. The Euclidian text has been reproduced in facsimile, *Ein neuer Text des Euclides Latinus*, ed. M. Folkerts (Hildesheim, 1970).

161 See p. 195.

162 See p. 195. M_2 contains only Books 4-15.

163 This ends with Euclid III, def. 4.

164 A fragment. This includes III, 34-36; IV, 14-16. See note 166.

165 Ms. 24 in C.U. Faye, W.H. Bond, *Supplement to the Census of Medieval and Renaissance Manuscripts in the United States and Canada* (New York, 1962), p. 22. Mr. Honeyman has very kindly sent me photographs of his manuscript. In the meantime, it has been sold by Sotheby Parke Bernet & Co., London (see *The Honeyman Collection of Scientific Books and Manuscripts*, Part III, lot 1086).

166 There is on f.2, r-v: the end from III, 34 Heiberg (= 33 Adelard) according to Adelard II (only the words *-onem abscindere*); an addition as in the Pseudo-Boethian Geometria II after Euclid III, 7 (see Folkerts, *Geometrie II*, p. 130, lines 279f.: *Similes ... trigone sint*); III, 35 Heiberg (= 34 Adelard) according to

scripts M_1M_2HP; V shows similarities with the Boethius tradition more clearly than the remaining codices.167 This particular classification combines V with O so that both manuscripts possibly represent the same version.168

The mélange X was not only copied literally (OV) but was again adapted by another compilor (Y). Y has in many places altered the X text with the help of an Adelard codex and through conjecture. Y may be reconstructed from the texts contained in M (= M_1M_2) and P. Both texts are not straight copies but rather editorial modifications from text Y: the author of M hardly made changes of his own, but used an Adelard text, from which he took some propositions and inserted them in the corresponding place in the Boethian text. On the other hand P tried to adapt the Adelard text to the Greek-Latin tradition. For this purpose the writer designated the various irrationalities in Book 10 with expressions which are transcriptions of Greek *termini*.169 Therefore, OV is closest to Boethius and M to the Adelard text. P takes a middle place. The manuscript H is a direct copy of the corrected text in the manuscript P.170

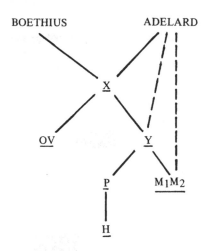

Adelard II; III, 7 beginning and addition as in the Geometria II (Folkerts, *Geometrie II*, p. 129, lines 272-279: *Si intra circulum ... perhibeatur*); III, 36 Heiberg (= 35 Adelard) according to Adelard II (broken off at the end of the page). On f.1, r-v: IV, 14-16 Heiberg (= 14-19 Adelard) according to Adelard II; an addition as in the Geometria II, after Euclid IV, 11 (Folkerts, *Geometrie II*, p. 132, lines 319-321: *Nam que ... terminum facientes); subscriptio: Euclidis philosophi liber quartus explicit. Incipit quintus viginti sex theoremata continens. et XVI^{cim} elementa.*

167 In V there are parts of the Boethius addition after Euclid III, 7 and IV, 11 (see note 166). P also contains a section of the addition after III, 7. M_1 and M_2 have nothing of the two additions.

168 The text in O breaks off before V begins.

169 For example, *diamese* for ἐκ δύο μέσων, *apotomese* for μέσης ἀποτομή, *dionimum* for ἐκ δύο ὀνομάτων, *longisimetrum* for σύμμετρον ᾗ μήκει, *longiretum* for εὐθεῖα ῥητή. These Greek terms had already appeared in Martianus Capella.

170 H shows all the errors of the corrected manuscript P and even particular mistakes of its own. The following fact especially speaks for a direct copying: because of the page ending, P left out the conclusion of proposition I, 47 so that instead of *lateribus continetur*, there is only *la*. H has correctly added *lateribus* but not the necessary *continetur*.

Through a careful consideration of the seven manuscripts mentioned above we have taken a look at the transformation which the 12th century mathematics went through as a result of the Euclid translations from the Arabic. We saw to what extent the Greek-Latin and the Arabic-Latin tradition within a few decades emerged and became fruitfully interwoven. We could trace the existence of two mutually independent collections of the *Euclides Latinus* (MOVP and Luneburg), the first of which had further revisions within a few years. Various attempts to revise the Euclid text are reflected in the various levels of the original mélange. The scholars took the trouble, with the help of codices by Adelard (Y, M) or through sources which were similar to the Greek (P), to improve their text. This three-fold symphysis of the original Boethian text with the help of Adelard's treatise is similar to the many redactions to which Adelard's translation was submitted in the course of the 12th and 13th century; the most important at that time was the Campanus version which likewise stands in the Adelard tradition. ·

2.5 The Geometries in the Renaissance.

After the two Geometries had long fallen into oblivion in the 13th and 14th centuries, the interest of the humanists turned to the Geometry I and II, which they doubtlessly interpreted as the authentic writings of Boethius. In the libraries of eight scholars from the 15th and 16th centuries there is found with certainty a copy of the Geometria II: Conrad Peutinger, Willibald Pirckheimer, Frans Nans, Jean Nicot, Pierre Pithou, Jacques-Auguste de Thou, Jacob Rehdiger and Marquard Freher. In addition to some of these humanists the following also possessed manuscripts of the Geometria I: Jacques Bongars, Hans Jakob Fugger, Hermann and Hartmann Schedel, Pieter Schrijver. Among the existing manuscripts of both Geometries there are still seven humanistic codices; these originate, for the most part, from Italy.[171] One of these manuscripts (Cesena, Plut. sin. XXVI.1), which originated in Cesena after 1450, is the text upon which the *editio princeps* is based.[172]

The first printing of the two Geometries was published in 1492 by Johannes and Gregorius de Gregoriis in Venice.[173] It is part of the first printed collection of Boethius' works; it contains the Geometria II in its entirety but the Geometria I with only the first two books.[174] It is not clear who edited these geometries.[175] The edition of the Geometria II stands out because of its numerous good conjectures. Often the editor corrected accumulated errors of his text and without the help of other codices, reconstructed the original text. The editor's conjectures show his abilities: aside from a few cases where he changed the text unnecessarily, often in places where the other manuscripts made a faulty transmission, he arrived at a logical text. For the Euclidian section he seems to have used a Greek Euclid text. In some places he even corrected the author of the Geometria II whose text was very faulty. Nevertheless, it is the editio Veneta that regardless of its inadequacies was the best text available until 1867.

All of the subsequent editions are derived from this *editio princeps*, as the new printing of Venice, 1499,[176] the edition contained in the Boethius collection by Heinrich Loriti (Gla-

---

171 See pp. 192, 194, and 195.
172 See Folkerts, *Geometrie II*, p. 56.
173 *Gesamtkatalog der Wiegendrucke*, no. 4511.
174 Presumably the rest has not been printed, because the Euclid excerpts, which make up the greatest part of books 3-5, are also present in the Geometria II.
175 In the printed edition there is a letter from the Venetian Nicolaus Judecus to Donatus Civalellus as preface. It does not inform us who was the editor of the Geometries.
176 *Gesamtkatalog der Wiegendrucke*, no. 4512.

reanus) (Basil, 1546), and its reprinting (1570) as well as the excerpts which were published by Henri Estienne and Simon de Colines (Paris, 1500 and after).[177]

2.6 Conclusion.

The role played by the two geometries attributed to Boethius in the Middle Ages can only be understood in the context of the evolution of mathematics. Because geometry had acquired a stable position in the *quadrivium* and no better geometrical text had come from antiquity, even these inadequate compilations survived. But more than that, because of the famous name „Boethius" and the fact that these texts contained otherwise inaccessible information (the Euclidian excerpts) both writings underwent, up until the 12th century, a certain blossoming and wide circulation as is evidenced by the number of manuscripts and the citations by other writers. Scholars also strove for an understanding of the more obscure parts; for this purpose there grew up scholia which are found in a few manuscripts of the Geometria II.[178] The significance of both geometries diminished as the fixed scheme of the *artes liberales* gradually became more flexible and new ideas in the medieval learning forced their way in.[179] The diminishing importance of the Geometries became particularly obvious after the 11th century as learned texts were continually translated from the Arabic. Computations on the abacus were gradually replaced by the Indic-Arabic methods; many new geometrical writings, also, were available in Latin translation. Especially, scholars now had the opportunity to use a complete and reliable Euclid text. Under these circumstances there was hardly any place left for the obsolete and faulty geometries of the Pseudo-Boethius. In the 12th century a few scholars tried to preserve the Euclidian part by combining it with the Latin Arabic tradition, but this was only a transitory phase. After the 13th century the Geometries fell into oblivion until they appeared again in the Humanistic context.

---

177 In addition to Sacrobosco's *De Sphaera* there are printed only the Euclid excerpts, taken from the Geometria II. Further information about these and other editions may be seen in Folkerts, *Geometrie II*, pp. 41-49.
178 These will be evaluated in another place.
179 For this one may consult the collection *Artes Liberales: Von der antiken Bildung zur Wissenschaft des Mittelalters*, ed. J. Koch (Leiden, Köln, 1959) and especially the essays by H.M. Klinkenberg, „Der Verfall des Quadriviums im frühen Mittelalter," and O. Pedersen, „Du quadrivium à la physique."

Addendum

[pp. 201, 205–208] In the mean time the origin of the version of Euclid, here referred to as "Version II by Adelard of Bath" or "Adelard II", and its relation to the other Latin Euclid texts of the 12th century has become clearer. We now know that "Version II" is not a translation from the Arabic, but a reworking which is based upon Adelard I (i. e. the translation that Adelard of Bath made from the Arabic), the translation of Hermann of Carinthia from the Arabic and the Boethius tradition; the author of this version was not Adelard of Bath, but almost certainly Robert of Chester. For details on this text see the edition of Version II (M. Folkerts, H. L. L. Busard (eds.), *Robert of Chester's (?) Redaction of Euclid's* Elements: *the so-called Adelard II Version*, Basel / Boston / Berlin 1992) with remarks about the origin of this version and its relation to other Latin texts of the *Elements* on pp. 12–31. Therefore, in all places where the name "Adelard" occurs, it should be replaced by "Robert of Chester".

Die Altercatio in der Geometrie I des Pseudo-Boethius.
Ein Beitrag zur Geometrie im mittelalterlichen Quadrivium

Die deutschsprachige Fachprosa des Mittelalters, die von den Germanisten lange vernachlässigt wurde, hat erst jetzt vor allem durch Gerhard Eis' Arbeiten einen festen Platz im Lehr- und Forschungsbetrieb gewonnen. Auch die in Latein verfaßte Fachliteratur der Artes steht noch immer oft im Schatten des dichterischen Schrifttums: Zwar sind durch den Aufschwung, den die Wissenschaftsgeschichte seit der 2. Hälfte des 19. Jahrhunderts genommen hat, viele Texte des Gebrauchsschrifttums im Mittelalter bekannt geworden; trotzdem liegen in den Bibliotheken noch viele Abhandlungen in lateinischer Sprache, die noch nicht beachtet oder nur in groben Umrissen ausgewertet wurden.

Einer dieser Texte wird im folgenden erstmals in einer kritischen Edition gedruckt vorgestellt. Obwohl er vom Inhalt her nicht zu den Meisterleistungen des mittelalterlichen Fachschrifttums gehört, lohnt es sich, ihn der Vergessenheit zu entreißen, da er zu den wenigen mathematischen Schriften gehört, die wir aus der vor- und frühkarolingischen Zeit besitzen, und er außerdem bis zum 15. Jahrhundert oft gelesen wurde. Es handelt sich um die sogenannte 'Altercatio duorum geometricorum', ein Wechselgespräch zwischen Lehrer und Schüler, das in der im folgenden als 'Geometrie I' bezeichneten Schrift des Pseudo-Boethius den größten Teil des 5. Buches ausmacht. Dieser Dialog ist der einzige Abschnitt der Schrift, der noch nicht ediert ist[1].

Zum besseren Verständnis sollen zunächst die für die 'Geometrie I' wesentlichen wissenschaftlichen Vorarbeiten genannt werden (Kapitel 1). Da sich die Geometrie I durch Blattversetzungen, Zusätze und Auslassungen in einem äußerst korrupten Zustand befindet, wird dann kurz auf Entstehung, Aufbau und Inhalt der Schrift hingewiesen (Kapitel 2). Aufgrund der Überlieferung ergeben sich für Umfang und Stellung der 'Altercatio' innerhalb des Werkes besondere Probleme, die im Kapitel 3 dargelegt werden. Kapitel 3 geht ebenfalls auf den Inhalt der 'Altercatio' ein, Kapitel 4 auf die Überlieferung. Den Schluß bildet die kritische Edition, die alle bekannten Handschriften berücksichtigt.

---

[1] Lediglich der Schluß wurde von Lachmann abgedruckt; siehe unten S. 89. Geymonats Ausgabe (4) blieb ungedruckt.

Die Altercatio in der Geometrie I des Pseudo-Boethius

1. *Übersicht über den Stand der Forschung*

Zwei Geometrien werden in den Handschriften unter dem Namen des Boethius überliefert. Die kürzere, die sogenannte 'Geometrie II' in zwei Büchern, kann hier vernachlässigt werden. Sie ist ein Kompendium, das aus drei Teilen besteht (Euklidexzerpte, gromatische Texte, Abschnitt über den Abakus) und von einem unbekannten Autor in der ersten Hälfte des 11. Jahrhunderts in Lothringen verfaßt wurde. Durch das Kapitel über das Rechnen auf dem Rechenbrett, in dem die sogenannten arabischen Ziffern erstmals in einem zusammenhängenden Text in lateinischer Sprache erwähnt werden, fand diese Schrift große Beachtung bei den Wissenschaftshistorikern. Auch die längere Geometrie in 5 Büchern (im folgenden 'Geometrie I' genannt) stammt in ihrer jetzigen Gestalt nicht von Boethius. Während die Fragen, die mit Entstehungszeit und Abfassung der 'Geometrie II' zusammenhängen, heute im wesentlichen als gelöst gelten können, gibt die 'Geometrie I' noch manche Probleme auf, die dadurch erschwert werden, daß es noch keine kritische Edition des gesamten Textes gibt. Immerhin existieren einige zum Teil recht umfangreiche Arbeiten zur 'Geometrie I', die im folgenden kurz charakterisiert werden sollen.

Die 'Geometrie I' ist aus drei Elementen zusammengesetzt: Auszüge aus dem 'Corpus agrimensorum' befinden sich in Buch I, II (Anfang) und V, arithmetische Exzerpte in Buch II und geometrische in den Büchern III, IV und V. Die beiden ersten Bücher sind schon lange ediert: erstmals in der editio princeps des Boethius (Venedig 1492) nach einem Codex, der der Handschrift Cesena, Malat. sin. XXVI, 1 nahestand, dann praktisch unverändert in den Boethius-Ausgaben Venedig 1499, Basel 1546, Basel 1570. Ein Nachdruck der Basler Edition wurde von Migne veranstaltet (6).

Der erste, der sich näher mit der 'Geometrie I' beschäftigte, war K. Lachmann. Im Zusammenhang mit seiner Edition der Agrimensoren (1) gab er von der 'Geometrie I' den größten Teil des 1. Buches, ferner Buch III, IV und Teile von Buch V heraus. Allerdings ist seine Arbeit noch mit wesentlichen Fehlern behaftet: Er und sein Mitarbeiter F. Blume unterschieden die 'Geometrien I' und 'II', deren Inhalt sich zum Teil überdeckt, noch nicht genau genug, und von den 26 erhaltenen Handschriften der 'Geometrie I' benutzten sie nur vier, die zudem noch jüngeren Datums sind (b, m, a, R)[2].

Entscheidende Beiträge zur Erforschung der 'Geometrie I' lieferten um die Jahrhundertwende N. Bubnov (2) und C. Thulin (8). Ohne an eine kritische Ausgabe heranzugehen, sichteten sie die Handschriften und versuchten, die Schrift in die Überlieferung des 'Corpus agrimensorum' einzuordnen. Ihre Ergebnisse sind auch

---

[2] Die benutzten Handschriftensiglen beruhen auf den von Thulin gewählten Abkürzungen ([8], S. 30 f.). Sie sind im Kapitel 4 aufgelöst.

heute noch grundlegend. Bubnov kannte 17 Handschriften der 'Geometrie I' ([2], S. 180—188; Addenda 559), ohne ihre Abhängigkeit untereinander zu prüfen. Immerhin erwähnte er erstmals die beiden Hyparchetypi ([2], S. 472—488) und sah richtig die Zusammenhänge zwischen der 'Geometrie I' und den Schriften der Agrimensoren. Thulin führte Bubnovs Arbeiten weiter. Er kannte bis auf vier Codices (d, A, w, b^r) alle Handschriften der 'Geometrie I', versuchte, sie in Gruppen einzuordnen, verbesserte einige Fehler Bubnovs und gab eine relativ ausführliche Analyse der Schrift, daneben eine Beschreibung der einzelnen Handschriften und eine Übersicht über Inhalt und Aufbau ([8], S. 3—43).

Zwei weitere Arbeiten aus dieser Zeit betreffen Spezialprobleme der 'Geometrie I': J. L. Heiberg beschäftigt sich in seinem Aufsatz (5) vor allem mit den Zusammenhängen zwischen den Euklidexzerpten in den beiden Boethius zugeschriebenen Geometrien. Seine Ansichten über den ursprünglichen Aufbau der 'Geometrie I' ([5], S. 52 f.) sind zumindest in Einzelheiten anfechtbar. Darauf hat schon P. Tannery in seiner Arbeit (7) hingewiesen. Tannery stellt ferner eine Fehldeutung richtig, zu der Bubnov durch die falsche Einschätzung einer Inscriptio der Bamberger Handschrift b verleitet wurde. Bei seinem Versuch, die ursprüngliche Gestalt der Geometrie zu rekonstruieren, gelangt Tannery zu Ergebnissen, die in den Hauptzügen vermutlich zutreffen, auch wenn er nur zwei Handschriften näher untersuchte. Tannerys Verdienst ist es vor allem, auf den vielleicht ältesten Codex (p^1 = Paris, BN, lat. 13020) hingewiesen zu haben.

Während die Teile der 'Geometrie I', die bei Lachmann (1) und Migne (6) abgedruckt sind, nicht den Ansprüchen an eine moderne Edition genügen, sind jetzt wenigstens die Euklidexzerpte in der 'Geometrie I' kritisch ediert: Die Ausgabe in (3), S. 174—214, berücksichtigt alle vorhandenen Handschriften.

M. Geymonats Dissertation (4) behandelt in ihrem ersten Teil, der jetzt gedruckt vorliegt[3], die lateinischen Euklidexzerpte in einem Veroneser Palimpsest, die nach Geymonats Meinung auf die Übersetzung des Boethius zurückgehen. Im Teil 2, der nicht publiziert wurde ([4], S. 95—172), sind drei Exzerpte aus der 'Geometrie I' kritisch ediert, wobei auch die Parallelüberlieferung angemerkt ist, und zwar das Kapitelverzeichnis in Buch I und die beiden Dialoge in Buch II und V[4]. Geymonat zog alle Handschriften bis auf p^r, c und A heran. Der Abschnitt L',,Altercatio duorum geometricorum" ed altri excerpta della geometria pseudo-boeziana auf S. 160—172 geht auf Fragen der Entstehungszeit, Quellenbenutzung und Überlieferung dieser drei Teile ein. Geymonats Ergebnisse, die mir erst kurz vor Abschluß meiner Arbeit bekannt wurden, decken sich weitgehend mit meinen eigenen. Geymonats wohldurchdachte und fundierte Erörterungen sind in dem vorliegen-

---

[3] *Euclidis Latine facti fragmenta Veronensia*, Milano/Varese 1964 (= Testi e documenti per lo studio dell'antichità, 9).
[4] Also die Kapitel I.2, II.1 und V.7 nach der Einteilung auf S. 89.

den Aufsatz berücksichtigt; seine sorgfältige Edition ermöglicht eine Kontrolle meiner eigenen Kollationen.

Noch eine Arbeit der jüngsten Zeit bringt wesentlich neue Ergebnisse in der Frage der 'Geometrie I': B. L. Ullman hat in seinem Aufsatz (9) durch geistreiche Schlüsse neue Erkenntnisse zur Entstehung und frühen Überlieferungsgeschichte der 'Geometrie I' gewonnen. Ausgehend von den Handschriften der Agrimensoren und den ältesten Codices der Geometrie I, konnte er überraschende Zusammenhänge aufdecken, die nicht nur Licht auf jene dem Boethius zugeschriebene Geometrie werfen, sondern auch das Zentrum geometrisch-gromatischer Arbeiten im 7.—9. Jahrhundert erstmals erkennen lassen.

2. *Mutmaßliche Entstehung und Inhalt der Geometrie I*

Insbesondere die Arbeiten von Bubnov (2), Thulin (8) und Ullman (9) gestatten es, Aussagen über Entstehung und Inhalt der 'Geometrie I' zu machen. Diese Schrift, die in den Handschriften im allgemeinen den Titel '*Ars geometriae et arithmeticae*' trägt, entstand spätestens im 9., vielleicht schon im 8. Jahrhundert. Sie enthält viel Material aus dem 'Corpus agrimensorum', war aber nicht als Lehrbuch für angehende Feldmesser gedacht, sondern als mathematisches Schulbuch für den Unterricht in Geometrie und Arithmetik innerhalb des Quadriviums. Alles deutet darauf hin, daß das Werk in Corbie verfaßt wurde: Dort war im 8./9. Jahrhundert der Stammvater der jüngeren Rezension der Agrimensoren (Handschriften PG) vorhanden, ferner mindestens fünf Codices von Isidors '*Etymologiae*' und einer von Cassiodors '*Institutiones saecularium lectionum*' ([9], S. 275 f.), alles Schriften, die vom Kompilator der 'Geometrie I' exzerpiert wurden. Auch das im Mittelalter besonders seltene Werk des Columella wurde für die gromatischen Exzerpte, die sich in den meisten Boethius-I-Handschriften im Anschluß an die 'Geometrie I' finden, und für die Herstellung der 'Geometrie I' selbst (siehe Kapitel 3) benutzt: Nur zwei Columella-Handschriften sind vor dem 15. Jh. bekannt, eine aus Fulda (heute in der Mailänder Ambrosiana) und eine aus Corbie (heute in Leningrad). Beide wurden im 9. Jahrhundert geschrieben. Von den Handschriften der 'Geometrie I' schließlich stammen drei der ältesten aus Corbie (p^1, p^2, p^3, 9./10. Jh.)[2].

Ausgehend von Corbie, "the gromatic and geometric capital of the mediæval world" ([9], S. 283), verbreitete sich die Geometrie I schon recht bald im lateinisch sprechenden Europa: Ein Verweis im Reichenauer Katalog von 822 zeigt, daß diese Schrift damals schon in diesem Kloster vorhanden war ([2], S. 181; [9], S. 277); eine Handschrift, die vermutlich eng mit diesem Codex zusammenhängt, ist heute noch erhalten (m = Clm 560). Auch in Murbach, einer Gründung der Reichenau, dürfte die Geometrie im 9. Jh. existiert haben ([3], S. 34). Gerbert benutzte 983 in Bobbio eine Handschrift, die auch die 'Geometrie I' enthielt (vermutlich den jetzt in Neapel befindlichen Codex N), und schickte sie oder eine Abschrift davon um 997 an Adelbold von Lüttich ([2], S. 398. 475; [8], S. 5 f.). Auch die Handschrift b könnte über Gerbert nach Bamberg gelangt sein (s. S. 96). Nur wenig später, im Jahre 1004, wurde in Luxeuil die einzige genau datierte Abschrift der 'Geometrie I' angefertigt (bⁿ); einige Jahre darauf schenkte sie Bischof Werner der Straßburger Dombibliothek anläßlich des Dombaus, der 1015 begonnen wurde ([8], S. 12; [9], S. 268). Überhaupt stammen von den 26 Handschriften der 'Geometrie I' 15 aus dem 9.—11. Jahrhundert; davon dürften acht vor dem Jahre 1000 geschrieben sein[5].

Die erstaunlich rasche Verbreitung des Werkes ändert nichts an der Tatsache, daß die 'Geometrie I' ein recht dürftiges Machwerk eines Kompilators ist, der von dem Material, das er heranzog, nicht eben viel verstand. Dazu kommt, daß die Überlieferung zeitlich zwar fast bis zum Original zurückverfolgt werden kann, daß sie aber zu einem ältesten rekonstruierbaren Archetypus führt, der durch Umstellungen und Zusätze oft nicht verständlich ist[6]. Versuche, den Zustand wiederherzustellen, in dem sich das Werk vor Anfertigung dieser Abschrift befand, sind daher notwendig spekulativ und erfordern gewagte Konjekturen. Brauchbare Ansätze findet man bei Tannery ([7], S. 46—50) und Thulin ([8], S. 27—30). Weitere Aufschlüsse könnte vielleicht eine ausführliche Analyse der Schrift unter Berücksichtigung der gesamten handschriftlichen Überlieferung erbringen, obwohl auch sie wahrscheinlich nur zu Vermutungen führen wird. Leider dürfte Tannerys Urteil zutreffen ([7], S. 49):

> „Il est illusoire de prétendre le restituer tel qu'il a pu exister avant la division en cinq livres attribués à Boèce … La Pseudo-Géométrie nous apparaît dans un désordre irrémédiable, et nous ne pouvons faire que des conjectures très hasardées sur l'état des sources manuscrites dont elle dérive."

Daher soll im folgenden eine knappe Übersicht über den Inhalt der Schrift, wie sie sich aufgrund der erhaltenen Handschriften ergibt, gegeben werden. Die Einteilung in Abschnitte entspricht Thulin (8), S. 32—36. Angegeben ist ferner der ungefähre Umfang der Abschnitte (1 Zeile = 55 Buchstaben) und der Ort, wo die Texte abgedruckt sind.

Buch I:

1. Einleitung über Bedeutung und Nutzen der Geometrie; 'epistola Iulii Caesaris' mit verschiedenen Auszügen aus den Gromatici. 270 Zeilen. Lachmann (1), S. 393, 1—406, 25; Migne (6), S. 1352, 44—1358, 3.

---

[5] Nähere Angaben hierzu in Kapitel 4.

[6] Heibergs Vermutung ([5], S. 52 f.), die scheinbar sinnvollere Anordnung des Bambergensis sei die ursprünglichere, ist von Tannery mit Recht zurückgewiesen worden ([7], S. 46—49). Daß auch die Veränderungen der Handschriften Rfm (s. Kapitel 4) spätere Umformungen sind, beweist Thulin ([8], S. 31 f.).

Die Altercatio in der Geometrie I des Pseudo-Boethius

2. Kapitelverzeichnis, das aber nur teilweise den später behandelten Themen entspricht. 30 Zeilen. Migne (6), S. 1358, 5—36.

Buch II:

1. Dialog zwischen Δ(ιδάσκαλος) und M(αθητής) über Ursprung und Inhalt der Geometrie. 40 Zeilen. Migne (6), S. 1358, 37—1359, 18.

2. Auszüge aus der 'Arithmetik' des Boethius. 300 Zeilen. Migne (6), S. 1359, 19—1364, 51.

Buch III:

1.—6. Euklid, 'Elemente', Definitionen zu Buch I—IV, Propositionen von Buch I—II. 250 Zeilen. Lachmann (1), S. 377, 2—388, 19 (ohne die Beweise zu I 1—3); 391, 18—23; Folkerts (3), S. 176—204 = Z. 1—225.

Buch IV:

1.—4. Euklid, 'Elemente', Propositionen von Buch III und IV (z. T.) 70 Zeilen. Lachmann (1), S. 388, 20—389, 20; 390, 21—391, 16; 391, 24—392, 17; Folkerts (3), S. 204—210 = Z. 226—277.

Buch V:

1. *Mensura est* . . .: Auszüge aus Isidor. 20 Zeilen. Lachmann (1), S. 407, 1—408, 2.

2. *Ad dato puncto* . . .: Euklid III 7.9. 6 Zeilen. Lachmann (1), S. 389, 21—27 = 408, 3—9; Folkerts (3), S. 210—212 = Z. 278—282.

3. *Nam extremitatum genera* . . .: Gromatische Texte, die auf Balbus, Frontinus, Hyginus beruhen. 34 Zeilen. Lachmann (1), S. 408, 10—409, 17.

4. *Sunt et medii termini* . . .: Gromatische Exzerpte aus dem 'Liber regionum', 'De sepulchris' und Hyginus. 15 Zeilen. Lachmann (1), S. 409, 18—410, 7.

5. *Si duo circuli* . . .: Teile aus Euklid III. 23 Zeilen. Lachmann (1), S. 389, 28—390, 20; Folkerts (3), S. 212—214 = Z. 283—302.

6. *Stadius habet passus* . . .: Text über Maße. 2 Zeilen.

7. *Quoniam diversae formae* . . .: Längerer Schuldialog über geometrische Fragen zwischen dem In(terrogator) und R(esponsor). 180 Zeilen. Nur das Ende abgedruckt von Lachmann (1), S. 410, 8—412, 15. Dieser Teil ist am Ende der vorliegenden Arbeit ediert.

8. *Sex sunt ordines* . . .: Schlußworte. 6 Zeilen. Lachmann (1), S. 412, 16—21.

Die Schrift ist also aus drei Elementen zusammengesetzt: arithmetischen, gromatischen und geometrischen. Der arithmetische Abschnitt nimmt den größten Teil des 2. Buches ein (II.2); er ist willkürlich aus verschiedenen Kapiteln von Boethius' 'Arithmetik' zusammengefügt. Die gromatischen Auszüge bilden Buch I.1 und Teile des 5. Buches (V. 1.3.4.6—8); auch der Dialog II.1 beruht auf Material des 'Corpus agrimensorum'. Auch hier erfolgte die Auswahl recht wahllos ([8], S. 37—39). Relativ geschlossen ist der geometrische Teil, der Exzerpte aus Euklid I—IV (ohne Beweise) enthält. Er nimmt Buch III und IV der 'Geometrie I' ein, wobei Teile, die ursprünglich ins Buch IV gehörten, durch Blattversetzung ins

5. Buch gelangten (V.2.5). Diese Euklidauszüge, die auf Boethius' Übersetzung zurückgehen, machen hauptsächlich den Wert der Schrift aus. Rechnet man die schlecht überarbeiteten Exzerpte aus bekannten arithmetischen und gromatischen Schriften ab, die den größten Teil des Werkes einnehmen, so bleiben an interessanten Stücken nur noch die beiden Dialoge II.1 und V.7 übrig. Gerade das längere Gespräch V.7 wurde (abgesehen von Geymonat) bislang immer aus den Untersuchungen ausgeklammert. Ihm gilt der folgende Abschnitt, wobei zunächst die Frage zu behandeln ist, welche Teile des sehr verderbten Buches V dem längeren Dialog zuzurechnen sind und wie dieses Buch überhaupt aufgebaut ist.

3. Die 'Altercatio' im 5. Buch der 'Geometrie I'

Die handschriftliche Überlieferung macht nicht eindeutig klar, wo die 'Altercatio' beginnt und wo sie endet, da sich die Überschrift *Incipit altercatio duorum geometricorum de figuris, numeris et mensuris* in den Codices an verschiedenen Stellen des 5. Buches findet: Der Hyparchetypus X^{II}, dem die Mehrzahl der Handschriften angehört, darunter auch die aus Corbie stammenden Codices p^1, p^2, p^3 (siehe Kapitel 4), bringt diese inscriptio am Anfang des 5. Buches, während die andere Gruppe X^I die Überschrift erst vor V.7 (siehe Kapitel 2) folgen läßt. Ein Explicit fehlt durchweg. So empfiehlt es sich, zunächst die Bestandteile des besonders verderbten Buches V zu analysieren und dann zu versuchen, aufgrund inhaltlicher Kriterien Umfang und Funktion der 'Altercatio' zu bestimmen.

Das Buch beginnt mit einem Abschnitt über Längen- und Flächenmaße, die bei der Feldmessung auftreten (Kapitel V.1). Dieser Text ist aus Isidor, 'Etymologiae' XV 15, entnommen; er begegnet uns unter dem Titel 'De mensuris agrorum' fast wörtlich auch an anderer Stelle im 'Corpus agrimensorum'[7]. Vom Inhalt her paßt dieser Teil zur 'Altercatio', die ja auch von Maßen handeln soll. Äußerlich allerdings trägt er kaum Kennzeichen eines Dialogs: Das in V.7 übliche vorgesetzte In(terrogator) bzw. R(esponsor) fehlt hier, und die einzigen Worte, die auf einen Dialog hindeuten[8], scheinen später eingefügt zu sein, um Dialogform zu geben.

Der folgende kurze Abschnitt V.2 wird fortgesetzt durch V.5. Es handelt sich um Auszüge aus Euklid, Buch III, die ursprünglich dem 4. Buch der Geometrie I angehörten (zwischen IV.2 und IV.3) und später durch Blattversetzung der Vorlage an diese Stelle gelangten.

---

[7] In der Agrimensoren-Handschrift G und in den gromatischen Exzerpten, die in den meisten Handschriften der 'Geometrie I' folgen; siehe (8), S. 26. Gegenüber dem Kapitel 'De mensuris agrorum', das Lachmann abdruckte ([1], S. 367, 9—368, 18), ist der Abschnitt V.1 in der 'Geometrie I' am Ende umgeformt und verkürzt.

[8] *Sed ut ad rem primae artis geometricae veniamus, quod pedali mensura comprehenditur edicere non tarderis*: (1), S. 407, 12—14.

Die Altercatio in der Geometrie I des Pseudo-Boethius

Die zusammengehörigen Euklidexzerpte V.2 und V.5 sind durch zwei geo-metrisch-gromatische Stücke unterbrochen: V.3 beginnt mit einer Definition der Begrenzungslinien (*extremitates*) nach Balbus (98, 5—9 La.). Dann werden die bei-den möglichen Arten *rigor* und *flexus* erklärt. Der Text geht von den *extremitates* über zur Vermessung derartig begrenzter Stücke und erläutert vor allem, wie bei nichtebenem Gelände die schiefe auf die horizontale Fläche reduziert werden kann (*cultellare*). Auch dieser Kommentar beruht zum großen Teil auf bekannten Texten[9]; wie in V.1 findet sich auch hier ein Einschub, der den Anschein eines Dia-logs geben soll[10]. — Der Abschnitt V.4 trägt die unklare Überschrift '*De interna ratione et non recipiendos limites*'. Er handelt von besonderen Grenzsteinen und von Grenzproblemen, die durch Bergbäche entstehen können. Der vorliegende Text ist zusammengestoppelt aus den völlig verschiedenartigen Exzerpten 'Liber regionum' 213, 9—14 La.; 'De sepulchris' 272, 5—8 La.; Hyginus 128, 11—19 La., die wörtlich wiedergegeben sind; er handelt keineswegs *de figuris, numeris et men-suris*.

Nach den Euklidexzerpten V.5 folgt ein zweizeiliger Text über die Maße *sta-dius, miliarium, centuria*[11]. Er steht in sachlichem Zusammenhang mit dem Isidor-Exzerpt V.1 und kann entweder aus Isidor XV 15 entnommen sein (368, 11—15 La.) oder aus der geometrischen Einleitung zum 5. Buch des Columella (V 1, 6 f.).

Der anschließende lange Text V.7 enthält den Schuldialog zwischen dem Inter-rogator und Responsor. Dem eigentlichen Wechselgespräch geht ein kurzer Ab-schnitt voraus (Z. 3—14 der Edition in Kapitel 5), der sich in zwei Teile gliedert. Die erste Hälfte ist ein fast wörtliches Zitat aus dem Vermessungskapitel zu Beginn des 5. Buches in Columellas *De re rustica* (V 1, 13—2, 1). Es handelt von den ver-schiedenen Arten der Äcker (quadratische, längliche, dreieckige, runde, halbkreis-förmige, bogenförmig begrenzte) und bildet den Zusammenhang zu dem in der Geometrie I vorangehenden Text V.1.6, der ebenfalls auf Columella beruht.

Es folgen die Definitionen 1.2.4.5 aus Buch I von Euklids Elementen (Definition von Punkt, Linie, Strecke, Fläche) mit einem Zusatz nach Definition 1, der eine weitere Erklärung des Wortes *punctum* enthält[12]. Dieser Zusatz findet sich wört-lich auch vor den Euklidauszügen in Buch III der 'Geometrie I'[13]; da er den Punkt als *Principium mensurae* bezeichnet, handelt es sich um eine aus gromatischen Quellen stammende Erläuterung. Der Text der Definitionen aus Euklid stimmt mit den gleichartigen Exzerpten am Anfang des 3. Buches der Geometrie I fast

---

[9] Hyginus 128, 18 f. La.; Frontinus 27, 3—5 La.; Hyginus 109, 16—20 La.

[10] *Sed ut haec plene scias, breviter insinuamus, quod doceas*: (1), S. 408, 16 f.

[11] *Stadius habet passus CXXV, VIII stadia miliarium faciunt. Centuria habet in se iugera CC.*

[12] *Principium mensurae punctum vocatur, cum medium tenet figurae.*

[13] Siehe Lachmann (1), S. 377, 2; Folkerts (3), S. 176, 1.

wörtlich und mit Isidor III 12, 7 wörtlich überein. Isidor ließ die Definition 3, die an dieser Stelle fehlt, ebenfalls aus und fügte auch die einleitenden Worte *Prima autem figura huius artis* und *Secunda* hinzu, die im Buch III dieser Geometrie fehlen. Der Kompilator scheint hier also Isidor abgeschrieben zu haben; der Zusatz nach Definition I stammt dagegen vermutlich aus Buch III der Geometrie I.

Den Anschluß an die Euklidexzerpte nach Isidor bildet ein langer Auszug aus Augustinus, 'De quantitate animae' VIII—XII (Z. 15—114 dieser Edition), der von Geymonat erstmals identifiziert wurde[14]; bisher galt dieser Teil als selbständige Leistung des Kompilators[15]. Augustin gibt in seiner Schrift über die Größe der Seele einen wohl fiktiven Dialog mit seinem Freund Evodius wieder, der in den Kapiteln VIII—XII über Eigenschaften und Vorzüge von Punkt, Linie und Figuren handelt. In seinem Versuch, eine Werthierarchie unter den geometrischen Figuren aufzustellen, lassen sich pythagoreische Elemente erkennen. Die mathematischen Kapitel in Augustins Schrift bezwecken, das Verhältnis von Seele und Leib zu veranschaulichen; sie führen zur Erkenntnis, daß sich die Ordnung aus dem Zusammentreffen von Unregelmäßigkeiten ergibt. Dieser Teil wird in den Übersetzungen und Kommentaren zu 'De quantitate animae' nur gestreift[16] und in mathematikgeschichtlichen Darstellungen völlig übersehen, obwohl er für die Einschätzung der Mathematik in der Spätantike wichtig ist.

In diesem Zusammenhang kann es jedoch nicht Aufgabe sein, den mathematischen Abschnitt aus Augustins Schrift zu würdigen; hier geht es nur um die Frage, wie der Kompilator der 'Geometrie I' seine Vorlage exzerpierte. Der Auszug beginnt mit Kapitel XI, 18 aus Augustins Werk, das zunächst bis XII, 21 ausgeschrieben wird (Z. 15—87). In diesem Teil gibt es nicht nur in den beiden Handschriftenklassen X¹ und Xᴵᴵ (siehe Kapitel 4) Lücken von einer oder mehreren Zeilen, sondern manchmal weist schon der Archetypus derartige Auslassungen auf, die ohne Kenntnis der Vorlage sinnentstellend sind[17]. Nach der Erörterung über die drei Dimensionen Länge, Breite und Höhe folgen zwei weitere Exzerpte aus 'De quantitate animae' (IX, 14 und VIII, 13), die in den Zusammenhang nicht recht passen, bevor Z. 110—114 der erste lange Augustin-Text wieder aufgenommen wird. Ohne Übergang schließt sich dann ein kurzer Auszug aus einer anderen Schrift des Augustinus an (Z. 117—119 = Augustinus, 'Soliloquia' I, IV, 10), der den Unterschied zwischen Kreis und Gerade verdeutlichen soll.

---

[14] Im nicht publizierten Teil seiner Dissertation (4), S. 169.
[15] So Tannery (7), S. 48.
[16] z. B. von Josef M. Colleran, Westminster/Maryland 1950; Pierre de Labriolle, Paris 1951; Carl Johann Perl, Paderborn 1960.
[17] z. B. Z. 23—26. 41—43. Es läßt sich nicht mehr klären, ob diese Auslassungen schon dem Kompilator anzulasten sind. In der Edition wurden diese Lücken ergänzt, also als Versehen des Archetypus-Schreibers gewertet.

Die Altercatio in der Geometrie I des Pseudo-Boethius

Die Erwähnung von *sphera* und *linea* veranlaßt den Kompilator, auf den *ager sphericus* und *quadratus* hinzuweisen (Z. 119—121). Dadurch ist der Übergang in die gromatische Terminologie vollzogen, und folgerichtig finden wir im Anschluß daran Auszüge aus dem 'Corpus agrimensorum', die rein formal in die Gestalt eines Wechselgesprächs gebracht sind (Z. 122—184). Dieser Text, den bereits Lachmann abdruckte[18], ist hauptsächlich Balbus entnommen; er behandelt die Arten der Linien, Begrenzungen, Flächen und Winkel. Auch hier ist der Text durch Lükken manchmal bis zur Unkenntlichkeit entstellt.

Immerhin zeigt der Textvergleich innerhalb des Balbus-Textes, daß enge Beziehungen zwischen der 'Geometrie I' und der Agrimensoren-Handschrift G bestehen. Dies paßt gut zur Ansicht Thulins ([8], S. 6), G sei die Quelle der mit „Boethius" verbundenen gromatischen Exzerpte gewesen. Für die 'Geometrie I' müssen wir diese Feststellung allerdings leicht modifizieren, da die Erklärung der *extremitates* ('Altercatio', Z. 139—146) unter den Balbus-Exzerpten der Klasse PG nicht vorkommt[19]. Der Autor der 'Geometrie I' benutzte also eine Balbus-Handschrift der Klasse PG, die vollständiger als die erhaltenen Codices dieser Gruppe war[20].

Vermutlich entnahm der Kompilator aus dieser Handschrift auch die außerordentlich verderbten Auszüge aus Hyginus ('Altercatio', Z. 167—174). Sichere Aussagen sind hier allerdings nicht möglich. Dieser Text wird nämlich nur im Arcerianus B und in den *codices mixti* der EF-Klasse überliefert[21], weicht im Wortlaut dort aber stark von der 'Geometrie I' ab. Hier sieht man, wie bruchstückhaft unsere Kenntnis von der Textgeschichte des 'Corpus agrimensorum' ist.

Diese Bemerkungen zu den Quellen des Kompilators dürften genügen, um die Entstehung des 5. Buches hinreichend zu erklären. Zunächst mußte die Legende zerstört werden, daß wir in der 'Altercatio' ein besonders wertvolles, weil altes Stück aus dem Mathematikunterricht des frühen Mittelalters vor uns haben: Auch dieser Teil des Werkes ist aus anderen Schriften übernommen. Wir erkennen, daß das Buch V in ganz ähnlicher Weise aus mehreren Vorlagen zusammengeschrieben wurde wie die übrigen Bücher und vor allem Buch I.

Das 5. Buch besteht also aus drei teilweise voneinander getrennten Blöcken: den Texten aus Isidor (V.1.6), den Euklidexzerpten (V.2.5) und der eigentlichen 'Alter-

---

[18] (1), S. 410, 8—412, 21.
[19] Dieser Teil wird nur in Abschriften des Arcerianus B überliefert.
[20] Daß Vorfahren der Handschriften PG im 8. Jahrhundert in Corbie zur Verfügung standen, ist sehr wahrscheinlich: P stammt nach Ullmans Meinung ([9], S. 274) aus Corbie, G aus dem benachbarten Saint Omer.
[21] Obwohl dieser Text in EF fehlt, wird seine Zugehörigkeit zu dieser Gruppe gesichert durch den Codex Nansianus (= London, BM Add. 47679). Siehe Menso Folkerts, in: Rhein. Museum für Philologie, NF 112 (1969), S. 59, 65.

catio' (V.7). Die Auszüge aus Euklid gehörten ursprünglich nicht dem Buch V an, und auch die dazwischenliegenden gromatischen Texte V.3 und V.4 machen innerhalb des Kontextes einen fremdartigen Eindruck. Mir scheint, daß es erweiterte gromatische Glossen sind, die bei einer Abschrift in den Text gerieten; auch die Euklidexzerpte in Buch III und IV der 'Geometrie I' sind oft durch Zusätze agrimensorischen Inhalts erweitert. Die Abschnitte V.2—5 dürften also mit dem eigentlichen Gespräch über geometrische Fragen nichts zu tun haben.

Dagegen scheint es, daß die übrigen Teile des 5. Buches, also V.1.6.7.8, vom Kompilator mit der Absicht zusammengefügt wurden, ein eigenes Ganzes zu bilden; ihm wurde dann der Titel *Altercatio duorum geometricorum de figuris, numeris et mensuris* zugelegt[22]. Jedenfalls gehören die Teile V.1, V.6 und V.7 Anfang eng zusammen: V.1 und V.6 sind Auszüge aus Isidor bzw. aus Columella, wobei V.6 den Text V.1 fortsetzt. Der Zusammenhang zwischen V.6 und V.7 wird durch die Columella-Exzerpte am Anfang von V.7 sichergestellt, während der unmittelbar anschließende Text nach Euklid die Überleitung zum eigentlichen Wechselgespräch bildet, das mit der Frage des Interrogators nach den Bausteinen einer Geraden beginnt.

Der Kompilator hat also auch Buch V aus verschiedenen geometrischen Texten zusammengesetzt. Auszüge aus Isidor, Columella und Euklid bilden den Anfang. An diese fast wörtlich übernommenen Texte schließen sich die langen Passagen aus Augustinus' 'De quantitate animae' und das kurze Zitat aus seinen 'Soliloquien' an. Geometrische Auszüge aus Balbus und weitere verderbte Exzerpte agrimensorischen Ursprungs folgen. Die Schlußworte des Buchs V und gleichzeitig der ganzen Schrift knüpfen an den Dialog in Buch II an, der ähnlich endet.

So hat die Quellenuntersuchung des 5. Buches zu zwei Ergebnissen geführt: Einmal ließ sich die Konzeption des Autors erkennen, der sich bemühte, aus ihm vorliegenden älteren Texten ein Buch zusammenzustellen, das über Maße und Figuren handelte. Den Kern bildete das lange Exzerpt aus Augustinus, dessen Dialogform übernommen wurde; seine äußere Gestalt wurde auf die übrigen ausgewählten Texte übertragen, die entweder (wie am Schluß) zu einem Gespräch umgeformt wurden oder (wie am Anfang) durch eingefügte Elemente den Eindruck eines Dialogs wecken sollten. So macht das 5. Buch — vernachlässigt man die vermutlich später dorthin gelangten Auszüge V.2—5 — einen unter den gegebenen Voraussetzungen relativ homogenen Eindruck. Das zweite Ergebnis ist die negative Erkenntnis, daß der Optimismus eines Tannery[23] oder Cantor[24] hinsichtlich einer

---

[22] Die Auslassung des Abschnittes V.1 und die geänderte Stellung der Überschrift in der Handschriftenklasse X^1 scheint sekundär zu sein, wie Thulin richtig vermerkt: (8), S. 31.

[23] Ce morceau, réellement curieux, paraît bien original, sans qu'il soit facile d'indiquer soit l'époque, soit l'école à laquelle il appartient: (7), S. 48. Siehe auch (7), S. 49 unten.

Die Altercatio in der Geometrie I des Pseudo-Boethius

Ursprünglichkeit der '*Altercatio*' nicht berechtigt ist: Auch dieser Abschnitt der 'Geometrie I' ist lediglich aus fremden, uns erhaltenen Schriften zusammengesetzt, ohne daß von eigenen Zutaten des Kompilators die Rede sein kann. Wir können nur die Liste der exzerpierten Werke erweitern um Augustins Schriften '*De quantitate animae*' und '*Soliloquia*', die also im 8. Jahrhundert in Corbie oder Umgebung vorhanden gewesen sein müssen.

In Anbetracht dieses Sachverhalts kann man mit Recht bezweifeln, ob eine kritische Edition der ganzen 'Geometrie I' noch notwendig ist. Damit aber wenigstens alle Teile dieser Schrift gedruckt vorliegen, erschien es mir sinnvoll, den einzigen noch nicht herausgegebenen Teil, eben die Abschnitte V.7 und 8, zu edieren. Die erhaltenen Handschriften bezeugen, daß die 'Geometrie I' trotz ihrer Dürftigkeit vor allem im 10. und 11. Jahrhundert als mathematische Quellenschrift weit verbreitet war. So ist es wohl berechtigt, alle Codices zusammenzustellen und dabei nach Möglichkeit Provenienz und frühere Besitzer anzugeben. Dies geschieht im folgenden Kapitel. Genaue Beschreibungen erübrigten sich im allgemeinen, da der Inhalt der meisten Handschriften bei Thulin und Bubnov angegeben ist. Dagegen fehlen bisher noch ernsthafte Versuche, die Codices im einzelnen zu klassifizieren. Im Hinblick auf die geistesgeschichtliche Wirkung der Schrift im Mittelalter habe ich versucht, auch hier zu Aussagen zu kommen, soweit es unter Auslassung der Bücher I und II möglich ist.

4. Überlieferung

Von der Geometrie I des Pseudo-Boethius existieren 26 Handschriften, die (mit einer Ausnahme) auch die *Altercatio* enthalten. Sie sind im folgenden chronologisch geordnet. Die von Thulin gewählten Siglen ([8], S. 30 f.) wurden beibehalten und ergänzt.

1) **N** = Neapel, Biblioteca Nazionale V A 13, Anf. 10. Jh. Enthält auf Bl. 1ʳ—15ʳ die 'Geometrie I'. Es folgen gromatische Exzerpte und ab Bl. 33 astrologische und chronologische Auszüge aus Isidor, Cassiodor und Beda. Genaue Beschreibung in (8), S. 9—12, und (2), S. 476—79, Einl. S. 49 f., 75. Vermutlich diesen Codex benutzte Gerbert 983 in Bobbio und sandte ihn später an Adelbold von Lüttich (siehe S. 88). Der ursprüngliche Text in N wurde von Gerbert oder einem seiner Schüler vor 1004 an vielen Stellen verbessert (N¹)[25].

---

[24] M. Cantor, *Vorlesungen über Geschichte der Mathematik*, 1. Band, Leipzig ³1907, S. 580 f.: „Im fünften Buche, welches im Drucke noch nicht herausgegeben ist, zeigt sich neuerdings ein Allerlei, aus welchem sich ein sehr interessantes, den Begleitstücken in keiner Weise ähnelndes Fragment '*Altercatio duorum geometricorum*' hervorhebt, ein katechetisches Zwiegespräch, dessen Ursprung in tiefstes Dunkel gehüllt ist."

[25] Thulin (8), S. 5 f.

2) b = Bamberg, Staatliche Bibliothek, Ms. class. 55 (früher H. J. IV 22), Anf. 10. Jh.
Frankreich. Der Codex kam vielleicht über Johannes Scottus und Gerbert in die
Bibliothek des Bamberger Doms, zu deren Erstausstattung er gehört haben dürfte
(kurz nach 1000)[26]. Die 'Geometrie I' steht in eigentümlicher Anordnung[27] auf
Bl. 1ᵛ—16ʳ; es folgen astrologische, gromatische und chronologische Auszüge. Näheres
siehe Friedrich Leitschuh, *Katalog der Handschriften der Königlichen Bibliothek zu
Bamberg*, 1. Band, 2. Abteilung, Bamberg 1895—1906, S. 61—66; Thulin (8), S. 21 f.
Auch hier ist der Text der 'Geometrie I' von einer gleichzeitigen oder nur wenig jün-
geren Hand korrigiert (b¹).

3) p¹ = Paris, Bibliothèque Nationale, lat. 13020, 9. oder 10. Jh.[28], Corbie, seit 1638
Saint-Germain-des-Prés, seit 1795/96 in der Bibliothèque Nationale. Die 'Geometrie I'
steht am Ende des Codex auf Bl. 59ᵛ—83ʳ; voraus geht Boethius' Musik. Beschreibung
bei Thulin (8), S. 23.

4) p² = Paris, Bibliothèque Nationale, lat. 13955, 9. oder 10. Jh.[28], Corbie, seit 1638
Saint-Germain-des-Prés, seit 1795/96 in der Bibliothèque Nationale. Neben der 'Geometrie I' (auf Bl. 107ʳ—123ᵛ) sind noch verschiedenartige Texte vorwiegend philo-
sophischen, astronomischen und gromatischen Inhalts vorhanden. Siehe Thulin (8),
S. 17 f.; Bubnov (2), Einleitung S. 61 f.

5) p³ = Paris, Bibliothèque Nationale, lat. 14080, 9. oder 10. Jh.[28], Corbie, seit 1638
Saint-Germain-des-Prés, seit 1795/96 in der Bibliothèque Nationale. Die Blätter sind
in falscher Reihenfolge eingebunden. Der 'Geometrie I' (auf Bl. 65ʳ—72ᵛ, 80ʳ—87ᵛ,
79ʳᵛ, 73ʳ—77ᵛ) geht Boethius' 'Musik' voraus; es folgen wie in p² Exzerpte aus dem
'Corpus agrimensorum'. Vergleiche Thulin (8) S. 18.

6) pʳ = Prag, Universitätsbibliothek (Univerzitní Knihovna), lat. 1717 (IX.C.6), 10. Jh.
Enthält neben der 'Geometrie I' (auf Bl. 47ʳ—63ᵛ) noch Boethius' 'Arithmetik', 'Musik' und astronomische Texte. Näheres siehe Joseph Truhlár: *Catalogus codicum manu
scriptorum latinorum qui in C. R. bibliotheca publica atque universitatis Pragensis
asservantur*, II, Prag 1906, S. 13; Thulin (8), S. 23.

7) m = München, Bayerische Staatsbibliothek, Clm 560, 10. Jh., Vorbesitzer: Hermann
Schedel, Hartmann Schedel (15. Jh.), Hans Jakob Fugger (ab 1552), seit 1571 in der
Hofbibliothek. Nach B. Bischoff[29] wurden die Bll. 89—149ʳ im 9. Jh. auf der Reichenau geschrieben. Dieser Teil enthält vor der 'Geometrie I' (diese auf Bl. 122ʳ—149ʳ)
noch Arats 'Phaenomena' mit Scholien. Die Handschrift ist beschrieben im *Catalogus*

---

[26] Ludwig Traube, *Paläographische Forschungen*, 4. Teil, Abh. Kgl. Bayer. Ak. Wiss.,
III. Kl., 24 (1904), S. 10. Siehe auch Rudolf Blank, *Weltdarstellung und Weltbild in Würzburg und Bamberg vom 8.—12. Jahrhundert*, Bamberg 1968, S. 113—119.

[27] Heiberg (5), S. 50—55; Tannery (7), S. 44—49; Thulin (8), S. 21.

[28] Die landläufige Ansicht, die Codices p¹, p², p³ stammten aus dem 10. oder frühen
11. Jahrhundert, dürfte nicht zutreffen, wie Bernhard Bischoff nachgewiesen hat: *Hadoard
und die Klassikerhandschriften aus Corbie*, in: Mittelalterliche Studien, I, Stuttgart 1966,
S. 60. Bischoff setzt die drei Codices in das 2. oder 3. Viertel des 9. Jahrhunderts.

[29] *Die südostdeutschen Schreibschulen und Bibliotheken in der Karolingerzeit*, Teil I,
Wiesbaden ²1960, S. 262, Anm. 3.

codicum latinorum bibliothecae regiae Monacensis, Tom. I pars I, München ²1892, S. 154 f., und bei Bubnov (2), Einleitung S. 41. Vgl. auch Heiberg (5), S. 50—53.

8) **T** = Cambridge, Trinity College, 939 (R. 15.14; 491), 10. Jh., Canterbury, gelangte durch George Willmer († 1626) in die Bibliothek des Trinity College. Der Codex enthält nach der 'Geometrie I' (Bl. 4ʳ—43ᵛ) noch geometrisch-gromatische Auszüge, die bei Thulin (8), S. 13 f. aufgezählt sind. Vergleiche auch die Beschreibung bei M. R. James, *The western manuscripts in the library of Trinity College,* Cambridge 1901, II, S. 349—354.

9) **e** = Einsiedeln, Stiftsbibliothek, Cod. 298, 10./11. Jh., Reims oder Trier (?). Der Codex kam wie der folgende (E) höchstwahrscheinlich mit dem Hl. Wolfgang von Trier nach Einsiedeln. Der 'Geometrie I' (auf S. 1—22) folgt Boethius' 'Musik'. Beschreibungen bei Gabriel Meier, *Catalogus codicum manu scriptorum qui in Bibliotheca Monasterii Einsidlensis servantur,* I, Einsiedeln/Leipzig 1899, S. 272, und Thulin (8), S. 24.

10) **E** = Einsiedeln, Stiftsbibliothek, Cod. 358, 10./11. Jh., Reims oder Trier (?). Auch hier geht die 'Geometrie I' (auf S. 1—20, 23—37) Boethius' Musik voraus. Vgl. G. Meier, a. a. O., S. 322 f., und Thulin (8), S. 24.

11) **bⁿ** = Bern, Burgerbibliothek, Ms. lat. 87, gemäß der subscriptio auf Bl. 17ᵛ im Juni 1004 vom Priester Constantius in Luxeuil geschrieben. Bischof Werinharius (1002—27) schenkte den Codex der Kirche St. Maria in Straßburg. Um 1600 gelangte er an den in Straßburg residierenden französischen Diplomaten und Humanisten Jacques Bongars (1556—1612), dessen Bibliothek 1632 nach Bern kam. Die 'Geometrie I' steht auf Bl. 1ᵛ—8ᵛ; es folgen die üblichen gromatischen Exzerpte. Siehe H. Hagen, *Catalogus codicum Bernensium,* Bern 1875, S. 104—108; Bubnov (2), S. 476; Thulin (8), S. 12 f.

12) **Bⁿ** = Bern, Burgerbibliothek, Ms. lat. 299, 11. Jh. Auch dieser Codex kam über Bongars nach Bern, dürfte aber entgegen Thulins Angaben nicht der Straßburger Dombibliothek angehört haben. Neben der 'Geometrie I' (auf Bl. 1ʳ—14ʳ) enthält er gromatische Auszüge, geometrische und arithmetische Texte. Vgl. Hagen, a. a. O., S. 316—318; Bubnov (2), S. 483 f., Einl. 22 f.; Thulin (8), S. 14—17.

13) **g** = St. Gallen, Stiftsbibliothek, Ms. 830, 11ᴵ. Jh., St. Gallen. Der Codex weist handschriftliche Noten Ekkeharts IV. († um 1060) auf. Die Handschrift enthält neben der 'Geometrie I' (auf S. 283—309) noch philosophische Schriften des Boethius und Cicero, aber keine weiteren gromatischen Exzerpte. Vgl. Gustav Scherrer, *Verzeichnis der Handschriften der Stiftsbibliothek von St. Gallen,* Halle 1875, S. 281 f.; Thulin (8), S. 25.

14) **d** = Oxford, Bodleian Library, Ms. Douce 125, 11. Jh., England (Canterbury?), im 15. Jahrhundert in Winchester, später im Besitz von Francis Douce (1757—1834). Die 'Geometrie I' befindet sich auf S. 1—67; sie nimmt den ganzen Band ein. Beschreibung in *A summary catalogue of western mss. in the Bodleian Library at Oxford,* IV, Oxford 1897, S. 529 (Nr. 21699); O. Pächt und J. J. G. Alexander, *Catalogue of illuminated manuscripts in the Bodleian Library,* 3.: English school, Oxford 1973, Nr. 31.

15) **W** = Wien, Österreichische Nationalbibliothek, Cod. lat. 55, 11. Jh., Süddeutschland. Im 15. Jh. im Dominikanerkloster zu Ofen (Buda), spätestens seit 1755 in der Wiener

Hofbibliothek. Der Codex enthält neben der 'Geometrie I' (Bl. 1ʳ—14ᵛ, 22ʳᵛ, 15ʳ—21ᵛ) noch Boethius' Arithmetik und Musik sowie das Hugbald zugeschriebene Werk 'Musica enchiriadis'. Vgl. S. Endlicher, Catalogus codicum philologicorum latinorum bibliothecae Palatinae Vindobonensis, Wien 1836, S. 254 f., Nr. CCCLXIV; Tabulae codicum manu scriptorum ... in Bibliotheca Palatina Vindobonensi asservatorum, I, Wien 1864, S. 8; Thulin (8), S. 24 f.

16) vᵇ = Vat. Barb. lat. 92, um 1100, Westdeutschland. Der Codex enthält neben der 'Geometrie I' (Bl. 19ᵛ—22ᵛ, 38ᵛ—44ʳ) auch die 'Geometrie II' sowie weitere gromatische und astronomische Exzerpte. Siehe Bubnov (2), S. 484—488, Einl. 71 f.; Thulin (8), S. 22 f.; Folkerts (3), S. 7; Sesto Prete, Codices Barberiniani Latini. Codices 1—150, Rom 1968.

17) c = Chartres, Bibliothèque de la Ville 498, 12ᴵ. Jh., Chartres. Dieser Codex bildet zusammen mit Chartres 497 das 'Eptateuchon' des Thierry von Chartres, eine Enzyklopädie der sieben artes liberales, die im 2. Viertel des 12. Jh.s entstand[30]. Thierry übernahm nicht die gesamte 'Geometrie I', sondern — bezeichnenderweise — nur die Kapitel 1.6—8 des Buches V, also die 'Altercatio'[31].

18) A = London, British Museum, Add. 47679, 12. Jh. Aus mehreren Quellen zusammengesetzte Agrimensorenhandschrift, die in ihrem 1. Teil auf eine Vorlage der Klasse EF zurückgeht, während der 2. Teil neben der 'Geometrie I' (auf Bl. 74ʳ—101ʳ) die in diesen Handschriften üblichen gromatischen Auszüge aufweist. Siehe Folkerts (3), S. 27—29, und Rhein. Museum f. Philologie, NF 112 (1969), S. 53—70.

19) v° = Vat. Ottob. lat. 1862, 12ᴵᴵ. Jh., Ostfrankreich. Die Handschrift enthält den Euklidtext nach Adelhard, Gerberts 'Geometrie', „Boethius" 'Geometrie II', die 'Geometrie I' (Bl. 37ʳ—44ᵛ) und weitere gromatische Texte. Siehe Thulin (8), S. 17; Folkerts (3), S. 14 f.

20) f = Florenz, Biblioteca Medicea Laurenziana, Plut. 29.19. Diese Handschrift wurde um 1250 für Richard de Fournival aus Amiens vielleicht nach einem Corbier Codex abgeschrieben; über Gerard de Abbeville gelangte sie 1271 an die Sorbonne[32]. Der 'Geometrie I' (Bl. 1ʳ—27ʳ) folgen weitere gromatische Auszüge, die bei Thulin (8), S. 20 angegeben sind. Der „Boethius"-Text ist in f weitgehend umgestellt[33].

21) a = München, Bayer. Staatsbibliothek, Clm 4024ᵃ, 13. Jh., Frankreich. Die Handschrift enthält heute auf Bl. 3ʳ—9ᵛ nur noch einen Teil aus Buch I und II der 'Geometrie I'; der Rest, also auch die 'Altercatio', ist durch mechanischen Verlust nicht mehr vorhanden. Siehe Thulin (8), S. 20 f.; Folkerts (3), S. 29—31.

---

[30] Zur Handschrift 497/498 und zu Thierrys Person siehe die Literaturangaben bei Folkerts (3), S. 8 f.
[31] Thulins Annahme (8), S. 18, der Rest des Werkes sei in der Lücke der Handschrift verlorengegangen, trifft nicht zu, da vor den Auszügen der 'Geometrie I' noch das Ende von Euklids 'Elementen' in der Fassung II des Adelhard vorhanden ist.
[32] Ullman (9), S. 279—282.
[33] Thulin (8), S. 32.

Die Altercatio in der Geometrie I des Pseudo-Boethius

22) R = Rostock, Universitätsbibliothek, Ms. phil. 18 (= IV 111,4), 13. Jh. Der 'Geometrie I' (S. 1—17) folgen gromatisch-geometrische Texte. Näheres bei Thulin (8), S. 18—20; Folkerts (3), S. 31 f.

23) w = Wien, Österreichische Nationalbibliothek, Cod. lat. 2269, 13. Jh. Der Codex enthält verschiedene Schriften Alkuins, ein astrologisches Werk, Schriften des Boethius zum Trivium, seine 'Arithmetik' und 'Musik', Hyginus, '*De astronomia*', und die 'Geometrie I' (Bl. 220ʳ—223ʳ). Vgl. Bubnov (2), S. 555, Einl. 89; *Tabulae . . .*, a. a. O., II, Wien 1868, S. 44 f.; *Aristoteles Latinus. Codices. Supplementa altera.* Brügge/Paris 1961, S. 58 f.

24) bʳ = Breslau, Universitätsbibilothek (Biblioteka Uniwersytecka), Rhedig. 55, 15. Jh., Italien. Der Codex ist eine Abschrift der Rostocker Handschrift R (Nr. 22). Die 'Geometrie I' steht auf Bl. 2ʳ—26ʳ. Siehe Folkerts (3), S. 32 f.

25) C = Cesena, Biblioteca Malatestiana, Plut. sin. XXVI.1. Der Codex wurde um 1450—65 in der Malatestianischen Schreibschule kopiert. Er enthält Schriften zur Arithmetik, Musik und die 'Geometrien I' und 'II' des „Boethius", erstere auf Bl. 221ʳ—249ʳ. Siehe Folkerts (3), S. 20 f.

26) y = New York, Columbia University, Ms. Plimpton 164 (= Boncompagni 90), 16. Jh. Die Handschrift ist in den Bll. 1—34 eine Abschrift von N (siehe Nr. 1); die 'Geometrie' steht auf Bl. 2—16ᵛ. Vgl. die Beschreibung bei Bubnov (2), S. 480, Einl. 73—75; Thulin (8), S. 12.

Die Frage, welche Beziehungen zwischen diesen 26 Handschriften bestehen und welche Bedeutung sie demnach für die Überlieferungsgeschichte der 'Geometrie I' haben, ist bislang nicht ausreichend untersucht. Übereinstimmung besteht darin, daß alle Codices auf einen Archetypus (im folgenden nach Thulin mit X bezeichnet) zurückgehen, jenen ältesten rekonstruierbaren Überlieferungsträger, dem schon die in Kapitel 2 erwähnten Blattversetzungen und Zusätze eigen waren. Wie Bubnov (2), S. 476—488, als erster erkannte, müssen von diesem Textzeugen zwei Abschriften existiert haben: X' (bei Bubnov: TBbbaa) und X'' (bei Bubnov: TBbbab). Von X' leiten sich die Handschriften NTbⁿy ab; alle übrigen gehören X'' an. X' und X'' unterscheiden sich deutlich dadurch, daß in X' der Anfang von Buch V (Kapitel V.1) wohl absichtlich fortgelassen ist und das Ende von Buch II durch Ausfall eines Blattes fehlt, während in X'' mehrere Lücken in der '*Altercatio*' (V.7) den Text zum Teil unverständlich machen. Thulins Ansicht trifft zu ([8], S. 31), daß X'' zwar älter als X' ist, daß aber X' den Text getreuer wiedergibt.

Hierin erschöpfen sich die bisherigen Forschungen: Thulins Bemerkungen über Handschriftengruppierungen innerhalb der Klasse X'' sind recht oberflächlich und nicht immer richtig ([8], S. 31 f.), und Ullmans Versuch (9), die Wirkungsgeschichte der einzelnen Handschriften zu untersuchen und dadurch die Überlieferung aufzuhellen, führt nur in Einzelfällen weiter. Neue Ergebnisse sind nur durch den

Textvergleich sämtlicher Codices zu gewinnen. Dies ist für Euklidexzerpte und die '*Altercatio*', also für den größten Teil der Bücher III—V, geschehen. Die Untersuchungen erlauben zwar noch nicht, mit letzter Sicherheit ein Stemma zu erstellen — hierfür wäre es nötig, auch die Bücher I und II einzubeziehen —, jedoch lassen sich Thulins Angaben präzisieren und erweitern:

Zunächst können von den erwähnten 26 Handschriften sechs eliminiert werden, da es sich um Abschriften erhaltener Codices handelt, und zwar (in Klammern die jeweilige Vorlage): b^r (R), p^3 (p^2), v^b (b), v^o (B^n), w (b), y (N). Von den übrigen 19^{34} Handschriften ist keine die Vorlage einer anderen gewesen.

Anhand des Textes der '*Altercatio*' und der Euklidexzerpte[35] lassen sich folgende Aussagen über Verwandtschaft und Eigenarten dieser 19 Codices gewinnen:

1) Nb^nT bilden die Gruppe X^I, die sich durch einen relativ guten Text auszeichnet. Nur wenig Sonderfehler von X^I können gefunden werden, z. B. 'Altercatio', Z. 16.43f.44; Euklidexzerpte, Z. 2.8.14.37.57(2x).88.98.111f.122.124.126(2x).137 usw.

2) Dagegen sind diejenigen Versehen Legion, die in allen übrigen Codices vorhanden sind; sie beweisen die Existenz einer gemeinsamen Vorlage X^{II}. Einige Fehler aus der 'Altercatio' mögen hier stellvertretend erwähnt werden: Z. 18(−c).23(−fc).30(−fc). 35—39(−fc).85.130[36].

3) p^2 nimmt eine Sonderstellung ein: Diese Handschrift gehört zwar zu X^{II}, ist aber mit X^I kontaminiert. Die X^I-Lesarten, die zunächst vom Schreiber in einem späteren Zeitpunkt hinzugefügt wurden, sind im weiteren Verlauf der Abschrift oft schon in den Text geraten, so daß p^2 im 'Altercatio'-Teil des Buches V eine reine X^I-Handschrift ist.

4) In X^I weichen Nb^nT nur selten voneinander ab. b^n ist mit N nahe verwandt, aber keine Abschrift davon, da N (wie auch Tb^n) Sonderfehler aufweist.

5) Die zahlreichen Korrekturen in N, die, wie erwähnt, auf den Gelehrtenkreis um Gerbert zurückgehen (N^1), sind teils sinnvolle eigene Konjekturen — meist handelt es sich um überlegt eingefügte Zusätze —, teils entstammen sie einer mit A verwandten Handschrift. Dies zeigen die Stellen 'Altercatio', Z. 44.119.128; Euklidexzerpte, Z. 68. 102.173.178.182(2x).183.

6) X^{II} gliedert sich in die beiden Klassen $\alpha = p^1p^r(p^2)B^nAfc$ und $\beta = beWEmdCgR$:
   a) Bindefehler von α und gleichzeitig Trennfehler gegen β sind etwa die Stellen 'Altercatio', Z. 64.79.151 (−fc); Euklidexzerpte[37], Z. 46(−f).87(−f).114.131(−f)[36].
   b) Bindefehler von β und gleichzeitig Trennfehler gegen α sind 'Altercatio', Z. 6.74 (−bm).79.169(−e, +f); Euklidexzerpte, Z. 95(+f).102(−b).155 (−g).293(−R, +f).

---

[34] Der Kodex a (= Nr. 21) enthält nicht die '*Altercatio*'.
[35] Diese werden nach den Zeilen der Edition von Folkerts (3), 176—214 (Edition Mc) zitiert.
[36] Zur Sonderstellung der Codices fc siehe unten S. 101.
[37] In c sind die Euklidexzerpte nicht vorhanden (s. S. 98).

Die Altercatio in der Geometrie I des Pseudo-Boethius

7) p^1 gibt den Text außerordentlich sorgfältig wieder. Er ist der zuverlässigste Codex der Gruppe X^{II}.

8) Auch p^r hat nur selten versucht, den Text zu verbessern. Aber p^r kopierte nicht ganz so gewissenhaft wie p^1 die Vorlage. Insbesondere finden wir einige durch Flüchtigkeit entstandene Lücken, z. B. 'Altercatio', Z. 27—30.160 f.

9) Die Handschriften $B^n Afc$ gehen auf eine Vorlage γ zurück, wie zahlreiche gemeinsame Versehen zeigen[38]: 'Altercatio', Z. 7($-A$).15($-A$).28($-Af$).44.56($-A$).83($-Af$). 115.117.117($-A$); Euklidexzerpte, Z. 36($-A$, $+g$).37.97($-B^n$).126.150.175.184($+C$). 199[37].

10) Es gibt eine Reihe von Bindefehlern für B^n und A, die es wahrscheinlich machen, daß beide Codices von ein und derselben Vorlage abstammen[39], z. B. 'Altercatio', Z. 16.70. 94.151.167($+p^1$); zahlreiche weitere Stellen aus dem Euklidteil lassen sich anführen.

11) B^n ist von einer etwa gleichzeitigen Hand (B^{n1}) nach einem Codex der Klasse X^I durchkorrigiert, so daß die meisten der dieser Klasse eigenen Fehler von B^{n1} verbessert wurden. Von den zahllosen Stellen seien nur 'Altercatio', Z. 79.130 als Beleg angeführt.

12) f und c haben neben der X^{II}-Vorlage auch eine Handschrift der Klasse X^I benutzt, mit deren Hilfe die folgenden wesentlichen X^{II}-Versehen behoben wurden: 'Altercatio', Z. 23.30.35—39.40.48.57($+m$).60f.61.67f.74.105.132.155f.156f.158.161.162.163. 165. An all diesen Stellen hat auch B^{n1} den richtigen Text nachgetragen. Die Varianten Z. 61[40] und 16.18[41] scheinen darauf hinzudeuten, daß nicht bereits eine (hypothetische) gemeinsame Vorlage von fc mit einem X^I-Codex kontaminiert war, sondern daß f und c unabhängig voneinander eine X^I-Handschrift heranzogen.

13) A entfernt sich durch viele eigenwillige und oft unnötige Änderungen stark vom überlieferten Text. Auch f schaltet sehr frei mit dem Text, wie schon Thulin erkannte ([8], S. 32). Die Handschrift c, die auf Thierry von Chartres zurückgeht, zeichnet sich durch zahlreiche gute Korrekturen aus, die sogar manchmal den fehlerhaften Text des Kompilators verbessern (z. B. 'Altercatio', Z. 23.33.147.159.163.164. 171). Die Varianten 'Altercatio', Z. 3.4 scheinen darauf hinzudeuten, daß c die Columella-Auszüge zum Vergleich heranzog[42]; die Stelle 'Altercatio', Z. 165 und vielleicht auch Z. 150 beweist, daß c auch mit Hilfe von Balbus-Exzerpten[42] den korrupten Text nicht ungeschickt verbesserte.

14) Einige Fehler der Handschriften $beWEmdC$ scheinen darauf hinzudeuten, daß diese Codices auf eine gemeinsame Vorlage δ zurückgehen: 'Altercatio', Z. 47($-m$, $+p^r$).73

---

[38] Da jeder der drei Textzeugen A, f, c seine Vorlage stark bearbeitete und f und c vermutlich auch eine Handschrift der Klasse X^I herangezogen (s. oben), ist es verständlich, daß oft nicht alle vier Codices übereinstimmen.

[39] Wie f und c zu dieser Vorlage stehen, muß angesichts der Kontamination dieser Handschriften ungewiß bleiben.

[40] $B^{n1} f$ haben den richtigen X^I-Text, c den X^{II}-Zusatz.

[41] $B^{n1} c$ weisen gegenüber f den richtigen Wortlaut nach X^I auf.

[42] Sie befinden sich im Codex c unter den Exzerpten im Anschluß an die 'Geometrie I' des Ps. Boethius.

$(+p^1p^rB^n)$; Euklidexzerpte, Z. $62(-C,+p^1).63(-C,+p^1).111(+g).219(-m).219.221.$
269.

15) Innerhalb der Klasse δ weisen die Handschriften *beW* oft gemeinsame Versehen auf: 'Altercatio', Z. $108(+E).145(+d).160(2x).170(-e).174$; Euklidexzerpte sehr oft, z. B. Z. $4.62.64(+b^n).87.96.99.100.105$. In der Mehrzahl der Fälle haben b^1 und e^1 diese Fehler verbessert. Dabei benutzte b^1 offenbar einen mit B^nA verwandten Codex, wie die folgenden typischen Übereinstimmungen zeigen: 'Altercatio', Z. $9.70.79$; Euklidexzerpte, Z. $38.116.126.128$ und öfter.

16) Alle drei Handschriften dieser Gruppe haben ihre Vorlage ε sorgfältig abgeschrieben. Sieht man von den schon in ε vorhandenen Fehlern ab, so bieten *beW* neben p^1p^r den reinsten Text der Gruppe XII[43].

17) Auch *Em* hängen eng zusammen. Zwar ist in der 'Altercatio' nur ein Fehler ihres gemeinsamen Vorfahren ζ anzutreffen (Z. 105), aber im Euklidteil machen zahlreiche Stellen diesen Sachverhalt evident, z. B. Z. $27.49.73.74.75.78.79.86$. Dabei zeigt die Variante Z. 78 *eiecta*] $\frac{eiecta}{egesta}$ E, *eiecta egesta* m, daß E ursprünglicher als *m* ist. Überhaupt gibt E seine Vorlage getreu wieder, während sich *m* durch eine Unmenge an Varianten (oft eigene Konjekturen) auszeichnet. *m* gehört neben A, f, c zu den Handschriften, die im Gegensatz zu den übrigen sehr frei mit dem Text umgegangen sind. Die Stellen 'Altercatio', Z. $57.78.82$ lassen vermuten, daß *m* auch einen X^I-Codex heranzog. Ganz besonders auffällig ist im Euklid-Text der Zusatz der Proposition III 23 nach III 19 (Z. 299), die außer in *m* in keiner anderen Handschrift der Boethius-Tradition zu finden ist. Woher *m* diesen Text nehmen konnte, ist völlig ungewiß.

Abgesehen von einigen Unsicherheiten[44] reichen die genannten Aussagen aus, um mit einiger Wahrscheinlichkeit das folgende Stemma zu erstellen:

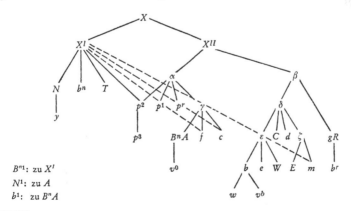

B^{n1}: zu X^I
N^1: zu A
b^1: zu B^nA

---

[43] In der 'Altercatio' weist *b* lediglich einen Sonderfehler auf, nämlich die Lücke Z. 20—22.

[44] z. B. Aufschlüsselung von $p^1 p^r$; Zugehörigkeit von R (wohl zu g, aber auch zu e möglich) und g, Stellung von d und C in der Gruppe δ.

Die Altercatio in der Geometrie I des Pseudo-Boethius

5. Edition

In Kapitel 3 wurde nachgewiesen, daß die Abschnitte V 1.6—8 vom Kompilator der 'Geometrie I' aneinandergefügt, überarbeitet und mit der Bezeichnung 'Altercatio' versehen wurden. Von daher müßte eine Edition der 'Altercatio' eben diese Texte enthalten. Aus äußerlichen Gründen erschien es jedoch unnötig, den Anfang (V.1.6) abzudrucken: Dieser Teil wurde nämlich schon von Lachmann ediert und ist im übrigen ein fast wörtlicher Auszug aus Isidor. So ist hier nur das Kernstück der 'Altercatio' wiedergegeben (V.7.8), das bis auf den Schlußteil von Lachmann nicht berücksichtigt wurde.

Die Fehler, die zum Teil schon im Autographen vorhanden gewesen sein müssen, erschweren die Textkonstitution nicht unbeträchtlich, da Versehen im Archetypus nicht immer verbessert werden dürfen. In den Fällen, wo die Vorlage des Kompilators bekannt ist, mußten Fehler, die dort und in allen „Boethius"-Handschriften vorhanden sind, beibehalten werden; im Apparat wird auf diese Tatsache hingewiesen. Wenn dagegen Sonderfehler der Codices gegenüber den exzerpierten Texten auftreten oder wenn sich die Vorlage nicht sicher bestimmen läßt, bleibt es oft Ermessenssache, welche der in allen Handschriften vorhandenen Versehen als Fehler des Autors (die folglich beizubehalten sind) oder des Archetypus (die also behoben werden müßten) anzusehen sind. Im allgemeinen wurde an diesen Stellen der handschriftlich überlieferte Text aufgenommen, da jede Verbesserung willkürlich gewesen wäre[45].

Die Orthographie folgt im allgemeinen der handschriftlichen Überlieferung, wobei die besten Codices (N und p^1) maßgeblich berücksichtigt wurden. In Kleinigkeiten (z. B. Assimilation der Konsonanten, Varianten *ti* : *ci* u. ä.) wurde die Schreibweise vereinheitlicht. Eckige Klammern bezeichnen Tilgungen, spitze Zusätze der Herausgeber.

Der kritische Apparat beschränkt sich nicht auf die Abweichungen der Hyparchetypi X^I und X^{II}, sondern erwähnt alle wesentlichen Varianten. Dadurch wird der Wirkungsgeschichte der 'Geometrie I' im Mittelalter Rechnung getragen und die Möglichkeit gegeben, Handschriften einzuordnen, die spätere Benutzer in Händen hatten. Vernachlässigt wurden nur unbedeutende Verschreibungen (meist Orthographica) und Korrekturen in den Codices, die für deren Klassifizierung unerheblich sind.

---

[45] Derartige Stellen sind vor allem am Schluß der 'Altercatio' in den Exzerpten aus Balbus und Hyginus anzutreffen. Insbesondere die Hyginus-Auszüge befinden sich in einem so korrupten Zustand, daß man kaum annehmen kann, diese Entstellungen seien in der kurzen Zeit zwischen Autograph und Archetypus entstanden. Vielmehr muß hierfür eine besonders verderbte Vorlage verantwortlich sein. Somit darf in diesem Teil der lückenhafte Text nicht verbessert werden.

Hinzugefügt wurde ein Similienapparat. Er gibt darüber Auskunft, welche Werke der Kompilator direkt benutzte. Da diese Frage im edierten Teil stets beantwortet werden kann, war es überflüssig, andere gleichzeitige oder frühere Texte ähnlichen Inhalts anzugeben.

Conspectus siglorum

A   = Londinensis, Addit. 47679, f. 99v—101v (s. XII)
b   = Bambergensis, Ms. class. 55, f. 13v—15v (s. X)
b^n = Bernensis 87, f. 7v—8v (a. 1004)
B^n = Bernensis 299, f. 12v—14r (s. XI)
c   = Carnutensis 498, p. 281—283 (s. XII)
C   = Caesenas, Plut. sin. 26.1, f. 247r—249r (s. XV)
d   = Oxoniensis, Douce 125, p. 61—67 (s. XI)
e   = Einsidlensis 298, p. 21—22 (s. X/XI)
E   = Einsidlensis 358, p. 34—37 (s. X/XI)
f   = Laurentianus, Plut. 29.19, f. 8v—11r (s. XIII)
g   = Sangallensis 830, p. 306—309 (s. XI)
m   = Monacensis 560, f. 146r—149r (s. X)
N   = Neapolitanus V A 13, f. 13r—15r (s. X)
p^1 = Parisinus, lat. 13020, f. 81r—83r (s. X)
p^2 = Parisinus, lat. 13955, f. 121v—123v (s. X)
p^r = Pragensis, lat. 1717, f. 62r—63v (s. X)
R   = Rostochiensis, Ms. phil. 18, p. 15—17 (s. XIII)
T   = Cantabrigiensis, Coll. S. Trinit. 939, f. 38r—43v (s. X)
W   = Vindobonensis 55, f. 18v—21r (s. 11)
X^I = consensus codicum Nb^nTp^2
X^{II} = consensus codicum $p^1p^rB^nAfcbeWEmdCgR$
α   = consensus codicum $p^1p^rB^nAfc$
β   = consensus codicum beWEmdCgR
γ   = consensus codicum B^nAfc
δ   = consensus codicum beWEmdC
ε   = consensus codicum beW
ζ   = consensus codicum Em
N^1, p^{21} etc. = correctio in libro N (p^2) ipsius scribae vel alterius cuiusdam aetate scribae proximi manu facta
X^I (-N) = familia X^I sine codice N, etc.
⟨ ⟩ includunt voces textui archetypi addendas
[ ] includunt voces e textu archetypi delendas
Aug. = AUGUSTINUS, *De quantitate animae* et *Soliloquia* (ed. Migne, Patrologia Latina 32)
Balb. = BALBUS, *Ad Celsum expositio* (ed. Lachmann, Grom. vet. I, 91—108)
Col. = COLUMELLA, *De re rustica* (ed. Forster-Heffner)
Hyg. = HYGINUS (ed. Lachmann, Grom. vet., I, 108—134, et Thulin)
Isid. = ISIDORUS, *Etymologiae* (ed. Lindsay)

Die Altercatio in der Geometrie I des Pseudo-Boethius

INCIPIT ALTERCATIO DUORUM GEOMETRICORUM DE FIGURIS, NUMERIS ET MENSURIS.

Quoniam diversae formae agrorum veniunt in disputationem,
sed tamen cuiusque generis species subiciemus, quibus quasi
5 formulis utemur.

Omnis ager aut quadratus est, aut longus, aut triquetrus,
aut rotundus, aut semicirculus, aut minus quam semicirculus,
velut arcus.

Prima autem figura huius artis punctus est, cuius pars
10 nulla est. Utique principium mensurae punctum vocatur, cum
medium tenet figurae. Secunda linea, praeter latitudinem longitudo. Recta linea est, quae ex aequo in suis punctis iacet.
Superficies vero, quod latitudines et longitudines solas
habet.

15 ⟨IN.⟩ Nonne tibi tale aliquod videtur etiam illud, unde
linea ducitur etsi figura nondum sit, cuius medium intelle-
gamus? Illud enim dico lineae principium, a quo incipit longitudo, quod volo sine ulla longitudine intellegas. Et si
longitudinem sine puncto atque sectione intellegis, nequa-
20 quam profecto intellegis, unde ipsa incipit aut ubi finit
longitudo.

---

3 Quoniam — *8* arcus] *Columella V 1,13—2,1*
9 Prima — *14* habet] *Isidorus III 12,7*
10 Utique — *11* figurae] *"Boethii" 'geometria I', lib. III in.*
15 Nonne — *87* cupiebam] *Augustinus, 'De qu. an.' 1045,51—1047,28 Mi.*

---

1 sq. inscr. (*A:* Incipit quintus de figuris, numeris et mensuris) *ante* Mensura (*p. 407,3 La.*) hab X^{II}(-cb), om. cb / 1 numeris] lineis m / 2 mensuris] versuris R / 3 ante Quon-iam add. A Disputatio geometricalis per interrogationem et responsionem, add. f Item quante sunt figure agrorum? R. / Sed quoniam c Col. / 4 sed tamen om. c Col. / cuiuscumque NTp^2 / generis] gentis m / subiecimus p^r / 5 formulas p^1, formulus b(?)W, corr. $p^{11}b^1$ / uteremur $B^{n1}c$, utemur B^nrell. / 6 est om. Col. / longus] aut cuneatus add. Col. / triquetrus Col., triquertus vel -cus β, triquestus $X^{1}p^{1}p^{r}$, triquestrus $B^n f(?)$, triquetus cb^1, triquester A / 7 aut (2.) — 8 arcus] aut etiam semicirculi vel arcus Col. / rotundus aut semicirculus] semiculus aut rotundus $B^n f c$ / quam semicirculus] a semiculo A / 9 artis] partis $X^{1}p^{1}p^{r}B^{n1}$, om. B^n / punctum Ab^1 / cui m / 10 est] est ve R / Utique] Itaque A / principium om. d / 11 Secunda linea] Secundo loco est linea, scilicet A / 13 quod] est quae A / longitudines et latitudines m Isid., latitudinem et longitudinem A, latitudines et longitudines rell. / 15 tale tibi $B^n f c$, tibi om. E / aliquid tale Aug. / aliquid $b^n A f e E C g R$ Aug. / 15sq. linea unde E / 16 etsi] et $B^n A$, corr. B^{n1} / nondum $X^{1}B^{n1}c$ Aug., non X^{II}(-c) / cui X^1 / 17 ante Illud add. Geym. R. / principium linee f / 18 quod — intellegas $X^{1}B^{n1}c$ Aug., om. X^{II}(-c) / Et si ⟨INT.⟩ Etsi Geym., Nam si Aug. / 19 sine — sectione om. Aug. / 20 aut — finit] aut finit f, om. Aug. / aut — 22 accipit om. b, add. b^1

⟨R.⟩ Tale omnino, [quia linea lineam accipit.]

⟨IN.⟩ Punctum vocatur, cum medium tenet figurae; ⟨si autem
principium lineae est vel lineis aut etiam finis, vel cum om-
25 nino aliquid notat quod sine partibus intelligendum sit, nec
tamen obtineat figurae⟩ medium, signum dicitur. Est ergo
signum nota sine partibus. Est autem punctum nota mediam
figuram tenens. Ita fit, ut omne punctum etiam signum sit,
non autem omne signum punctum videatur, quamquam plerique
30 punctum appellant non quod omnis figurae medium, sed quod
solius circuli vel pilae tenet. Vides certe etiam, quantum
valeat punctum. Nam ab ipso incipit linea, ipso terminatur.
Figuram utique rectis lineis nullam videmus fieri posse,
nisi ab ipso puncto angulus claudatur. Deinde undecumque
35 secari linea potest, per ipsum secatur, cum ipsum omnino
nullam in se admittat sectionem. Nulla linea lineae nisi per
ipsum copulatur. Postremo cum ceteris planis figuris eam
praeponendam ratio demonstraverit, quae circulo clauditur
propter summam aequalitatem, quae alia ipsius aequitatis
40 moderatio est quam punctum in medio constitutum? Talia omnino et multa de huius potentia dici possunt, ⟨sed adhibeo
modum et tibi ipsi cogitanda plura permitto.

R. Sane ut videtur. Non enim me requirere pigebit,⟩ si
quid fuerit obscurius. Cerno autem mediocriter, ut puto, ma-
45 gnam huius signi esse potentiam.

---

22 Tale X^Ib^I, Talem $X^Ib^I(?)$, hoc est *add.* A / omni *g* / quia — accipit *om. Aug., qui*
alium textum hab. / quia] cum *add.* A / accipit γb^Img, accepit $B^nIrell.$ / **23—26** ⟨INT.⟩
Punctum vocatur, cum medium tenet figurae, medium signum? ⟨R.⟩ Dicitur: est *Geym.*
/ Punctum — tenet X^IB^nIfc, *om.* $X^{II}(-fc)$ / cum] Cum vero *C* / figurae] Fugere p^IA,
Fungere p^r / si — **26** figurae *om. codd., add. Folk. ex Aug.* / **26** ducitur B^nc /
27 nota (2.) — **30** medium *om.* p^r / nota mediam] medium nota B^n, *corr.* B^nI / **27**sq.
medium figurae *m Aug.* / **28** tenens] obtinens *R*, habens B^nc, *corr.* B^nI / **29** *post* videatur
add. Aug. aliqua verba / plerumque *A*, multi *add.* p^IB^nA (multa) $bWdCg$, multi *del.* B^nI /
30 punctum] puncta cum sint, non tamen *A* / appellant — medium X^IB^nIfc, medio in
loco sit punctum $X^{II}(-fc)$ / appellent *Aug.* / **31** pilae] medium *add.* A / *ante* Vides *add.*
Aug. quaedem, *add. Geym.* ⟨INT.⟩ / etiam certe *f* / **32** punctum *om. Aug.* / **32**sq. terminatur figura. Utique $X^{II}(-A)$ / **33** utique *om. Aug.* / nullam *post* utique *A*, figuram
add. c / **34** illo *R* / ab ipso puncto] ipso *Aug.* / **35** linea secari b^n, linea *om.* R / cum —
39 aequalitatem X^IB^nIfc, *om.* $X^{II}(-fc)$ / **36** Nulla] Nullam *c* / **38** demonstravit *f* / **39** quae
X^IB^nIAfc, qua $p^Ip^rB^nb^Im^IdCg$ *Aug.*, quia $\varepsilon\zeta R$, ergo *add.* A / aequitatis] aequalitatis
Aug. / **40** quam X^IB^nIAfcR *Aug.*, quia *b*, qua $b^Irell.$ / in *om.* f / Talia X^IB^nIfc, Tale
$X^{II}(-fc)$ / **40**sq. Talia omnino et *om. Aug.* / **41** sed — **43** pigebit *om. codd.; lacunam a*
Geym. indicatam suppl. Folk. ex Aug. / **43**sq. si — obscurius *om.* X^I / si quid] sed quod *f* /
44 fuerint εE, *corr.* e^I / obscurus *c* / *ante* Cerno *add.* A $\Delta S.$, *add.* N^I In.; *spatium album*
rel. c / autem $X^{II}(-A)Aug.$, haud N^IA, *om.* X^I / ut puto] puncto B^nIc, me *add.* $X^{II}(-\gamma)$,
me *del.* b^I / magnam — **45** esse] inesse huius signi magnam γ / **45** huic signo *A* / potentiam esse p^r

Die Altercatio in der Geometrie I des Pseudo-Boethius

IN. Nunc ergo illud attende, cum et quid sit signum et quid sit longitudo et quid sit latitudo perspexeris, quid horum tibi recte videatur alterius indigere.

R. Video quod latitudo longitudine indigeat, sine qua
50 prorsus intelligi non potest. Rursum longitudinem cerno latitudine quidem non indigere, ut sit, sed sine signo illo esse non posse. Illud autem signum per semet ipsum esse et nullius horum indigere manifestum est.

IN. Ita est, ut dicis. Sed diligentius considera, utrum
55 latitudo undique secari queat an aliunde nec ipsa sectionem possit admittere, quamvis plus admittat quam linea. Certe etenim latitudinem sic intellegas, ut cogitatione tua de altitudine nihil usurpes.

R. Sic omnino.

60 IN. Accedit igitur huic latitudini altitudo. Et responde iam, utrum etiam aliquid accesserit, quo magis undique secari queat.

R. Mire omnino admonuisti. Nunc enim video non solum desuper aut ex inferiore parte, sed lateribus quoque posse ad-
65 mitti sectionem.

IN. Quoniam tibi, si non fallor, et longitudo et latitudo

---

46 IN. *add. Folk., om. codd.* (*-A*), MΓ. *A, R. N¹Geym., spatium album rel.* c / et quid *X¹Bⁿ¹f,* (a)equis *X¹¹*(-*pʳAfc*), aequus *pʳ,* quid *Ac* / et (2.) *om.* ƒ / **47** sit (2.) *om. pʳA* / pr(a)espexeris *pʳεEdC, corr. b¹* / **48** recte *om.* m *Aug.,* recta *Bⁿ,* rectius *A, corr. Bⁿ¹* / alterius indigere *X¹Bⁿ¹fc, om. X¹¹*(-*fc*), alterius et cuius indigere, sine quo esse non possit *Aug.* / **49** R. *add. Folk., om. codd.* (*-A*), ΔS. *A,* IN. *N¹Geym.* / longitudinem *p¹pʳBⁿeWζd, corr. Bⁿ¹*(?)*W¹d¹*(?) / longitudine — **50** Rursum *om.* b, *add.* b¹ / **50** intelligi *om.* d / potest] valet *R* / Rursus *pʳfgR Aug.* / cerno] certo *d* / latitudinem *p¹pʳBⁿb¹*(?)*eζd,* latitudine *Bⁿ¹b*(?)*rell.* / **51** quidem non] equidem *A* / sed *X¹,* et *X¹¹* / signo *om. ε, add. b¹e¹* / **52** potest *pʳg,* posse *pʳ¹rell.* / **53** indigere *om.* d / **54** IN. *add. Folk., om. codd.* (*-A*), MΓ. *A, R. N¹Geym., spatium album rel.* gc / ut dicis est c / Sed] Si c / **55** latitudo] per puncta *add. A,* vere *add. Aug.* / secare *C* / an] in *m,* aut ƒ / alicunde *Aug.* / ipsa] ipsam *m,* ne ipsa *A* / **55sq.** possit sectionem ƒ / **56** admittere *X¹yb¹*(ammittere *bⁿA*), amittere *rell.* / plus *ante* quam *Bⁿfc* / admittat *X¹yb¹*(ammittat *bⁿA*), amittat *rell.* / *ante* Certe *add. Geym.* ⟨INT.⟩, *add. quaedam Aug.* / etenim] velim *A,* enim *Aug.* / **57** sic] si ƒ / intelligis *Aug.* / de altitudine *X¹Bⁿ¹fcm, om. X¹¹*(-*fcm*) / **58** nihil *post* tua (*v.57*) *Geym.* / **59** R. *om. mg,* ΔS. *A,* In. *N¹* / **60** IN. *add. Folk., om. codd.*(-*A*)*Geym.,* MΓ. *A, R. N¹, spatium album rel.* ceg / Accidit ƒeER, Accedat *m Aug.,* Accedet *C¹,* Accedit *b¹C rell.* / igitur *om. pʳ* / latitudini] ut audivi *C* / Et] Sed *c* / **60sq.** responde iam, utrum *X¹Bⁿ¹fc,* respondeam quod *X¹¹*(-*fc*) / **61** aliquid *X¹Bⁿ¹Af,* aliquot *m,* aliquod *rell.,* si *add. p¹Bⁿce¹*(?)*mdCg,* sic *add. pʳεER,* sibi *add. A,* si *del. Bⁿ¹* / quo *X¹,* quod *Bⁿ¹AfcεER,* quid *b¹e¹rell.* / magis *om.* m / **63** R. ƒ, *om. rell.* (*-A*), ΔS. *A,* In. *N¹Geym., spatium album rel.* cg / Mire] Ire *d,* Iure *d¹* / videro ƒ / **64** inferiori *X¹¹*(-*gR*) *Aug.* / sed *X¹αb¹,* sed e *vel* sede *β,* sed de *C¹,* sed a *Aug.* / amitti *p¹BⁿcδR,* ammitti *bⁿA* / admitti posse *Aug.* / **64sq.** sectionem amitti posse *C* / **65** *post* sectionem *add. quaedam Aug.* / **66** IN. *Folk.,* MΓ. *A, R. N¹, om. rell. Geym., spatium album rel.* cg / tibi] igitur *add. Aug.* / si non] nisi *A,* non *m* / **66sq.** latitudo et altitudo] altitudo et latitudo *C,* latitudo *pʳf*

et altitudo nota est, quaero, utrum possint deesse duo priora,
ubi aderit altitudo.

R. Utique altitudo sine longitudine esse non potest. Alti-
70 tudo autem sine latitudine potest.

IN. Redi ergo ad cogitationem latitudinis, et si eam quasi
iacentem animo figuraveris, erigatur in quodlibet latus, tam-
quam si eam velles per tenuissimam rimam, ubi seclusae ianuae
iunguntur, educere. An nondum intellegis quid velim?

75　R. Intellego quid dicas, sed nondum fortasse, quid velis.

IN. Illud scilicet, ut respondeas, utrum sic erecta latitudo videatur tibi migrasse in altitudinem et latitudinis iam
nomen descriptionemque amisisse, an adhuc maneat latitudo,
quamvis ita sit collocata.

80　R. Videtur mihi altitudo esse facta.

IN. Recordare, obsecro, quemadmodum altitudinem definieramus.

R. Recordor utique et me iam sic respondisse pudet. Nam
etiam hoc modo quasi erecta altitudo sectionem per longum
versus non admittit. Quare nulla in ea possunt interiora co-
85 gitari, quamvis medium et extrema cogitentur. Nulla omnino
est altitudo, ubi nihil intus cogitari potest.

---

67 nata $p^r B^{n1} cgm$, natura d, nota B^n rell. / est om. p^r / quaero — 68 altitudo $X^1 B^{n1} fc$,
om. $X^{11}(-fc)$ / priora] superiora Aug. / 69 R. om. mg, et A / Utique — 70 potest] Sine
longitudine quidem video esse non posse altitudinem, sine latitudine autem potest Aug. /
Altitudo utique c / longitudine] latitudine c / Altitudo (2.)] latitudo A / 70 autem]
etiam A, post latitudine m, om. Geym. / latitudine] altitudine A, esse non add. $B^n Ab^1$,
corr. B^{n1} / 71 IN. Folk., om. codd. Geym., spatium album rel. c / etsi Geym. / 72 figura-
beris $p^1 B^n cbWm$ / quolibet $p^1 B^n \varepsilon EdR$, quotlibet m / latere R / 73 velles $X^1 Aug.$, vellis
$p^1 bW$, bellis m, velis rell. / pro m / rimam $X^1 B^{n1} AfcgR$, ripam rell. / seclusae] se clausae
c Aug. / 74 iungunt $EdCgR$ Aug., iugunt eW / educere $X^1 B^{n1} Afc$, edice b, edicere rell. /
An — velim $X^1 B^{n1} fc$, an non $X^{11}(-fC)$ / 75 R. fc, ΔS. A, In. $N^1 Geym.$, om. rell., spatium
album rel. ceg / sed] si c / sed nondum] secundum f / velis $X^1 \alpha b^1 R$ Aug., velim $\beta(-R)$ /
76 IN. Folk., MΓ. A, R. N^1, om. rell. Geym., spatium album rel. cg / Illud] enim add.
$X^{11}(-A)$ / sic] se f / erectae d / 77 migrare f / latitudinis] altitudinis f / 77sq. nomen iam
R / 78 descriptionemque $X^1 Aug.$, discriptionemque m, discretionem qu(a)e fbW, discre-
tionemque rell. / 79 ante quamvis add. A ΔS., add. N^1 In. / ita $X^1 B^{n1}$, ista m, iuxta αb^1,
iusta $\beta(-m)$ / collata R, cum locata p^r / 80 R. f Geym., om. rell., spatium album rel. c /
81 IN. om. mgR, MΓ. A, R. N^1 / Recordarisne Aug. / 82 R. om. $b^n mg$, ΔS. A, In. N^1 /
utique] plane Aug. / sic om. $X^{11}(-m)$ / 83 quasi] quia si m / erepta C, est add. $X^{11}(-A)$ /
altitudo quasi erecta A / altitudo] latitudo p^{21} Aug. (recte), om. $B^n c$ / sectione (?) g /
longum] deorsum add. Aug. / 84 non admittit] nāmittit d / amittit $p^1 B^n(?)c\beta$, ammittit
$b^n B^{n1}(?)A$ / ante Quare spatium album rel. c / in ea nulla C / cogitare $p^1 B^n$, corr. B^{n1},
INT. add. Geym. / 85 ante quamvis add. $X^{11}(-f)$ Geym. Utique, add. f itaque / extremum
X^{11} / cogitantur $eWE(?)R$, corr. W^1 / cogitentur, nulla Geym. / 85sq. est omnino p^r

Die Altercatio in der Geometrie I des Pseudo-Boethius

IN. Recte dicis, et sic te omnino meminisse cupiebam. Potest esse figura, quae quattuor solummodo lineis continetur, an et tribus lineis invenitur?

90 R. Et quattuor et tribus. Nam est ager quadratus, est et trigonus, sicut infra monstravi.

IN. Melior est igitur figura, quae quattuor lineis rectis paribus quam quae tribus constat?

R. Certe melior, siquidem magis in ea aequalitas valet.

95 IN. Quid ergo ista, quae quattuor rectis paribus lineis continetur? Potest fieri, ut non anguli omnes in ea pares sint, an non potest?

R. Nullo modo prorsus.

IN. Quid si rectis lineis tribus et imparibus figura con
100 stet? Possunt etiam in ista pares esse anguli, an aliud intellegis?

R. Omnino non possunt.

IN. Recte dicis. Sed dic, quaeso, quaenam tibi figura melior videatur et pulchrior, ea quae paribus an quae impari
105 bus lineis constat.

R. Quis dubitet eam esse meliorem, in qua aequalitas praevalet?

IN. Ergo inaequalitati aequalitatem praeponis?

R. Nescio, utrum quispiam non praeponat.

---

92 Melior — 97 potest] *Augustinus, 'De qu. an.' 1043,22—27 Mi.*
98 Nullo — 109 praeponat] *Augustinus, 'De qu. an.' 1043,2—13 Mi.*

---

87 IN. *Folk.*, MΓ. *A*, R. *N¹Geym.*, *om. rell.*, *spatium album rel.* ceg / te *om.* f / cupiebam] cogitabam *R* / *ante* Potest *add. A* ΔS., *add. rell.*(-cmg) *Geym.* IN(T)., *spatium album rel.* g / 88 solismodo *p¹* / continentur c / 89 et tribus] etribus *pⁿεE*, et in tribus *A*, *corr.* e¹ / 90 R. *om.* mg, MΓ. *A*, *spatium album rel.* g / est ager] ager est f, ager *m* / est (2.) *om.* feECR / 91 monstrabimus *A* / *post* monstravi *quattuor versus albos reliquit* c / 92 IN. *om.* εζdR, ΔS. *A* / figura] id est (idem C) locus *add.* X^{II}(-fc) / quae] et *add. R* / rectis *om. A* / 93 quamque p¹(?)p² , quamquam c, *corr.* p¹¹ / 94 R. *om.* m, MΓ. *A* / Melior certe *Aug.* / in ea aequalitas] inequalitas f / qualitas BⁿA, *corr.* Bⁿ¹ / 95 IN. *om.* m, ΔS. *A* / istam *Aug.* / rectis *om.* g / lineis rectis paribus f, rectis lineis paribus cR / 96 continetur? Potest] confit, censesne posse etiam ita *Aug.* / 98 R. *om.* m, MΓ. *A* / prorsus modo *Aug.* / 99 IN. *om.* m, ΔS. *A* / Quod *R* / et] sed *m Aug.* / 100 etiam — 102 possunt *om.* p¹, *add.* p¹¹ / pares in ista pʳ / 102 R. *om.* m, MΓ. *A* / 103 IN. *om.* fm, ΔS. *A* / quaeso] te *add. A* / 104 videtur g / ea, quae] eaque d, eane quae *Aug.* / an] an ea pʳ, aut *A* / an quae] anque p¹, atque *Bⁿ*, *corr.* Bⁿ¹ / 105 lineis X^{I}Bⁿ¹fc *Aug.*, *om.* X^{II}(-fc) / constet ζ, *corr.* E¹ / 106 R. *om.* m, MΓ. *A* / praevalet *om.* m / 108 IN. *om.* m, ΔS. *A* / inaequalitate p¹pʳm, inequalitati / f / inaequalitati aequalitatem] εE, *corr.* e¹ / praenis W, praepisunt b, *corr.* b¹ / 109 R. *om.* m, MΓ. *A*, Respondero *add.* f / praeponet m

110   IN. Dic ergo, quaeso te, numquid vera linea est, quae
per longum secari potest, aut verum signum, quod ullo modo
secari potest, aut vera latitudo, quae, cum erecta est, sectionem per longum deorsum versus non admittit?

    R. Nihil minus.

115   IN. Nonne duos angulos haberi posse pariter licet?

    R. Pariter omnino.

    IN. Quid? Linea et sphera unumne aliquid tibi videatur,
an quicquam inter se differunt?

    R. Aliud linea, aliud sphera. Nam est ager sphericus
120 extra lineam qui est rotundus, et est quadratus, qui linearum ratione constringitur.

    IN. Ergo linearum genera sunt tria?

    R. Utique.

    IN. Quae?

125   R. Rectum, circumferens, flexuosum.

    IN. Quid rectum?

    R. Recta linea est quae aequaliter suis signis posita est,
quae in planitie positae flexae in utraque parte non concurrunt.

---

110 Dic — *114* minus] *Augustinus, 'De qu. an.' 1047,31—35 Mi.*
117 Quid — *119* sphera] *Augustinus, 'Soliloquia' IV 10*
122 Ergo — *135* sunt] *Balbus 99,3—10 La.*
128 quae — concurrunt] *Balbus 98,16—99,2 La.*

---

110 IN. *om. m, ΔS. A* / numquidnam *Aug.* / 111 longum] vel latum *add. $p^{2l}B^{n1}f$,*
latumque *add. c* / aut — 112 potest *om. $B^{n}Aε\zeta$, add. $B^{n1}e^{1}$* / nullo *Geym.* / 112 longitudo *f* /
erepta *C*, recta *mg* / est] ut diximus *add. Aug.* / 113 pro *m* / seorsum *C* / non *om. Aug.* /
amittit *$p^{r}B^{n}f^{1}cR$ Geym.*, ammittit *$b^{n}A$*, admittit *f rell. Aug.* / 114 R. *om. m, MΓ. A* /
115 IN. *om. m, ΔS. A* / pariter posse *γ* / 116 R. *om. m, MΓ. A* / 117 IN. *om. m, ΔS. A* /
sphera *$Ab^{1}e^{1}Eg$*, sera *b*, sfera *C*, fera *eWd*, spera *rell.* / unum nec *$p^{1}p^{r}$* / tibi aliquid
$B^{n}Ac$, tibi aliquod aliquid *f* / aliquod *Geym.* / videntur *$B^{n}fc$ Aug. Geym.*, videtur *b^{n}*,
videantur *R*, videatur *b^{n1} rell.*, esse *add. Aug.* / 118 quiquam *$p^{1}f$*, quicquid *C* / 119 R.
om. m, MΓ. A / linea] est *add. $N^{1}A$* / sphera *$Ab^{1}g$*, sfera *eEC*, fera *bWd*, spera *rell.* /
sphericus *$Ab^{1}E^{1}(?)g$*, sfericus *eWdC*, fericus *b(?)*, spericus *rell.* / 120 qui (1.)] que *f* /
121 rationem *T* / 122 IN. *om. m, ΔS. A* / 123 R. *om. m, MΓ. A* / 124 Int. *$N^{1}f$, ΔS. A,
om. rell.* / Quae *om. $X^{1}ceEdgR$, del. $B^{n1}b^{1}$* / 125 R. *$N^{1}f$, MΓ. A, om. rell.* / eflectuo-
sum (?) *f* / 126 IN. *om. m, ΔS. A* / 127 R. *om. m, MΓ. A* / suis] in suis *c* / 128 quae — con-
currunt] *textus corruptus. Legendum esset* ⟨Ordinatae rectae lineae sunt⟩ quae in ⟨eadem⟩
planitie positae et eiectae in utraque parte ⟨in infinitum⟩, non concurrunt *(cf. Balbum)* /
posita *c*, et *add. $N^{1}A$* / flexae] et eiectae *Balb.*, electae *Balbi codd.PG* / parte *om. p^{rc}*

Die Altercatio in der Geometrie I des Pseudo-Boethius

IN. Quid circumferens?

130   R. Cuius incessus a conspectu signorum suorum distabit.

IN. Quid flexuosum?

R. Flexuosa linea est multiformis, velut arborum aut signorum aut fluminum; in quorum similitudine et arcifiniorum
extremitas finitur, et multarum similiter, quae natura inae-
135   quali linea formata sunt.

IN. Ecce de figuris et lineis quae infra agrum sunt nobis
enuntiasti: de extremitatibus agrorum quae sunt ad conclusio-
nem nobis edicito.

R. Extremitatum genera sunt duo.

140   IN. Quae?

R. Unum quod per rigorem observatur, et aliud quod per
flexus.

IN. Quid flexus? Quid rigor?

R. Rigor est quicquid inter duo signa vel in modum lineae
145   rectum perspicitur. Flexuosum est quicquid secundum naturam
locorum curvatur.

IN. Quantae sunt summitates?

R. Summitatum genera sunt duo, summitas et plana. Summitas
est secundum geometricam appellationem quae longitudinem et la-
150   titudinem habet tantummodo. Summitatis finis lineae. Plana
summitas est quae aequaliter rectis lineis est posita.

---

139 Extremitatum — 146 curvatur] *Balbus 98,5—9 La.*
148 Summitatum — 151 posita] *Balbus 99,11—14 La.*

---

129 IN. *om. m*, ΔS. *A* / 130 R. *om. m*, MΓ. *A* / incessus a X^lB^{nl}, incensura $X^{ll}(-fc)$,
incessura a *f*, ineisura (?) a *c* / conspectus $X^{ll}(-B^nfc)$ / suorum] pariter *add. Geym.* /
131 IN. *om. m*, ΔS. *A* / 132 R. *om. m*, MΓ. *A* / multiformis X^lp^lAc *Balb.*, multis formis
$p^{ll}X^{ll}(-Ac)$ / arbor *d* / aut $X^lB^{nl}fc$, *om.* $X^{ll}(-fc)$ / signorum *om. p^r* / **133** similitudine
X^l *Balbi cod.G*, similitudinem X^{ll} *Geym.* / **134** multarum] rerum *add. Geym.* / similiter]
similitudinum *A* / 134sq. inaequali linea X^l *Balb.*, in(a)equa linea p^lp^rAbWdC, inaequali
inea *m*, in(a)equalia $B^nfceEgR$ / **136** IN. *om. m*, ΔS. *A* / lineas *b*, corr. b^l, *om. A* /
137 conclusionem] nunc *add. f* / **139** R. *om. m*, MΓ. *A* / 140 IN. *om. m*, ΔS. *A* / 141 R.
om. m, MΓ. *A* / pro (bis) *m* / 142 flexos $p^lεζ$, fixos *d*, corr. b^l / 143 IN. *om. m*, ΔS. *A* /
flexis *d* / 144 R. *om. m*, MΓ. *A* / Rigoris *m* / velut *f* / 145 perspicitur $X^lB^{nl}Afb^lgR$,
prospicitur $p^lp^rB^nc$, praespicitur $e^lζC$, respicitur $εd$ / **147** IN. *om. m*, ΔS. *A* / Quantae]
Quot *c* / 148 R. *om. m*, MΓ. *A* / et] est *g* / plena p^2m, corr. p^{2l}, Planities $A_;$ summitas
add. c / **149** et latitudinem *om. b^n* / latitudinem $X^lp^rB^nf$, altitudinem B^{nl} *rell.* / 150 fines
c Balbi cod.G, finis *rell. Balbi cod.P* / lineae] sunt *add. c* / **151** est (1.) *om. p^r* / qualiter
B^nA, corr. $B^{nl}A^l$, in *add. A*, a *add. c* / lineis *om.* $p^lp^rB^nA$, post est (2.) *add. B^{nl}* / est
(2.) *om. c*

IN. Quanta sunt genera angulorum?

R. Genera angulorum rationalium sunt tria, id est rectus,
hebes et acutus. Haec habent species novem: rectarum linea-
155 rum tres, rectarum et circumferentium tres, circumferentium
tres. Rectarum linearum species angulorum tres: recta, hebes,
acuta. Rectus angulus est ethigrammus ex rectis lineis comprehensus, qui Latine normalis appellatur. Quotiens autem
rectam super rectam lineam trans ordinem angulos pares fe-
160 cerit, ut singuli anguli recti sint, et linea perpendicularis fuerit iuncta, efficiet triangula recto angulo. Hebes
angulus est plus normalis, hoc est excedens recti anguli
positionem, et quia si triangulus secundum hanc positionem
constitutus fuerit, perpendiculare extra finitimas lineas
165 habebit [inde trigonum]. Rectus angulus est normalis, hebes
angulus est plus normalis, acutus angulus est minus normalis.

Omnem locum defixum rigoribus cuiusque observatur. Nam

---

153 Genera — *166* normalis] *Balbus 100,5—101,11 La.*
167 Omnem — *172* est] *Hyginus 128,19—129,6 La.* = *92,2—12 Th.*

---

152 IN. *om.* m, ΔS. A / sint p^r / **153** R. *om.* m, MΓ. A / rationabilium *f* / rectum A /
154 ebes $p^2p^1\delta$ / et *om.* p^r / acutum A / Haec X^I, et X^{II} / **155sq.** circumferentium tres.
Rectarum X^IB^n1fc, ebes (hebes p^rgR, hebetum A) a (ad p^r, ac B^nA) circumferentium
$X^{II}(-fcm)$, Ebes arcum ferentium m / **156sq.** recta, hebes, acuta X^IB^n1fc, *om.* $X^{II}(-fc)$ /
156 hebes] habes p^2 / **157** ethigrammus X^I, eotigrammus $p^1p^rB^nWdC$, eotrigrammus $\varepsilon\zeta R$,
cotigrammus (?) *b*, etigrammus (?) b^I, eutigramus *f*, eutigrammus Acg^I, eutigrammis *g* /
ex] vel *f* / **158** qui] a *add.* A / Latine] latitudine $X^{II}(-fceER)$ / Quotiens — **161** angulo]
textus valde corruptus emendandus non erat (cf. Balbi codd.PG: p. 100,11-101,2 La.) /
autem X^IB^n1fc, *om.* $X^{II}(-fc)$ / **159** rectam (1.) $X^{II}(-Ac)$ *Balbi cod.G*, rectum X^I, recta
bene coni. A Geym., recta linea *c,* lineam *add.* g / rectam (2.) *om.* c / linea A, *om.* g /
stans *bene coni. Geym., sed Balbi codd.GJR habent* trans / ordinem $X^Ip^r1B^nfcm, Balb.$
(G), ordinarie A, ordine *rell. Geym.* / fecerit pares *f* / **160** anguli *om.* ε, *add.* b^1e^1 /
recti — **161** triangula *om.* p^r / sunt c, si ε, *corr.* b^1e^1 / et] si *add. f* / propendicularis m /
161 efficietque A / triangulum *f* / recto X^IB^n1fc, recti B^n, recta $X^{II}(-\gamma)$ / rectiangula A,
recta angula *d* / Hebes $b^np^rAfcdgR$, Ebes *rell.* / **162** angulus] angulis *d* / plus X^IB^n1fcg,
om. $X^{II}(-fcg)$ / **163** et - positionem (2.) X^IB^n1fc, qui si A, cum C, *om. rell.* / et] Sed
B^n1 / quia X^IB^n1 *Balbi cod.G (false),* qui *bene coniec. cf Geym.* / **164** perpendiculare
X^IB^n1 *Balb. (G),* perpendicularis $X^{II}(-c)$, perpendicularem *bene coniec. c Geym.* / extra]
extrema m / **165** inde trigonum *vel glossa quaedam esse videtur vel quod reliquum est ex
definitione acuti anguli.* c *habet* Accutus angulus est compressior recto. Qui si a recta
linea quae ̤sedis loco fuerit rectam lineam in occursum exceperit, efficiet triangulum qui
perpendicularem intra tres lineas habebit *(i. e. Balbi definitio acuti anguli e posteriore
codicis parte, ut videtur, sumpta)* / normalis X^IB^n1fc, normaliter $X^{II}(-fc)$ / ebes $p^1B^n\delta$ /
166 angulus est (**1.**)] et angulus A / minus $X^IB^n1A^1fcm$, minor $X^{II}(-fcm)$ / **167** Omnem — **174** deficiunt] *textum valde corruptum e Hygino corrigere Geym. conatus est.
Haec verba habet f post* continet *(v. 178)* / Omne *f* / cuiuscumque *f,* quibusque A / ob-
servant c, observatis A / Nam — **168** vicinos (**169** facit A) *ante* Omnem *pos.* p^1B^nA,
corr. $p^{11}B^n1$

Die Altercatio in der Geometrie I des Pseudo-Boethius

aliquando unus qui rigor in multos vicinos agri fundos finem
facit. Sunt et rigores declinantes ad locum vicinos per pla-
170 nitiem. Aut marginibus cepti finitique intus agrum invenire
possunt. Similitudinibus quod fuerit, nequid noxia nobis
fieri videatur, sed ab extremis rigoribus servandum est.

Ideo sunt supercilia, id est colles tumentes, quae ad
agrum declinantes deficiunt.

175 Sed quaecumque particulas dimensionis lineae intra circulum erunt, acutos angulos faciunt generis sui, quos tamen
omnes in circumferentia cludet, sicut de duobus agris subterius pictura continet.

Sex sunt ordines in opere demonstrationis artis geometri-
180 ae, id est propositio, dispositio, descriptio, distributio,
demonstratio, et conclusio. Quod primum est in propositione,
fundum; in dispositione, linearum genera; in descriptione,
anguli; in distributione, figurae; in demonstratione, summitas; in conclusione, extremitas.

---

173 Ideo — *174* deficiunt] *Hyginus 128,15—16 La. = 91,19 Th.*
175 Sed — *178* continet] *Balbus 102,2—4 La.*
179 Sex — *184* extremitas] *"Boethii" 'geometria I', 1358,19—21 Mi. et 1359,16—18 Mi.*

---

168 unus *om.* R / unus quis *f*, unusque *d*, et *add.* A / 169 faciunt *f* / Sunt] Sicut *f* /
locos *Ac* / vicinos *om.* *fβ(-E)*, *add.* *b¹* / pro planitie *m* / 170 coepti *BⁿAeEC* / finiti
quae *m* / intur *bW*, inter *Bⁿ¹(?)R*, intra *Ac*, intro *T*, intus *b¹ rell.* / inveniri *A* / 171 quod]
quibus oportunum *A* / noxae *A*, noxium *c* / 172 observandum *A* / 173 est] ad *add.* *f* /
174 agrum] agitur *ε*, *corr.* *b¹e¹* / declinantes *om.* *f* / 175 quaecumque] per *add.* A / particula *c*, particulares *f*, particularis *Geym.* / dimensiones *f* / 176 Acutis *bW* / generi *bW*,
corr. *b¹* / 177 omnes *om.* *f*, omni *T* / in *om.* A / claudet *mc* / 178 continet] Explicat liber
quartus geometricorum *add.* R / 179 Sex *usque ad finem om.* *pʳ*, *alio loco habet* *f*, *alia
manu scr.* *p¹* / Sex] Sed *c* / opere] et *add.* *m* / demonstratione *m* / artis *om.* *cg* / geo-
metric(a)e *p¹BⁿAcb¹R* / 180 propositio *X¹γ*, *om.* *p¹*, praepositio *N¹T¹ rell.* / 181 et *om.* R /
Quod] Quorum A / propositione *γ*, praepositione *rell.* / 182 fundus A, *om.* R / 184 in]
Et in *m*

Die Altercatio in der Geometrie I des Pseudo-Boethius

Abgekürzt zitiertes Schrifttum

(1) F. Blume, K. Lachmann und A. Rudorff, *Die Schriften der römischen Feldmesser,* I—II, Berlin 1848—52, Neudruck Hildesheim 1967[46].

(2) Nicolaus Bubnov, *Gerberti postea Silvestri II papae opera mathematica,* Berlin 1899, Neudruck Hildesheim 1963.

(3) Menso Folkerts, *„Boethius" Geometrie II, ein mathematisches Lehrbuch des Mittelalters,* Wiesbaden 1970.

(4) Mario Geymonat, *Studi intorno alla tradizione di testi geometrici latini.* Università degli studi di Milano, Facoltà di lettere e filosofia, Tesi de laurea, Anno accademico 1962—63.

(5) J. L. Heiberg, *Euklid's Elemente im Mittelalter (Beiträge zur Geschichte der Mathematik im Mittelalter,* II), Zeitschrift für Mathematik und Physik 35 (1890), historisch-literarische Abtheilung, S. 48—58, 81—98.

(6) Migne, *Patrologia Latina,* LXIII, Paris 1847, Sp. 1352—1364.

(7) Paul Tannery, *Notes sur la Pseudo-Géométrie de Boèce,* Bibliotheca mathematica, 3. Folge 1 (1900), S. 39—50 (wiederabgedruckt in: *Mémoires scientifiques,* Bd. 5, Toulouse/Paris 1922, S. 211—228).

(8) Carl Thulin, *Zur Überlieferungsgeschichte des Corpus agrimensorum. Exzerptenhandschriften und Kompendien.* Göteborgs Kungl. Vetenskaps- och Vitterhets-Samhälles Handlingar, Ser. IV, 14, Göteborg 1911.

(9) B. L. Ullman, *Geometry in the mediaeval quadrivium,* in: Studi di bibliografia e di storia in onore di Tammaro de Marinis, IV, Rom 1964, S. 263—285.

---

[46] Vgl. neuerdings auch Hans Butzmann [Hrsg.], *Corpus agrimensorum romanorum,* Leiden 1970 (= Codices graeci et latini photographice depicti ..., 22).

The *Geometry II* Ascribed to Boethius[*]

There are two geometrical writings attributed to Boethius. Several questions arise about them, of which the answers are important for the history of mathematics. The problems which are connected with the second of these texts are here stated and, where possible, solved[1].

Two testimonies from the 6th century[2] show that Boethius translated Euclid into Latin. The translation belongs to his early writings and was made probably before 507. The work has not come down to us in its original form. But there are two different redactions of a geometry attributed in the manuscripts to Boethius:

One of these writings, which we here call *Geometry I*, is in most of the manuscripts divided into five books. This work is made up from three parts characterized by three different types of content: excerpts from the *agrimensores* (i. e. the Roman land surveyors), arithmetic and geometry. Book 1 and the so-called *Altercatio duorum geometricorum* in book 5 form the *agrimensores* part. In book 2 are excerpts from the *Arithmetic* of Boethius. Books 3, 4 and the beginning of book 5 contain geometrical material: the definitions, postulates, axioms and most of the enunciations of the theorems in the first four books of Euclid's *Elements*, though without proofs. In this section the pages of the ancestor of all existing copies were evidently mis-ordered; and the consequent chaos appears in all manuscripts.

The other geometry attributed to Boethius, which is here called *Geometry II*, comprises two books. After a short preface, book 1 presents mostly the same excerpts from Euclid as are found in *Geometry I*. There follow the propositions of Euclid I,1–3, with proofs, and mathematical excerpts from the *agrimensores*. Book 1 ends with a section on the abacus. The greater part of the second book consists of *agrimensores* texts on

---

[*]This is the English translation, with some changes, of the article "Das Problem der pseudo-boethischen Geometrie", *Sudhoffs Archiv* 52 (1968), 152–161. A more detailed analysis is given in Folkerts, 'Geometrie II', pp. 69–107.

[1]For the so-called *Geometry I* see the short notes below and Folkerts, 'Importance' (reprinted in this volume, item VII).

[2]By Cassiodorus; see Folkerts, 'Geometrie II', pp. 69f.

the geometry of triangles, quadrilaterals, regular polygons and the circle. The work ends with a section on fractions.

The following remarks are intended to identify the sources of the various parts of *Geometry II*: the section on the abacus, the excerpts from the *agrimensores*, and the extracts from Euclid's *Elements*.

First the section on the abacus. In this part there are symbols that are similar to the modern Hindu-Arabic numerals. If Boethius were really the author of this section, it would follow that these numerals were already in use about 500 A.D. in western Europe. Thus it is no wonder that this chapter provided the impetus in the 19th century to investigate the genuineness of the whole work. G. Friedlein was the first to maintain that the work was not genuine; but this did not convince M. Cantor, who vigorously defended Boethius' authorship. N. Bubnov was the first to bring irrefutable proof against genuineness; but since his most important work on the subject[3] appeared only in Russian, his arguments and results were almost unknown outside Russia.

The most important deductions about the author's sources in the abacus section can be made from the text of the rules on multiplication and division. A comparison of these rules with the surviving works of the older writers on the abacus shows that the "Boethius" text is similar to Gerbert's *Regulae de numerorum abaci rationibus*[4], but not to the others. There are two redactions of Gerbert's work[5]: the shorter version (*A*), written by Gerbert about 980; and the longer version (*B*), which was written after Gerbert's death (1003). The rules for multiplication in "Boethius" agree with the corresponding rules in version *B*, but not with those in *A*. The six rules of division, too, show considerable similarities to the extended Gerbert text (*B*)[6]: in the "Boethius" text of the sixth rule there is a small inexactitude, and this is also to be found in Gerbert's version *B*. Both texts connect the division of the hundreds by "interrupted" hundreds (i. e., hundreds with units, but no tens) with the division of the thousands through "interrupted" hundreds, although the second division is essentially more complicated than the first. Again, the strange statement by "Boethius" that the difference in the division is placed *above* the *articuli*, but *beneath* the *digiti*, can be explained by a clumsy gloss which is transmitted in all manuscripts of the Gerbert *B* text. Thus the connection between Gerbert *B* and "Boethius" is evident.

---

[3]Bubnov, 'Abak i Boécii' (The abacus and Boethius).

[4]Edited in Bubnov, 'Gerberti opera mathematica', pp. 1–22.

[5]See Folkerts, 'Geometrie II', p. 84.

[6]See Folkerts, 'Geometrie II', pp. 85f.

The "Boethius" text of the rules 2–5 for divison is expressed so shortly that it cannot be understood without knowing the procedure of division on the abacus. Only when one reads the corresponding detailed rules of Gerbert *B*, can one understand the sense of the "Boethius" text.

These considerations lead to the conclusion that the author of the abacus section used the extended Gerbert text (*B*) and hence that this section was written not before the 11th century. Since the two oldest manuscripts were written in the middle of the 11th century, the work must have been compiled between 1000 and 1050.

In the abacus section there is a illustration of a calculation board. It is a table which is divided by horizontal and vertical lines into rows and columns. In the two oldest manuscripts there are six rows and twelve columns. In the first row are the names of the Arabic numerals which were used by writers on the abacus in the 11th and 12th centuries: *igin, andras, ormis, arbas, quimas, calctis, zenis, temenias, celentis, sipos*[7]; the second row is occupied by the nine Arabic numerals and a sign that looks like a zero[8]. The remaining cells (i. e. the rectangles created by the vertical and horizontal lines) are labelled with Roman numerals. In row 3 (from right to left) are 1, 10, 100, 1000 etc.; in row 4 the halves of the corresponding numerals in row 3; in row 5 the halves of the corresponding numerals in row 4; in row 6 the halves of the corresponding numerals in row 5. It was not uncommon that, as a mnemonic aid, not only the powers of 10, but also their halves, quarters and eights were marked on the abacus table[9]. It should be noted that the special Roman symbols for the fractions, too, were written on this illustration. Other manuscripts have similar, but not identical, illustrations[10]. The diagram of this table was not created by the author of the text, because there are other, and sometimes older, representations of the abacus table in several manuscripts[11].

The author of the *Geometry II* reports that a certain "Architas" had adapted (*accommodatam*) the abacus for the Roman world. The name "Architas" is artifical[12]: the abacists of the 11th and 12th century considered Gerbert as the inventor of the abacus table. The compiler

---

[7]Strictly speaking, *sipos* is not the name of a numeral. See Folkerts, 'Names and Forms', p. 10.

[8]This sign was used as a marker on the abacus; see Folkerts, 'Geometrie II', p. 87, and Folkerts, 'Names and Forms', p. 10.

[9]See Folkerts, 'Geometrie II', p. 88.

[10]Facsimiles of the relevant pages of all manuscripts are given in Folkerts, 'Geometrie II', tables 3–21.

[11]See, for instance, the reproductions in Folkerts, 'Names and Forms'.

[12]See Folkerts, 'Geometrie II', p. 89.

wanted to ascribe the work as a whole to Boethius. It would accordingly have made no sense to attribute the invention of the abacus to Gerbert who lived 400 years after Boethius, and thus the author had to find another name. Archytas was a reasonable choice, since he is mentioned in Boethius' *Arithmetic* and *Music* as a great mathematician.

In most of the manuscripts there is a row in the abacus table for the Roman fractions. In another part of the text, at the end of *Geometry II*, the author also treats fractions[13]. This part seems to be an original contribution – perhaps the only one – of the compiler of *Geometry II*: he brings a series of twelve names of fractions, some of them being measures of length, some of weight and some of time – a mixture otherwise unknown in antiquity or the Middle Ages. The relation between the values of the fractions is not very clear. Instead of the traditional signs for the fractions, the author introduces the first twelve letters of the Latin alphabet, associating each with one of the fractions. He also introduces a table with 12 rows and 12 columns, whose first row is the twelve letters A ... M. It is not clear how this helps calculations with fractions. In fact it makes matters worse by obscuring relations between the fractions and by introducing Roman numerals for 1, 10, 100, ... without any indication of the purpose behind it. It is possible to make sense of the fraction table by rotating it by 90 degrees clockwise and attaching it to the abacus table[14], but also then the advantage of the table is not evident. Altogether, it is not surprising that this table was forgotten later in the Middle Ages: the Roman fractions continued to be used as they had been.

Part of book 1 and almost the whole of book 2 are in the *agrimensores* tradition. As we shall see, the analysis of this part[15] supports the results deduced from the abacus section.

The rules for finding the area and dimensions of plane figures, which occupy most of book 2, are also to be found in the treatise of Epaphroditus and Vitruvius Rufus and also in the anonymous *Liber podismi*, two geometrical writings which belong to the oldest part of the *Corpus agrimensorum*[16]. A number of mistakes and unusual formulations in *Geometry II* may only be explained by influence from the text of Epaphroditus or the *Liber podismi*. Three examples must suffice here.

---

[13]For the part on fractions see Folkerts, 'Geometrie II', pp. 90–94.

[14]See the reconstruction in Folkerts, 'Geometrie II', p. 92, which is based upon Bubnov.

[15]For the sources of this part see Folkerts, 'Geometrie II', pp. 95–104.

[16]A concordance between these two text and *Geometry II* is given in Folkerts, 'Geometrie II', p. 98.

1) In the problem of the right-angled triangle the author includes in his procedure division by 5 instead of taking the square root. The result is by chance correct because the number to be divided is 25. We can explain this as derived from a false conjecture in the *Liber podismi* text of manuscript class X: in the archetype of the *Liber podismi* the words *sumo latus* ("I take the side" [i. e., square root]) had been omitted; the scribe of the common ancestor of the X class manuscripts had noted that there was an omission and supplied it with the unfortunate *quinta pars*. The same words appear in *Geometry II.*

2) The diameter 2ρ of the in-circle of a right-angled triangle is given by Pseudo-Boethius, as by Epaphroditus, correctly as $2\rho = b + c - a$ (a is the hypotenuse, b, c are the other two sides). In Epaphroditus the example is taken with hypotenuse = 15 feet, the other two sides = 8 and 12 feet, the diameter = 6 feet. But a triangle with sides 8, 12 and 15 is obtuse-angled and not right-angled. Clearly, the 8 in the Epaphroditus text was wrongly written for 9, for the triangle with sides 9, 12 and 15 is indeed right-angled and the diameter of its in-circle is 6. The author of *Geometry II* noticed that there was a mistake in the text he had before him and made a correction, but he changed the wrong number: he corrected 5 feet into 6 feet, so that the formula $2\rho = b + c - a$ was satisfied. But after this 'correction', the triangle was no longer right-angled.

3) The areas of regular polygons were calculated in *Geometry II*, as in Epaphroditus, with the help of polygonal numbers (the formulae found with these numbers always give too great a number for areas). According to this approach, found in several *agrimensores* texts, the areas of a pentagon and hexagon are

$$F_5(a) = \frac{3a^2 - a}{2}, \quad F_6(a) = \frac{4a^2 - 2a}{2}.$$

But Epaphroditus has these formulae with the mistake of adding instead of subtracting:

$$F_5(a) = \frac{3a^2 + a}{2}, \quad F_6(a) = \frac{4a^2 + 2a}{2}.$$

We find the same mistake in Pseudo-Boethius. He made the further mistake of forgetting the 2 in the formula for $F_5(a)$.

These examples (and there are many others) show that the author, in the section on the calculation of areas and lenghts of geometric figures, used the treatise by Epaphroditus and Vitruvius Rufus and the *Liber podismi*, both in manuscript class X. The compiler took over mistakes from these works. Sometimes he tried to correct the text, but in so clumsy a way that he made new mistakes.

As we have seen, in the abacus section of *Geometry II* the name "Architas" apparently comes in the place of "Gerbert". This could also be the case for the chapters based on the *agrimensores*: in this section Architas is mentioned three times – in passages which were taken from Epaphroditus or the *Liber podismi*[17]. In the X class manuscripts the *Liber podismi* and Epaphroditus were almost always transmitted anonymously, but in one of the oldest manuscripts of this class, Naples V.A.13 (N), Gerbert is given, in a late 10th or 11th century hand, as author of the *Liber podismi*. If the compiler of *Geometry II* had this (or a similar) manuscript in front of him, he must have assumed that Gerbert was the author; and replaced Gerbert's name by "Architas" in this section as he had done in the abacus section.

The third section of *Geometry II*, which constitutes about a third of the work, contains excerpts from Euclid's *Elements*[18]. This section is here called *Md* and is related to three other groups of excerpts from Euclid, called *Ma, Mb, Mc*:

Ma = Euclid excerpts in manuscripts that transmit the third recension of Cassiodorus' *Institutiones*

Mb = Euclid excerpts in some manuscripts of the later redaction of the *agrimensores*

Mc = Euclid excerpts in books III–V of Pseudo-Boethius' *Geometry I*.

Common to all four groups are the definitions, postulates and axioms of book I of the *Elements*. In addition, *Ma* contains the definitions of book V, *Mb* propositions 1 to 3 of book I (with proofs), and both *Mc* and *Md* most of the definitions and propositions of books I–IV (but without proofs)[19]. It is clear from common mistakes that all four groups depend on a common source (M)[20]. The source of M was Boethius' translation of Euclid. In the 8th century M was in northern France, probably in Corbie, at that time the "gromatic and geometric capital of the mediaeval world"[21]. There the definitions, postulates and axioms of book I and the

---

[17]See Folkerts, 'Geometrie II', p. 104.

[18]For the sources of this part, see Folkerts, 'Geometrie II', pp. 69–82.

[19]For details, see the list in Folkerts, 'Geometrie II', pp. 70f.

[20]An edition of the text of *Mc* as well as a reconstruction of M, which is based upon the readings of *Ma, Mb, Mc* and *Md*, is given in Folkerts, 'Geometrie II', pp. 173–217.

[21]Ullman, 'Geometry', p. 283.

definitions of book V were taken over from M into manuscripts of the third recension of Cassiodorus' *Institutiones*. No doubt it was also in Corbie that Mb and Mc were extracted from M or a derivative of it. Mc and Md are similar in the extracts selected and in the mistakes copied; Md is rather late (see below). Of all the four groups of extracts, Mb presents the best text and Mc/Md the worst: the latter have within the Euclid text extra material which comes from the *agrimensores* tradition, and the copies we have were made from a codex in which the folios were out of order. It is accordingly unfortunate that the excerpts from books II to IV were transmitted only in Mc and Md.

The relationship between Mc and Md may be characterized by the following facts[22]:

1. There is a large number of additions common to Mc and Md.
2. Through disordering of pages some propositions from Euclid III which belong to book 4 of *Geometry I* (Mc) appear in book 5. In Md these propositions are missing.
3. In Md there are ten propositions that are not in Mc.
4. A comparison of Mc and Md shows that Mc often has a faulty text, but made almost no conjectures to mend the passages; Md on the contrary tried to correct the mistakes, but in most cases without success.
5. There are about 30 places in which Md carries a better text than Mc. In some of these there is no obvious mistake in Mc of which the correction could have led to the text in Md (so Md had a better text to copy from than Mc).

Thus it is clear that neither of Mc and Md were copied from the other, but that both were copied from some lost common source.

Summary and conclusions

Geometry II is a mathematical compendium in two books, which is partly geometrical and partly arithmetical. The major part is geometrical; only at the end of each of the two books is arithmetic treated, rules for calculation on the abacus.

It is clear what sources the author used: in the Euclid excerpts a derivative of Boethius' translation of the *Elements*, which was similar to the excerpts found in *Geometry I*; in the *agrimensores* section a manuscript of the X group of the *Corpus agrimensorum* similar to a manuscript now in Naples; in the abacus section redaction B of Ger-

---

[22] For details, see Folkerts, 'Geometrie II', pp. 74–80.

8   The *Geometry II* Ascribed to Boethius

bert's treatise on the abacus, which did not originate before Gerbert's pontificate (999–1003).

Thus, *Geometry II* was written not before 1000. Since the oldest manuscripts are from the middle of the 11th century, the work must have been written in the first half of the century.

The compiler very probably lived in Lorraine: the redaction B of Gerbert's treatise on the abacus was later incorporated in the collection of the oldest abacists made about 995 in Lorraine. At that time Liège, the "Athens of Lorraine"[23], was the centre for writers on the abacus[24]. The X-manuscript N or a manuscript closely related to it was at Liège in the 11th century: Gerbert used this manuscript in Bobbio in 983 and later sent it to Adelbold of Liège. The supposition that the author of *Geometry II* was from Lorraine is supported by the fact that all the oldest extant manuscripts were written in western Germany or eastern France.

The name of the scholar who compiled this work from two or three sources, is not known. We can, however, characterize style, method of work and ability to make corrections: the introductory and connecting passages often are artificially elaborate. The reader is constantly struck by the efforts of the author to introduce variations in expression with no real difference in meaning. These remarks apply not only to one section of the work, but to all, and show that the whole was written by only one compiler. On stylistic grounds, if no other, we may suggest that Franco of Liège could well have been the compiler[25].

The mathematical abilities of the compiler were meagre. When he tried to correct mistakes in his source, he was seldom successful. On the contrary, he often made matters worse in an already bad text. Some errors he failed to notice altogether. He copied large pieces of text without understanding them. In the section on fractions – the only part the compiler appears to have written himself – he suggested a reformation of the fractions which could not be put into practice because it contradicted the traditional nomenclature and relations between the fractions. Thus the text has many mistakes and some incomprehensable passages, though the transmission in general is good and may be traced almost to the author's time.

---

[23] Bubnov, 'Selbstständigkeit', p. 15.

[24] Herigerus, Adelbold, Wazzo, Radulf and Franco came from Liège. Bernelinus wrote in his treatise on the abacus: "Lotharienses ... quos in his cum expertus sum florere".

[25] An edition of Franco's treatise on squaring the circle is reprinted in this volume, item X.

The *Geometry II* attributed to Boethius is typical of the state of mathematics in the 11th century – at a time when mathematics was at a very low level. The historical value of the work lies in its being one of the earliest in which the abacus table and the Hindu-Arabic numerals were presented and in its preservation of parts of the translation of Euclid's *Elements* by Boethius.

Bibliography

N. Bubnov, *Gerberti postea Silvestri II papae opera mathematica*, Berlin 1899.

N. Bubnov, "Abak i Boécij. Lotaringskij naučnyi podlog XI veka", *Žurnal ministerstva narodnogo prosveščenija, novaja serija*, nos. 7, 8, 16, 17, 19–21, St. Petersburg 1907–1909.

N. Bubnov, *Arithmetische Selbstständigkeit der europäischen Kultur*, Berlin 1914.

M. Folkerts, *"Boethius" Geometrie II, ein mathematisches Lehrbuch des Mittelalters*, Wiesbaden 1970.

M. Folkerts, "The Importance of the Pseudo-Boethian *Geometria* During the Middle Ages", in M. Masi (ed.), *Boethius and the Liberal Arts*, Bern, Frankfurt a. M. 1981, pp. 187–209. (Reprinted in this volume, item VII.)

M. Folkerts, "The Names and Forms of the Numerals on the Abacus in the Gerbert Tradition" (In this volume, item VI).

B. L. Ullman, "Geometry in the Mediaeval Quadrivium", in *Studi di bibliografia e di storia in onore di Tammaro de Marinis*, vol. 4, Rome 1964, pp. 263–285.

X

A TREATISE ON THE SQUARING OF THE CIRCLE BY FRANCO OF LIÈGE, OF ABOUT 1050
[Part I]

Menso Folkerts, Berlin and A. J. E. M. Smeur, Dorst/Breda

REMARKS ON THE EDITION

The Latin text of the treatise by Franco which is printed below is based on the Edinburgh (*e*) and Vatican manuscript (*v*). In order to reconstruct the illegible parts in *v* the codex Plimpton 250 (*n*) served as reliable source. As the greatest part of the text is based on only two manuscripts it was impossible to reconstruct with accuracy the original text by Franco in all parts. For this reason the editors confined themselves to inserting only the absolutely necessary corrections in the text of the manuscripts *v* and *e*, which resulted in a comprehensible text in all parts. Some orthographical peculiarities of the manuscripts (missing or superfluous aspiration, for example) were removed or the spelling standardized. Proper names are written in the current spelling of the Middle Ages. Obliterations by the editors in the text are denoted by square brackets [ ], supplementary notes by the editors concerning the manuscripts *e*, *v* and *n* are denoted by braces ⟨ ⟩. These corrections of the transmitted text are not listed once more in the critical apparatus. The arrangement of the books in sections was made by us. The page-numbers of the three manuscripts *e*, *v*, *p* are mentioned in the text.

As four manuscripts had to be taken into consideration we could not dispense with a critical apparatus. It includes all deviations of the manuscripts *e*, *v*, *p* from the printed text, excepting the following: variants *ae-e, y-i, d-t* (e.g. *haut*), *p-pp* (e.g. *repperire*), *h-ch* (e.g. *nichil*), additional or missing *h*, assimilation, variants like *quoties-quotiens*, different methods of writing numbers (e.g. *VIIII-IX*). Variants of the copy *n* were not mentioned, if errors are concerned which can only be found in *n*. The codices *e*, *v* and *p* have a number of oversights which later on were corrected by the scribe himself or another person. We have decided not to mention these corrections for the following reasons: they are mistakes which must have obviously been made in transcribing and which are insignificant for the reconstruction of the text. In many cases it would have been impossible to reconstruct the original text in the manuscripts because parts of it have been scratched out or covered by the overhead correction. On the other hand the indication of the mentioned corrections would have at least doubled the size of the apparatus. For this reason corrections within the manuscripts are only mentioned in the apparatus when the original text proved to be the better one or the transmission of the text was more complicated. In these cases exponents (v^1, v^2) signify corrections in the corresponding manuscript either made by the scribe himself (v^1) or by another person later on (v^2). But we have to point out that at times it proved to be difficult to exactly distinguish between v^1 and v^2.

We think it superfluous to list the deviations of our edition from Winterberg's.

60

In the latter the pronunciation is mostly given incorrectly or the abbreviations were wrongly expanded by Winterberg. Thus he never read the abbreviation for *est*; and in consequence of this *est* is always lacking. Instead of *etenim* Winterberg wrote *et*. And for the abbreviation *unde* he found out different meanings: *IIII* (*v*, fol. 89 r), *numerus* (89 r), *tamen* (105 r), *tandem* (107 r), *unum* (108 r). Some abbreviations which were strikingly wrongly expanded are compiled in the following whereby the original text is followed by the variants of Winterberg which are given in brackets:

quia non quoslibet (*requiro num quilibet*, 86 r)

quae talis est (*quantitatis*, 86 v)

terminus (*numerus*, 87 v)

operae pretium est (*quare procuravi*, 87 v)

XII (*rursus*, 88 r)

econtra (*etiam*, 88 r)

domini (*diutius*, 88 v)

tantundem (*latitudine*, 97 v)

mutilata (*minuta lato*, 99 r)

immerito (*minuto*, 99 v)

cudentibus malleis (*cum dentibus malleis*, 100 v)

nunc igitur consistamus (*negotii stamus*, 103 r)

modo ter modo quater (*modo tantummodo quantum*, 104 r)

deinde (*demitur*, 104 r)

facticio (*easque*, 104 v)

tricies (*tocies*, 106 r).

The second apparatus indicates the treatises from which the quotations mentioned in the text are taken and reveals which passages had already been edited by other scholars. Remarks concerning the contents were dispensable in most cases since the introduction (which will appear in the next issue) gives detailed information to this effect.

FRANCO LEODIENSIS, DE QUADRATURA CIRCULI

Meaning of the sigla and abbreviations:

e = Edinburgh, Royal Observatory, Ms. Crawford 1.27 (13th century), f. 25 r–41 v (complete)

v = Vat. lat. 3123 (12th century), f. 84 v–108 v (complete)

p = Paris, BN lat. 7377 C (about 1100), f. 1 r–4 r (incomplete: begins with book 6, line 185)

n = New York, Columbia University, Ms. Plimpton 250 (15th century), f. 129 r–155 v (complete; copy of v)

e^1, v^1, p^1, n^1: corrector of e, v, p, n (the same hand as the writer)

v^2:   corrector of v (a later hand)

[ ]   text which is transmitted in $ev(p)n$ but must be obliterated

⟨ ⟩   text which is missing in $ev(p)n$ but must be supplemented

†   corrupt text

add.   = *addidit* (added)
corr.   = *correxit* (corrected)
del.   = *delevit* (obliterated)
in marg. = *in margine* (in the margin)
om.   = *omisit* (omitted)
pos.   = *posuit* (placed)
supr.   = *suprascripsit* (superscribed)
ut vid.   = *ut videtur* (presumably)

   The *Apparatus criticus* and the *Apparatus locorum similium* are to be found at the end of the Latin text.

(*e, f. 25r v, f. 84v*) INCIPIT PROLOGUS IN PRIMUM LIBRUM DOMNI FRANCONIS DE QUADRATURA CIRCULI

   Ex quo, mi papa, praesulum decus, corona totius per orbem cleri, nullius umquam immemor honestatis, ex quo liberalitatem tuam gratuito in me
5 confirmasti, ex eodem iam tempore nulla hora sollicitus esse non potui, cuius officii ratione tantam gratiam promereri deberem. Quippe pecuniarum nihil erat, non cava superbi cornipedis ungula, non Minervale pretium operosae arteque elaboratae vestis, non fusilis metalli admirandis figuris opus incusum, non peregrini lapidis multiplex ac varii coloris aspectus. Quam-
10 quam ego, si eiusmodi rerum abundantia pollerem, haud umquam in animum subrepere auderet tantae nobilitati tantaeque dignitati ex eisdem munuscula offerre. Haec enim cum principum largitione distribuuntur in vulgus, indigentiae et necessitati subvenitur; cum vero eorum oblatione ipsorum principum benevolentia comparatur, procul dubio avaritiae illorum occulto quo-
15 dam elogio exprobratur. Neque enim redimi muneribus possent, si non eos, quod nimirum accidit, ex illo avaritiae monstro ipsa munera delectarent. Quis autem nescit omnem illam idolorum servitutem ab arce animi tui procul extrusam et velut in infima praecipitatam? Quod si non ita se habet, ubi illae facultates, quas affluit tibi cum amplissimis rebus [divitis et] ditis Colo-
20 niae tantus hereditarius et imperialis fiscus? Has tu profecto, ut ita dicam, profligare nullo tempore desisti iis, quicumque eis indigeant hilariter erogando. Quidni? Nempe aurum in massam contrudere diabolicum putas, distribuere autem humanum et proximum deo. Quapropter iure metuerem talibus eulogiis tuis conspectibus praesentari, cum et tu opum praecipue (*v, f. 85r*)
25 contemptor existas, et undique circumspecta prudentia oculata – infirma[m] scientia animalis ante et retro – ista fortassis exprobrationi suae cupiditatis assignaret. Sed tandem post multas et varias deliberationes occurrit animo nihil esse, in quo aeque devotio mea perclaresceret, quam si procurarem iuxta vires ingenii munus aliquod liberale. Nempe hoc munus summa illa
30 ingenuitas in rebus honestissimis et nata et omnem aetatem versata hoc ita avide amplexatur, ut infelici avaritiae longe minus blandiantur ignavae opes. Atque ita hac via adductus sum, ut ederem libellum de circuli quadratura,

imperitia plurimum renitente, sed ad temerarios ausus perurgente devotione.
Hoc igitur opus quod editum est, te ad studia nos beneficiis invitante, alterius
35 opus nisi tuum esse non poterit. Quamobrem certus esto omnem eius diligentiam, sive penitus reprobandum sit sive in parte corrigendum sive ex toto
provehendum, id tua maxime interesse. Ergo quid facto opus sit, exemplo
consultus eris Augusti. Habes mihi quamplures et Tuccas et Varos, quibus
minime sit moris alienis studiis invidere. Horum fidei, horum caritati laborem
40 nostrum committes, ut non tantum superflua resecent, sed erratis quoque
adhibeant te probante correctionem. Itaque hoc opus suae emendationis
opera castigent, ut non sit indignum, quod tuo nomini, veluti cuidam numini,
debeat consecrari.

DOMNI FRANCONIS LIBER PRIMUS INCIPIT DE QUADRATURA CIRCULI

45 (A)
  Quadratura circuli inter occultas rerum adeo est abstrusa naturas, (*e, f. 25v*)
ut de eius ratione nemo hodie vel dubitaret, nisi Aristoteles, quem etiam
inventorem ferunt, ipsius mentionem Praedicamentis suis indidisset. Eius
itaque scientiam haud dubium ferunt usque ad Boetium perdurasse, illo
50 autem sublato ipsa quoque omnis simul interiit praeter solam dubitationem,
quae talis ac tanta est, ut in ea omnes Italiae, Galliae atque Germaniae defece-
rint (*v, f. 85v*) sapientes, siquidem hanc noster Adelboldus, hanc maximus
doctorum Wazo, hanc ipse studiorum reparator Gerbertus multique alii
studiose investigabant. Qui si effectu potiti fuissent, num id ab illis profectos,
55 quorum aliqui adhuc supersunt, universos lateret? Et Gerberti quidem geometricus libellus habetur aliaque eiusdem scripta, a quibus, ni fallor, numquam exclusisset, siquidem eius diligentiae super hac scientia compertum
fuisset. Quamobrem dementis esset in tanta difficultate perfectam cognitionem
polliceri.

60 (B)
  Nihil ergo volumus promittere praeter studium et laborem. Qui primo
sudabit in illa quaestione, quae plurimum etiam fatigavit maiores nostros, de
comparatione videlicet angulorum. Quorum est
  (a) una quidem divisio secundum propriam speciem in rectum, hebetem et
65 acutum;
  (b) altera secundum positionem, cum alius exterior, alius interior appellatur,
videlicet quod hic intra figurae terminos comprehensus sit, ille vero extra
appositus, ad hunc modum.

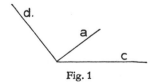

Fig. 1

Est enim interior angulus sub ABC, exterior sub ABD. Neque enim illis
70 credendum est, qui nihil volunt interius aut exterius dici, nisi aliquid intel-
ligatur intra aut extra, siquidem hi tali utentes figura rectum angulum ita
collocant, ut sit intra hebetem et extra acutum.

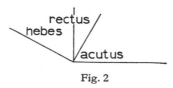

Fig. 2

Ad quem hebetem referentes exteriorem appellant, videlicet quod magis a recto
extra acutum inveniatur, atque ad eundem acutum comparantes interiorem
75 iudicant, quod intra hebetem magis contineatur a recto. Sed hic magnus est
error nihilque aliud fuit, quod impedisset eos, qui conati sunt approbare
triangulum III interiores angulos aequos habere duobus rectis. Equidem, si
bene comprehensum fuisset, quid interior angulus accipi deberet, nihil esset
reliqui, quod eis posset obstare. Quare sciendum est omnem angulum exte-
80 riorem et interiorem iuste vocari, prout se habebit circa propriam figuram aut
extra aut intra, et exteriorem dici ad comparationem eius, qui fuerit intra
ipsam figuram, interiorem ad illum, qui extra positus fuerit, referri; quomodo
ulterior Gallia non eius comparatione, quae ultra sit, sed ad citeriorem Galliam
ulterior (v, f. 86r) nuncupatur et ulterior Hispania ad Hispaniam dicitur
85 citeriorem. Igitur de his angulis et lineis ipsorum inter eos, qui curam habent
geometricae disciplinae, difficillimae quaestiones versari solent, qui anguli
quibus angulis comparentur. Unde etiam haec nata est quaestio, cuius nunc
quaerimus rationem, quam nos saepe tractatam, nondum vero perfectae
solutionis adapertione reseratam, in praesenti loco ne patiamur indiscussam.
90   Nam demonstrata in diversis angulis aequalitate spatiorum probabile erit
eandem aequalitatem reperiri in figuris, licet inter se qualitate formae diversis.
Sit ergo propositum, ut III interiores angulos aequos ostendam duobus rectis.
Describo in primis duos rectos hoc modo: Iacente in plano AB linea recta
aliam rectam, id est (e, f. 26r) CD, lineam superinpono, D puncto ab A et B
95 punctis aequaliter distante.

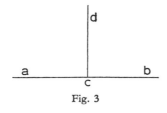

Fig. 3

Sunt igitur recti duo; et quia non quoslibet III interiores, sed diffinitae
triangulae formae III interiores angulos propositos teneo, hos angulos non

per se constituo, sed circa triangulos, quorum anguli sunt, requiro. Itaque triangulum aequilaterum constituo, cuius anguli sunt AEF, BEF, CEF; 100 hos compono II rectis, et aequale spatium invenio.

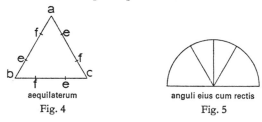

aequilaterum
Fig. 4

anguli eius cum rectis
Fig. 5

Idem accidit, si quis in ortogonio eandem imperet probationem.

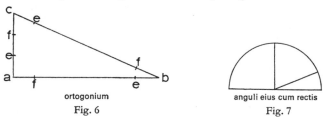

ortogonium
Fig. 6

anguli eius cum rectis
Fig. 7

Dato quoque ambligonio eiusdem rei gratia nihil diversum proveniet.

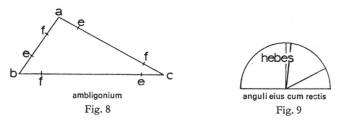

ambligonium
Fig. 8

anguli eius cum rectis
Fig. 9

Et postremo, cum III existant triangulorum genera, nullum esse poterit, cuius anguli huic comparationi dissentiant, sicut harum descriptionum proba-105 tur exemplis. Et hic est interiorum angulorum consensus ad rectos. Exteriores etenim supra aequalitatis modum longe exuberant. In hac vero demonstratione dominus Wazo ascribit figuram hanc A, magister Adelmannus hanc B, Ratechinus hanc Γ, et praeter hos alius quidam hanc Δ, et alii alias.

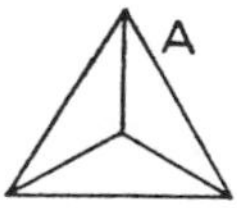

Fig. 10

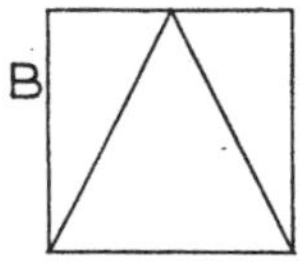

Fig. 11

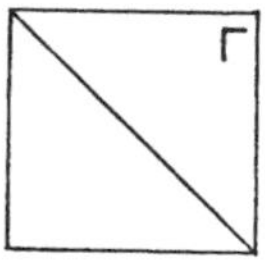

Fig. 12

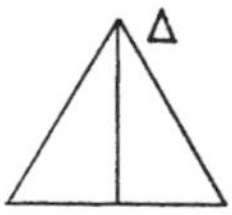

Fig. 13

Sed nos potius animum proposito operi commodemus.

110 (C)

Igitur quadratura circuli reductio quaedam videtur esse ipsius circuli in quadratum vel adaequatio figurae ad se (*v, f. 86v*) invicem utriusque.

(a) Hanc quidam ita constituunt, ut a puncto diametrum in VIII dividant portiones demptaque portione VIII. latus quadrati ducant.

115 (b) Sunt, qui rursus diametrum a medietate quater partiantur et eiecta IIII angulos statuant quadrati.

(c) Praeterea existunt, qui ambitum circuli in IIII distrahunt partes, ex quibus quadratum struunt, quem aiunt illi circulo aequalem.

Sed hi omnes a veritate longe absunt eo, quod ubi et qualiter investiganda
120 sit non attendunt. Nam quicumque demonstrare voluerit formarum quarumlibet aequalitatem, hunc primo advertere oportet, ubi illa versetur aequalitas.

(D)

Omnium enim figurarum aequalium

(a) aliae solo numero coaequantur,
125 (b) aliae spatio tantum,

(c) aliae utroque.

Ergo cum sint figurae circulus et quadratus, necesse est aut primo aut II° aut III° modo aequalitatis recipiant comparationem.

Sed numero solo nequeunt aequari. Nam quaecumque solius numeri servant
130 aequalitatem, ut XXXVI triangulus idemque tetragonus, in illis areas numquam eiusdem reperies quantitatis. Sed aequales sunt areae quadrati et circuli. Non igitur aequantur numero solo. Neque vero numero et spatio has quisquam formulas probabit aequales. Hoc autem ita probo. Quaecumque enim aequalitatis hunc retinent modum, in his communis numerus secundum
135 regulam utriusque figurae potest inveniri, ut in his, quarum una IIII^or habet in hoc latere, in illo IX, altera vero in omni latere VI gestat. In his enim et IIII per novenos et VI per VI multiplicatis XXXVI invenitur, qui utriusque figurae communis est numerus. Sed in quadratum et circulum non cadit, ut videlicet communis numerus propria inveniatur regula utriusque. Quam
140 propositionem etiam probamus hoc modo. Est enim communis numerus tam circuli quam quadrati aequalis circulo CLIIII. Hunc autem facile est invenire iuxta regulam circuli, quae talis est: Triplicata diametro adiectaque VII. diametri fit numerus, qui circulus sive circuli ambitus appellatur. Cuius medietate in medietatem diametri ducta provenit (*v, f. 87r*) numerus, qui pro ipso circulo
145 reputatur. Est autem diametros CLIIII circuli XIIII. Qua triplicata (*e, f. 26v*) et omnibus, quae regula docet, deinceps observatis CLIIII circulum secundum rationem circuli reperimus. Sed eundem secundum quadrati rationem nullo modo possumus invenire. Est enim ratio quadrati, ut ex qualibet summa in se ipsa multiplicata concrescat. Hoc autem modo CLIIII non creatur.
150 Nam si ab aliqua summa in semet ducta procrearetur, id profecto fieret vel a XII vel a XIII. Sed XII, si duodecies sumantur, X minus quam CLIIII invenies; quod si XIII, XV amplius habebis. Non igitur CLIIII, cum sit communis numerus et circuli et quadrati, propria colligitur utriusque ratione.

Quare, si in omnibus figuris, quaecumque spatio et numero sunt aequales,
155 communis numerus propria utriusque regula colligitur, CLIIII vero communis
numerus circuli et quadrati propria utriusque non concrescit, assero non esse
aequales spatio et numero quadratum et circulum.

Adhuc aliud argumentum pono: Si circulus et quadratus in spatio et
numero aequalitatem reciperent, facile ac sponte altera in alterius formam
160 transiret. Hoc enim in omnibus aliis pervidetur, quaecumque numeri et
spatii retinent aequalitatem. Exemplum aliquod dare placet: Sit item figura in
hoc latere IIII pedum, in illo IX, cui superponatur altera illa profecto, quae
ex omni latere senario metitur. Dico in his figuris palam esse, qualiter una in
alteram transformetur. Quod ea de causa putamus contingere, quia nec solo
165 numero nec solo spatio comparationem habent aequalitatis. Quod idem ipsum profecto in circulo et quadrato accideret, si eodem modo inter se com-
parabiles existerent. Sed neque circulus in quadratum neque rursus quadratus
in circulum, nisi cum summa difficultate et nisi reperta qua id fieri possit
artis cuiusdam facultate, quam deo praestante tradituri postmodum sumus,
170 neuter umquam in neutrum transire possunt. Causam vero quare non possint
eam quoque posterius, si divinitas permiserit, monstrare conabor. Nunc igitur concludendum est, quod his ita se habentibus non recte utroque iu-
dicentur aequales quadrata forma (*v, f. 87v*) et circularis.

(E)
175 Adhuc aliud: Si essent saepe dictae figurae iuxta numerum et spatium
aequales, numerus communis, ut puta CLIIII, non solum circulus, sed etiam
quadratus esset. Sed non est quadratus CLIIII. Hoc ita probo.

(a) Omnis tetragonus, sicut etiam arithmeticorum regula docet, ex imparium
coacervatione generatur. At vero CLIIII, coacervatis super se ipsos ab uno
180 quousque velis imparibus, nusquam occurrit. Non est igitur CLIIII quadratus. Eadem lex alios quoque circulares numeros includit, neque umquam
poterit terminus reperiri, qui, cum circuli proprietate nitatur, quadrati
quoque ratione participet.

(b) Sed fieri poterit, ut aliquis dicat CLIIII, etsi minime numero, minutiis
185 tamen quadrari posse. Operae pretium est investigare et hoc. Ergo, quamvis
contra naturam videatur, ut CLIIII in quadraturam redigatur, incipiamus
tamen ipsique naturae iura inferentes ad uncias quoque animum vertamus,
an forte illis appositis numero totaque illa summa per eundem numerum et per
ipsas uncias dimensa CLIIII tetragonare valeamus. Et quia hic ipse numerus,
190 si ab aliqua summa in se ducta crearetur, id vel a XII vel a XIII, ut supra dictum est, fieri oporteret, XII in primis assumatur. Huic autem adiciatur aut
triens aut quincunx. Sed quincunx exuberat; nam XII in se et XII quincunx,
item XIIes quincunx, postremo quincunx in quincuncem, supra CLIIII sextantem et dimidiam sextulam apponunt. Triens autem minus (*e, f. 27r*) reddit
195 eisdem CLIIII; namque XIIes XII^cim et XIIes triens, item XII triens, ad
extremum triens in trientem, CLII accrescente uncia et duella accumulant.
Quibus adhuc desunt ad perfectionem CLIIII assis, dextans et II duellae.

Quamobrem, quincunce exuberante, triente vero ab integritate plenitudinis refugiente, nonne dementiae est vel illo minorem vel illo sumere potiorem?
200 Illud tamen adhuc fieri potest, ut eisdem XII et trienti minutias apponamus diligentius investigantes, an vel sic per aequalem multiplicationem ad ęam quam quaerimus summam pertingamus. Erunt autem (v, f. 88r) hae minutiae, quae apponentur: semuncia, sicilicus, sextula. Quippe aliae aut plus componunt aut multo minus; hae autem solae ad ipsos CLIIII tam proxime accedunt,
205 ut, exceptis partibus unius oboli tertia, IIII., IX., nihil excrescat, nihil excedat, nihil exuberet. Quod apertius ostendi valet abaco quam stilo computando potius quam disputando. Neque id quicquam proficiet, si quis minutias confingat a numero denominatas, quomodo solent calculatores ad minutissimum aliquid, quod iam nomine careat, divisione perducta. Qui enim nesciat,
210 quantam partem aut scripuli aut oboli aut ceratis aut novissimi calci assumat, nonne frustra pro huius aequalitatis perquisitione operam insinuet? Et certe quota pars accipienda sit, qua poterit nosse disciplina? Omnis enim tetragonus quotum ab unitate locum obtinet in ordine tetragonorum, totam ad sui multiplicationem expostulat summam. Hac arte cuiuslibet quadrati
215 longissime et ultra mille milia positi tetragonicum latus facile et cito pervestigamus. Quod in CLIIII vel alio quolibet circulari numero, cum ipsi minime tetragoni sunt, quis valeat invenire? Nimirum inter tetragonicum latus ipsosque tetragonos haec quasi pactio firmata est, ut tam latus nisi ex loco et ordine ipsorum nequeat deprehendi, quam ipsi per multiplicationem
220 creari non possunt praeter lateris libratam dimensionem. Sed quomodo XII nihil addere potes, ut CLIIII in quadratum provehatur, ita econtra XIII numero nihil auferre, ut idem quadratus aequis ex omni parte lateribus construatur. Quamobrem non oportet ulterius cum ipsa concertare natura, quam nulla vis a suo statu inflectere valet. Et his demonstratum sit CLIIII quadrari
225 non posse, quod minime mirabimur, si ad alios numeros, quicumque naturaliter quadrati non sunt, considerationem vertamus. Nam nullus eorum in quadraturam reduci potest, neque integris terminis in se multiplicatis, neque si uncias vel minutias adiciamus, quod mox in II° libello monstrabimus.

PROLOGUS SECUNDI LIBRI

(v, f. 88v) Etsi nullus omnium quantalibet sit felicitate praestantior, haud vereor tamen, mi Caesar, dedeceret hoc veluti quoddam diadema, quod tuo capiti fabricare molimur. Si quantum nobilis materia, in tantum artifex
5 sapiens esset, verum perpendas oportet, quoniam aurum non tam ex arte placet quam ex propria virtute, neque ita pretiosum caelatura quantum naturali praestantia iudicatur. Sed forte aliqui dicant geometricalis scientiae curam a sanctitate praesulari (e, f. 27v) alienam existere. Nimirum, qui ita putaverint, hi minime recolunt sanctum Moysen quam maxime huius disci-
10 plinae habuisse peritiam; hi permensam diluvii arcam Egyptiis cubitis nequaquam retractant; ipsique non reputant omnem terram repromissionis funiculis geometricalibus distributam. Ad haec Salomonem tum ipsum templum, tum porticus, tum atrium templi, postremo quicquid ad templum

68

respiceret, convenientibus ordinasse mensuris. Praeterea apud Ezechielem
15 virum, cuius erat species quasi species aeris, totum aedificium illud civitatis
multaque in eo praeter numerum linea mensurali calamoque designasse.
Quod, si ita est, quae religio sit etiam sacerdotes domini a tanto studio pro-
hibere? Nonne Ezechiel sacerdos? Nonne vir ille, cuius species aeris, ipse
Christus summus et maximus sacerdos? Sed quae iam supra distulimus, ingre-
20 diamur ostendere.

PROLOGUS EXPLICIT. INCIPIT LIBER II

(A)
   Omnium numerorum
  (a) alii sunt sponte et naturaliter tetragoni,
25 (b) alii minime.
   Quicumque ergo natura tetragonorum tenentur, hos si quis quadrare
voluerit, nulla difficultas impediet, ut si forte in quadrati figuram disponere
velit sive quaternarii summam sive novenarii sive quorumlibet reliquorum
naturali tetragonorum ordine subsequentium.
30   Sin autem ceterorum aliquem, qui a natura tetragonorum separantur, in
eorum formam (*v, f. 89 r*) reducere contendas, id, quantum ad se ipsos, nullum umquam consequeris effectum. Verumtamen circa superficies spatiorum
idem fieri possibile est. Nihil enim prohibet eiusmodi spatia inveniri quadrati
formam habentia, quorum aliud duobus pedibus constet, aliud trium in
35 se mensuram retineat, aliud quinis, aliud senis, aliud septenis vel octonis vel
etiam pluribus ultra pedibus teneatur, atque in hunc modum omnes numerorum quantitates circa subiectas corporum materias, quominus debeant quadrari, nullatenus recusant. Ceterum in se ipsis et seorsum extra considerationem mensurabilium spatiorum sola animi speculatione perceptis, vacuum
40 quisque consumet laborem, si eosdem numeros redigere curet ad quadratorum rationem. Potest tamen compositis ex adiectione minutiarum aequis
lateribus ad eorum summam proxime accedi, ut parum invenias aut deesse aut
ad perfectionem et integritatem superabundare. Et aliis quidem, ut certe
binario latera creantur, ex uno et triente, semuncia, duella, dimidia sextula,
45 unde aut binario paulo minus conficitur, vel item ex uno et quincunce, unde
paulo amplius eodem binario concrescit, aliis vero aliae unciarum aut solae
aut cum minutiis pro tetragonicis [et exquatis] lateribus constituuntur, ut
quisque calculandi peritus per semetipsum intelligere valet. In quibus omnibus, ut dictum est, perfecta integritas numquam investigari poterit, sed id
50 vitii semper adest, ut licet parum, tamen aliquid aut superabundet aut desit.
Nihil igitur mirum, si circularis numerus, ut puta CLIIII et quilibet eius
generis, quadrari non potest, quando quidem et aliorum omnium nullus
potest praeter naturales tetragonos.
   De qua re ea de causa longius traximus disputationem, quia sunt nonnulli,
55 qui putant quadraturam circuli in numero per minutias constitui posse. Scio
Werenboldum hac opinione inductum XXXVIII semis, id est aream XXII
circuli, in quadratum, (*e, f. 28r*) ut sibi visum (*v, f. 89v*) est, redegisse multi-

plicata in se senarii summa et adiectis minutiis, quae sibi visae sunt ad rem
pertinere. Quem hoc quidem fefellit, quod per senarium tum senario multi-
60 plicato tum etiam minutiis, ipsas tandem minutias, sicut per omnes quadratos
fieri oportet, in sese multiplicare neglexit.

(B)
    Sed ut ad praeposita revertamur, cum CLIIII aliique circulares quadrari
non possint, manifestum est non esse aequales in comparatione numeri et
65 spatii quadratum et circulum. Restat ergo aut solo spatio comparationem
habere aut prorsus aequari non posse. Sed quis hoc dicat, cum nulla sit figura, quae alteri per aequalitatem conferri non possit? Quare, ut etiam istae
aequandi facultatem non habeant, stultum est arbitrari. Restat ergo, ut hanc
exaequationem in spatio solo quaeramus. Nam supra diversis conclusionibus
70 extortum est, ne ultra vel in numero solo requiratur vel in utroque.
    Sed omnium figurarum, quae solo spatio sunt aequales,
    (a) aliae per se demonstrari possunt,
    (b) aliae non per se, sed per alias probantur aequales.
    Per se monstrari dico, quae ipsae in aequales suas absque mediae alicuius
75 interpositione resolvuntur. Per se non posse dico, quae nisi per medias resolutionem non capiunt. De hac resolutione postea tractabimus.
    Ad praesens vero sciendum, quod in hac parte quadratum in comparatione
circuli ponimus, quae non per se ad aequalitatem reduci possunt. Semper enim
media quadam figura opus est, ut vel circulus in quadratum vel quadratus
80 reducatur in circulum. Haec autem figura unde nascatur, in hoc libello
monstrare propono.
    Nascitur sane ex ipso circulo et vel per diametrum eius vel per resolutionem
invenitur. Sit igitur a quolibet imperatum, ut sibi quadratum producam aequale scilicet ABΓΔ circulo. Quem cum dederit, diligenter attendo possitne
85 statim a principio aequale [a] circulo qudratum produci. Video non posse,
sicut in priore libello satis est approbatum. Nondum tamen desisto. Utor
illo Terentii consilio, quia hac via non processit, aliam aggredior et rursus
perquiro, num saltem valeat aliquod spatium IIII laterum a proposito circulo
procreari, ut vel per illud adinventi- (v, f. 90r) onem quadrati peraccedam,
90 quo ipse circulus nullum mihi per se aditum aperiebat. Et hoc iure. Nam
omnis quadrilatera figura quadrato proxima et quasi consanguinea existit,
cum praesertim eodem modo rectis nitatur angulis. Quare nihil mirum, si per
illam ad hanc familiarius accedi potest. Huiusmodi autem figura facillima est
inventu: Denique proposti circuli diametrum in partes XIIII divido, deinde
95 quadrilateram figuram constituo hac nimirum ratione,. ut per duo latera
ipsam diametrum totam, item per alia duo latera XI eiusdem particulas
disponam. Quo facto habeo formam eidem circulo prorsus aequalem spatii
quantitate dumtaxat. Haec igitur figura ABCD litteris per IIII angulos
designata in hunc modum describatur, ut a lateribus ad latera et per transver-
100 sum et per longum pro numero partium diametri, id est XI et XIIII, lineae

ducantur, quatinus ipso intuitu manifestum sit, quot in tota figura partes includantur, hoc est CLIIII, ad hunc modum.

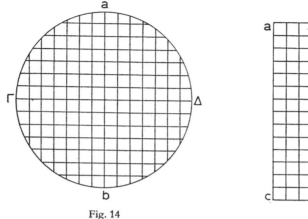

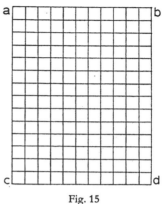

Fig. 14   Fig. 15

(*e, f. 28v*) Dico igitur quadrilaterum ex appositi circuli diametro productum esse. Undecies enim XIIII CLIIII fiunt, eadem videlicet summa, quam area
105 circularis includit.

(C)

Sed hic aliquis fortassis obiciet: Quis umquam scire potuit, quantum area circuli comprehendat, ac per hoc quis valet scire, quid circulo vel sit vel non sit aequale? Cuius enim nescias propriam quantitatem, quid extra valeas
110 existimare ad ipsius aequalitatem? His ita obiectis quibus obnitar rationes non desunt. Peritia, inquam, geometricae disciplinae (*v, f. 90v*) de inveniendo circuli embado eiusmodi regulam describit, ut medietas diametri in medietatem circuitus debeat protendi. Quare, cum huius circuli diametros XIIII, circuitus autem XLIIII, horum autem dimidium VII et XXII existat, non potest
115 incertum esse habere circulum CLIIII; siquidem hanc summam conficit VII et XXIIorum multiplicatio. Quibus ita se habentibus verissime dictum est illud aequilaterum esse circulo aequale, quod ex tota diametro et eius XI partibus constat esse productum, ob hoc nimirum, quia quantum VIIes XXII, tantundem conficiunt undecies XIIII. Sed adhuc instabit et his contradictio-
120 num validius impugnabit telis. Nulla est auctoritas regulae, nisi, quod ipsa docet, ita se in rebus habere aut animo perspiciatur aut sensu teneatur; alioquin fidem praestare non oportet. Num soli regulae credatur? Nonne omnis regula ex subiectis rebus accepta et praescripta est? Certe de quadrila-tero, cum per II latera undenario metiatur, in aliis vero duobus XIIII gerat,
125 qui, ut dictum est, in se multiplicati CLIIII reddunt, non est dubium, quin in tota areae suae latitudine CLIIII non plus nec minus comprehendat. Quod etiam in superiore figura promptum est pervidere. Sed in circulo cui id ip-

sum copia est perspicere? A quo potest circulus aliquando simili modo veluti quadrilaterum in CLIIII partes aequaliter distribui, ut hoc nobis videre liceat
130 et sic tandem fidem adhibere? Numquid nos in rebus visibilibus credere oportet, quod nequaquam oculis videri potest? An negat quisquam figuras geometricas visibiles esse, cum omnes circino aut regulae ad dimensionem subiaceant? Ergo CLIIII inesse spatio circulari aut visibili argumento, cum nimirum res ipsa visibilis existat, probandum est, aut non credendum. Sed
135 probari non potest circumferentis lineae prohibente natura. Igitur, ne etiam credi oporteat, validissimae sunt huiusmodi obiec- (v, f. 91 r) tiones. Quid ergo dicemus? Faciendum ergo quod exigimur, et visibili, ut ita dicam, argumento utendum, verisimillimum ego quidem puto. Cum studiosi geometricae disciplinae diuturna dubitatione turbarentur quid existimare de quantitate
140 circularis embadi deberent, neque sicut areas angulis rectis inclusas, ita quoque comprehensionem circuli manifeste deprehendere potuissent ullo mensurarum genere, neque unciis, neque pedibus, neque aliis quibuslibet, quibus omnis dimensio sive in longum sive in latum sive etiam in altum proficiscitur, quibus porticus, quibus miliaria, quibus stadiorum longitudo,
145 quibus agrorum fluviorumque latitudo, quibus parietum montiumque altitudo comparatur, cumque nullatenus pateretur circumferentis ambitus curvatura huiusce dimensionis rationem, ipsi autem nihilominus certissime rescire vellent, cuius aestimationis esset planities circularis, animadverterunt quod linea eius, etsi curva ac circumferens (e, f. 29r) esset, aequaliter tamen undique versum a puncto medietatis distaret. Quo deprehenso (res etenim manifesta est)
150 divisere totum circuitum in quot eis visum est partes, utpote in IIII et XL, quas totidem punctis designavere; et a punctis ad medium totius circuli punctum lineas duxere adhibitoque sectionis labore XLIIII protulere partes, quarum unaquaeque VII mensuras, id est medietatem diametri, in longitudinem
155 habebat (perlatum vero XIIII inveniebatur eiusdem diametri); nec mora partes partibus adiungentes acumen latitudini obverterunt. Quo facto inventae sunt undenae et undenae sibi invicem congruentes, factaeque sunt duae formulae ex XLIIII, quarum utraque XXII illarum XLIIII, id est medietate circuli, constabat, habens VII in uno latere, XI vero in altero. Quibus sibi eo modo
160 copulatis, quo plurimum (v, f. 91v) posset latitudini respondere longitudo, egressa est forma binis lateribus undenas, binis autem XIIII gerens unitates.

Quorum exemplo uti res manifestior fiat, datum circulum in IIII partes divido, ductis diametris altera ab A in B, altera a Γ in Δ. Quo facto singulas circuli quartas per XI distribuo, sicut haec figura demonstrat.
165 Facta autem resolutione XI partes XI partibus intermiscendo compono, versis aliarum extremitatibus ad capita aliarum secundum huius descriptionis exemplum. Hoc modo resoluto circulo et partibus eius hac arte dispositis nonne producta est quadrilateri figura? Quae profecto, dum XI in latitudine, in longitudine vero XIIII gerat, qui in se multiplicati CLIIII componunt,
170 manifeste ostendit, quantum intra se spatii circularis ambitus comprehendat. Unde iam nulli potest dubium esse, (e, f. 29v) quin ex hac resolutione subtiles geometrae regulam illam collegerint, qua praecipiunt ad inveniendam aream

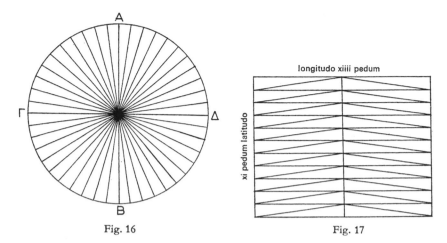

Fig. 16   Fig. 17

circuli aut diametri medietate medietatem circuitus extendi, aut totam diametrum IIIIᵃ parte circuitus, aut totum circuitum diametri IIIIᵃ, aut quolibet
175 alio modo ad idem ipsum proficiente. Quam regulam idcirco compendii
causa inventam putamus, ut non semper necesse esset circulum resolvere, quotiens vellemus arealem eius quantitatem scientia tenere, cum praesertim cura
ea subiecta saepe considerandum scirent, quae nullatenus resolutionem pate-
rentur. Haec enim nisi in membranis et pelliculis et si qua sunt (*v, f. 92r*) eius-
180 modi, exerceri commode non valet. Nunc igitur obiectionibus supra memoratis satis factum arbitramur simulque demonstratum, qualiter illa media
figura, per quam circulo quadratus aequalis esse probatur, ex ipso circulo vel
per diametrum eius per resolutionem generatur. Iam igitur huius scientiae
quasi fundamento quodam firmati consequenter certis incerta et notis ignota
185 perquiremus, ac deinceps quemadmodum ex eodem quadrilatero, in quod
circulum resolvimus, quadrati species promoveatur curabimus intimare. Hoc
autem faciemus, si prius confragosa oratio alicuius prooemii interpositione
levigetur.

PROLOGUS TERTII LIBRI

Si tu is esses, praesul eximie, cuius suae laudes animum delectarent et pasce-
rent, nulla facundia satis esset his, quae tum de te ipso, tum de avis, de
proavis, deque universis maioribus tuis non modo Romanorum, sed Graiorum
5 quoque principum illustrissimis, de horum nobilitate, de dignitate, de gloria,
de potentia, de copiis, de amplitudine rerum, de praeclarissimis factis, de
virtute, de sapientia, de meritis ipsorum iactari valerent. Virgilius cupiens a
parentibus magnificare Augustum, Aeneida XII libris conscripsit. Quanto
pluribus opus esset, si quis vellet colligere, quae primus, quae secundus,
10 quaeque tertius gessit Ottonum, quorum primus ab Henrico patre suo suscepit
regnum, sed filio reliquit imperium; pater apud Theutones primus regnavit,

filius apud ipsos primus imperavit. Et quibus, nisi illis, Germania debet,
quod sibi cum tanto orbe ipsa exsolvit tributum Italia? Per quos alios nostri
imperatores Romani sceptri facti sunt successores? Vellem mihi diceres, o
15 Maro: quid tale contulit vestro Latio ille tuus, ille pius, ille magnus, ille deo
natus Aeneas? Agnosce, Auguste, quanto sit infra tuum genus piis et praecla-
ris Ottonibus comparatum. Sed, esto, fuerit talis Aeneas ille; quid tua refert?
Tu enim Aeneam vix millesimus attingas nepos. Putasne (v, f. 92v) igitur
Ottonum nepos, si qualem tu Virgilium haberet, qui eum extolleret a laude
20 parentum, putasne tuam famam quanta gloria offuscaret? Ita quidem, utpote
qui non millesima, sed prima tam praeclari sanguinis proles existat. Verum
haec universa, sancte ac religiose pontifex, prudenter aspernaris, sicut omnia
in te piae humilitatis signa et opera attestantur, bene memor, quia neque
nobiles neque sapientes, sed ignobilia et contemptibilia mundi elegit deus.
25 Quare cautus fui ita scribendo nequaquam attingere, quae animum cupidum
(e, f. 30r) laudis insanum redderent, tuis vero auribus suis favoribus merito
infestis offensionem incuterent. Itaque verti stilum ad ea potius dictanda, quae
sensum instruerent, ingenium acuerent, sollertiam excitarent, haud ignorans
in tantum tibi placere rationabilium studiorum utilitatem, quantum minime
30 diligere vales inanis gloriae ac ventosae iactantiae odiosam vanitatem.

EXPLICIT PROLOGUS. INCIPIT LIBER TERTIUS

(A) Superiori libello qualiter a circulo quadrilateri species exiret ostendimus.
Nunc autem ex eodem quadrati producere formam aggredimur, opus quidem
difficile et laboriosum nec minus ipsa circuli quadratura intemptatum et in
35 quo, nisi geometricae sublevemur auxiliis, necessario arbitrer deficiendum.
Quis enim adhuc quadrandi tradidit rationem? Et quare ab omnibus praeteri-
ta, nisi propter difficultatem eius invictam? Quam ob causam non parvis in
hoc loco angustiis torquemur, et quocumque transire temptamus, veluti clauso
repellimur limine. Si enim illud quo longitudo latitudinem superat dividamus,
40 dimidium latitudini demus, dimidium vero relinquamus longitudini, nonne
velut angulari subducto trunca quaedam nascitur figura? Hoc [si] quidem in
apposita pervidetur descriptione. Quodsi per obliquum sumere velimus eius-

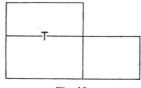

Fig. 18

dem spatii quantitatem, quo temperamento fiet quatenus, ne infra sit nec
modum excedat? Puto manifestam (v, f. 93r) esse difficultatem. Nunc autem
45 potest nobis demonstratio dari brevissima.

Quadrilateri latus, scilicet AB, diminuatur III partibus ad E signum; similiter CD latus ad signum F totidem partibus sit recisum. Igitur spatium inter EF et BD lineas dividi oportet, quod fiet hac arte: Ducetur angularis ab E usque G, ad cuius mensuram inter ipsum E et B assignetur H punctum, 50 sumptaque distantia eiusdem H ab A secundum ipsam HK et BI lineae fiant. Quo facto si spatium quod includitur de medio subtrahatur figuraeque ad latitudinem apponatur, dico ex quadrilatero productum esse quadratum.

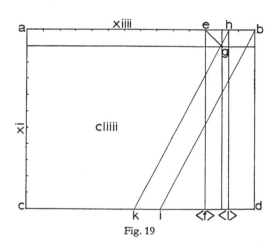

Fig. 19

Et hoc modo illud, quod ex circulo prodit, in quadraturam redigitur. Ad finem vero arreptae disputationis in omne quadrilaterum universalem qua- 55 drandi dabimus rationem. Hoc autem nomine appello quicquid excepto quadrato IIII latera et totidem rectos habeat angulos. Sed hanc, quam modo produximus, quadrati figuram aequalem asserimus circulo. Nam si quadrilatero unde processit aequalis, aequalem circulo quis neget, qui quadrilatero existit aequalis? Sed quadrilatero nimirum aequalis, cuius videlicet nulla 60 pars relicta sit extra. Haud dubium igitur, quin et circulo omnimodis quantitate conferatur aequali. Verum aequalis utrique spatio solo, non etiam numero. Non autem hoc dico, (v, f. 93v) quin idem numerus omnibus his insit figuris, cum ut prima, sic et secunda, sic et tertia CLIIII pedibus conste[n]t aut assibus. Nec tamen quadratum numero aequale dicimus, quia numerus 65 (e, f. 30v) iste non ita recipere potest aequalem multiplicationem laterum, sicut circuli sive quadrilateri proprietate constitui.

(B)
Sed rursus hic aliquis dicet in minutiis hoc fieri posse. Licet dudum probaverim, placet tamen adversus hanc opinionem aliam hoc in loco introducere 70 argumenti rationem. Dico enim si quadrato aequa latera constituuntur in minutiis, nihilominus et in numero aequalia constitui. Nam quicquid multiplicant minutiae, idem numero quoque multiplicatur integro.

(a) Sit autem exemplo quadratus minutiarum, cuius sunt latera unitas et semis, eo quod omnibus iste videatur apertior. Hic vero tali constituitur modo:
75 multiplico I in se, nascitur unus; deinde semissem per eandem unitatem. His propter II latera duco. Rursum mihi provenit unus. Quo iuncto superiori habeo II. Deinde semissem in se ipsum duco, egreditur quadrans. Qui appositus II quadratum efficit supradictum continentem in tota superficie sua quadrantes IX. Similiter IX invenio, non iam quadrantes sed asses, si unita-
80 tem semissem, quod erat latus huius quadrati, duplicavero. Quae duplicatio III producit, qui in se ductus novenarium generat quadratum, nihil a superiore, in eo videlicet, quod IX partibus constat, differentem. Ad hunc modum et quadratum circuli, si minutiis constaret, integro reperiri posse arbitramur. Sed, ut manifestum est, impossibile est in integris eum reperire. Ne igitur
85 requiras in minutiis, nisi frustra fatigari malueris et inutili labore consumi.

(b) Adhuc aliud dicam. Si sint II° numeri, alter cum minutiis, absque minutiis alter, ille autem cum minutiis multiplicetur in se, alter vero ratione, quae in circulo servatur, accrescat: quam comparationem ad se habuerint summulae (*v, f. 94r*) in hunc modum productae, eadem etiam servabitur in illis summis,
90 quae eodem modo productae fuerint utroque multiplicandorum secundum eandem proportionem integro sumpto. Sit ergo numerus cum minutia VIS. Item sit absque minutia VII. Rursus sint alii duo, uterque integer, eadem proportione servata, id est XIII et XIIII. Assero ergo XIII in se, XIIII vero iuxta regulam circuli ducto, qui inde nascuntur numeri eandem inter se
95 comparationem servare, quam et illi habent, quorum alter a VIS, alter a VII, utroque iuxta suam rationem multiplicato, concrescit. At vero nullos reperimus integros numeros, qui ad invicem conservent comparationem circuli et aequalis sibimet quadrati. Quid igitur in minutiis requirere labores, quod numquam sinit invenire natura?
100   Sed etsi ratione calculandi abacique peritia subtilissima illic reperiri convinceretur, quid hoc mensoribus et studio geometricali conferret? Quis enim redigeret sub mensuram tum obolum, tum ceratem, tum calculum et harum particulas infinitas? Quod nos quidem ita impossibile rati sumus, ut facilius concedamus quadraturam per minutias circulo aequari, quam easdem minutias
105 sub dimensionem venire posse. Iure igitur circa has illam quaerere recusans artem potius geometricam, qua metiendo componi posset, pervestigare curavi. Neque hoc temere, proprioque arbitrio, ac praeter exemplum fecisse me sapiens quisquam redarguat, qui sciat geometras in multis mensura tantum contentos esse, neque se inquisitione numerorum occupare, ut puta cum
110 interiores tres anguli duobus rectis comparantur, aliaque permulta quae longum est enumerare.

(C)
    Quis hanc aequalitatem in numero ac non potius in spatio et mensura aestimet requirendam? Neque hoc in aequalibus dumtaxat faciendum, verum
115 (*e, f. 31r*) in duplis, in triplis, in quincuplis, et in his quae sexcuplam vel septuplam vel octuplam habent comparationem, ut nulla earum in minutiis aut

in numero requiratur, magnopere servandum. Dicet aliquis: quomodo
(*v, f. 94v*) probas duplum in minutiis simplici quadrato non conferri? Superiore quidem ratione hoc quoque probo. Nam in qua comparatione erunt
120 duo numeri, quorum alter ex integro, alter vero productus est ex eo latere,
quod minutiae componunt, eadem servabitur in illis quoque summulis,
quarum latera omnino sunt integra, sic tamen, ut unam habeant proportionem. Ut si sint duo numeri IIII et VI quadrans, quorum alterius latus est duo,
alterius duo S. Item sint duo numeri XVI et XXV, alter IIII alter V iuxta
125 proportionem superiorum laterum, id est duorum item duorum S, ipsi quoque latera habentes. Dico in qua proportione reperiuntur IIII et VI quadrans, in eadem proportione etiam XVI et XXV reperiri. Uterque enim
minorem in se habet minorisque medietatem et ipsius medietatis octavam. Ad
hunc modum itaque, si comparatio dupli et simplicis quadrati in minutiis
130 constaret, nihilominus in integris terminis eandem facile esset invenire. Sed
quis umquam in numeris deprehendit dupli et simplicis quadrati comparationem? Nam simplo habente latus, quod numero sit dimensum, numquam in
latere dupli numerum invenies; ut si quaternarius, cum sit quadratus, duo
habeat in latere, quo numero dupli eius latus notabitur? Quodsi in latere dupli
135 secundum integrum numerum facta est dimensio, tum nimirum latera simplicis iuxta numerum impossibile est metiri. Utputa XVI, duplicis quadrati ad
octo, quaternario metimur latus, quomodo putas latera simpli partiemur?
Quare, cum omnis comparatio, quae ex minutiis creatur, in integris etiam
numeris existat, cumque dupli relatio ad simplicem quadratum in nullis um-
140 quam numeris sit reperta, hinc audaciam arripimus hoc demum concludendi,
quod ne per minutias quidem latera illorum valeant comparari, quamquam
doctissimus vir Regimboldus asserat in latere dupli quincuncem et latus
simplicis contineri. Quod et ipse et cum Gerberto noster Racechinus fatetur.
Hoc autem aliter in re esse tali probamus (*v, f. 95r*) argumentatione.
145   Omne latus dupli eiusdem mensurae omnino est cuius et diagonium
simplicis quadrati. Omnis autem linea diagonii tantum solummodo maior esse
debet laterali linea, ut dum haec simplum, illa aequaliter duplum efficiat
quadratum, cui nihil desit nihilque accedat. Hoc cum omnes geometricae disciplinae peritissimi attestentur, tum etiam visibili argumento, si quis artem
150 metiendi non nesciat, ita rem esse facillime comprobabit. Quod cum ita se
habeat, sumatur exempli gratia XXV quadratus; is retinet latus V, de quo
perficiamus diagonium eiusdem lateris quincunce sibimet adiecto. Est autem in V assibus duo asses et uncia proportio quincuncis. Hanc ipse quinarius
recipiat, erunt ergo VII et uncia. Hoc igitur est diagonium proposti qua-
155 drati quantum ad magistri Regimboldi sententiam. Nunc ergo, cum omne
diagonium nihil plus minusve nisi duplum aequaliter provehat, videamus, sicut
V in se ducti congregant XXV, an eodem modo VII et uncia in VII et
unciam ducti secundum tetragonicam multiplicationem quinquagenarium
accumulent, qui nimirum ad XXV dupla proportione confertur. Minime;
160 septies namque VII XLVIIII multiplicant. Inde septies uncia et item septies
uncia, novissime uncia in unciam assem, sextantem et dimidiam sextulam

coacervant. Quorum assis XLVIIII adiectus quinquagenariam (*e, f. 31v*) pluritatem complet; superat quantitas sextantis et dimidiae sextulae, id ipsum nihilominus integro numero licet approbare. Et quoniam ipse con-
165 firmat diagonium quadrati et latus eiusdem in tali proportione reperiri, qualis est inter XII ac XVII, quorum duodenarium claudit et eius insuper V duodecimas, quod appellant quincuncem, utrosque in se, et XII duodecies et XVII decies septies congregemus. Quae summa proveniet? Nimirum ex priore CXLIIII, ex posteriore CCLXXXVIIII colligentur. Sed diagonium
170 (*v, f. 95v*) nihil a duplo nec minus nec amplius creare debet. At vero CCLXXXVIIII unitate excedunt duplam proportionem ad C⟨X⟩LIIII per comparationem reducti. Unde absque dubitationis scrupulo constat diagonium non habere in se quincuncem et latus sui quadrati. Quare, si in duplo quadrato mensura lateris ab hoc diagonio nihil distat, manifestum est ita, ut
175 nullatenus refutari possit, quod ne ipsum quidem latus dupli minoris latus eiusque quincuncem possit includere. Sed his tandem qualicumque modo finitis illud volumus intimare quadrati formam vix tantis difficultatibus repertam. Si quis noverit aequaliter collocare ita sane, ut quanto circulus quadrati angulis, tanto quadrati latera circuli excessionibus superentur, is
180 profecto habet pretiosissimam illam et a doctis viris saepe ac diu perquisitam, quam vocat geometricae peritia circuli quadraturam. Atque haec collocatio quomodo etiam per artem fiat, monstrare curabo. Sed hoc postmodum fiet; nunc vero istic conquiescam.

PROLOGUS QUARTI LIBRI

Cum singulae rerum, Tullio in Topicis auctore, multas habeant causas, eaque inter omnes principalis existat, praeter quam ceterae stolidae sunt et nihil agentes, quomodo materia et instrumentum nihil agunt, si non manus artificis
5 fuerit adhibita, tu modo, decus pontificum, solus es, si quid in hoc opere placeret, primaria ipsaque efficiens causa, ego vero qualecumque instrumentum iure videar, quo tu sis abusus ad huius operis effectum. Abusus, inquam, quia inutile instrumentum. Sitne instrumentum utile opus, ⟨quod⟩ non esse bonum non posset, cum praesertim et materia sit integerrima et auctor ut
10 melior nullus? An non auctor ille, qui incitator? Ille utique auctor. Numquid enim operarii mercede conducti Romam condidisse dicuntur posthabita Romuli memoria, qui illos forsitan operarios mercede conduxit? Sed illorum potius obsoleta est omnis (*v, f. 96r*) memoria. Quis enim caementa confecerit, qui lapides comportaverint, quis iecerit fundamenta, quis illum murum compe-
15 gerit, nemo recordari, nemo dicere valet. Sin etiam de divinis exemplum petatur: Cuius illa apud Mattheum vinea, patrisne familias an eorum, qui denarium accipiunt? Distinguit hoc Mattheus ipse his verbis: conducere, inquit, operarios in vineam suam; suam ait, non eorum. Sed apertius ipse pater familias suam esse testatur, dum ait ad illos: Ite et vos in vineam meam;
20 in meam ite, non in vestram. Ergo tu quoque huius operis auctor. Quare nisi tibi, qui pigrum et stolidum ingenium beneficiis impulisti, si quid in hoc

opere gratum existeret, nulli debetur. Quisquis enim in nos retorqueret, is
quasi limae, quasi serrae, quasi asciae aut securi scriberet laudem quam artifex
meruisset. (*e, f. 32r*) Sed unum est in quo hoc ipsum opus tibi magis addicere
25 queas, quippe si digneris illud maiestatis tuae patrocinio tueri. Rarus enim
scriptor sua auctoritate ita numquam fretus fuit, qui⟨n⟩ principum defensione
non indigeret. Horatius Maecenatis praesidio gaudet, amat Pollio Virgilium,
utrique Augustus favet. Terentius concitante aemulorum invidia odium pro
gratia incurrisset, nisi Calliopius callidis argumentis accusatoribus obstitisset.
30 Manlius Torquatus Boetius, vir cunctarum artium perfectione et consulari
dignitate praecipuus, cum in latinos thesauros insignis quadruvii pretium
transtulisset, quod a Pytagora omnium ducum veteris philosophiae iudicio
probatum et excoctum esset, frequenter tamen et humiliter orat, quatinus
Patricii Symmachi paterna gratia labor ille provehatur. Quodsi tanti nominis
35 viri praeclarissimo labori suo absque favore principum diffisi sunt, multo
magis ego, quem neque fortuna neque scientia commendat, sine tuo praesidio
nihil habeo spei. Quo ego si fuero potitus, tanto securior ero Horatio, Marone
ac ceteris, quanto meliore (*v, f. 96v*) patrono defensus. Tu enim Patricio
melior, tu maior Augusto. Qui si de me senseris bene, haud vereor male
40 aestimare quisquam pradsumat. Certe, ubi radius splendidae auctoritatis tuae
velut ipsius solis effulserit, necessarium erit ignorantiae meae tenebras non
apparere. Composita circuli quadratura, cur circulo quadratum principaliter
et per se ipsum etiam in mensura non comparetur, iam placet continuam disputationem adiungere. Sed de ea altius ordiendum est tali principio.

45   EXPLICIT PROLOGUS. INCIPIT LIBER QUARTUS

(A)
       Omnium aequalium formarum (de his enim agitur)
(a) aliae sunt eiusdem formae,
(b) aliae diversae.
50   Eiusdem formae sunt ut triangulus et triangulus, quadratus et quadratus,
circulus et circulus. Porro diversae, ut triangulus a quadrato, aut triangulus
a circulo, aut item quadratus a circulo. Nam triangulus alias uno obtuso et
II acutis, alias uno recto, cum II° reliqui sint acuti, alias omnibus acutis
constat. Quadratus autem IIII semper lateribus et totidem rectis angulis
55 continetur. A quorum utroque circuli figura distat, cum eadem et angulos
ignoret et unius lineae circuitione ambiatur. Tales itaque communem mensuram non recipiunt, nec umquam valet ipsarum quantitas arealis eodem
circino deprehendi. Quamobrem saepe contingit dubitationem creari, dum
inter se comparantur, utrum altera tantundem spatii comprehendat necne.
60 Neque enim videtur proposito circulo et quadrato sive item triangulo et
circulo tantundem spatii continere; eo quod haec in acutos angulos coartetur,
altera in rectos latius porrigatur, III angulorum nullis quasi faucibus obnoxia
liberiori spatio dilatetur. Quae dubitatio tamen maxime potest excludi certo
scientiae fine, si haec in illius redigatur figuram, ut videlicet [vel] a triangulo,

65 si forte de ipso dubitetur, provehamus circulum vel quadratum vel certe
longilaterum, aut aliam quamlibet formarum, prout opus fuerit, in aliam
reducamus. Sed sciendum hanc reductionem nullatenus ad (*v, f. 97r*) effectum
perduci, nisi praecognito quod harum figurarum aliae minus, aliae plus a se
distant. Minus distant, quae (*e, f. 32v*) vel longitudine sola vel latitudine
70 concordant. Inter illas vero habetur maior distantia, quae utroque sunt
diversae. Sciendum quoque, quod minus distantes amplius distantium in
medio versantur. Quamobrem ad illas, nisi per has, nulla suppetit facultas
transeundi. Et quomodo ad diversissima, nisi per minus diversa, pervenias?
Nullo id pacto natura permittente contingat. Sic a bono ad malum nisi per
75 id, quod neque bonum neque malum est; sic a iusto ad iniustum nisi per
illud, quod indifferens vocant; sic a nigro ad album nisi per venetum, pallidum
aut rufum aut per alia huiuscemodi non pervenitur. Idem etiam specialissimo-
rum a generalissimis distantia peraffirmat, ubi, nisi per subalternas species et
genera, neque ab his ad illa neque ab illis ad ista descendes sive conscendes.
80 Quid, cum etiam mollissimus aquarum liquor in lapideam cristallorum
duritiam solidatur, num id fieri potest, nisi prius in glacialem rigorem
constipetur? Infinita sunt in rebus extrariis huius transitionis exempla.
Verumtamen, ut idem indubitanter in eisdem figuris agnoscas, de ipsis potius
aliquid exempli causa proferemus, si prius harum ipsarum formarum pressiore
85 usus fuero divisione hoc modo.

Omne quod habet dimensionem et est eiusdem quantitatis et aequalis
spatii, aut aequale est aut maius aut minus. Quod nemo miretur, quomodo
eiusdem spatii esse possit, quod sit maius vel minus. Quod enim sic maius
est, si contingat, ut in omnibus dimensionibus sit maius, nequaquam per
90 omnes partes maius existet, sed per aliquas partes amplius habebit et longitudinis et latitudinis et altitudinis, per aliquas minus, et tanto minus in aliis
partibus, quanto in aliis amplius retinebit, quod posterius manifestum erit.
Eadem ratio est de minore. Unde contingit, ut quae sunt vel maiores vel
minores, etiam si forte (*v, f. 97v*) eveniat, ut omnibus dimensionibus vel
95 excedant vel aliquid debeant, quia tamen tanto minus habent in aliis partibus
quanto in aliis amplius et econtra tanto in aliis amplius quanto in aliis minus,
non omnino ab aequalitatis comparatione recedant. Tantundem enim in his
quantum et in illis reperitur.

Sed cum sint III dimensiones, id est longitudo, latitudo, altitudo, hae
100 collectae cum aequali et maiori et minori quibuscumque modis colligi possunt,
aequalium figurarum XX et unum faciunt modos. Quia vero de planis
figuris tractatus incidit, quod in altum non excrescunt, altitudine reiecta,
quot modis [AB] complecti possint longitudine et latitudine potius require-
mus; harumque complexionum rationes et naturas, quomodo vel diminui
105 vel augeri debeant et altera in alteram resolvi, diligentius exequemur, ut his
ad cognitionem deductis, cur primo circulus in quadrilaterum, dein in quadratum redigatur neque per se principaliter valeat quadrari, manifestius intelliga-
tur. Hos complexionum modos circa litteras A, B, C loco aequalis, maioris
minorisque appositas ostendemus, D et E acceptis ad longitudinem latitu-

110 dinemque significandam. Horum autem VIIII existere arbitramur. Quod ita
probamus: ad A utrumque copulamus, id est (*e, f. 33r*) D et E, inde D solum,
posthac E. Fiunt ergo complexionum III figurae. Plures autem qui fierent?
Nam cum sit A solum, cui tum D, tum E, tum utrumque comparetur, aut
utrumque habebit aut unum aut alterum; neutrum enim non potest habere.
115 Quomodo enim A, id est aequale, esset, si neque D neque E aequale haberet?
Tot modis etiam B, signum maioris, ad easdem litteras, id est D, E, potest
componi, et C minoris totidem nota. Quarum coniunctionum ratio nihil
habet diversum.

i	*a*	*de*
ii	*a*	*d*
iii	*a*	*e*
iiii	*b*	*de*
v	*b*	*d*
vi	*b*	*e*
vii	*c*	*de*
viii	*c*	*d*
ix	*c*	*e*

Fig. 20

Quodsi quis dicat maius neutro esse maius aut minus neutro minus, hoc
120 minime procedit ipsique (*v, f. 98r*) naturae repugnare convincitur. Ergo
merito comprehendendum est cum aequali, cum maiori, cum minori VIIII
extare modos complectendi secundum longitudinem et latitudinem altitudinis
dimensione reiecta, quod in ipsis potius videamus figuris.

Omnis enim eiusdem spatii figura ad aliud extra relata aut aequalis est et
125 longo et lato, aut aequalis longo, aut aequalis lato, aut maior et longo et lato,
aut maior longitudine tantum aut latitudine tantum, aut certe minor utroque,
aut minor longitudine dumtaxat, aut mensura latitudinis sola, quod subiectae
declarant figurae; quippe circa quas secundum diversam positionem acceptas
omnes hi complexionum VIIII modi demonstrari non abnuunt.
130 Nunc igitur, cum modos ipsos numero comprehenderimus, necnon exemplis omnia ad intelligentiam patescant, quae sit ratio ipsorum augendi minu-
endique et quam prorsus habeant resolutionis naturam, expedire curemus.
(a) Omnium, quae tam longo quam lato sunt aequales, altera nullatenus in
alteram resolvitur. Cuius rei non alia causa videtur, nisi per omnia conveniens

135 aequalitas. Ergo hic modus a ratione auctionis et diminutionis excluditur. Quid enim ad explendam aequalitatem alii, quod prorsus aequale est, addas? Sed neque quod auferre debebis. Sic enim aequalitatis destrues comparationem.

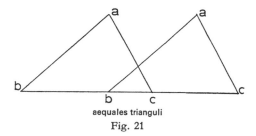

aequales trianguli

Fig. 21

(b) Restat, ut ad alias transeamus, et primo dicendum de illis, quae longo tantum aut lato aequales existunt. Quibus omnibus (*v, f. 98v*) huius rationis
140 communio inhaerere videtur, ut quotiens oporteat eas vel augeri vel minui, una tantum resolutio fieri debeat; ut si ABC trianguli latitudinem recidas inter AB et inter CB per medium et adiungas B ad C, dico ex triangulo quadrilateri formam procreatam esse.

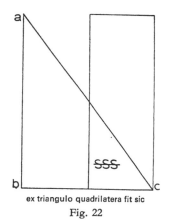

ex triangulo quadrilatera fit sic

Fig. 22

Sin autem quadrilateri formam, id est ABCD, ab angulo C usque ad E, quod
145 est medium BD, partiaris et D iuxta B apponas, rursus ex IIII laterum spatio triangularem formam procreasti, unius scilicet resectionis detrimento necnon unius auctionis additamento, quadrilateri quidem longitudine diminuta, trianguli vero latitudine recuperata.

Videsne familiarem huiusmodi figurarum inter se transitionem? Quodsi
150 iacente quadrilatero inter AC et inter CB eadem (*e, f. 33v*) trigoni superficies longitudine curtetur, ad signum DC autem B apponatur; rursus hoc modo ortogonium ex tribus ad IIII^{or} latera reduxisti, quod item poteris reparare, si ab angulo B inter CD quadrilaterum dividas ad signum E, et D colligas

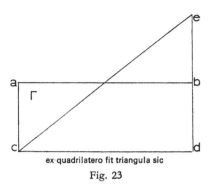

ex·quadrilatero fit triangula sic

Fig. 23

ad C; et his exemplis, quae sunt aequales aut longitudine tantum aut latitudine,
155 altera in alterius compositionem transfiguratur.

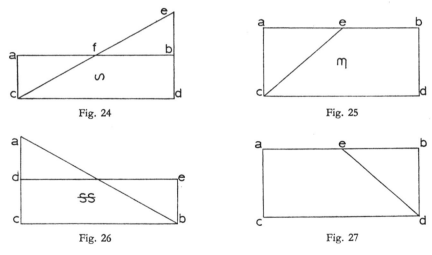

Fig. 24

Fig. 25

Fig. 26

Fig. 27

(c) Nunc de his, quae sunt vel longitudine vel latitudine maiores minoresve,
consideremus. Harum III.ᵒʳ sunt modi, qui eo a superioribus distant, quod
illi quidem vulgares existunt, utpote qui per se ipsos facillime pateant et a
quovis imperito artis geometricae natura tantum monstrante et augeantur
160 et (v, f. 99r) minuantur. At hi IIII modi numquam valent absque artis ad-
miniculo administrari, et necesse est a geometra petas, quotiens aliquid
spatiorum vel in agro vel in mensa vel in pagina vel ubicumque fieri potest
destinaveris aequare. Ut si quadrilaterum, sive stet, sive iaceat, quadrare
volueris, quomodo vel eius longitudinem vel eius latitudinem aequaliter
165 poteris diminuere, si non artem didiceris? Item, si tetragonum vel longiorem
vel latiorem reddere desideres, ut sic tibi quadrilaterum restituatur, id etiam,
praeter artis peritiam, quonam modo perficies? Probet unusquisque, cui hoc
facillimum videatur, quare hi IIII ab illis distant. Ceterum unius resectionis

dampnum sine dampno cum eisdem patiuntur et unius adiectionis accipiunt
170 emolumentum. Nam quod superat in longitudine, idem demptum secundum
suam rationem, si quis eam non nesciat, latitudinem emendat; vel si latitudo
superat, ipsa quoque mutilata aequalitatem longitudini restaurat. Cur autem
una omnes in se resectione transformentur, ratio nobis videtur esse eo, quod
in omnibus una dumtaxat reperiatur aut longitudinis aut latitudinis distantia.

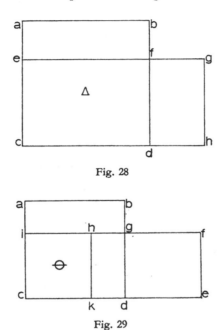

Fig. 28

Fig. 29

175   (d) Sed iam de his disputatione, prout potuimus, ad finem perducta nunc
ad reliquas duas, quae adhuc de IX supersunt, complexiones animi vertamus
intentionem. Hae igitur tanto difficilius ad aequalitatem reducuntur, quanto
se invicem et latitudine et longitudine excedunt, ipsaeque solae in aequales
suas non resolvuntur nisi per medias quasdam figuras. Cuius rei est causa,
180 quantum nobis conicere fas est a deo, magna distantia utriusque dimensionis
tam in lato quam in longo. Quae distantia id sane efficit, ut ad cognationem
aequalitatis non perveniant, nisi secunda vice mutilentur, si forte fuerint
maiores, vel econtra duplici augmento fulciantur, si minores extiterint. In
quibus etiam hoc est animadvertendum, quod maiores non sunt omnino
185 maiores, (v, f. 99v) sed dum aliis partibus exteriores habeantur, aliis partibus
interiores existant, quamobrem maiorum partium quantitas truncatur, unde
minorum dampna, quibus haec forma ab illa videtur disparari, resarciantur;
et truncantur quidem secunda vice, quia nimirum partim longitudine,
partim latitudine auctiores (e, f. 34r) habentur, quatinus latum de lato emen-
190 detur et longum de longo compleatur. Eadem causa est, cur minores secunda

vice augeri expostulant, quia videlicet dum aliqua pars longitudini, aliqua latitudini deficiat, tam grande dampnum nisi secunda auctione restaurationem non capit. Sunt igitur in talibus exercendi, si qui hoc, quod oratione significamus secundum aridi sermonis possibilitatem, subtilius intelligere non 195 dedignentur. Non enim res ex oratione, sed oratio ex rebus illustratur.

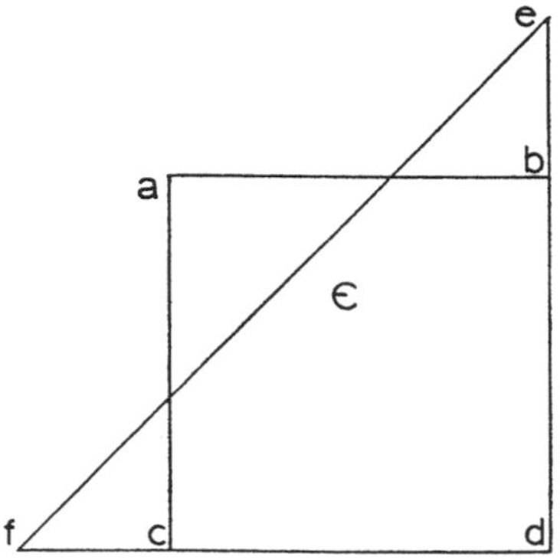

Fig. 30

(B)

Nunc igitur ad circuli revertamur expositionem dicturi, cui praecipue inter IX complexiones deputetur. Nimirum quadrilatero comparatus vel illi, quod longitudine tantum est aequale, vel illi, quod latitudine tantum, non 200 immerito deputari valet. Habet enim XIIII in diametro tam per latum quam per longum, quantum etiam quadrilaterum per II° latera habet. Unde facillime, si quis adverteret, in hanc formam posset resolvi. Sed ex quo tempore obsoleta est et antiquata geometricae facultatis exercitatio, quae scabies fere ante ipsum Boetium omnem philosophorum gregem ipso conquerente invasit, 205 nullus id existimare praesumpsit, videlicet quia nulli videbatur credibile, ut circumferentis lineae curvitas illatenus extendi posset et ad rectitudinem aliquatenus reduci; idem autem circulus relatus ad quadratum, dum distet ab eo tam lato quam longo, difficilius in id ipsum resolvitur, quia videlicet ab eodem duobus dimensionibus disparatur. Quamobrem etiam primo resolutionem 210 capit in quadrilaterum, unde non amplius quam una dimensione distat, ut eo reciso (v, f. 100r) et illa distantia iam privato regulare quadratum per se refulciatur, et his rationibus demonstratur, cur circulus in quadratum per se reduci non possit. Hoc in loco liber iste suo fine claudatur.

PROLOGUS QUINTI LIBRI

Haud me ipsum latet, praesul clarissime, nostrum hoc studium non usque quaque ad dignam subtilitatem extenuatum, multaque nimis in eo reprehensioni patere in ipsa quoque compositione verborum, nedum in rerum veritate.

5 Quod partim accidere ex inculto ingenio ac raro scribendi usu (dum magis
procurando corpori quam stilo exercendo dediti sumus) minime recuso,
partim autem evenire ex nimia difficultate subtilissimarum rerum, in quibus
etiam perfectos viros aliquando falli necesse est, nemo qui refellat. Accedunt
his alia pleraque impedimenta quam maxima. Primo officii cura tum provincia
10 familiaris rei etsi parvae; deinde crebra inaequalitas corpusculi mei; postremo
interioris hominis quamplures absque numero molestiae conceptae ex novis
et insolitis rebus, quas repente emergere nostris diebus ad dissipationem
omnium honestarum rerum nemo tam ferreis praecordiis dolere non potest.
Hae igitur res veluti contra unum hominem legio armata, ita contra eam,
15 cuius tibi debitor sum devotionis assistunt, ut non tam indignandum sit, si
quod in rebus aut verbis peccatum offendat, (e, f. 34v) quantum illud est
animadvertendum posse quemquam ad scribendum inter tot curas tantasque
omnium rerum perturbationes voluntatem applicare, quando flere quam
studere potius libet. Quod certe fieri non posset, nisi praecipua vis tuae
20 dulcedinis hanc in nobis amaritudinem, veluti mel absinthium, temperaret.

EXPLICIT PROLOGUS. INCIPIT LIBER QUINTUS

(A)

Quoniam igitur in perquisitione circuli quadraturae ex circulo quadrilaterum, ex quadrilatero quadratum produximus, quaeret forsitan aliquis,
25 contrane a quadrato ad eandem quadrilateri (v, f. 100v) formam indeque
perveniri possit ad circulum. Quod cui sit dubium, cum non aliter in figuris
contingat atque solet evenire in qualibet massa plumbi, ferri, aeris, argenti,
auri ceterisque id genus, quorum unumquodque valet eodem pondere con-
servato a quadrati forma, si forte quadratum existat, cum in trigonam, tum
30 in pentagonam, tum in aliam, quamcumque maluerit artifex, speciem converti
et rursus ab ea specie ad pristinam qualitatem cudentibus malleis retorqueri?
Quod ipsum natura cereae molis potissimum non recusat, quae ignei caloris
inattacta facile, amissa sui rigoris duritia, nunc in longissimum tenorem
distendi, nunc in modum pilae eadem sese patitur rotundari, et nunc ab hac
35 figura in illam traduci, nunc ab illa in hanc non est difficilis reformari.
Nihilominus exempli ratione tractabilis argillae nos instruit Livius, unde Horatius protulit ita: argilla quidvis imitaberis uda. Adhuc in liquoribus eandem
naturam copia est perspicere. Nam cyathus aquae, vini, mellis aliorumque id
genus in vase rotundo, ipse quoque rotundam conservat figurationem. Eius-
40 dem autem in vas oblongum transfusio necesse est longitudinis productionem
suscipiat; a quo valeat item priori formae restitui, si reddatur eiusdem vasis
rotundae capacitati. Ergo non est dubium ad similem modum omne spatium
eiusdem quantitatis mutua transformatione variari, nec minus a quadratura
ad circulum redire, quam a circulari compositione IIIIᵒʳ angulorum cum
45 totidem lineis figuram accipere. Verum difficillimum est eiusmodi negotium,
nec tam sine labore invenitur quomodo fiat, quam facile cognoscitur, quod
ita fieri debeat. Nam dato quadrato cuiuscumque magnitudinis, ut par eidem

circulus producatur, unde noverit quis imprimis sitne lateri addendum aliquid
an potius aliquid auferendum, deinde quod demi convenit, qua ratione
50 sumendum? Huius rei cum forte ostendemus, qualiter omnes figurae qua-
drilaterae in quadraturas redigantur, mani- (v, f. 101 r) festa erit et transformatio
et reformatio. Ad praesens autem qua arte quadrati forma[m] in illud modo
quadrilaterum, unde processit, reduci oporteat, paucis ostendam, cum nulla
difficultas impediat, licet inventio plurimo sit involuta labore.
55   Proposito igitur quadrato ABCD, latera superius et inferius in partes
XIIII disperties et ex his III per lineam EF excludes atque ad ipsius lineae
caput ex ipsis XIIII quadrabis unam, iuxta quam [id est ad mensuram] eius
quadrati diagonum posito supra F circino affiges punctum H sub illo, quod
est C, a quo ducetur linea ad oppositum (e, f. 35r) sibi inferius I punctum,
60 excluso spatio quod inter ipsam HL et AC lineas includitur; quod si demp-
seris a latitudine figurae et eodem modo inserueris ad reparandam longitudinem, quo supra ablatum meministi; scilicet per obliquum dico quadratum
in formam quadrilateram, ut propositum fuerat, redactum esse, veluti figura
monstrat subiecta.

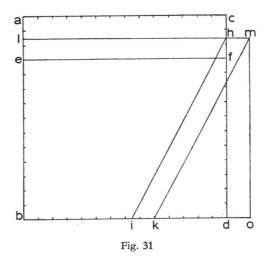

Fig. 31

65   Aequale est enim ACHL spatium HDMO spatio. Est autem totum quadrilaterum LMBO. Cuius si dividas LB, minus latus, in XI partes, invenies
easdem et III insuper eiusdem quantitatis in LM, (v, f. 101v) latere maiore.
Quare non est dubium reformationem hanc integre perfectam esse, cum
eandem profecto latera quam ante habuerant repraesentent comparationem.

70 (B)
    Huius itaque rei demonstratione completa nunc iam constitutio quadra-
turae circuli describatur. Haec autem iuxta hanc regulam fieri debet: Dato ad
quadrandum circulo, ductis per medium centrum lineis duabus, primo

secetur ipse in aequas IIII portiones. Quo facto utraque linea XIIIIes dividatur.
75 Deinde autem iuxta centrum circuli quadratum fiat, cuius singula latera ad
unius XIIII^{mae} mensuram sint protensa; hoc vero quadratum diagonio
partiatur. Porro a diagonio dematur latus et quod restat in geminas particulas
scindatur; harum sumatur una. Deinde a puncto utriusque lineae sive diametri sexta pars undique versum numeretur et huic illa particula adiungatur
80 atque ille locus diligenter punctis assignetur. Deinde per ea puncta lineae
ducantur, quoad sibi invicem concurrant. Igitur hoc completo inventa est
procul dubio circuli quadratura.

   Sed huius rationis dabimus exemplum, ut sicubi alicuius dubitationis
obscuritate caliget, exempli luce palam fiat. Sit item propositus ABΓΔ
85 circulus et sint ductae per medium centrum lineae duae, altera ab A in B,
porro altera a Γ in Δ, et sit utraque linea in partes XIIII distributa, et sit
constitutum quadratum iuxta ipsum centrum, quod habeat per singula latera
unam XIIII^{am} totius diametri, et sit quadratum illud diagonali linea divisum,
quae est AC. De hac autem dempta sit mensura unius lateris AB, remanet
90 pars diagonii, quantum est inter BC. Hoc autem per medium dividatur, et sit
ibi E punctum. Erit igitur medietas EC. Ea vero ad VI^{tam} a centro diametri
partem undique versum adiuncta sit, ubi lineae per singula latera dirigantur,
quousque concurrant. Dico ergo constitutam circuli quadraturam. Hanc
autem rationem subiecta figura demonstrat, in qua appel- (*v, f. 101 v finis*) lamus
95 illud quattuordecimae quadratum clavem circuli quadraturae eo, quod per
hanc constitutio eius obserata fit et reserata.

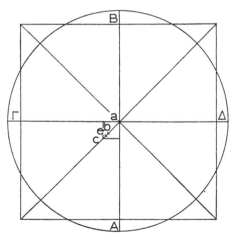

Fig. 32

(C) (*e, f. 35v*)
   Quod a medio puncto IIII lineas produximus, ubi latera quadrati concurrunt, idcirco factum est, ut anguli quadrati apertius coniungerentur. Sed
100 cum possibile ac facile sit arte cognita ita circulare quadratum veluti circuli

quadrare quadriformam, cur potius quadratura circuli quam circulatura quadr⟨at⟩i dicatur quam brevissime valeam libet ostendere. Quicquid enim ceterae figurae metiantur, plerum fallit, quia et ipsae nisi per circulum dimensae praeter aliquod vitium casu fortasse aliquando, ratione vero num-
105 quam poterunt invenire. Nec solum figurae, sed etiam figurarum anguli per ipsum regulantur, ut ex primo valet intelligi libello, ubi medietatem circuli, ut apertius demonstretur aequalitas angulorum cuiusque trianguli, rectis circumscripsimus angulis. Unde quis [non] iure fateatur universitatis creatri-cem omnipotentiam unicae illius stelliferae sperae convexa ad huius exempli
110 similitudinem incurvasse, quod etiam ipsum egregie praeter omnes formas unitum est singulare, cum inter tot species triangulorum, inter tot diversitates omnium formarum quarum pluralitas finem non capit, circulum dumtaxat sola una linea includat. Qui, quod nihilominus mirum est, angulorum re-cessus ignorat itaque est totus aequabili compositione perpolitus, ut vere in
115 illo et primum medio conveniat et medium imo consentiat. Quod privilegium ne ipse quidem quadratus meruit possidere, quamquam et ille †praerogatura quadra naturae† reliquo formarum generi praeluceat, videlicet quod in illo solo et aequa latera aequis lateribus respondent et recti anguli rectis angulis opponuntur. In quo etsi anguli inter se et item latera velut amicitiae foedere
120 copulentur, dum tamen anguli lateribus et rursus angulis latera dissimilitudine pugnent, merito sprevit consultissima divinae mentis prudentia iuxta hoc exemplar informare illum caelestem globum, qui futurus esset et voluntate celerrimus et inter res corporeas aeternitate, ut ita dicam, permanentissimus.
Denique circulus non modo terrarum spatia, sed [modo] etiam caelestia
125 metitur: Per hunc ascensus et descensus tum solis, tum lunae, tum VII planetarum, tum omnium siderum metiri valent; per hunc eorum inter se distantiam comprehendi, per hunc crementa et immutationes dierum ac noctium, per hunc meridianam lineam quam ceterorum climatum principa-liter indicant positionem promptum est inveniri. Sed aliis omissis huius lineae
130 inventionem praeterire nolo propter eos, qui putant excepto aequinoctiali die meridianum tempus comprehendi numquam posse. Unde etiam arbitran-tur horologia solaria aequinoctiali tantum die collocanda hoc modo. Expec-tant enim aequinoctii diem ortuque et item occasu solis observato ita meri-dianam horologii lineam vertunt, ut altersecus pari videatur intervallo distare,
135 arbitrantes ad hunc modum horologio collocato tam meridiem quam omnes horas inventu facile esse: (e, f. 36r) quasi vero per ortum meridies recte perquiratur et non potius ortus a meridie investigari debeat. Nam haec eodem modo semper se habet, ortus vero et occasus cottidie variantur. Sed quod inconstans et varium est, incertum esse probatur; incerta vero quid
140 afferant certi? Non igitur meridies ab ortu investiganda est, sed inventa meridies ortum probat. Quamobrem hic magnus est error, cui illud iungatur incommodum, quod affert aequinoctialis diei expectatio tam longa. (v, f.102r) Qui si forte matutinum solem nubibus obtexit, quid, nisi aliud aequinoctium, expectandum? Et si item illo die tempestas obvenerit, quid, nisi rursus usque
145 ad tertium vel ad quartum vel, si ita contigerit, ad XXX aequinoctium dilatio

facienda? Cuius impedimento tollendo quo pacto unoquoque die, sole
lucente, meridies investigetur monstrare curemus. Sit enim positus gnomo in
medio quolibet circulo. Is autem ita temperatus esto, ut umbra illius paulo
ante meridiem intra circinationem sese recipiat, et item paulo post meridiem
150 eandem excedat, fixisque punctis, ubi umbra vel ingressa est vel egressa,
spatium illud inter II puncta per medium dividatur. Ad quam medietatem ab
eo puncto, ubi gnomo affixus est, ducatur linea. Nimirum hac arte centrum
solis meridianum deprehensum affirmamus, iuxta quam horologia regulare
potes sicut aequinoctiali, ita et in ceteris quibuscumque totius anni dierum
155 curriculus. Cum igitur tanta sit dignitas circuli, iure non ipse ad quadratum,
sed potius ad ipsum referri videtur quadratus.

(D)

His etenim finitis tandem ad excessuras veniendum, quarum in tantum
difficilis videtur ad inveniendum mensura, ut plerique resectas particulas in
160 trutinam mittant, libraeque lancibus examinent aequalitatem. Dico igitur XX
et VIII partes, qualis est centesima LIIIIᵃ, si ad quadrilaterum respicias, ab
excessuris contineri, quarum dimidia pars excessuris circuli, dimidia vero
quadrati deputatur excessuris. Hoc autem ita probabis.

Quadratura circuli constituta tale circa gyrum circuli quadratum ascribe,
165 quod partes eius attingat extremas. Id profecto in singulis lateribus XIIII
mensuras iuxta diametrum circuli necessario retinebit. Hoc autem quadratum
per singulos angulos X semis partibus excedit aream circularem. Cuius rei
(v, f. 102v) probatio in promptu est tibi. Nam quater decies XIIII CXCVI
reddunt, quae est circumascripti continentia quadrati. Sed embadum circulare
170 CLIIII continebat. Quos superant CXCVI, XLII. Hos autem XLII in IIIIᵒʳ
partes aequaliter divide: fiunt X et semis. Iure ergo dictum est per singulos
angulorum X semis particulas inveniri. Igitur hoc ita probatum priori
dubitationi argumento duc, et quanto angulum interioris quadrati exterioris
angulus vincat perquire. Hoc autem fiat ita. Ab angulo minore illud quo
175 maior superat praecides, idque ipsi minori comparabis, et bis tantum invenies.
Quare minoris anguli continentia tribus semis particulis constat. Sed III
semis quater ducti XIIII apponunt. (e, f. 36v) Hi autem duplicati, ut cum eis
excessurae circuli annumeratae sint, XXVIII restituunt. Ergo, si hoc feceris,
probatum est, uti reor, XX et VIII partes in illis angulis contineri, quibus
180 se alterutrum excedunt circulus et quadratus.

(E)

Hoc ita demonstrato illud etiam volumus explicare, quanto superentur ab
invicem tetragonica forma et item quadrilatera, quod monstratu difficillimum
est. Est autem distantia haec XI pedes vel asses et totidem septunces paulo
185 plus. Quanto vero plus, diffinite monstrari non potest calculando; cadit enim
in minutias eius modi, quae nulla valeant comprehendi ratione. Sed hic forte
aliquis in animi rationem ducatur, cum his figuris circulus adaequetur, cur
ille et quadratus minus se invicem excedant vel cur amplius istae sese supervadant, quando omnium eadem quantitas invenitur. Quod facile poterit

190 adverti, si quis aliarum respiciat configurationem formarum. Sint ergo III
aequales formae, quarum prima IIII°ᵉ pedibus in longitudine, tribus autem
in latitudine constet, at vero secunda senis in longum pedibus porrigatur,
duobus autem in latum tendatur; porro IIIᵃ uno dumtaxat pede in minore
latere, dimensa XII extendatur in longum, et hae omnes super se ad unum
195 idemque punctum col- (*v, f. 103r*) locentur hoc modo. Quo facto quis non

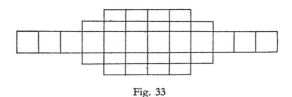

Fig. 33

videat aequales esse formas, verumtamen non esse aequales excessuras? Nihil
ergo mirum, si distantia quadrati a quadrilatero maior est ea, qua ab eodem
quadrato circulus differt (idem autem distantiam appello quod excessuram):
illa enim plus septemdecim, haec autem XIIII habet. Sed de his satis; nunc
200 igitur consistamus.

PROLOGUS SEXTI LIBRI

Cum multotiens ad memoriam reduco in levioribus studiis perfectissimi
viri quam plurimos annos attriverint, vereor nimis, ne a pluribus intentionem
meam non videntibus pro tam arduo incepto temeritatis accuser. Quid est
5 enim, obsecro, Thebais vel Aeneis illa Papinii Statii Sursuli, id est sursum
canentis, illa Virgilii Maronis, doctissimi poetarum, quid igitur nisi leve
quoddam ac inane poeticae fictionis inventum? Et tamen, dum in utraque
sudatur, his VI annorum revolutiones explicantur. Quod si ita est, putasne
quantis sit opus indutiis ad integram expositionem circuli quadraturae?
10 Utique, si difficultas musicorum sonorum eo usque Pytagoricae subtilitatis
acumen evicit, ut eorum investigationi frustra studium applicasset, nisi
tandem diligentiam illius mallei fabrorum diutissime desiderata ratione
instruxissent. Multo magis laboriosam huiuscemodi perquisitionem necessario fateri debemus. Nempe, ut de ceteris reticeamus, sola dumtaxat pro-
15 positarum formarum divisio quanto divisionem monochordi artificio non
vincat. Quare fortassis reprehensores non deerunt, qua ego quidem animatus
temeritate ad tam difficillimam rem prorumpere non extimuerim. Ignosce
pietas ignosce, date veniam, quaeso, quicumque non estis ignari trahi quem-
que omnium hominum voluptate sua. Voluptate et ego, siquidem peccatum
20 est, peccavi, ea ni- (*e, f. 37r*) mirum voluptate, qua satis superque delector in
amore et gratia Maecenatis mei, Patricii mei, Caesaris quoque sive potius
Augusti mei. Valida causa est amor sanctus. Quamobrem (*v, f. 103v*) quicquid
fiat, id temeritati sapiens ascripserit numquam. Sed nunc animum labori
reddamus.

25   EXPLICIT PROLOGUS. INCIPIT LIBER VI^{tus}

(A)

Ad aequalitatem circuli et quadrati seu constituendam seu approbandam
nihil ita prodesse puto, quam si utrique figurae ars foret reperta divisionis.
Quapropter hanc etiam sed interim in solo quadrato, altera adhuc a sensu
30 nostro remota, ut potero, investigare conabor ingredien[te]s a speculatione
communi compendiosum tractatum.

Omnis divisio formarum

(a) aut naturalis est

(b) aut arte perficitur.

35   Naturalis est, quam per se quisque facile novit. Illam vero ad artem dicimus
pertinere, quae nulli est obvia, nisi aut disciplina aut studio et exercitatione
comparata. Sed naturalis duobus modis perfici videtur: aut enim a latere
ad latus oppositum lineae ducuntur, quibus figurae distribuuntur vel in
duas vel in III vel deinceps consequentes partes, aut a puncto medium
40 tenente locum eiciuntur. Uterque fieri potest in his formulis, quae et punctum
in medio claudunt, et ipsae IIII lateribus et rectis angulis includuntur. Sunt
huiusmodi, quae lateribus carent, ut circulus, et sunt item, quarum medietas
puncto designari non valet, ut trigonus ysosceles vel etiam ambligonius; in
quo quis ita in medio punctum constituat, ut ab omni angulo omnique latere
45 aequali differat intervallo? Quare huius formae triangulus ab utraque remotus
est divisione.

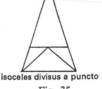

circulus a puncto divisus   isoceles divisus a puncto   ampligonius divisus
Fig. 34   Fig. 35   Fig. 36

At vero circulus, quia punctum eius ab omni circumferentia aequaliter distat,
eam recipit, quae fit a puncto exeuntibus lineis, divisionem, quia vera latera
recusat ⟨et⟩ a laterum divisione alienus existit. Porro superficies quadrati,
50 cum sit rectiangula et IIII lateribus inclusa, a quibus aequali distantia punctum
in medio fixum invenitur, utrumque sectionis non refugit modum. Sciendum
autem hanc divisionem, cum aequas recipit partes, nullam in partibus proportionem habere ad exemplum videlicet numerorum, (v, f. 104r) qui vel in
duos binos partiuntur, ut IIII, vel in tres, ut seni, vel in IIII, ut VIII, et
55 ceteri; cum vero inaequales habuerit partes, illarum posse aliquam inveniri
comparationem. Quam si quis forte nosse voluerit, summam aliquam numeri
secundum eas, quarum habitudinem requirit, partes distribuat, earumque
partium adi nvicem relatione perspecta facillime comprehendet; ut si forte
medietatis ad III partem comparationem requirat, senarium bis itemque ter

60 partiatur. Erit igitur medietas III, tertia vero duo. Sed III ad II sesqualtera
cognatione iunguntur. Sesqualteram ergo rationem habent tertia pars et
medietas. Eodem modo XII, qui III et IIII habent modo ter, modo quater
divisis, tertiae quartaeque partis ad se relationem invenies. Nam quomodo se
habent III ad IIII, eadem ratione III^a pars respicit IIII^{am}, atque hanc insi-
65 stentes viam omnium partium proportiones absque ulla reperiemus offensio-
ne. Haec autem dicta sunt pro divisione secundum naturam.

At vero divisio artis inaequalibus lineis perficitur. Quarum in quadrato
diagonium maxima reperitur, reliquae autem, quanto magis angulo accedunt,
magis magisque breviantur. Porro in circulo diametros maximum habet
70 modum. Quae vero ad ambitum propinquiores existunt, maiore contra-
huntur brevitate. (*e, f. 37v*) Cuius quantae sint utilitates, ego quidem enumerare non possum. Hoc tamen dicam, quot pedibus circulum quadratus excedat, nulla melius deprehenditur ratione. Nunc igitur qualiter area propositi
quadrati, quem circulo aequavimus, per hanc dividi possit, investigare curabo.

75 (B)

Primo etenim videndum, quot naturales includat tetragonos a pari numero
productos. Naturales dico, quia sunt et facticii arteque compositi, velut ipse,
quem nunc habemus in manibus. Quot enim naturales erunt, tot etiam lineas
invenies certas. Habet autem, ut a maximo incipiamus, CXLIIII, deinde C,
80 post hunc LXIIII, subinde vero XXXVI, hinc autem XVI et novissime
quaternarium opere et actu quadratum. Sed quot sunt isti? VI. Sex igitur
lineas in (*v, f. 104v*) CLIIII tetragono facticio certissimas habes. Quid tum?
Quid de reliquis faciendum? Ullamne monstrare possibile est? Puto, quia
aliquas adhuc investigare copia erit. Nam latus omnis dupli diagonium est
85 simpli. Ergo XII, cum sit latus CXLIIII tetragoni, faciamus ex eo diagonium.
Eodem modo X, cum sit latus C quadrati, diagonium fiat. Item VIII, VI,
IIII, II, omnia haec latera pro diagoniis accepta, lineae fiant ad eandem
divisionem idoneae. Et harum quidem atque superiorum, quot pedes unaquaeque contineat, brevissime succingam. Ille enim dimidiam sui quadrati,
90 harum vero quaeque partem eiusdem includit IIII, suorum vero proprie
quadratorum similiter ut ille dimidiam. Quorum item lateribus sumptis ad
diagonia eadem consequitur ratio. Et hac arte invenies omnes simpli ad
duplum cognationem habentes.

Sed hinc exceptas qua ratione perquiremus? De quo genere est quidem
95 III, V, VII, XI et quaecumque ad illam dupli et simpli non pertinent rationem.
De quibus omnibus satagendum est nobis, ut et eas metiendi ars non desit,
habita prius consideratione de III, an ulla proportione iungatur ad altrinsecus
positas. Nam IIII ad primam existit dupla. Inter quas II et III numquid ad
ipsas et inter se aliqua proportione conferuntur? Et hanc quam esse dicamus?
100 Ex illis quippe, quae in usu habentur quasque diligentissimi numerorum
inspectores memoriae posterorum scriptis tradiderunt, quis tantae subtilitatis
hic aliquam ostendat?

(a) Annon manifestum est huiusmodi proportionem neque in multiplici

genere neque multiplici superparticulari neque multiplici superpartiente in-
105 veniri? Si enim II^a ad primam multiplex existeret, et item III^a ad II^{am},
IIII^{a}que ad III^{am}, profecto duplex ex III proportionibus multiplicis necessario constaret. Est enim IIII^a ad primam dupla. Sed hoc falsum esse quis
nesciat? Cui sit ignotum minimam esse omnium multiplicium duplicem proportionem? Quare ipsam impossibile est ex multiplicibus constare. Quodsi
110 non ex istis, multo minus ex aliis supra memoratis, quia et illis duplex proportio multo minor invenitur.

Restat ergo, ut super- (*v, f. 105r*) particulari proportione ad se conferantur,
aut superpartiente, aut nulla.

(b) Primo superparticularem attemptemus. Invenimus enim multiplicem
115 ex tribus superparticularibus constitutum, id est VI ad III comparatos, inter
quos IIII et V, ut omnibus notum est, sesquitertiam, sesquiquartam ac
sesquiquintam proportionem efficiunt. Quamobrem putet aliquis (*e, f. 38r*)
in divisione quadrati inter I^{am} lineam et IIII^{am}, quae se invicem duplici
collatione respiciunt, talem secundam debere constitui, quae primae III^{am}
120 habeat partem, talem deinde III^{am}, quae secundae contineat IIII^{am}, cuius
etiam III^{ae} IIII^a linea suscipiat quintam. Sed nihil minus verum. Hoc autem ea
ratio probat, quod VIII a I esse duplicem secundae oportet. Quippe quaternario praecedit IIII. Omnes autem quaternario se praecedentes duplices erunt
ad illas relatae, quae se iuxta numeri naturalem ordinem consequuntur,
125 quemadmodum mox II^a sequitur I. Quare necesse erit, ut ad ipsam secundam
duplex inveniatur VIII, sicut ad I^{am} quarta. Quae, si illud constiterit, maior
erit quam dupla. Impossibile enim erit prioribus IIII ita se ad invicem
habentibus, posteriores etiam IIII consequentes proportiones alterutrum non
conservare, videlicet ut linea V^a sesqui VI^a sit ad IIII^{am}, VI^a ad V^{am} sesqui
130 VII^a, VII^a ad VI^{am} sesqui VIII^a, VIII^a ad VII^{am} sesqui IX^a. Unde amplius
conficitur quam proportio dupla.

Quod ut liquido patefiat, numero ad demonstrationem utamur. Sint itaque
III, quia III habet⟨ur⟩ quasi prima linea, tum IIII^{or} velut II, tum V loco
III^{ae}, postremo VI pro linea IIII. Dico, si illud confirmatum fuerit, VIII^{am}
135 quod non oportet ultra duplum maiorem esse secunda. Quae enim erit
secunda? Quattuor. Quae octava? X. Nam X a tribus VIII° distat loco. At
vero X ad IIII^{or} relati amplius quam duplum reddunt, quantum est ipsius
quaternarii pars dimidia. Non igitur primae lineae pars tertia sumenda est,
nec prorsus inter ipsam et IIII^{am} tales sunt proportiones requirendae, quales
140 inter III et VI duplam faciunt (*v, f. 105v*) comparationem. Quid igitur? Ubi
deinceps in omni genere superparticulari alias invenies, quarum solae tres
ad duplum aspirent, id prorsus fieri non potest. Itaque sicut multiplices et
ceterorum, ita etiam ista inaequalitatis proportio ab hac quam tractamus
divisione separanda est.

145 (c) Restat ergo aut in superpartiens genus incidere, aut minime iuxta
proportionum procedere rationem. Sed cunctis etiam superpartientibus per-
tentatis nusquam tales repperimus, qui continuatim huic apti sint divisioni.
Neque enim secunda linea duas tertias I^{ae}, aut III quartas aut IIII V aut V VI

aut deinceps plures iuxta proprietatem superpartientis accipiet partes. Quare
150 et ipsum ab hac divisione excludendum. Quid igitur agendum? Numquid
credendum cunctis his remotis generibus rem ullam comparari?

Sed forte erit, cui haec comparatio per adiectionem et diminutionem fieri
videatur hoc modo: Constituta linea, scilicet I^a, si ablata VI^a eius, deinde ipsa
per medium dividatur adiectaque medietate augeatur, secundae lineae quanti-
155 tatem inventam esse. Sin autem ab hac X^a pars recidatur et post haec ipsa
quater secetur quartaeque suae deinceps adiectione concrescat, III^{am} lineam
inveniri; a qua rursus XIIII^a subtracta VI^{a}que residui inventa atque integre
suae quantitati adiecta, IIII^{am} lineam dimensam esse. Atque hoc ordine
observato, ut videlicet in diminutione a VI incipientes eam semper auferamus
160 partem, quae denominata est a summa quater-(*e, f. 38v*)nario auctiore (ut X
a VI, XIIII a X quaternario auctiores existunt), rursus autem in adiectione a
II^a parte progredientes illas semper adiciamus, quae a paribus appellationem
trahunt continuo ordine se sequentibus: eo modo omnes procul dubio quadrati lineas inveniri.
165 Quod utrumque procedat in lineis, numeri pro lineis accepti demonstrabunt.
Sit ergo quasi prima linea $\overline{\text{I}}$CCVIIIIDC, ab his aufero sextam, id est $\overline{\text{CC}}$IDC,
remanent $\overline{\text{I}}$VIII, horum sumo medietatem, id est $\overline{\text{D}}$IIII, hos addo primae
summae, hoc est $\overline{\text{I}}$CCIXDC, fiunt $\overline{\text{I}}$DCCXIIIDC, qui pro secunda linea depu-
tentur. A quibus rursus decimam tollo, videlicet $\overline{\text{CLXXI}}$CCCLX, remanent
170 $\overline{\text{I}}$DXLIICCXXXX. Quibus subduco IIII., quae est CCCLXXXVDLX, eam-
que $\overline{\text{I}}$DCCXIIIDC (*v, f. 106r*) adiungo, fiunt $\overline{\text{IIXCIX}}$CLX. Hi autem pro III^a
linea computentur. A quibus item XIIII^{am} abstraho partem. Haec XIIII^a in
CXLIXDCCCCXL invenitur, qua subtracta remanent $\overline{\text{I}}$DCCCCXLIXCCXX.
Quos item sexies divido et invenio [VI] $\overline{\text{CCCXXIIII}}$DCCCLXX, atque his
175 illi summae coniunctis, quae pro III^a linea erat, id est $\overline{\text{IIL}}$XXXXIXCLX,
exeunt mihi $\overline{\text{II}}$CCCCXXIIIIXXX. Hos igitur pro IIII^a linea appono. Sed
IIII^{am} primae duplicem esse oportet. Hi autem ad primum numerum comparati tantum exuberant supra duplum, quantum est summa $\overline{\text{IIII}}$DCCCXXX.
Quod idem vitium in lineis nihilominus evenire quis dubitet? Quapropter
180 illarum dimensio secundum adiectionem et diminutionem procedere non
valet. Sed rursus hic quod obiciatur deesse non puto. Dici enim potest, ut
totam istam exuberantiam in eam quantitatem dispertiamus, quae colligitur
ex his summulis, per quas ipsa diminutio progressa est. Quae sunt igitur
istae summulae? Nimirum VI, X et XIIII. Sed quaenam ex his (*p, f. 1r*)
185 quantitas colligitur? Utique XXX. Per hos ergo distribuamus $\overline{\text{IIII}}$DCCCXXX,
quo numero dupla proportio primae a IIII^a linea superabatur, egrediturque
pars XXX^a, CLXI; quam deinde primo sexies multiplicemus, fiunt
DCCCCLXVI. Quos statim ab illo numero, qui secundae lineae retinet vicem,
id est $\overline{\text{I}}$DCCXIIIDC, subtrahamus. Relinquitur $\overline{\text{I}}$DCCXIIDCXXXIIII. Quo
190 ita diminuto, ad XXX^{am} revertamur, quae est CLXI. Rursusque non modo
per VI, sed etiam per X, hoc est sedecies, multiplicetur, unde concrescunt

IIDLXXVI. Quorum subtractione illud, quod superat in IIIᵃ linea, quae et suo abundat et IIᵃᵉ lineae vitio, emendemus. Supersunt ĪIXCVIDLXXXIIII. Et his ita ad legitimam mensuram redactis rursum XXX per VI, per X, per
195 XIIII, id est tricies, in augmentum ducamus. Ex qua multiplicatione tota illa summa consurgit, qua primae lineae numerus a quartae lineae numero superabatur. Haec autem est ĪIIIDCCCXXX, qua tandem subtracta ab ipso IIIIᵃᵉ lineae numero, id est ĪICCCCXXIIIIXXX, relinquuntur ĪICCCCXVIIII CC, quem numerum manifestum est dupla (v, f. 106v) relatione conferri ad
200 ĪCCVIIIIDC. Hos numeros, quorum omnium radix est VI, propter confusionem tollendam ita distinximus, uti in subiecta formula patet. Quibus tam promotis necessario usi sumus eo, quod non sunt inventi minores, qui tum integri in partes VI et X ac XIIII dividerentur, tum diminuti IIᵃᵐ, IIIIᵃᵐ, VIᵃᵐ quoque reciperent portionem, quorum etiam exuberantia tricies auferri
205 non recusaret. (p, f. 1v) (e, f. 39r)

Partes linearum			
Sexta	Decima	Quarta Xᵃ	
CCĪDC	CLXXĪCCCLX	CXLVIIIIDCCCCXL	
Priores lineae ablatione superfluitatis corrigendae			
ĪCCVIIIIDC	ĪDCCXIIIDC	ĪIXCVIIIICLX	ĪICCCCXXIIIIXXX
Remanentia linearum ablatis partibus			Superabundantia linearum
ĪVIII	ĪDXLIICCXXXX	ĪDCCCCXLVIIIICCXX	ĪIIIDCCCXXX
DIIII	CCCLXXXVDLX	CCCXXIIIIDCCCLXX	
Medietas	Quarta	Sexta	tricesima superabundantiae
Partes remanentium			CLXI
Sex tricesimae	XVI tricesimae	XXX tricesimae	
DCCCCLXVI	ĪIDLXXVI	ĪIIIDCCCXXX	
Lineae ablatione superfluitatis de superioribus correctae			
ĪCCVIIIIDC	ĪDCCXIIDCXXXIIII	ĪIXCVIDLXXXIIII	ĪICCCCXVIIIICC

Fig. 37.

Quoniam igitur de IIᵃ linea VI, de IIIᵃ vero VI et X, porro de IIIIᵃ VI et X et XIIII, id est omnibus tricesimis, eo quod et suo et secundae ac IIIᵃᵉ vitio laborabat, abstractis, IIIIᵃᵐ lineam primae invenimus duplam, sicut earum exigit legitima dimensio, forte commodum videbitur et omnino rationabile,
210 ut iuxta illam debeant metiri rationem. Quae nimirum opinio in errorem adducit exercitatis etiam ingeniis vix intelligibilem. Ad quem deprehen- (p, f. 2r) dendum, ne quando ab illo decipiamur, animum libet intendere. Ubi consi- (v, f. 107r) derandum, quod prima linea eius quadrati latus existit,

cuius secunda diagonium invenitur. Haec autem comparatio non servatur in
215 superiore descriptione numerorum a secundo ad primum, id est a $\overline{\text{ICCVIIIIDC}}$
ad $\overline{\text{IDCCXIIDCXXXIIII}}$. Nam utroque in se multiplicato quadratus ille, qui
ex maioris multiplicatione concrescit, eum quadratum, qui ex minore produ-
citur, plus quam duplo excedit, cum II^a linea duplum aequaliter reddat, prima
vero simplum, si utraque fuerit in latere sui quadrati constituta. Unde mani-
220 festum est, quod hae lineae sicut iuxta proportiones [V] inaequalitatum, id
est multiplicis et superparticularis, tum ceterorum metiri non possunt. Ita
etiam non procedere, ut per adiectionem et diminutionem mensurari consen-
tiant.

Quare aut eorum mensuram necesse est ignorare, aut rursus ad geometricae
225 facultatis confugere adiumenta, nec poterit quisquam obtendere quod per
minutias debeant conferri, demonstrato iam superius, quod omnis proportio,
quae in minutiis constat, possit etiam ad integros numeros transferri.

(C)

Quare prorsus ab arithmeticis supputationibus recedentes proportionem
230 hanc sive potius mensuram per artem geometricam, quam non usque quaque
arithmeticae putamus addictam, immo in quam pluribus propriam exerceri
considerationem, perquirere studeamus. Ad quem ergo tendimus finem?
Quid fieri oportet? Scilicet quemadmodum illae dudum repertae sunt per
(e, f. 39v) quadrata naturalia sumentibus nobis diagonia singulorum, ut
235 tandem hae quoque simili reperiantur modo. Quadrabis enim eiusmodi VI
particulas, quales continet quaternarius quadratus aut novenarius aut quilibet
aliorum in hac figura, et hac arte linea III^a inventa erit. Item, ut V invenias,
quadrabis X. Septimam quoque invenies eo, qui est duplus ad septenarium,
in quadratum redacto. Et si ubique eandem rationem observes, poteris reperire

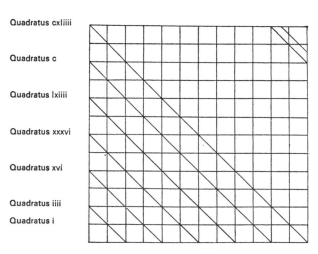

Quadratus cxliiii

Quadratus c

Quadratus lxiiii

Quadratus xxxvi

Quadratus xvi

Quadratus iiii

Quadratus i

Fig. 38

240 ad eundem modum reliquas omnes. Denique, si in III aut IIII aut V aut alias
     deinceps partes quadrati spatium ab angulo partiri volueris, eadem te ubique
     ratio quadrandi (*p, f. 2v*) iuvabit, quod manifestum erit eadem arte cognita,
     quae ita monstrabitur, ut in ipso limine sistam volentem ingredi. Figura vero
     (*v, f. 107v*) subiciatur prius, ut, quae de ea dicta sunt vel dicenda, in ea clare-
245 scant.
        Hinc iam ostendemus subiecto prius quadrato, de quo hactenus tractatum
     est, ut eo sub oculis posito ratio eius clarior fiat. In hac igitur figura ab angulo
     sinistro sursum ordinavimus lineas naturalium tetragonorum a pari numero re-
     iectis imparibus eo, quod medietates non habeant ex integro numero deno-
250 minatas. Et lineam quadrati duo habentis in latere primam posuimus, quae
     inter se et angulum centesimas quadragesimas quartas, quas brevius appel-
     lamus pedes vel asses, continet duas. Quadrati vero habentis IIII^{or} in latere
     secundam posuimus, quae asses includit VIII. Eius vero, qui VI in latere,
     III^{am} claudentem XVIII. Itemque eius, qui VIII, quartam, continentem
255 XXXII. Deinde eius, qui X, V^{am}, sub qua sunt asses L. Postremo qui XII,
     VI^{am}, cui subiciuntur LXXII. Has autem omnes diximus notas et per has
     alias inveniri posse ignotas. Quod ita est. Nam lateris, quod pertingit usque
     ad VI, si posita fuerit mensura inter IIII et V, habebis lineam XXXVI in-
     cludentem. Eius vero lateris, (*p, f. 3r*) quod summitas V attingit, in medio
260 III^{ae} et quartae mensura ponatur, et veniet tibi linea XXV sub se continens.
     Item eius mensura, quod ad caput quartae copulatur, si ducta fuerit inter alte-
     ram et III^{am}, erit linea, cui subiciantur XVI. Rursus illa pars lateris, quae
     ascendit usque ad III^{am}, ipsa quoque ponatur inter eandem III^{am} et alteram,
     et facta est linea habens sub se pedes VIIII. Adhuc longitudo lateris secundam
265 attingens (*e, f. 40r*) inter ipsam signetur et I^{am}, et haec linea contine- (*v, f. 108r*)
     bit IIII. Denique et illa lateris particula, ubi summitas primae lineae finitur,
     eidem subiecta unum dumtaxat includit pedem. Igitur hoc ordine omnis haec
     ratio procedit, ut principales lineae ipsa tetragona dividentes medietatem,
     lineae vero a lateribus sumptae IIII^{am} eorum semper includant partem.
270   His ita in hoc angulo demonstratis alteram in opposito angulo ostendamus
     artem, ut, ubi ista defecerit, iuvemur ab illa. Hoc modo, si diagonium primi
     assis metiaris eaque mensura notes utrimque figurae latus, ducasque lineam a
     nota B usque ad notam A, habes quantitatem eiusdem assis artis divisione
     repertam. Et hoc modo constituta linea prima accipe dimensionem eius eaque
275 sicut antea utrimque designa latus et a signo C trahe lineam usque ad signum
     D, quae profecto II^{a} erit, duos asses angulariter includens, eadem quoque in
     angulo altero reperta I° quadrato diviso per medium, ut notum fiat esse inter
     has quarum mensurae existat et alius modus. Sed iam ventum est ad III^{am}. Hic
     standum, quia arte qua nunc et prius usi sumus nihil hic proficere possumus.
280 Quid ergo agendum? An cedendum difficultati? An a labore cessandum?
     Minime. Quaerenda enim ars quadrandi, per quam non solum III, sed V et VI
     et VII et quaecumque adhuc repertae non sunt, metiri possunt.
        Haec autem ars ita se habet. Accipies de proposita VI asses, unde per se
     figuram compones hoc modo.

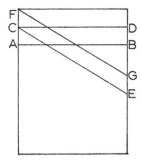

Fig. 39

285 Huius igitur figurae minus maiori latus demes ad punctum A, (*p, f. 3v*)
trahesque lineam ab eo usque B, et spatium quod supra est divides per
medium CD linea. Quo facto ad mensuram lateris, quod est sub C, trahe
lineas CE, FG. Erit igitur inter has spatium aequale ABCD spatio; spatium
dico aequale, sed non per latum. Fac ergo, ut et latitudo sit aequalis, et eo
290 reciso a longitudine latitudinique adiecto quadrasti figuram. Cuius absque
dubio, si diagonium accipias, III^{am} lineam invenisti, quae pedes descripti supra
tetragoni tenebit III. Ad hunc quoque modum (*v, f. 108v*) quadrato denario
invenies V. Et quia impares non se aptos exhibent ad quadrandum, semper eis
duplicatis quadraturam eodem modo perficies. Et quaecumque lineae requi-
295 rantur, sic poteris invenire.

His finitis magnum nobis videtur, quod hucusque pervenire potuimus.
Quare disputatio conticescat hoc loco. Libet autem in fine inventam circuli
quadraturam aliquot versiculorum quasi coloribus depingere, quatinus modis
omnibus adornata, ut fieri a nobis potest, digna fiat, quae ante oculos tuae re-
300 verentiae apparere mereatur. (*p, f. 4r*)

LIBER FRANCONIS DE QUADRATURA CIRCULI EXPLICIT

Respicit hic spatiis quadratum circulus aequis.
Nil, quicumque probes, plus minus invenies.
Primus Aristoteles fertur sociasse figuras.
305 Dum studium viguit, regula nota fuit.
Quae tulit Anicium, mors artibus ipsa sepulchrum.
Hanc, pater, hanc †partem†, tu revocas tumulo.
Laus, decus et probitas, lux, gloria, splendor, honestas,
Regum pignus ave, nobile munus habe.

APPARATUS CRITICUS

BOOK I

1–2 inscr. om. e | *5* non *om. e, supr. e*¹ | *8* arteque *v*] atque *e* | *9* inclusum *e* | *13* principium *v* | *15* exprobatur *e* | *18* in *om. e* | se ita *e* | *21* desisti iis] desistis *v*, desisti *e* | eis] eius *e* | *24* elogiis *e* | *25* infirmam *v*, in formam *e* | *26* scientia] sci *ev* | *31* minus *en*] munus *v* | *32* ita *om. e* | via *v*] vita *e* | *33* perunguente *v* | *34* quoniam edictum *v* | *37* quid] quidem *v* | *38* et Tuccas et Varos *v*] varos et tuccas *e* | *40* committes *v*] comites *e* | *44 inscr. om. e* | *46* obstrusa *(ut vid.) e*, abtrusa *v* | *47* aristotiles *ev* | *48* ferunt] c̄ *(ut vid.) add. e* | Cuius *e* | *49* sciencia *e* | *50* dubitacionem solam *e* | *53* Gerebertus *e* | *54* studiose *om. e* | investigarunt *v* | *55* Gereberti *e* | *62* quae *e*] quod *v* | *64* hebetem] etiam *add. e* | *66* altera *e*] alia *v* | *68* positus *e* | *68.72 figurae in cod. e pictae valde differunt* | *71* aut *v*] vel *e* | hi] hii *e*, hic *v* | *72* et *om. e* | *75* quod *v*] quidem *e* | a *om. e* | est] hic *add. v* | *84* Hispania] hyspani *e* | *87* est *om. v* | *88* vero *om. v* | *89* adapertione *v*] ad apercionem *e* | *90* probabilis *v* | *92* ut *v*] in *e* | *93* Iacente *v*] Placento *e* | recta linea *e* | *95 litteras iuxta figuram positas om. e* | *96* quilibet *ev* | *97* angulos *(2.)*] angulo *v* | *99* sunt anguli *e* | CEF *v*] CGF *e* | *100–108 litteras vecesque iuxta figuras 4–13 positas om. e* | *100 fig. 5 om. e* | *101* eandem – *102* ambligonio *om. v* | *fig. 7 om. e* | *102 fig. 9 om. e* | *103* cum *v*] est *e* | III] VI *ev* | *108* Razenus *e* | Γ] T *ev* | quidem *e* | Δ] a *ev* | *fig. 10:* A] M *v* | *fig. 12:* Γ] T *v* | *fig. 13:* Δ] a *v* | *109* comendemus *e* | *113* ut *v*] ita ut *e* | *117* existunt *om. e* | *120* sit *v*] sunt *e* | monstrare *e* | *123* Omnis enim figura *e* | *129* coequari *e, corr. e*¹ | *131* eiusdem *om. e* | *133* probabit *v*] probat *e* | *137* et *(1.)*] in *add. e* | *138* Sed *v*] Si *e* | et] in *add. e* | *139* invenitur *e* | *140* etiam probamus *v*] improbamus *e* | enim *om. e* | *145* diametrus *e* | *149* multiplica *e* | modo *om. e* | *150.151* a *v*] aū *e* | *151* sumatur *e* | quam *e*] a *v* | *158* Adhuc aliud *v*] Aliud autem *e* | *160* providetur *e* | *161* aliquo *e* | *162* supponatur *e* | *164* ea de causa putamus *v*] eo deputamus *e* | *165* idem *v*] id *e* | *166* si *e*] si in *v* | *168* et – *169* facultate *om. v* | *170* umquam] inquam *ev* | *171* permisit *e* | *172* utroque *v*] uterque *vel* utrique *e* | *177* Hoc *om. e* | *182* qui, cum *e*] quicumque *v* | *186* quadratura *e* | *187* iura *om. e* | *190* vel *(1.) om. v* | ut *v*] eciam *add. e* | *193* item XIIes quincunx *om. v* | sextantem *v*] sex tandem *e* | *194* opponunt *e* | autem *v*] aut *e* | *195* XIIes XII^cim *v*] XII XII^cies *e* | *198* quinquncem *e* | *200* et trienti *om. e* | opponamus *e* | *203* opponentur *e* | *207* quicquam *v*] quiquam *e* | *208* a *v*] aut *e* | *209* producta *e* | *210* obuli *v* | *211* insumet *(ut vid.) e* | *212* certe *e*] certet *v* | qua *e*] quo *v* | *221* potest *v*

BOOK II

1 Prohemium secundi libri. *e* | *6* quam *e*] Quoniam *v* | *9* quam *e*] quoniam *v* | *11* retractant *v*] ///tractant *e* | non *om. v* | *12* geometricabilibus *v* | distributam *om. e* | *21 inscriptio*] liber secundus *e* | *24* sunt *om. v* | *26* ergo *om. e* | tenentur *ev*¹] teneatur *v*, teneantur *n* | *30* a] ab *ev* | separentur *e* | *32* consequetur *e* | *36* etiam *om. e* | *45* aut *e*] a *v* | *47* cum *om. e* | minutis *v* | exequatis *e* | *49* perfecta integritas *v*] perintegritas *e* | investigare *v* | *56* hanc opinionem *e* | id est] idem *ev* | *59* pertinere – *60* etiam *om. e* | *65* aut *v*] a *e* | comparatione *v* | *74* in aequales] in(a)equales *ev* | *77* in *(1.) v*] d'm *e* | *80* in circulum reducatur *e* | *82* ipso *om. e* | *84* aBΓA *e* | *88* a *v*] aū *e* | *90* per se *om. e* | *91* quadrilatera *e*] quadrata *v* | *92* nitatur *v*] invitatur *e* | *96* ipsam *om. v* | item *v*] Idem *e* | *97* eiusdem *v* | *99* et *om. e* | *101* quot *e*] quo *v* | *102 litteras iuxta figuras positas om. e* | *108* vel *(2.) om. e* | *112* regula *e* | *113* tendi *e* | *117* esse *om. e* | *121* prospiciatur *ev* | *122* fidem *om. e* | *126* nec *v*] neque *e* | *129* quadrilatāi *v*] *133* circuli *e* | *134* nimium *(ut vid.) e* | *136* oporteat] oportet *v*, oporteret *e* | *137* ergo *v*] igitur *e* | *139* estimare *e* | *143* sive etiam in altum *v*] etiam *e* | *145* fluviorum *e* | montisque *e* | *146* comparatur *v*] invenitur *e* | *147* racione *e* | *151* in *(2.)*] III *(ut vid.) v* | *155 sq.* partibus partes *e* | *159* altero *e*] latere *v* | *163* Γ] T *e* | Δ] a *e* | *164 litteras vocesque iuxta figuras positas om. e* | *165* Facta *v*] Acta *e* | *166* aliarum *(1.)* – capita *bis habet e* | *169* multiplicata *ev* | *172* collegerunt *e* | *175* Qua *e* | *177* cura *e*] circa *v* | *179* Haec *e*] Nec *v* | eiusmodi *om. e* | *181* factum *v*] sanctum *e* | *182* vel *e*] et *v* | *183* eius *e*] vel *add. v* | sciencie huius *e* | *184* consequentur *v*

100

BOOK III

1 Prohemium libri tercii. *e* │ *2* laudis *e* │ *3* tum *(1.) om. e* │ avis] tum *add. e* │ *5* eorum *e* │ *6* factis *v*] sanctis *e* │ *8* Augustum *post* Virgilius *(7) pos. e* │ *10* quaeque *v*] quae *e* │ octonum *v*, imperium *(ut vid.) supr. v*² │ *13* exsolvit *e*] ^{ex}esoluit *v*, exoluit *n* │ *14* facti *v*] sancti *e* │ mihi] ut *add. e* │ *15 sq.* deo natus *v*] dampnatus *e* │ *17* octonibus *v* │ *18* attingis *e* │ *19* octonum *v* │ *27* infestis *v*] in fectis *e* │ *30* diligere *e*] dediligere *v* │ *31 inscriptio*] liber tercius *e* │ *33* eodem – aggredimur *e*] eo *v* │ *36* tradit *e* │ omnibus *v*] hominibus *e* │ *39* lumine *e* │ *40* dimidium *(1.)*] dimidii *e* │ dimidium latitudini demus *om. v* │ *43* ne *v*] nec *e* │ *44 sq.* autem potest nobis *e*] ut a nobis potest *v* │ *45* dari *e*] detur *v* │ *46* III] in III *v* │ ad E signum *v*] adde signum E *e* │ *47* sit *e*] fit *v* │ *49* assignetur *e*] affigetur *v* │ *51* spatium] id est superficies HKIB *add. v*² *in marg.* │ includunt *e* │ *52 litteras iuxta figuram positas lineamque HL om. e* │ *53* illud *om. e* │ *54* omni *v* │ *56* et – *57* figuram *om. e* │ *56* rectis *v* │ *57* quadrilatero] nimirum aequalis cuius videlicet *add. v* │ *59* aequalis *(1.)*] utrique spatio *add.. v* │ quadrilacio *e* │ *63* constet] constent *v*, constans *e* │ *64* aut *e*] ut *v* │ *68* hoc] hec *add. e* │ *70* argumenti rationem *v*] argumentacionem *e* │ quadrato equa latera *e*] quadratum (quadrato *v*¹) aequilatera *v* │ *73* quadratus] quadratis *ev* │ *79* Similis *ev* │ *81* a *v*] autem *e* │ *83* integro *v*] in integro *e* │ *85* volueris *e* │ *86* absque *v*] sine *e* │ *87* minutiis *om. v* │ *89* eadem – *90* productae *om. v* │ *94* inter *v*] in *e* │ *95* VIS, alter *e*] VI *v* │ *96* concrescunt *v* │ *97* servent *e* │ *98* requirere *v*] querere *e* │ *101* quid *e*] quod *v* │ conferretur *e* │ *102* mensura *e* │ obulum *v* │ calculum et *v*] calcum tum *e* │ *104* quadratura *e* │ aequari *v*] perequari *e* │ *105* circa *v*] curo *(ut vid.) e* │ *107* praeter *v*] potest *e* │ *109* contemptos *v* │ se *v*] si *e* │ *110* III interiores *e* │ rectis *v*] fectis *v* │ *113* ac *e*] hac *v* │ *122* habent *e* │ proportionum *e* │ *123* sint *v*] sicut *e* │ quadrans] V *v*, quinque *e* │ *124* S *e*] quinque *v* │ sint] sunt *ev* │ *125* S *e*] Sed *v* │ *127* proportione *om. e* │ *128* minorisque *v*] minoris *e* │ *129* itaque modum *e* │ *130* in *om. v* │ *131* umquam *om. v* │ *133* dupli *e*] quo dupli *v* │ *137* octo *v*] IX *e* │ *138* quae *v*] qui *e* │ *139* existat – *140* numeris *om. e* │ *142* asserit *e* │ in latere *en*] illatere *v* │ *143* et *(1.) v*] ait *(ut vid.) e* │ Gereberto *e* │ Zacheus *e* │ *145* Omne *v*] Pbone *(ut vid.) e* │ *146* linea diagonii *e*¹] diagonii linea *e*, linea triagonii diagoni *v* │ *147* dum hec simplum *v*] di exemplum *e* │ efficiet *v* │ *152* diagonum *v* │ autem *del. (ut vid.) v*¹ │ *153.154* et *om. v* │ *154* est igitur *e* │ *155* Rechinbaldi *e* │ *156* proveat *e* │ *157* congregant *om. v* │ et *(bis) om.. v* │ *159* accumulant *v* │ XX *v* │ *160* Inde *v*] In *e* │ *161* unciam *v*] uncia *e* │ *162* quinquagenarium *v* │ *166* XII *e*] X *v* │ quorum *v*] XVII *add. e* │ *168* XVII *e*] VII *v* │ *169* CXLIIII *e*] C quinquaginta IIII *v* │ colliguntur *e* │ *170* a *v*] autem *e* │ *171* CCLXXXIX *e*] ducentum sexaginta XXXVIIII *v* │ ad *e*] et *v* │ CXLIIII] CLIIII *e*, CLIII *v* │ *172* constet *v* │ diagonum *e* │ *177* tantis] stantis *v* │ *178* aequabiliter *v* │ circulus *om. e* │ *179* angulis *e*] circulus *v* │ is *e*] his *v* │ *181 sq.* haec collocatio *e*] excollocatio *v*, hec *supr. v*¹ │ *183* quiescam *v*

BOOK IV

1 Prohemium libri quarti *e* │ *2* habent *e* │ eaque *e*] ea quae *v* │ *6* placerent *v* │ efficiens *e*] effugiens *v* │ *8* Sitne] Sive *v*, Si enim *e* │ *12* forsan *e* │ *13* scementa *v* │ cumfecerit *e* │ *14* comportarint *v* │ *17* hoc *v*] hic *e* │ *20* huius *e*] cuius *v* │ *21* qui *erasit e*¹ │ *22* is *e*] his *v* │ *23* ascie] aciae *v* │ securi] aut *add. e* │ scriberem *v* │ quam] quoniam *v* │ *24* meruisset *om. e* │ opus *om. v* │ *25* Rarus *e*] Ratus *v* │ *26* umquam *v* │ principum *n*¹] principium *evn* │ defensionem *e* │ *27* Oracius *e* │ *28* aemulorum *v*] eorum *e* │ *29* calliopus *e* │ restitisset *v* │ *30* Mallius *v* │ consulari *en*¹] consolari *vn* │ *32* pithagoria *e* │ *33* esset *v*] fuisset *e* │ *34* Patricii *om. e*, *supr. (ut vid.) e*¹ │ simachi *e*, Seminachi *v* │ tanta *e* │ *35* principium *v* │ *37* oracio *e* │ *38* patriciis *ev* │ *39* augustis *e* │ *40* aestimare *om. e* │ *43* non *om. v* │ *44* alterius exordiendum *e* │ principio tali *e* │ *45* inscriptio] liber quartus. *e* │ *50* sunt *om. v* │ *52* idem *e* │ *53* sunt *v* │ *57* non *om. e* │ *60* enim *om. e* │ *63* liberiorum *e* │ tamen *v*] tum *e* │ *64* illius *ev*¹] illorum *v*, illam *n* │ *65* de ipso *om. v* │ *69* longitudine – latitudine *v*] longitudinem *e* │ *70* quae *e*] quod *v* │ sint *e* │ *72* versentur *v* │ *73* ad *om. v* │ *75 sq.* nisi per illud *om. e* │ *77* huiusmodi *e* │ *78* a *v*] aut *e* │ *82* exemplum *e* │ *83* agnoscat *e* │ *87* Quod – *88* minus *om. e* │ *89* in *om. e* │ *90* existit *v* │ et *om. v* │ *91* altitudinis *om. v* │ *93* quae *v*] queque *e* │ vel *(1.) om. e*, *supr. e*¹ │ *95* debeant *e*] dehabeant *v* | *98* in *om. v* | *101* XX et unus *v*, XXI *e* | *103* possunt *ev* | *105* alteram *v*] altera *e* | *106* deinde *e* | quadrato *v* | *108* litteras *v*] lateras *e* | ARC *v* | *109* appositis *e* | *110* signandam *e* | arbitramus *v* | *113* D *v*] b *e* | tum *(3.) om. v* | *115* si *v*] sed *e* | *117* C] E *ev* | Quorum *e* | *119* neutro *(2.)*] esse *add. e* | *120* natura *e* | *121* merito comprehendendum *v*] comprehensum *e* | equali *en*] equalis *v* | *122* altitudinis *e*] altitudinem *v* | *123* dimensionem *e* | *124* eiusdem *v*] eius *e* | aequalis] equaliter *ev* | est − *125* lato *(2.)*] lata *e* | *125* longo *(1., 2.)*] longa *v* | lato *(1.)*] lata *v* | aut *(2.)*] ut *v* | aequalis *(2.)*] aequaliter *v* | lato *(2.)*] lata *ev* | *126* tantum *(1.) om. v* | *127* latitudinis *v*] latitudine *e* | *130* comprehendimus *e* | *131* quae *e*] quod *v* | *132* quam] quia *e*, quoniam *v* | *134* alteram *v*] altera *e* | *135* modis *v* | actionis et diminutiis *v* | *136* Quid *v*] Quod *e* | alii *v*] illi *e* | *137* quid *e* | *137–195 figurae 21–30 in codd. ev valde differunt; litteras vocesque om. e* | *138* ad *om. e* | *140* eas *e*] cẽas *v* | *141* recidas *e*] rescidas *v* | *142sqq. litterae in textu cod. v scriptae ad litteras iuxta figuras positas non quadrant* | *147* longitudinem *e* | *151* longitudinem *e* | curretur *v* | *152* reduxistis *v* | *153* dividas *om. e* | *154* quae sunt aequales *om. e* | aequalis *v* | *157* a *v*] aut *e* | *160* valeant *v* | *164* longitudinem vel eius latitudinem *v*] longitudine vel latitudine *e* | *165* diminuare *(ut vid.) e* | didisceris *e* | *167* cui *om. e* | *168* refectionis *e* | *169* sine *v*] se ne *e* | *170* emolimentum *e* | idem *e*] id est *v* | *172* superat *v*] super a *e* | mutilata *v*] mutilia *vel* inutilia *e* | *174* distanciam *e* | *175* nunc *v*] non *e* | *176* supersunt *e*] super *v* | *178* in aequales] in(a)equales *ev* | *179* causa est *e* | *182* perveniat *e* | *183* augmento *e*] argumento *v* | exteterint *v* | *189* auctores *v* | *193* quod *om. e* | *199* aequale *n*] equalis *ev* | illi *om. e* | quod *(2.)*] quae *v*, quam *e* | latitudinem *e* | non *om. e* | *202* hac forma *e* | ex *om. v* | *203* obseleta *v*, absoleta *en*] nullus *en*] nullo *v* | *206* curvitatis nullatenus *e* | *207* distat *v* | *208* id *om. v* | *209* dimensibus *e* | *210* dimensio *e* | ut *v*] in *e* | *211* resciso *v* | *211sq.* per se refulciatur *e*] fulciatur *v* | *212* demonstratur] demonstrato *ev*; *fortasse* demonstratum

BOOK V

1 Prohemium libri quinti *e* | *2* Haud *v*] Aut *e*, haut *e*[1] *in marg.* | *2sq.* usquam quamque *v* | *3* comprehensioni *e* | *5* accideret *e* | *8* refellat *e*] te fellat *v* | *10* parvae] rei *add. e* | *11* novis *v*] nocius *(ut vid.) e* | *12* in solutis *e* | *13* omnium *om. v* | *15* si *om. v* | *16* quantum illud *om. e* | *17* quemquem *e* | *20* dulcedini *e* | *21 inscriptio*] liber quintus *e* | *23* quadraturi *v* | *26* a circulo *v* | *29* tum *e*] nam *v* | *31* specie] spem *ev* | *32* cereae *v*] certe *e* | *33* nunc *v*] non *e* | *34* redundari *e* | *35* est difficilis *e*] exdifficilis *v* | *36* argille *e*] argilli *v* | Livius] limus *(ut vid.) ev* | Horatius *v*] or. *e* | *37* imitaberis uda *v*] imitabilis *e* | *38* prospicere *ev* | sciatus *v* | *40* transfuso *e* | *41* valet *v* | valeat − formae] priori forme valeat *e* | substitui *e* | *44* a circulari *e*] articulari *v* | *47* eiusdem *e* | *49* quod *v*] quid *e* | convenit *om. v* | *50* cum forte *e*] conforte *v* | *51* quadraturas *v*] quadrilateras *e* | *52* formam quadrati *e* | illo *v* | *53* praecessit *v* | *54* involuta sit *e* | *56* ad *om. e, supr. e*[1] | *57* quam] quae *ev* | *58* diagonium *v*] affigies *v* | illo *e*] illud *vn*, illa *(ut vid.) v*[1] | *60* HL] HI *ev* | demiseris *v* | *61* et *e*] ad *v* | inferveris *v* | *65* ACHA *v* | HDMO] HLMN *v*, BLMN *e* | *66* IMBO *v* | IB *ev* | *67* in LM] in IM *(ut vid.) e*, MIM *v* | *68* non *e*] enim *n*; *v legi non potest* | *69* eundem *e* | repraesentem *v* | *71* nunc iam *v*] Nam nunc *e* | *72* autem] siquid *e*, siquidem *v* | *73* quadrandum *e*] quadratum *v* | *75* centrum *v*[2]] punctum *ev* | *76* diagono *v* | *79* hinc *(ut vid.) e* | *80* signetur *e* | *84* ABΓΔ] ABΓA *v*, ab ΓH *e* | *86* porro] ab *add. e* | a *v*] ab *e* | Δ] A *v*, H *e* | *91* centro *v*[2]] puncto *ev* | *94* /lamus − *142* longa *desunt in cod. v (uno folio omisso)* | *96 fig. 32 habet v in fine tractatus (v, f. 108v)* | *100* sit] sic *e* | *112* circulum] circulura *e* | *116sq. textus corruptus. Fortasse praerogata quadratura legendum est* | *117* preliceat *e* | *124* sed] in *e* | modo *(2.)*] *fortasse* momina *vel* motus? | *127* immuniciones *e* | *128* quam] quod *e* | *134* videantur *e, corr. e*[1] | *143* Qui si *e*] Quasi *v* | obtexeret *v* | quid *v*] quod *e* | *145* ad *(2.) om. e* | *146* Cuius *e*] Cum *v* | *147* gnomo *om. e* | *148* ut *om. e* | *149* intra − meridiem *(2.) om. e* | *150* est *om. e* | *152* hac *v*] hoc *e* | *153* deprensum *e* | orologia *v* | *154* in *om. v* | *155* ipse *v*] posse *(ut vid.) e* | *158* tantem *e* | in tantum *v*] iterum *e* | *159* refectas *e* | *163* probabis ita *v* | *166* retinebis *e* | *167 in ima pagina figuram pinxit v* | *168* promptum *e* | *171* divides *e* | Iure − *172* semis *om. v* | *172* probant *e* | *173* dubitacionem *e* |

figura ad v. 167 pertinens

174 fiet *v* | *175* praescides *v* | idque *e*] id quod *v* | *176* continentiam *v* | *177* ponunt *v* | eis *om. e* | *178* annumerati *v* | *179* XX et VIII] XXIX *e*, X et VIII *v* | *182* ita *e*] itaque *v* | *184* hec distancia *e* | *185* Quantum v | *187* racione *e* | adequatur *e* | *190* ergo *om. v* | IIII *e* | *192* in longum *om. e* | *193* dumtaxat *e*] tantum *v* | in *(2.) om. e* | *195 post* modo *add. ev* V *vel* Y | *figura a scriba codicis v male picta et per errorem in f. 102r posita hic retractata correctaque est. e praebet tres figuras* | quis *v*] si quis *e* | *197* ergo *v*] igitur *e* | si *e*] sic *(ut vid.) v* | quadrilatero – est] quadrilatera maiorem *e* | eodem *en*] eadem *v* | *199* enim *e*] cum *v* | haec *e*] hoc *v* | *200* constamus *e*

BOOK VI

1 prohemium libri sexti. *e* | *2* Cum *v*] Quod *e* | *3* attri///uerint (b *in rasura?*) *v*, attriverunt *en* | *4* non videntibus *e*] dividentibus *v* | accuset *v* | *5* surtuli *e* | id *v*] hoc *e* | *8* subdatur *e* | *10* pictagorice *e* | *15* monocordi *e* | *18* non *om. e* | *19* omnium hominum *om. e* | Siquidem *e*] si quid *v* | *21* mei *(1.) om. e* | *22* ob quam rem *e* | *23* fit *e* | *25 inscriptio*] liber sextus *e* | *28* utriusque *v* | *29* a sensu *v*] assensu *e* | *31* compendioso tractatu *ev* | *35* quisque *v*] quodlibet *e* | *40* Utraque *e* | *31* Sunt] in *add. v* | *43* valent *v* | ysocheles *e* | *44* ita *om. e* | *46 voces iuxta figuras scriptas om. e; figuram quadrati add. e* | *48* recipiat *v* | *49* existat *e* | *50* cum] dum *ev* | *52* partibus] ysocheles *add. e, del. e*[1] | *54* paciuntur *e* | ut IIII *om. e* | ut *(2.) v*] vel *e* | in *(2.) v*] III *e* | et *om. e* | *55* inaequales *v*] in equalis *e* | illarum *om. e* | aliqua *e* | inveniri – *56* aliquam *om. e* | *60* paciatur *e* | III, tertia vero *v*] tercia tercia. duo *e* | *61.62* haberent *v* | *63* tertiae *v*] tres *e* | *66* autem *v*] utique *e* | sint *e* | *68* reliqua *e* | *73* Nunc *v*] Non *e* | *74* quem *v*] quoniam *e* | *76* naturalis *e* | *77* fasticii *v* | *78* quem *v*] quoniam *e* | *79* ut *v*] ita ut *e* | *86* sit *v*] fit *e* | quadratum *(ut vid.) e* | VIII *v*] IX *e* | *88* atque *v*] itaque *e* | *94* hinc *e*] huic *v* | *95* VII *v*] VI[a] *e* | *97* porcione *e* | *99* dicemus *v* | *103* est *om. v* | proporcionem *ev*[2]] proportione *vn* | *104* neque *(1.)* – superpartiente *e*] partiente *v* | *106* IIII[a] que] quarta is *(ut vid.) e* | *111* minor *om. e* | *112* ut *e*] aut *v* | superparticulari – *115* tribus *om. e* | *116* ac *om. v* | *118* duplici *v*] dupplicet *e* | *119* debere *e*] habere *v* | *120* habent *e* | quae *om. e* | *122* a I *v*] I[a] *e* | *124* numeri iuxta *v* | *129* conservare *v*] est servare *e* | *130* VIII[a] *(2.) om. v* | *132* Sit *v* | *133* III habet *v*, terciam habet *e* | *134* IIII *v*] linea VII[a] *e* | *135* ultra] lineam *add. e* | *139* perquirende *e* | *140* Ubi *v*] Y *e* | *141* tres *om. v* | *145* generis *e* | *146* pertemptatis *e* | *147* reperimus *e* | sunt *e* | *148* aut *(1.) om. e* | *149* quare *e*] Quae *v* | *151* rem *om. e* | *153* Constituta linea *vn*] Linea constituta *v*[1] | scilicet linea *e* | *155* rescidatur *v* | *156* adiectionem *e* | *157* VIque *e*] VI quae *v* | residue *e* | *158* dimensam *v*] diversam *e* | *159* ut *om. e* | *160* quae *e*] quod *v* | auctiore] id est excedente *supr. v*[2] | *161* VI, XIIII a X *v*] V *e* | *162* illa *v* | *163* modo *e*] in *v* | *165* utrum *v* | *167* remanent] in *add. e* | *168* est *om. e* | $\overline{\text{I}}$ *(2.)*] in *(ut vid.) v* | *170* $\overline{\text{I}}$] in *e* | CCXXXX] CCIIII *ev* | $\overline{\text{CCCLXXXV}}$] CCCLXXXV *e* | *171* $\overline{\text{II}}$] II *e* | *172* abstraho *e*] ab ho *v* | *172* sq. in $\overline{\text{CXLIX}}$ DCCCCXL] $\overline{\text{I}}$CXLI$\overline{\text{X}}$DCCCCXL *e* | *174* DCCCLXX] DCCCCLXX *e*, DCCCLXXX *v* | *175* illi *ev*[1]] illae *v* | $\overline{\text{II}}$LXXXXIX] $\overline{\text{II}}$DCCCCIX *v*, $\overline{\text{II}}$DCCCXIX *e* | *177* primum *e*] unum *v* | *178* DCCCXXX] DCCCCXX *e* | *180* et diminutionem *om. e* | *183* igitur *om. e* | *184* et *om. e* | *185* quantitas: *hic incipit p* | ergo *ep*] igitur *v* | *186* primae–linea] IIII[(a)e] ac prim(a)e line(a)e *epv* | egreditur *ev* | *187* CLXI *pv*] CLXIX *e* | *188* quae ... retinent *v* | *189* in $\overline{\text{DCCXIII}}$ *ev* | *190* diminucio *e* | *191* increscunt *e* | *193* (h)abundabat *ep* | *196* a *pv*] que *e* | *197* superabat *e* | *198* $\overline{\text{II}}$CCCCXXIII *v*, $\overline{\text{II}}$CCCCXXIIII *e* | *204* reciperetur *e* | *205 figura 37:* (7) $\overline{\text{IVIII}}$] $\overline{\text{IVIIII}}$ *v*, $\overline{\text{IIX}}$ *e* | $\overline{\text{DXLII}}$] $\overline{\text{DLXII}}$ *v* | CCXXXX] CCIIII *epv* | DCCCXXX] DCCCCXXX *v* | (8) DCCCLXX] DCCLXX *(ut vid.) e* | (9) tricesima superabundantiae *om. e* | tricesimus *v* | (10) CLXI

om. e | *(13)* ablate *e* | *(14)* DCXXXIIII *ep*] XXXIIII *v* | II̅X̅CVI *ep*] IIIIX̅CVI *v* |
II C̅C̅C̅ X̅IX̅ CC *v* | *206* IIII³] III *e* | *207* id est] idem *e* | eo *om. v* | *209* legitima exi-
git *v* | forte] forme *e* | *211* adducit] adhuc *e, corr. e*¹ | etiam] et *e* | vix] iuxta *e* | *213*
quod] qu(a)e *ep(ut vid.) v* | *214* comparati *e* | *215* id est a] idem autem *e* | *216* D̅C̅C̅X̅II̅]
D̅C̅C̅II̅ *e* | *221* tum *ep*] cum *v*, et *v*¹n | *222* adminucionem *e* | *224* aut *(2.)* – *225* nec
om. e | *230* usque quaque] usquam *e* | *231* exercere *e* | *232* ergo] igitur *e* | *233* dudum
ep] dubium *v* | *234* quadrata *ep*] quadratura *v* | *235* VI eiusmodⁱ *v* | *236* contineat *e,
corr. e*¹ | *238* qui est] quod *e* | *239* eandem] si in ut *(?)* eam *e* | *243* quae *p*] quod *ev* |
245 diagonia et verba iuxta figuram posita om. v* | *246* Hinc *v*] Hanc *ep* | tractatum *e*] tctatum
(ut vid.) p, epactarum *v* | *247* igitur *ep*] *om. v* | *251* anguli *e* | quinquagesimas *epv* |
252 habentes *e* | *253* secundum *v* | vero *om. e* | *254* Item *ep* | *256* duximus *e* | et
– *257* ignotas *om. e* | *261* quarte caput *e* | alteram] secundam *e* | *263* alteram] II°ᵐ *e* |
265 et *(1.) om. v* | *268* principalis *epv* | *270* hoc *om. e* | *271* modo *e*] ergo *pv* | *272*
eaque] ea quae *e* | *274* Et *ep*] Ad *v*, At *n* | *275* utrique *e* | *276* II° *ep*] duc *v* | duos]
duo *v* | singulariter *v* | *277* reperta *ep*¹] reperto *pv* | *278* Sed *ep*] Si *v* | est *om. e* |
279 et prius] plus *e* | hic *om. e* | *280* An *(2.) om. e* | *281* enim] est *add. e* | *282* pos-
sint *pv* | *283* opposita *e* | *284* figuram *om. v* | *285* Huius] Suius *e* | *288* ABCD *ep*]
AB *v* | *290* resciso *v* | latitudini *e* | *291* super *e* | *296* His] Sic *e* | quod hucusque
ep] quousque huc *v* | *299* quae *p*] quod *ev* | *301* subscr.] Explicit *e, om. p* | *302sq. om. v* |
302 quadratus *e* | *304–309 om. e* | *v. 304 sq. post v. 309 pos. v* | *304* aristotiles *pv*

APPARATUS II

The names of authors of scientific works refer to the bibliography which will follow in the
next issue at the end of the introduction.

BOOK I
    3   *Ex* – 16 *delectarent:* edited by A. Mai, p. 346 f., and Migne, PL 143, col. 1373–75
   32   *Atque* – 63 *angulorum:* edited by A. Mai, p. 347 f., and Migne, PL 143, col. 1375 f. In both is
         lacking the sentence 37 *Ergo* — 38 *Augusti*
   38   Plotius Tucca and Varius Rufus edited the 'Aeneis' by order of Augustus against the will
         and after the death of Virgil
   46   *Quadratura* – 58 *fuisset:* edited by Bubnov, p. 384 f.
   48   *Praedicamentis:* Aristoteles, *Categoriae* 7 b 31–33 Bekker; comp. *Aristoteles Latinus* I 1
         (1961), p. 21, 14 f.; I 2 (1961), p. 61, 14 f.
   49   *Boetium:* Boethius, *Comm. in Aristotelis praedicamenta,* edited by Migne, PL 64, col. 230 f.
   52   *Adelboldus:* comp. Adelboldus, *Ep. ad Gerbertum,* p. 302–309 Bubnov
   53   *Wazo:* The experiments of Wazo concerning squaring of the circle are not mentioned
         elsewhere
   53   *Gerbertus:* We suppose that there is a reference to the letter of Adelbold to Gerbert. There
         is not known any work by Gerbert himself on squaring the circle
   55 f. *geometricus libellus:* Geometry by Gerbert (p. 48–97 Bubnov). The anonymous continuation
         of the geometry deals with squaring the circle (p. 346, 9–16; 354, 9–17; 356, 9–12 Bubnov)
   69 f. *Neque-est:* Comp. Regimbold, edited by Tannery (d), p. 526, 16–527, 1; Gerbert, *Geometria,*
         p. 68, 12–19 Bubnov
  178   *arithmeticorum regula:* Boethius, *De arithmetica,* p. 117, 26–28 Friedlein

BOOK II
    9   *Moysen:* Exodus 25–27
   10   *permensam – cubitis:* Genesis 6, 15
   11 f. *terram – distributam:* Numeri 34–35, 5
   12   *Salomonem:* 3. Regum 6
   14   *apud Ezechielem:* Ezech. 40–42
   87   *Terentii consilio:* Terence, Andria 670

104

144f. comp. "Boethius" Geometry II, l. 560f. Folkerts
149f. comp. "Boethius" Geometry II, l. 404–406, 412f. Folkerts
171f. *subtiles geometrae:* e.g. Epaphroditus 31 Bu. (p. 546, 10–14 Bubnov); *Geometria incerti auctoris* IV 18 (p. 346, 12–16 Bubnov); letter of Adelbold to Gerbert (p. 304, 19–305, 3 Bubnov); "Boethius" Geometry II (l. 885–888 Folkerts)

BOOK III
142 *Regimboldus:* Correspondence, Tannery (d), p. 271. 278
164 *ipse* = Regimboldus (see l. 142)

BOOK IV
2 *Tullio in Topicis:* Cicero, Topica § 58–67
17 *Mattheus:* Matth. 20, 1
19 *ait ad illos:* Matth. 20, 4. 7
29 *Calliopius:* late ancient grammarian, editor of the works by Terence. Scholia in medieval manuscripts of Terence indicate that he was the speaker of the prologues by Terence. See F. Schlee, *Scholia Terentiana*, Leipzig 1893, p. 9f.; 79, 10; 94, 26; 140, 24
33 *orat:* introduction to Boethius, *De arithmetica*, particularly p. 5, 21f. Friedlein

BOOK V
36 *Horatius:* Epistulae 2, 2, 8

BOOK VI
5 *Sursuli:* The by-name *Sursulus* or *Surculus* is due to the fact that the poet Publius Papinius Statius was mistaken for the rhetor L. Statius Ursulus. The nonsensical etymology *Sursulus* = *sursum canens* can be found in the medieval *Accessus* to 'Achilleis' by Statius (comp. Paul M. Clogan, *The Medieval Achilleid of Statius*, Leiden 1968, p. 9; 21, 19f.)
18f. *trahi – sua:* Virgil, Ecloga 2, 65.

LIST OF PERSONAL NAMES AND NAMES OF PLACES MENTIONED BY FRANCO
The Roman numerals refer to the book, the Arabic ones to the lines of the present edition.

ADELBOLDUS from Utrecht (about 970–1027): I 52
ADELMANNUS, scholar of Fulbert from Chartres, 1st half of 11th century: I 107
AEGYPTIUS: II 10
AENEAS: III 16. 17. 18
AENEIS: see VIRGILIUS
ANICIUS: see BOETIUS
ARISTOTELES, Greek philosopher (384–322): I 47, VI 304
  Praedicamenta: I 48
AUGUSTUS, Roman emperor (63 B.C. to 14 A.D.): I 38, III 8. 16, IV 28. 39, VI 22
BOETIUS, Anicius Manlius Severinus, Roman philosopher (about 480 to 524): I 49, IV 30. 204, VI 306
CAESAR, Caius Iulius, Roman statesman (100–44): II 3, VI 21
CALLIOPIUS, late Roman grammarian: IV 29
CHRISTUS: II 19
CICERO, Marcus Tullius, Roman writer and author (106–43): IV 2
  Topica: IV 2

COLONIA: I 19
EGYPTIUS: see AEGYPTIUS
EZECHIEL: II 14. 18
GALLIA: I 51. 83
GERBERTUS, later on Pope Silvester II (about 940–1003): I 53. 55, III 143
GERMANIA: I 51, III 12
GRAII: III 4
HENRICUS = King Henry I (919–936): III 10
HISPANIA: I 84
HORATIUS, Quintus H. Flaccus, Roman poet (65–8 B.C.): IV 27. 37, V 36
ITALIA: I 51, III 13
LATIUM: III 15
LIVIUS, Titus, Roman writer and author (59 B.C.–17 A.D.): V 36
MAECENAS, Caius Cilnius (about 70–8 B.C.): IV 27, VI 21
MANLIUS: see BOETIUS
MARO: see VIRGILIUS
MATTHEUS: IV 16. 17
MOYSES: II 9
OTTONES: III 10. 17. 19

A TREATISE ON THE SQUARING OF THE CIRCLE BY FRANCO OF LIÈGE, OF ABOUT 1050

[Part II]

Menso Folkerts, Berlin and A. J. E. M. Smeur, Dorst/Breda

In 1831 A. Mai[1] published some parts of a 12th century manuscript, at present in the Vatican library, which started:

Incipit prologus in primum librum domni Franconis de quadratura circuli.

These parts were borrowed, unaltered, by J. P. Migne in 1853[2]. The complete text of the manuscript was published by C. Winterberg in 1882. His edition, however, has many faults and inaccuracies and is lacking the necessary elucidation which, consequently, makes it difficult to read and even incomprehensible in many places.

A smaller part of the text has also been published by N. Bubnov[3]. After Winterberg's edition Franco's treatise was mentioned in works on the history of mathematics but, obviously as a consequence of the imperfections of that edition it has received little attention. There is some information in C. Le Paige and M. Cantor and also, very briefly, in M. Clagett[4]. D. E. Smith only names the work[5]; D. J. Struik and A. P. Juschkewitsch on the other hand do not even mention it. J. E. Hofmann only gives a negative verdict[6]:

Der Tiefstand der geometrischen Kenntnisse wird noch deutlicher aus einer wissenschaftlich belanglosen Abhandlung des Franco von Lüttich über die Kreisquadratur (um 1050).

Others, less expert on the subject, are perhaps too previous in their praise, as for instance[7]:

His treatise, De quadratura circuli, means a true advance in mathematical thinking.

In 1968 Franco's text was re-edited by A. J. E. M. Smeur, provided now with an introduction. The Vatican manuscript, however, is in a very bad condition and in several places absolutely illegible. Recently M. Folkerts has found new manuscript-copies. These have led to the present new, and far better, edition.

The greater part of the introduction is a translation of the one from 1968. In some places important corrections were possible. This introduction contains: 1) some older records of Franco's squaring of the circle, 2) some data about the used manuscripts, 3) a short historical survey in connection with some names appearing in Franco's text, 4) some remarks on the mathematical knowledge which Franco had at his disposal. Then follows a discussion of the contents of Franco's treatise, which consists of 6 books, each opening with a '*Prologus*' not devoted to mathematics and which has not been given further attention. The proper mathematical text of each book is subdivided into paragraphs A, B, C and so on. Quotations have been provided with the page-number of the Vatican manuscript. These have also been indicated in the text-edition so that each quotation can be found again in its context.

INTRODUCTION

1.

The historiographer Sigebert (1030–October 5th, 1112) from the Abbey of Gembloux near Namur mentions in his *Chronica* in the year 1047[1]:

Franco scolasticus Leodicensium et scientia litterarum et morum probitate claret; qui ad Herimannum archiepiscopum scripsit librum de quadratura circuli, de qua re Aristoteles ait: Circuli quadratura, si est scibile, scientia quidem nondum est, illud vero scibile est.

This testimony can be regarded as reliable for Sigebert was but a little younger than Franco and lived not far away from him.

We meet it again, with some modifications, in later editions of Sigebert's chronicle, mentioning moreover that Franco dedicated his treatise[2]:

ad Hermannum Coloniae archiepiscopum, . . .

Franco's work is mentioned again in *Ly Myreur des Histors* by Jean d'Outremeuse (Liège, January 2nd, 1338–November 25th, 1400) but this time in the year 1070 and specifically in May; moreover, the text goes back to Sigebert's[3]:

L'an milh et LXX, en mois de may, commenchat et parfist et compoisat en escript, Franque, li scolastre del englise de Liege, qui astoit de scienche, de letres, de manere, de probeteit renommeis, a Hermain, archiepiscopum Bonneburgense, le libre de la quariuree de cercle, de la queile chouse Aristolt parolle, de cercle quareit, si est: Sciendum quidem meritum est; illud vero.

Johannes de Trittenheim mentions several works by Franco[4] and ends with:

Claruit sub Henrico imperatore tertio. Anno domini. M.LX.

The *Histoire littéraire de la France*, in which explicitly is referred to Sigebert, has[5]:

Francon Scholastique de Liège fit un traité du Comput, et un autre sur la quadrature du cercle, qui a toûjours donné, et donnera encore de l'exercise aux Mathématiciens.

The assertion that the work was still being studied (that is in the year 1867) is for the responsibility of the composers of the *Histoire*; at that time it was totally unknown to mathematicians.

In a publication by an otherwise unknown Falchalinus, can also be read that he helped Franco[6]:

De Falchalino Ludovici Discipulo.
Falchalinus etiam scholarum magister, ejus coaluerat disciplinis. Iste Franconi Scholastico sancti Lamberti in opusculis de quadratura circuli, . . . cooperatus est.

2.

The treatise concerning the squaring of the circle written by Franco from Liège is still available since four manuscripts are known which reproduce the text in part or as a whole.

The oldest codex is MS. latin 7377C in the Bibliothèque Nationale in Paris. This manuscript, made known thanks to the researches by P. Tannery[1], is incomplete: it contains only the end of the text on f. 1 r–4 r (beginning page 94, line 185 of this edition). An old foliation beginning with f. 23 gives evidence that 22 sheets are missing at the beginning of the manuscript. The table of contents on f. 48 r written by Nans reveals that the loss did occur only after 1600. We may assume that even the number 22 is too small as the missing part of the tractatus by Franco might have comprised a larger number of sheets. The peculiarity of the writing signifies that the codex was written about the end of the 11th century in Eastern France (Reims?). The following owners of the manuscript could be traced: Henricus Wierus from Treves (16th century); Marquard Freher, historian in Heidelberg (1565–1614); Frans Nans, philologist and lawyer in Dordrecht (1525–1599); Pieter Schrijver (1576–1660). Besides Franco's treatise this codex contains further mathematical works which were written or made known in Lorraine in the 11th century: parts of the correspondence between Regimbold and Radolf and the correspondence between Adelbold and Gerbert, the *Geometry II* of Pseudo-Boethius as well as the Geometry of Gerbert with parts of the anonymous supplement of it[2]. This fact gives also evidence of the close relation of the manuscript to Franco's original. The scribe faithfully copied the text and did not insert any corrections of his own. The more we have to regret that this oldest and best codex only reproduces a small part of the whole work.

The manuscript Vat. lat. 3123 in the Vatican Library which contains Franco's text nearly completely is of later date. This collection of mathematical and astronomical texts comprises 112 folios in the size of 112×197 mm. The sheets f. 65–111 belong together; this part was written in Western Germany in the 12th century, certainly before 1210. The written text has the size 71×148 mm with 30 lines per page. This codex, too, is damaged: a large number of sheets have spots due to mould and as a consequence of this the upper parts of some pages are illegible. One sheet is lost between f. 101 and 102: f. 101 v ends in the middle of the sentence '*In qua appel-*', whereas f. 102 r begins with a new chapter '*Quasi forte*'. The numeration of the pages which was added later on does not clearly indicate how many sheets have been lost, but the text preserved in the Edinburgh manuscript (see below) shows that this must have been one sheet. This manuscript, too, contains a number of texts, which were written in Lorraine in the 11th century. In addition to three short treatises which are too insignificant to be mentioned in this context the following three texts which originate from the scholarly group in Lorraine are contained in the second part of the codex: Ps. Boethius, *Geometry II* (f. 65 v–84 r), Franco's treatise (f. 84 v–108 v) and a letter of the 'Monk B' to Regimbold which stands in relation to the dispute on the angle (f. 108 v–110 r). Tannery printed this letter as No. 9 of the correspondence[3], whereas Winterberg erroneously took it for Franco's work[4]. The manuscript in the Vatican Library is one of the two older ones which, apart from the already mentioned missing page, reproduce Franco's text completely. The scribe, however, did not copy the text with accuracy: there are a number of errors which possibly did occur when copying the text. Part of these were corrected by the scribe himself or another person at the same time but most of these errors were not corrected and make the study of the text very difficult.

In addition to these two manuscripts which are independant from each other, a copy of the codex in the Vatican exists which was written in Italy about 1500. It was in the possession of Sir Thomas Phillipps (no. 11727) until 1908 and is now kept as MS Plimpton 250 in the Columbia University, New York[5]. Franco's treatise is written on f. 129r–155v. The scribe copied the Vatican text with extreme accuracy and corrected hardly any of the errors in the manuscript. Though this codex is only a copy of an available manuscript it is of special importance for the reconstruction of the text since this codex is the only expedient one to reconstruct the text in the Vatican library which is illegible due to mould. The lacuna in the Vat. 3123 after f. 101v is to be found in Plimpton 250 too; there is the reference '*hic deficiebat cartha*'.

Finally, there exists a fourth manuscript, containing the complete text of Franco's treatise. Our attention was drawn to this fact by Ron B. Thomson, Worcester College, Oxford, to whom we are much indebted. The manuscript in question is Edinburgh, Royal Observatory, Ms. Crawford 1.27, written in the 13th century and, therefore, of later date than Vat. lat. 3123.

The manuscript includes the following texts:

f. 1r–13v: Jordanus Nemorarius, *Liber phylotegni de triangulis*, a shorter version.
f. 14r–21v: Jordanus Nemorarius, *Liber de ratione ponderis*[6].
f. 21v: Ps. Archimedes, *De ponderibus*, proposition 4.
f. 22r–24r: Anonymus, *Liber de canonio*[7].
f. 24r: Anonymus, Fragment of a treatise on statics.
f. 24rv: Heron, Computation of the area of a triangle.
f. 25r–41v: Franco, *De quadratura circuli*.
f. 42r–52v: Gerard of Brussels, *Liber de motu*[8].

The manuscript is probably the same as the Codex 43, mentioned by Richard de Fournival in his *Biblionomia* (about 1250)[9]:

Jordani de Nemore liber philothegny CCCCXVII propositiones continens. Item eiusdem liber de ratione ponderum, et alius de ponderum proportione. Item cuiusdam ad papam de quadratura circuli. Item Gerardi de Bruxella subtilitas de motu. In uno volumine cuius signum est littera D.

Some years later this codex, like many others, came through Gérard d'Abbeville to the library of the Sorbonne. In the catalogue of the Sorbonne of 1338 it is described, with the signature LVI 30, as follows[10]:

Philoteny de Nemore continens LXIIII propositiones, liber eiusdem Jordani de ponderibus, subtilitas magistri G. de Brucella, ex legato predicto (= Gerardi de Abbatisvilla). Incipit in 2° fol. 'triangulo', in pen. 'una sit'. Precium X sol.

Afterwards the manuscript belonged to Pope Pius VI (1717–1799)[11]. Then, like many other manuscripts and books, this codex also came through the Leipsic bookseller T. O. Weigel into the possession of James Ludovic Lindsay, Earl of Crawford (1847–1913)[12]. In the two decennia from 1870 till 1890 the astronomer Lindsay acquired, by systematic purchase, a large scientific private library, to which also belonged the estates of Charles Babbage and Michel Chasles[13]. In 1888 this library was bestowed to the State of England and, from that time, it has been the foundation of the library of the Royal Observatory, Edinburgh.

In addition L. Thorndike mentions[14] a text from the 15th century which is available in the University Library in Cambridge (Ee III, 61, f. 176v–177v): *Franco scolasticus loadicensis ad Hermanum archiepiscopum scripsit hunc librum de quadratura circuli.* This attribution to Franco is unjustified as the text on these pages was not written by Franco. It is generally attributed to Campanus but this assumption probably lacks substantial ground[15].

3.

May we assume that the Vatican manuscript indeed contains the squaring of the circle, regarded by Sigebert as a work of Franco of Liège? We read that it is by a certain Franco; the indication Liège, however, is missing as is also the name of archbishop Herman of Cologne[1]. However, there are several indications which give sufficient certainty. To this end there follows a brief historical survey.

In 919, when the Carolingian dynasty died out, the Saxon duke Henry was elected king. His son, Otto the Great (912–973) was crowned king in Aix-la-Chapelle in 936 and crowned emperor in Rome in 962; all the monarchs mentioned hereafter also received the imperial crown. Otto I was succeeded by his son, Otto II, who died when his son Otto III (980–1002) was still under age. The latter's interests were looked after, among others, by archbishop Adalbero of Reims and his later successor Gerbert. This Gerbert also was one of the tutors of Otto III. Otto III made Rome his imperial residence. He had no direct successors and after his death Henry II (973–1024) was the last Saxonian emperor. After the three most important representatives, the above period has also been called the Otto dynasty.

With Conrad II (about 990–1039), elected in 1024, the Salic dynasty started. Under his son Henry III (1017–1056), already crowned king before Conrad's death and crowned emperor in 1046, the imperial authority reached its peak.

Under the Salic monarchs mentioned the most important person of the German empire was the archbishop of Cologne, Herman II, a grandson of Otto II, and totally devoted to the emperors[2]. He died on February 11th, 1056, shortly before Henry III.

The afore-mentioned Gerbert was born around 950[3]. After first lessons at Aurillac and a journey through Spain he became a teacher at Reims (972–982), then abbot at Bobbio in Italy, archbishop of Reims (991) and Ravenna (998) and finally, from 999 till his death in 1003, pope Silvester II. At Reims one of his pupils was Fulbert (about 960–1028) who started a school at Chartres[4] and of whom some pupils became well-known at Liège, such as Rudolf or Radolf, Adelman and Franco[5].

Since the 9th century, Liège had a very famous cathedral-school. Notger (Swabia, about 930 – Liège, April 10th, 1008), first prince-bishop of Liège (from 971), 'un des éducateurs de l'Europe', as G. Kurth calls him, had been, for some years, head of this school before he became bishop[6]. Some of his successors at the school were Wazo (1003)[7], afterwards bishop of Liège (1042–1048), Adelman, who became bishop of Brescia, a certain Gozechinus[8] and in 1066 Franco[9].

In his time Gerbert was the most influential mathematician in Western Europe[10]. Among his many mathematical papers there has been preserved a letter to his friend Adalbold, afterwards bishop of Utrecht (1010–1027)[11]. This Adalbold also studied at Liège.

From one of Fulbert's pupils at Chartres, Radolf of Liège, a correspondence has been preserved, dealing with mathematical subjects, with Regimbold, head of the cathedral-school at Cologne[12]. This Regimbold possibly was a pupil of Wazo at Liège and a friend of a certain Racechinus[13].

Most of the names mentioned can be found again in the Vatican manuscript, as Adalbold (85v), Adelman (86r), Gerbert (85v, 94v), Racechinus (86r, 94v), Regimbold (94v, 95r), Wazo (85v) and also Henry (92r) and the Otto-dynasty (92r and v). This, added to the evidence that Franco of Liège wrote on the squaring of the circle, does ascertain that the Vatican and Edinburgh text is indeed the one by Franco of Liège. As to Adelman, apart from what is mentioned in the manuscript, further witness that he practised mathematics is lacking. The name Werenboldus also appears (89r); P. Tannery suggests that this perhaps should be Regimboldus[14].

Although the name of Herman, the archbishop of Cologne, does not appear in the Vatican manuscript there are some indications that the treatise is devoted to him. The prologue of Book I opens with (84v):

mi papa, praesulum decus, corona totius per orbem cleri, . . .

and then, in connection with the obviously intended clerical dignitary, some lines further on the rich town of Cologne is mentioned. The prologue of Book III begins with (92r): '*praesul eximie*', eminent bishop, that of Book IV has (95v): '*tu . . . decus pontificum . . . es*' and that of Book V again begins with (100r): '*praesul clarissime*'. Finally, in the prologue of Book III reference is made to the three Otto's, German emperors and relations of Herman (92r):

. . . quae primus, quae secundus, quaeque tertius gessit Ottonum, quorum primus ab Henrico patre suo suscepit regnum, sed filio reliquit imperium; pater apud Theutones primus regnavit, filius apud ipsos primus imperavit. Et quibus, nisi illis, Germania debet, quod sibi cum tanto orbe ipsa exsolvit tributum Italia? Per quos alios nostri imperatores Romani sceptri facti sunt successores?

All this is in accordance with Sigebert's information that the work is devoted to archbishop Herman of Cologne and may even be seen as its confirmation. But the conclusion then has to be that the work could not have been written later than 1055.

Only little is known about Franco himself[15]. According to Sigebert his renown dates from 1047. So his treatise may be placed around 1050. By no means in 1060 or 1070, as some have done[16]. In 1066 Franco became head of the school of Liège and about 1083 he died. Apart from the squaring of the circle some other treatises are ascribed to him[17].

4.

For a well-understanding of Franco's treatise and a right appreciation of it, some idea of the level of mathematical knowledge in his days is wanted. This was very low. Some familiarity was possible with the works of Martianus Capella (about 450), Cassiodorus (480?–575?) and Isidorus of Sevilla (570–636), which were low in mathematical contents. More was known by Boethius although not all the writings circulating at the time under his name are his own. The *arithmetica* of Boethius has some number theory, borrowed from Nicomachus. The *geometria* attributed to Boethius contains a translation of axioms and propositions from the first four books of Euclid's *Elements*, lacking proofs. Furthermore something was known from the manuals of the Roman surveyors[1].

Therefore the preconditions for the solution of mathematical problems were extremely unfavourable in the 11th century. All the more we have to praise any endeavour to investigate mathematical contexts. Thanks to the researches by N. Bubnov and P. Tannery[2], one group of mathematicians which exercised a great influence on Franco's work in particular gained recognition. Their influence spread from Liège to Lorraine and beyond in the 11th century. This scholarly group is well-known for their works on the abacus but also for treatises concerning geometry. The most important works of these men are a large correspondence between the scholar Regimbold from Cologne and Radolf from Liège possibly from the year 1025, which deals with the angle sum in a triangle[3] and the so-called *Geometry II* of Ps. Boethius. The *Geometry II* which was possibly written in the period from 1025 to 1050, contains excerpts from Euclid, texts written by the Gromatici and a chapter dealing with the abacus[4]. The treatise by Franco stands in close relation to the two works mentioned above.

Franco's treatise shows that he mastered arithmetics very well, including fractions. At that time these still were the Roman fractions, subdividing the unity, the *as*, into 12 *unciae*. Because of the fact that multiples and parts of the *uncia* had their proper names, those occuring in Franco are given here[5]:

$\frac{1}{6}$ sextans	$\frac{1}{24}$ semuncia	$\frac{1}{288}$ scripulus
$\frac{1}{3}$ triens	$\frac{1}{36}$ duella	$\frac{1}{576}$ obolus
$\frac{5}{12}$ quincunx	$\frac{1}{48}$ sicilicus	$\frac{1}{1152}$ cerates
$\frac{5}{6}$ dextans	$\frac{1}{72}$ sextula	$\frac{1}{2304}$ novissimus calcus

As to plane geometry, however, neither Franco nor for instance Radolf, Regimbold and Adalbold had a right idea of what in fact was the meaning of the problems handed down to them. A great part of the correspondence between Radolf and Regimbold is on Boethius' communication[6]:

Scimus triangulum habere tres interiores angulos aequos duobus rectis.

The question now is not only how to prove this proposition but: what are interior angles and, consequently, what may be exterior angles. For Boethius also has[7]:

Omnium triangulorum et exterior angulus duobus angulis interius et ex adverso constitutis est aequalis, . . .

Radolf supposes interior angles to be in plane geometry and exterior angles in solid geometry. Regimbold on the other hand states that the vertex of an exterior angle differs from that of an interior one and starting from an equilateral triangle ABC with centre O he says that indeed the exterior angle ($\ast AOB$, so 120°) equals the sum of two interior angles ($\ast CAB$ and $\ast CBA$, both 60°)[8]. He does not prove his statement but has an annotation which is important to get an idea of with what one had to content oneself within geometry in those days[9]:

Quod si adhuc te latet nostrae veritas expositionis, cum circini probatione vel proportionali membranarum incisione cuncta quae dicimus vera esse poteris comprobare.

This is typical. Geometry was considered to be experimental science and propositions were verified with the help of a compass or a model.

Elsewhere Regimbold also has an exterior angle to be obtuse and an interior one to be acute[10].

The problem concerning the angle sum in a triangle is not definitely solved in the correspondence between Regimbold and Radolf, but only some time later it found a solution on an experimental basis. J. E. Hofmann showed that Franco knew of the solution of the dispute on the angle sum as it is transmitted in an anonymous text from Cues[11].

In the correspondence between Radolf and Regimbold the question also arises whether the proportion of the diagonal of a square to the side (the value of $\sqrt{2}$) equals $\frac{7}{5}$ or $\frac{17}{12}$[12]. This in connection with the problem of doubling a square. Geometrically, of course, no problem exists, taking the diagonal of the smaller square as side of the larger one. The conclusion, correct indeed, but unproved is that the proportion cannot be given in numbers.

Franco first takes $\frac{17}{12}$ as a value for $\sqrt{2}$ and easily verifies that this is too much. He succeeds in finding a better approximation, also for other irrational roots. He moreover states that such roots cannot be rational and tries to prove this.

Furthermore, he applies, not having the slightest idea that this again is an approximation, the value $3\frac{1}{7}$ for π when calculating the circumference of a circle. From this the area is calculated correctly: half the diameter multiplied by half the circumference, in accordance with the method of the Roman surveyors[13].

Franco's text stands also in relation to the *Geometry II* of Ps. Boethius: they have in common the rhetorically redundant, almost stilted style, which is mainly to be found in the introductions to the six books. Regarding the formulation and the number values, the calculation of the circle in Franco's treatise (p. 70, l. 111–116)[14] reminds of the corresponding passage in the *Geometry II* (line 885–888)[15].

The *porticus*, *miliaria*, *stadia* and *fluvii* are mentioned by Franco (p. 71, l. 144f.) as well as by Ps. Boethius (line 560f.). The glorification of Pythagoras and the

233

Patricius Symmachus in connection with Boethius (Franco, p. 78, l. 30–34) might refer to his *De arithmetica* (pp. 3, 1. 11. 13; 7, 21–26 of the edition by Friedlein) as well as to the *Geometry II* (see line 439–441. 447–453. 927–930. 2). There are some striking formulations in the text by Franco which strongly resemble passages in Ps. Boethius so that an influence between the authors of the two treatises must be considered[16]. On the assumption that Franco had knowledge of the *Geometry II* the date of origin of the latter text can be determined more precisely than up to now[17].

DISCUSSION OF FRANCO'S TREATISE

BOOK I

A.

Franco points to the origin of the problem, a statement in the *Categoriae* or *Praedicamenta* of Aristoteles, translated by Boethius into Latin. Boethius' quotation became well-known in the Middle Ages. To Aristotle the squaring of a circle was an example of a problem that can be solved and so can be the object of scientific investigation but of which the solution still was unknown[1]:

Nam si scibile non sit, non est scientia; scientia vero si non sit, nihil prohibet esse scibile, velut circuli quadratura si modo est scibilis, scientia siquidem eius nondum est, ipsa vero scibilis est.

According to Franco the problem has not yet been solved. He mentions Adalbold, Wazo and Gerbert who have tried to find a solution.

B.

Franco now introduces another difficulty, viz. the classification of angles. This can be done in two ways:

    a) right, acute and obtuse angles,
    b) interior and exterior angles.

As said before, b) has caused difficulties in the correspondence between Radolf and Regimbold. Franco's explanation is that an exterior angle lies outside the figure. This would be correct if he had drawn *BD* in a direct line with *CB* (figure 1 on f. 85 v)[2]. He dismisses the idea of an exterior angle being obtuse and an interior angle being acute. He has a right understanding of the assertion that the sum of the interior angles of a triangle equals two right angles (f. 85 v):

Sed hic magnus est error nihilque aliud fuit, quod impedisset eos, qui conati sunt approbare triangulum III interiores angulos aequos habere duobus rectis. Equidem, si bene comprehensum fuisset, quid interior angulus accipi deberet, nihil esset reliqui, quod eis posset obstare.

His proof, however, is experimental. First he has a figure of two right angles an then, transposing angles, he shows the correctness of the proposition for an equilateral, a right-angled and an obtuse-angled triangle. To him now the matter is finished (f. 86 r):

cum III existant triangulorum genera, nullum esse poterit, cuius anguli huic comparationi
dissentiant, sicut harum descriptionum probatur exemplis.

The meaning of the other figures on f. 86r is not entirely clear. Figure A (=
no. 10) of Wazo resembles Regimbold's one in his demonstration of the proposition
of the exterior angle[3]. Regimbold, most probably a pupil of Wazo's at Liège, could
very well have borrowed this figure from him[4]. Figure Γ (= no. 12), ascribed to
Racechinus, is also used by Radolf. Half of a square is a triangle with one right
angle and two acute ones, each being half of a right angle. For such a triangle
Radolf could demonstrate the proposition about the sum of the angles. If we take
it for known that a diagonal divides a rectangle into two equal parts (which is
easy to see) then the figures B (= no. 11) and Δ (= no. 13) could also demonstrate
the proposition just mentioned[5].

C.

For the squaring of a circle Franco mentions three methods known in his days.
   a) The side of the square equals $\frac{7}{8}$th of the diameter of the circle. This means
the value $3\frac{1}{16}$ for π, occuring already in an earlier Indian text[6]. Since no relation
to it is known, one has to assume that it was found once more in Europe.
   b) Start from a circle with two mutual perpendicular diameters. Divide the
radii into four equal parts and prolong each with its quarter to find the vertices of
the square. The sides now equal $\frac{5}{4}R\sqrt{2}$, so $\pi = 3\frac{1}{8}$, a good approximation,
known from Indian texts, practised by the Roman surveyors and occurring, too, in
a treatise by a monk B. of Regimbold's and Radolf's days[7].
   c) Take a quarter of the circumference as side of the square. This method also was
practised by the Roman surveyors. M. Cantor, by solving $\left(\dfrac{2\pi R}{4}\right)^2 = \pi R^2$,
coucludes that $\pi = 4$. But this obviously is incorrect for the circle and square are
isoperimetric so the square certainly is smaller in area than the circle, much smaller
even, for its area is about 2.46 R^2 [8].
   Franco dismisses these three methods[9], not so much because of their results (at
least on these he does not comment) but more so because of the method (f. 86v):

Sed hi omnes a veritate longe absunt eo, quod ubi et qualiter investiganda sit non attendunt.
Nam quicumque demonstrare voluerit formarum quarumlibet aequalitatem, hunc primo
advertere oportet, ubi illa versetur aequalitas.

D.

What is discussed now is the question on what the equality of two figures can
be based (equality in area, not congruency). At first Franco distinguishes between
three possibilities:
   a) The areas only equal in number and a construction to demonstrate the intended
equality is impossible.
   b) The exact opposite of a). Both areas do not wholly equal in number or the
equality at least cannot be calculated, but it can in fact be demonstrated by a
construction. This Franco calls an equality *spatio tantum*. This possibility may be

surprising but it should be considered that in those days a more or less evident verification was sufficient for a geometric proposition. This mostly meant an approximation which, however, was not regarded as such. Geometry was practical, Greek geometry was totally unknown.

c) The areas are equal in number, as well as by construction.

Franco then comes to a closer examination of these three possibilities.

To a). This is not acceptable in geometry where a construction is wanted. Franco's example of a square and a triangle, both with area 36, is not a suitable one for it is not difficult to show the equality by construction[10]. But the meaning is clear. A mere equality in number without any comment or indication as to the construction is insufficient to him.

To c). This possibility is of course the most complete. As an example Franco has a square with sides 6 and a rectangle with sides 9 and 4 both with area 36. The construction is lacking though very easy to perform, by subdividing both, either in 36 unit squares, or in larger parts[11].

As to the squaring of a circle, an equality in number is impossible, as Franco shows. The circumference is taken as $3\frac{1}{7}$ of the diameter. Here the proportion of circumference and diameter is put at $3\frac{1}{7}$, accepted as a right value in those days. The area then is found, correctly, as the product of half the circumference and half the diameter. The circumference of a circle with a diameter 14 equals 44 and the area 154, so more than 12^2 and less than 13^2, from which Franco concludes that the area of such a circle cannot equal a square in number (f. 87r):

assero non esse aequales spatio et numero quadratum et circulum.

So this method disposes of this possibility for the squaring of a circle and there remains a construction (which has to be an approximation). This construction is promised (f. 87r):

Sed neque circulus in quadratum neque rursus quadratus in circulum, nisi cum summa difficultate, et nisi reperta qua id fieri possit artis cuiusdam facultate, quam deo praestante tradituri postmodum sumus, neuter umquam in neutrum transire possunt.

E.

At the end of Book I Franco once more considers the fact that 154 is not a square number.

a) He points out that 154 cannot be found as sum of successive odd integers, no more than any other circle number like πR^2 (f. 87v):

Eadem lex alios quoque circulares numeros includit, . . .

He thus is familiar with the relation $\sum\limits_{1}^{n} (2k-1) = n^2$, already known to the Pythagoreans[12].

b) Is 154 perhaps a square number when calculating with fractions? Franco then calculates $(12 + \frac{5}{12})^2$, which is $\frac{1}{6} + \frac{1}{2} \times \frac{1}{72}$ too much, and $(12 + \frac{1}{3})^2$, which is $1\frac{5}{6} + 2 \times \frac{1}{36}$ too little[13]. He rightly dismisses the possibility of finding a correct value by augmenting $12\frac{1}{3}$ by smaller fractions (f. 87v):

diligentius investigantes, an vel sic per aequalem multiplicationem ad eam quam quaerimus summam pertingamus.

As to such a calculation, using smaller fractions, one can read (f. 88 r):

Quod apertius ostendi valet abaco quam stilo computando potius quam disputando

which proves him to be familiar with the abacus, the counting-board. The final conclusion is (f. 88 r):

Et his demonstratum sit CLIIII quadrari non posse,

to which he adds at once:

quod minime mirabimur, si ad alios numeros, quicumque naturaliter quadrati non sunt, considerationem vertamus. Nam nullus eorum in quadraturam reduci potest, neque integris terminis in se multiplicatis, neque si uncias vel minutias adiciamus, quod mox in II° libello monstrabimus.

BOOK II

A.

Franco classifies the positive integers in this way:

a) Some of them are square by nature, e.g. 4,9 and so on. The side of a square with an area of these values is known.

b) The others are not square. This does not mean that the side of a square with an area of such values is unknown but that it cannot be expressed in numbers. Only an approximation is possible, using fractions[1]. Franco then estimates $\sqrt{2}$, at first by $1 + \frac{1}{3} + \frac{1}{24} + \frac{1}{36} + \frac{1}{144}$ which is too small, and next by $1 + \frac{5}{12}$, being too large. Likewise the root of a circle number can only be approximated. Mentioned is Werenboldus' opinion that $38\frac{1}{2}$ (the area of a circle with a diameter of 7) equals the area of a square with sides 6 plus some fractional parts. We are not informed about those parts but Franco does say that Werenboldus, when multiplying, neglected to multiply the fractional parts. From another text, viz. that by monk B, these fractions are known. He took $6 + \frac{1}{5} + \frac{1}{200}$ as the side of the square with an area of $38\frac{1}{2}$ but he was aware that his result was too large[2].

B.

It is evident now that the area of a circle and a square cannot equal in number (f. 89 v):

manifestum est non esse aequales in comparatione numeri et spatii quadratum et circulum. Restat ergo aut solo spatio comparationem habere aut prorsus aequari non posse.

But the latter statement surely has to be rejected for, according to Franco, the areas of all figures can be compared.

The equality of a square and a circle can thus be based only on a geometric examination. For this two possibilities exist:

a) Of some figures the equality is clear at once[3].

b) In other cases the equality can be illustrated only by means of an intermediate figure. This holds for a circle and a square, the intermediate figure being a rectangle with sides 11 and 14 (in Franco's case).

C.

What is the origin of the rule for finding the area of a circle? At first Franco refers to former geometricians (f. 90 r and v):

> *Peritia, inquam, geometricae disciplinae de inveniendo circuli embado eiusmodi regulam describit, ut medietas diametri in medietatem circuitus debeat protendi.*

So the rule of the circumference equaling $3\frac{1}{7}$ times the diameter is not being discussed. Is it possible now to tell more about it; can the rule be illustrated? Franco's geometric argumentation is shown in both figures on f. 91 v. The circumference is divided into 44 parts with a length of 1 and the 44 sectors are fitted into a rectangle of sides 11 and 14[4]. The conclusion is (f. 91 v):

> *Unde iam nulli potest dubium esse, quin ex hac resolutione subtiles geometrae regulam illam collegerint, qua praecipiunt ad inveniendam aream circuli aut diametri medietate medietatem circuitus extendi, aut totam diametrum IIII^a parte circuitus, aut totum circuitum diametri IIII^a, aut quolibet alio modo ad idem ipsum proficiente.*

Worth mentioning are Franco's next words (f. 91 v–92 r):

> *Quam regulam idcirco compendii causa inventam putamus, ut non semper necesse esset circulum resolvere, quotiens vellemus arealem eius quantitatem scientia tenere, cum praesertim cura ea subiecta saepe considerandum scirent, quae nullatenus resolutionem paterentur. Haec enim nisi in membranis et pelliculis et si qua sunt eiusmodi, exerceri commode non valet.*

So the area can always easily be found by applying the rule while without it the area has to be found experimentally. But this can only be performed with parchment and membranes, by tearing, fitting together and remodelling, if necessary.

Having thus established the equality of the circle and the rectangle the latter has to be transformed into a square (f. 92 r):

> *ac deinceps quemadmodum ex eodem quadrilatero, in quod circulum resolvimus, quadrati species promoveatur curabimus intimare.*

BOOK III

A.

Franco's next step is to complete the squaring of a circle by transforming the rectangle into a square. The first step was an approximation, taking $\pi = 3\frac{1}{7}$. The transformation of a rectangle with sides in proportion 11 to 14 into a square is equivalent to finding a mean proportional. In this, however, Franco does not succeed, having no knowledge of the well-known Greek construction. His method is as follows: First cut off from the rectangle a maximal square. In the figure on f. 93 r $AC = 11$ and $AB = 14$. Take $AE = AC = 11$. After this the remaining strip $FDBE$ has to be divided in such a way that it fits like a gnomon to the square $CFAE$. Franco continues: $EH = EG$ (so $EH = \sqrt{2}$) and thus AH is the side of the square. The construction is not completed. But he has measured $HK = BI = HA$ and says, correctly, that $KIBH$ equals $LDBH$. So he is suggesting that the distance between HK and BI is $\sqrt{2}$ too and that the strip $KIBH$ fits on AH. In this construction $11 + \sqrt{2}$ is an approximation for $\sqrt{154}$, $(11 + \sqrt{2})^2 = 154.11$. Where he started from a circle with a radius of 7, now, after completing the squaring, his value for π can be calculated as $\dfrac{(11 + \sqrt{2})^2}{49}$ or $3.1451 \ldots$ so a good 0.1% too large[1].

In Franco's opinion, where $\sqrt{154}$ is irrational, the squaring of the circle can only be performed geometrically which means here: by an approximative construction.

B.

Franco returns once more to the irrationality of $\sqrt{154}$. In Book I he showed that $\sqrt{154}$ cannot be expressed by fractions. He still has a second argument, an indirect proof. Supposing, he says, that $\sqrt{154}$ could indeed be expressed by fractions, then the squaring of the circle could be calculated in integers too (f. 93 v):

Dico enim si quadrato aequa latera constituuntur in minutiis, nihilominus et in numero aequalia constitui. Nam quicquid multiplicant minutiae, idem numero quoque multiplicatur integro.

His argumentation is not wholly clear but what he means is something like this: if p denotes the least common multiple of the occuring denominators then all numbers become integers if the unit of length is taken p times smaller, and then a circle with radius $7p$ could indeed be squared in integers. Franco rejects this conclusion. He has two examples.

a) A square with sides $\frac{3}{2}$ has the area $\frac{9}{4}$: *quadrantes IX*. Change over to a unity twice as little and the same 9 is found from 3^2, so in integers.

b) Let $a_1 = 6\frac{1}{2}$, $b_1 = 7$ be line-segments of which one has a fractional part and let $a_2 = 13$, $b_2 = 14$ be their multiples without fractional parts. a_1 and a_2 are sides of squares, b_1 and b_2 diameters of circles. Now, as Franco shows by calculation, the relation

$$\frac{a_1^2}{\pi(\frac{1}{2}b_1)^2} = \frac{a_2^2}{\pi(\frac{1}{2}b_2)^2}$$

is true, so the proportion of the areas remains, when changing over from fractions to integers, that is to say: when changing the scale. Franco's conclusion is (f. 94 r):

Quid igitur in minutiis requirere labores, quod numquam sinit invenire natura?
Sed etsi ratione calculandi abacique peritia subtilissima illic reperiri convinceretur, quid hoc mensoribus et studio geometricali conferret?

The last sentence in fact weakens the preceding argumentation but is typical of the ideas on geometry at that time viz.: what does geometry profit by those very small fractional parts?

C.

In addition to what has been said before, Franco discusses the irrationality of $\sqrt{2}$, $\sqrt{3}$, $\sqrt{5}$ and so on. As to $\sqrt{2}$ he says (f. 94 v):

Ad hunc modum itaque, si comparatio dupli et simplicis quadrati in minutiis constaret, nihilominus in integris terminis eandem facile esset invenire. Sed quis umquam in numeris deprehendit dupli et simplicis quadrati comparationem?

The line of thoughts is as before. As an example he has: $a_1 = 2$, $b_1 = 2\frac{1}{2}$, $a_2 = 4$, $b_2 = 5$. If the proportion of the sides is kept then also that of the areas and if the latter has the value 2 then the proportion of the sides could be found in integers. But to find that proportion is impossible, even if Regimbold, Gerbert and Racechinus took it to be $1 + \frac{5}{12}$, which value Franco demonstrates to be incorrect. For in geometry $\sqrt{2}$ is known, viz. as a diagonal of a square with sides 1. This means to Franco that if the diagonal of one square is taken as side of another, then the area of the latter is double that of the former. Now take a square with side 5 and area 25 and another with side $7\frac{1}{2}$, which is 5 times $1 + \frac{5}{12}$, and area $49 + 1 + \frac{1}{6} + \frac{1}{144}{}^2$, then the latter appears to be more in area than double 25. This could be found too, says Franco, from $17^2 \neq 2$ times 12^2 which shows $\frac{17}{12}$ to be an incorrect value for $\sqrt{2}$.

BOOK IV

This book is a systematic dissertation of the ways in which areas of different figures can be compared mutually.

A.

Figures equal in area can have:

a) The same shape. This, and the condition of equal area, means congruency.

b) Different shapes, e. g. a square, a rectangle, a circle, a triangle. The dissertation only refers to b). Two figures differ minimally in shape if one dimension, length or width, is equal (f. 97 r):

Minus distant, quae vel longitudine sola vel latitudine concordant. Inter illas vero habetur maior distantia, quae utroque sunt diversae. Sciendum quoque, quod minus distantes amplius distantium in medio versantur. Quamobrem ad illas, nisi per has, nulla suppetit facultas transeundi.

What differences are possible? Two figures, equal in area, may differ only in length, in width, or in both, and in the latter case it even is possible that both length and width of the one figure are larger than those of the other[1]. Let a, b and c stand for equal, larger and smaller respectively, and d and e for length and width. Franco now distinguishes 9 different possibilities, given in a list on f. 97 v[2]. There are four groups, viz. a) 1; b) 2 and 3; c) 5, 6, 8 and 9; d) 4 and 7, which are discussed succesively and illustrated by figures.

a) Figure A (= no. 21). The two triangles are equal in length, width and area[3].

b) A rectangle and a triangle of equal length (height, figure B = no. 22), and of equal width (figure Γ = no. 23). The elucidation in the text refers to the letters in the figures, which are not always correct. In figure B the left part of the triangle can be put, upside down, on the right part to get the rectangle (f. 98 v):

dico ex triangulo quadrilateri formam procreatam esse.

Something similar is done in figure Γ (f. 98 v):

ex IIII laterum spatio triangularem formam procreasti⁴, . . .

c) Franco discusses the next four cases together (f. 98 v):

Nunc de his, quae sunt vel longitudine vel latitudine maiores minoresve, consideremus. Harum IIIIᵒʳ sunt modi, . . .

to which belong the figures Δ (= no. 28) and ⊖ (= no. 29). These cases treat the comparison of a rectangle and a square, equal in area, which has already been done in Book III. The subdivision is:

5, the rectangle is larger in length than the square,
6, the rectangle is larger in width than the square,
8, the rectangle is smaller in length than the square,
9, the rectangle is smaller in width than the square.

d) Two cases finally remain (f. 99 r):

nunc ad reliquas duas, quae adhuc de IX supersunt, complexiones animi vertamus intentionem.

Figure ∈ (= no. 30) belongs to case 4: the triangle is larger than the square in length as well as in width, while the reverse applies to case 7. These are the most difficult cases in which to show the equality in area. From the protruding parts something has to be cut off and to be fitted elsewhere. For this reason, says Franco, an intermediate figure is necessary.

B.

After this Franco returns to the squaring of the circle (f. 99 v):

Nunc igitur ad circuli revertamur expositionem dicturi, cui praecipue inter IX complexiones deputetur.

From the previous dissertation his intention is clear, the squaring of the circle has to be performed by an intermediate figure. For length and width of the circle (i.e. the diameter) are larger than those of the square. The book ends with the conclusion (f. 100 r):

et his rationibus demonstratur, cur circulus in quadratum per se reduci non possit.

BOOK V

This book contains some complements on the squaring of the circle.

A.

Is it possible to transform a square into a circle or, for only this is being discussed, can a square be transformed into a rectangle with sides in proportion 11 to

241

14^1? This is shown in the figure on f. 101 r: the transformation of a square $BDCA$ of sides 14 into a rectangle $BOML$. From the text it is clear that $FH = \sqrt{2}$ and $HI = AC = 14$. Then $HM = CH = 3 - \sqrt{2}$ and the sides of the rectangle are $11 + \sqrt{2}$ and $17 - \sqrt{2}$, in proportion 1.255 ... instead of $\frac{14}{11} = 1.273$ ... and the area is 193.5 ...[2].

B.

The construction of the square by an intermediate figure now being known, it is possible to draw at one go the equal square of a given circle. The text refers to a figure (f. 101 v):

Hanc autem rationem subiecta figura demonstrat.

The figure is missing here in the Vatican manuscript but it is the one on f. 108 v, at the end of Franco's treatise. The construction is performed as follows. In a circle with radius 7 are drawn two mutual perpendicular diameters AB and $\Gamma\Delta$. In a unit square, drawn at the centre, Franco successively constructs $\sqrt{2}$ *(AC)*, $\sqrt{2} - 1$ *(BC)* and $\frac{1}{2}\sqrt{2} - \frac{1}{2}$ *(EC)*. Mark on the four radii then a length 6 plus $\frac{1}{2}\sqrt{2} - \frac{1}{2}$ and draw the square with sides $11 + \sqrt{2}$[3].

C.

What follows next is beside the subject: some remarks on the importance of the circle, that is, to determine the exact noontide with the help of a gnomon[4]. The text ends (f. 102 r):

Cum igitur tanta sit dignitas circuli, iure non ipse ad quadratum, sed potius ad ipsum referri videtur quadratus.

D.

In figure 32 on f. 108 v[5] Franco attempts to find the area of the circle segments outside the square. Previously he remarks that this problem is so difficult that most geometers cut off those parts and weigh them (f. 102 r):

His etenim finitis tandem ad excessuras veniendum, quarum in tantum difficilis videtur ad inveniendum mensura, ut plerique resectas particulas in trutinam mittant, libraeque lancibus examinent aequalitatem.

This again is a good example of the ideas on geometry in those days. To Franco the area of such a segment, in a circle of radius 7, has the value $3\frac{1}{2}$. The reasoning is not quite clear. The circle and the subscribed square differ 42 in area, viz.: $14^2 - 154$. So one of the parts outside the circle is $10\frac{1}{2}$ in area. In the figure 40 this is part I. Franco next tries to find how much one wedge *(angulus)* of the larger *(maior, exterior)* square (so area I) is exceeding one of the smaller *(minor, interior)* square (area II) (f. 102 v):

Ab angulo minore illud quo maior superat praecides, idque ipsi minori comparabis, et bis tantum invenies. Quare minoris anguli continentia tribus semis particulis constat.

He says therefore: the area of I, reduced by II, equals double the area of II.

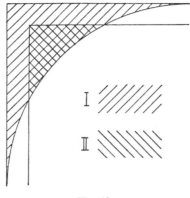

Fig. 40

From this indeed follows the value $3\frac{1}{2}$ for the area of II. But the reason why I — II equals 2 times II is not explained[6].

E.

Starting from a rectangle with sides 11 and 14 and a square with sides $11 + \sqrt{2}$ ($= 11 + \frac{17}{12}$), which are concentric and have sides, that are parallel two by two, the total area of the parts of the square outside the rectangle, as well as the total area of the parts of the rectangle outside the square, equals (f. 102v):

XI pedes vel asses et totidem septunces paulo plus.[7]

Replacing the rectangle by a circle of the same area, the parts of the circle equal 14, as found in D, instead of $17\frac{5}{12}$ (f. 102v):

Sed hic forte aliquis in animi rationem ducatur, cum his figuris circulus adaequetur, cur ille et quadratus minus se invicem excedant vel cur amplius istae sese supervadant, quando omnium eadem quantitas invenitur.

This is not surprising as Franco shows by an example, viz. one rectangle with sides 3 and 4, another one with sides 2 and 6 and a third one with sides 1 and 12, which are equal in area but have different mutual excesses, as easily can be seen in the figure. These excesses are 4 at the rectangles 3 by 4 and 2 by 6, 8 at the rectangles 3 by 4 and 1 by 12 and 6 at the rectangles 2 by 6 and 1 by 12[8].

BOOK VI

The maior part of this book is devoted again to irrational numbers. After a survey of various divisions of figures, Franco discusses the possibility of finding $\sqrt{n}$ (n of the form $2m^2$) geometrically. He ends with an approximative construction of transforming a rectangle into a square, so in fact a construction of $\sqrt{n}$.

A.

A division of figures can be:
a) Naturally.
b) Artificially.

To a). There are two possibilities, viz. by drawing a line from one side to the opposite one, or through the centre. In the latter case the two parts are equal and therefore, as Franco says, without proportion. In addition to this, something more is mentioned on proportions. If, for instance, half a figure is compared with its third part, the proportion is $\frac{1}{2} : \frac{1}{3}$ or $3 : 2$, '*proportio sesquialtera*'[1].

To b) This, for instance, happens by drawing in a square or circle lines not running through the centre; the parts are unequal.

Franco does not refer to this survey of divisons any more. Maybe it is intended as an introduction to what follows: the division of a square.

B.

Franco starts with a square with sides 12 and in this, with one vertex in common, squares with sides 10, 8, 6, 4 and 2 are drawn. Taking those sides as diagonals of new squares he finds the sides of squares with areas 2, 8, 18, 32, 50 and 72 and so he is able to construct $\sqrt{2}$, $\sqrt{8}$, $\sqrt{18}$, $\sqrt{32}$, $\sqrt{50}$ and $\sqrt{72}$[2]. Further to this he queries how to find $\sqrt{3}$, $\sqrt{5}$, $\sqrt{7}$ and so on. He confines himself to $\sqrt{2}$ and $\sqrt{3}$ and tries to investigate their numerical proportion to $\sqrt{1}$ and $\sqrt{4}$. To this, Franco now sucessively considers the three kinds of proportions distinguished at that time.

a) A multiple, '*proportio multiplex*'. It is clear that the successive proportions of $\sqrt{1}$, $\sqrt{2}$, $\sqrt{3}$ and $\sqrt{4}$ cannot be multiples. For from $\sqrt{2} = k \sqrt{1}$, $\sqrt{3} = m \sqrt{2}$, $\sqrt{4} = n \sqrt{3}$ follows $\sqrt{4} = nmk \sqrt{1}$[3]. But $\sqrt{4} = 2 \sqrt{1}$ so nmk should equal 2 which is impossible for 2 is the smallest multiple (f. 104v):

Quare ipsam impossibile est ex multiplicibus constare.

So two other possibilities remain, or else the proportion is irrational (f. 104v–105r):

Restat ergo, ut superparticulari proportione ad se conferantur, aut superpartiente, aut nulla.

b) A proportion like $\dfrac{m+1}{m}$ (m is a positive integer) is called '*proportio superparticularis*'[4]. It is named after the denumerator with the prefix '*sesqui*', for instance $\frac{3}{2}$, *proportio sesquialtera*, $\frac{4}{3}$, *proportio sesquitertia* and so on.

The fact that 2 is a product of the three successive proportions $\frac{4}{3}$, $\frac{5}{4}$ and $\frac{6}{5}$ leads Franco to the supposition that $\sqrt{2} = \frac{4}{3} \sqrt{1}$, $\sqrt{3} = \frac{5}{4} \sqrt{2}$, $\sqrt{4} = \frac{6}{5} \sqrt{3}$, for then indeed $\sqrt{4} = 2 \sqrt{1}$. But is this true? It would be easy to verify that $\sqrt{2} \neq \frac{4}{3}$ and $\sqrt{3} \neq \frac{5}{3}$ but Franco has a different – and faulty – reasoning. If the foregoing were correct then also, he says, should hold $\sqrt{5} = \frac{7}{6} \sqrt{4}$, $\sqrt{6} = \frac{8}{7} \sqrt{5}$, $\sqrt{7} = \frac{9}{8} \sqrt{6}$ and $\sqrt{8} = \frac{10}{9} \sqrt{7}$ and thus $\sqrt{8} = \frac{10}{4} \sqrt{2}$, which leads to a contradiction for $\sqrt{8} = 2 \sqrt{2}$ (from 8 = 4 times 2) and $\frac{10}{4} \neq 2$ (f. 105r):

Unde amplius conficitur quam proportio dupla.

Likewise his next example: let 3 be the length of the first line-segment, then the second one is 4 (viz. $\frac{4}{3}$ times the first) and so on, and the eighth is 10. But, as argued before, this should equal double the second one.

c) The only possibility remaining is a '*proportio superpartiens*', of the form $\frac{m+k}{m}$ (*m* and *k* are positive integers, $k \neq 1$, $k < m$)[5]. Also in this case of so many more possibilities no solution, as Franco says, can be found[6]. Nevertheless, he is going to discuss one of them in an elaborate way (f. 105 v):

Quid igitur agendum? Numquid credendum cunctis his remotis generibus rem ullam comparari?

Sed forte erit, cui haec comparatio per adiectionem et diminutionem fieri videatur hoc modo: . . .

What follows is surely the most remarkable part of Franco's treatise, viz. a recurrence relation for $\sqrt{n}$. We are not informed about its origin[7]. Franco successively calculates (f. 105 v):

$$\sqrt{2} = \left(\frac{1 - \frac{1}{6}}{2} + 1 \right) \sqrt{1}; \quad \sqrt{3} = \left(\frac{1 - \frac{1}{10}}{4} + 1 \right) \sqrt{2}; \quad \sqrt{4} = \left(\frac{1 - \frac{1}{14}}{6} + 1 \right) \sqrt{3}$$

Atque hoc ordine observato, ut videlicet in diminutione a VI incipientes eam semper auferamus partem, quae denominata est a summa quaternario auctiore (ut X a VI, XIIII a X quaternario auctiores existunt), rursus autem in adiectione a IIᵃ parte progredientes illas semper adiciamus, quae a paribus appellationem trahunt continuo ordine se sequentibus: eo modo omnes procul dubio quadrati lineas inveniri.

He thus suggests that the foregoing holds generally, so:

$$\sqrt{n+1} = \left(\frac{1 - \dfrac{1}{4n+2}}{2n} + 1 \right) \sqrt{n}$$

He has a numeric example, taking as first length, instead of $\sqrt{1}$, 1209600. Then

$$\ell_2 = \left(\frac{1 - \frac{1}{6}}{2} + 1 \right) 1209600 = 1713600, \quad \ell_3 = \left(\frac{1 - \frac{1}{10}}{4} + 1 \right) 1713600 = 2099160 \text{ and}$$

$$\ell_4 = \left(\frac{1 - \frac{1}{14}}{6} + 1 \right) 2099160 = 2424030, \text{ which should equal 2 times } 1209600 =$$

2419200. They differ by 4830 which Franco corrects diminishing ℓ_2 by $\frac{6}{30}$ of 4830, ℓ_3 by $\frac{16}{30}$ of 4830 and ℓ_4 by $\frac{30}{30}$ of 4830. These corrections are of proportion 6 : 16 : 30, chosen by Franco because of the numbers 6, 10 and 14 in the formulas for ℓ_2, ℓ_3 and ℓ_4; $16 = 6 + 10$, $30 = 6 + 10 + 14$. The corrected values are $\ell'_2 = 1712634$, $\ell'_3 = 2096584$ and $\ell'_4 = 2419200 = 2$ times ℓ_1. The whole calculation can be read on f. 105 v and 106 r and is given, once more, in a diagram on f. 106 v. Franco, too, motivates his start from the number 1209600 (f. 106 v):

Quibus tam promotis necessario usi sumus eo, quod non sunt inventi minores, qui tum integri in partes VI et X ac XIIII dividerentur, tum diminuti IIᵃᵐ, IIIIᵃᵐ, VIᵃᵐ quoque reciperent portionem, quorum etiam exuberantia tricies auferri non recusaret.

So he has taken $6 \times 10 \times 14 \times 2 \times 4 \times 6 \times 30 = 1209600$, to avoid fractions[3].

After all this, the result is verified. Does $(\ell'_2)^2$ indeed equal double $(\ell'_1)^2$ or is $\dfrac{1712634}{1209600}$ a correct value for $\sqrt{2}$? Franco has: $(\ell'_2)^2 < 2 \times (\ell'_1)^2$, so again he failed to find a rational value for $\sqrt{2}$[9] (f. 107 r):

Unde manifestum est, quod hae lineae sicut iuxta proportiones inaequalitatum, id est multi-plicis et superparticularis, tum ceterorum metiri non possunt.

In spite of this negative conclusion the recurrent relation is worth further examination. One can reduce:

$$\frac{\ell_{n+1}}{\ell_n} = \left(\frac{1 - \dfrac{1}{4n+2}}{2n} + 1 \right) = \frac{8n^2 + 8n + 1}{4n(2n+1)}$$

and next:

$$\left(\frac{\ell_{n+1}}{\ell_n} \right)^2 = \frac{n+1}{n} \times \frac{64n^3 + 64n^2 + 16n + \dfrac{1}{n+1}}{64n^3 + 64n^2 + 16n}$$

so the approximation:

$$\sqrt{\frac{n+1}{n}} \approx \frac{1 - \dfrac{1}{4n+2}}{2n} + 1$$

becomes closer for growing values of n[10].

The latter result corresponds with the well-known Babylonian approximation of roots: If b_1 is an approximation for $\sqrt{a}$, then $\frac{1}{2}(b_1 + \frac{a}{b_1}) = b_2$ is a next, and closer approximation and $\frac{1}{2}(b_2 + \frac{a}{b_2}) = b_3$ is a third one and so on. Writing $\sqrt{\frac{n+1}{n}}$ as $\sqrt{1 + \frac{1}{n}}$ and putting $b_1 = 1$, we get $b_2 = 1 + \dfrac{1}{2n}$ and $b_3 =$

$$= \frac{1 - \dfrac{1}{4n+2}}{2n} + 1 [11].$$

C.

Finally, having rejected the possibility of finding $\sqrt{n}$ numerically, as a rational number, Franco attempts to arrive at a geometrical solution (f. 107 r):

Quare prorsus ab arithmeticis supputationibus recedentes proportionem hanc sive potius mensuram per artem geometricam, quam non usque quaque arithmeticae putamus addictam, immo in quam pluribus propriam exerceri considerationem, perquirere studeamus.

As in B, he starts again from a square with sides 12. As can be seen in the figure 38 on f. 107v, he avails himself of the well-known method of halving a square by its diagonal, trying to choose the diagonals *(lineae)* in such a way that the areas of the triangles between them and the square-angle have definite values (f. 107 r):

si in III aut IIII aut V aut alias deinceps partes quadrati spatium ab angulo partiri volueris, . . .

Franco uses three similar methods. First (f. 107 v) in a square with sides 12 he assumes smaller ones with sides of even numbers ($a = 2$, 4, 6, 8, 10, 12) and which are halved by their diagonals. To side a corresponds a diagonal $d = a \sqrt{2}$ including a rectangular triangle with area $\frac{a^2}{2}$. This produces the areas 2, 8, 18, 32, 50 and 72.

Next (f. 107 v–108 r) he starts from diagonals equaling one of the 6 different square-sides, so being even in length. These diagonals d are inserted between the lineae previously drawn in the figure on f. 107 v. They represent diagonals of squares with irrational sides $a = \dfrac{d}{\sqrt{2}}$. Franco is right in saying that those diagonals include rectangular triangles with areas $\frac{d^2}{4}$, so 1, 4, 9, 16, 25 and 36 (f. 108 r):

Igitur hoc ordine omnis haec ratio procedit, ut principales lineae ipsa tetragona dividentes medietatem, lineae vero a lateribus sumptae IIII^{am} eorum semper includant partem.

Also in his third attempt he starts by dividing a square by its diagonal (f. 108 r) as shown at the upper right of the figure on f. 107 v[12]. The diagonal of the smaller square (with area 1) is taken as side $AE = EB$ of a square, so $AE = \sqrt{2}$. Then the diagonal AB of this square has a length of 2 and the triangle ABE an area 1, so the '*quantitas eiusdem assis*' (f. 108 r). He then takes AB as side $CE = DE$ of a next square. The inscribed rectangular triangle CED has an area of 2, so half of the first square at the lower left. Franco ends at the third square – to which would belong a triangle-area of value 4 – for again he does not arrive at the next integer (i. e. 3) (f. 108 r):

quia arte qua nunc et prius usi sumus nihil proficere possumus.

Thus Franco has not succeeded in expressing each positive integer n as the area of a half-square.

Finally a solution is introduced by (f. 108 r):

Quid ergo agendum? An cedendum difficultati? An a labore cessandum? Minime. Quaerenda enim ars quadrandi, per quam non solum III, sed V et VI et VII et quaecumque adhuc repertae non sunt, metiri possunt. Haec autem ars ita se habet.

So a construction is proposed to transform each rectangle into a square, the geometrical equivalent of finding $\sqrt{n}$.

AB is drawn so (figure 39 on f. 108 r) that $AB = AP$[13]. Let C be the centre of AF then CP is the side of the square. As in Book III, when squaring the rectangle of sides 11 and 14, $CE = CP$ and $FG = CE$ are drawn. Now $ABCD$ and $CEGF$ are equal in area. So what is intended is, that the strip $CEGF$ fits along CP and so the part above AB can be fitted, divided like a gnomon, around the square. It is clear that this does not hold for the proposed construction. If a and b are sides of the rectangle $(a < b)$ then Franco's squaring is equivalent to

$$ab \approx \left(a + \frac{b - a}{2}\right)^2$$

This equals $ab + \left(\dfrac{b - a}{2}\right)^2$, so the approximation becomes closer for smaller values of $b - a$.

Franco starts from a rectangle with area 6 and "constructs" a line segment with length $\sqrt{6}$. Taking this as diagonal of a square he finds $\sqrt{3}$. In a similar way he finds $\sqrt{5}$ from a rectangle with area 10. However, he does not mention the proportion of length and width of the rectangle he started from.

What Franco does, is in fact equivalent to the very old approximation:

$$\sqrt{a^2 + b} \approx a + \frac{b}{2a}$$

which is part of the general inequality[14]

$$a \pm \frac{b}{2a \pm 1} < \sqrt{a^2 \pm b} < a \pm \frac{b}{2a}$$

Both limits were used as approximation of $\sqrt{a^2 \pm b}$; they are quite correct for numbers differing but little from square-numbers.

With this Franco ends (f. 108 v):

Et quaecumque lineae requirantur, sic poteris invenire.

CONCLUSION

Apart from his attempt to perform the squaring of a circle, having first argued that this should be done by an intermediate figure, Franco's treatise is mainly devoted to the irrationality of square-roots. He tries to find a good numerical approximation. Besides that, he also gives an approximate construction.

His very elaborate treatise is of historical interest, dating from a period of which but a few mathematical texts are known. What he conveys is a very welcome addition to our knowledge of mathematics in that time.

Moreover, we think that Franco himself has to be praised. He appears to be very well acquainted with the mathematical knowledge of his time. He proceeds very systematically when solving the posed problems, making a graduation to find the best way, thus avoiding working at random. He also tries to solve the related problems, as the transformation of a square into a circle, and the calculation of the mutual excesses of a circle and a concentric equal square. He is well aware of the fact that $\sqrt{2}$, $\sqrt{3}$, $\sqrt{5}$, ... cannot be calculated as fractions but can be found geometrically. All this is surely no mean achievement for a man who could have no knowledge of Greek mathematics.

NOTES

(Prologue)

[1] A. Mai, pp. 346–348. For full titles see the list of literature.

[2] J. P. Migne, CXLIII, 1373–1376.

[3] N. Bubnov, pp. 384–385.

A treatise on the squaring of the circle by Franco of Liège, of about 1050 248

[4] C. Le Paige, pp. 9–11; M. Cantor, I, pp. 876–878. Their informations, however, are not complete, owing to the imperfectness of Winterberg's edition. M. Clagett (b), in a note on p. 15.

[5] D. E. Smith, I, p. 197.

[6] J. E. Hofmann (b), p. 83.

[7] F. Sassen, p. 51.

(Introduction)

1.

[1] G. H. Pertz, pp. 358–359. An almost identical text is in: *Germanicarum rerum* . . ., p. 125.

[2] C. Le Paige, p. 11.

[3] S. Bormans, p. 259. B. Lefebvre, p. 112, has: 'with exception of the date, the unknown seat of Hermanus, and partly Aristotle's quotation, it is a free translation of Sigebert's comment'. To this may be added: from as early as 1265, so also in Jean d'Outremeuse's days, Bonn was the residence of the clerical electors of Cologne, so of the archbishop. Although '*castra Bonnensia*' is the old Latin name of that town, '*Bonneburgensis*' may well relate to it.

[4] J. Trittenheim, p. 78 v.

[5] Part VII, p. 138.

[6] B. Pezius, IV, part III, book I, p. 22.

2.

[1] P. Tannery (d), p. 489.

[2] Exact description see M. Folkerts, pp. 5–7, and N. Bubnov, introduction pp. 55–58.

[3] P. Tannery (d), pp. 533–536.

[4] C. Winterberg, p. 138. Data which are far more indefinite are to be found on p. 141. The codex is described by E. Narducci, pp. 112–119 and by M. Folkerts, pp. 12–14.

[5] The codex is described by Samuel A. Ives, pp. 44–45.

[6] E. A. Moody and M. Clagett, pp. 174–227.

[7] E. A. Moody and M. Clagett, pp. 64–75.

[8] M. Clagett (a), pp. 73–175.

[9] A. Birkenmajer, p. 166 of the French edition.

[10] A. Birkenmajer, p. 166. The identity of the Sorbonne-manuscript with Crawford 1.27 is assured by the Incipits of the second and the last but one page.

[11] *Catalogue of the Crawford Library of the Royal Observatory, Edinburgh*, Edinburgh 1890, p. 495.

[12] Annotation within the manuscript.

[13] E. G. Forbes, pp. 7–13.

[14] L. Thorndike, column 569.

[15] M. Clagett (b), p. 583. From this mention we can assume that in the 15th century it was well-known that Franco wrote a treatise on squaring the circle.

3.

[1] M. Cantor, I, p. 877, questions the correctness of the comment about the dedication to the archbishop of Cologne.

[2] *Allg. Deutsche Biogr.* XII, pp. 130–131.

[3] M. Cantor, I, pp. 847 ff.

[4] F. Sassen, pp. 19, 50. A. Clerval (a), pp. 35, 38; on p. 124 also the comment that Fulbert was taught mathematics by Gerbert.

[5] A. Clerval (a), p. 63; on p. 90 also is explained that Franco very probably was a pupil of Fulbert.

[6] G. Kurth, pp. 282–285. J. Daris, pp. 279–314.

[7] R. Huysmans, pp. 40, 41.

[8] F. Sassen, p. 37. *Hist. lit. de la France VII*, pp. 499–505, mentioning Gozechin as Adelman's successor in 1050.

[9] *Biogr. nat. de Belg.* I, pp. 62–63; also the information that Franco was a pupil of Adelman. This is also mentioned in *Hist. lit. de la France VII*, p. 544. A. Clerval (a), p. 90, has: 'Francon . . . occupant en 1047 la chaire d'écolâtre après Gozechin et Valcher.' In that case Franco could

not have been Gozechin's immediate successor. Moreover, the year 1047 seems to be too early, others have 1066.

10 See N. Bubnov. He obtained his mathematical knowledge in Spain. He also found mathematical manuscripts in the library at Bobbio.

11 Edited by N. Bubnov, pp. 41–45. Cf. M. Cantor, I, p. 859. About Adalbold see also *Nieuw Nederlands biografisch woordenboek* IV, p. 12; C. Le Paige, pp. 7, 8; R. Huysmans, p. 40.

12 P. Tannery (d) and (e). According to A. Clerval (a), pp. 62, 63, Regimbold also studied at Chartres.

13 P. Tannery (d), pp. 492 and 531. The name may be Ratechitius or Razeginus as well. Is he perhaps identical with the named Gozechinus? G. Kurth, p. 285, mentions a Rasquin, pupil of Radolf's. Rasquin and Racechinus could be the same name but Regimbold was older than Radolf so a friend of his could not have been Radolf's pupil.

14 P. Tannery (d), p. 492. See also the coming discussion in Book II.

15 *Biogr. nat. de Belg.* VII, pp. 267–269.

16 C. Winterberg is quite wrong when writing, on p. 139: 'Traktat des Dominikaners Franco von Lüttich aus der Zeit Otto's III.' For Otto III died in 1002, so before Herman became archbishop of Cologne, and the Order of the Dominicans or Black Friars was founded only in 1216.

17 See *Biogr. nat. de Belg.* and A. Quételet, p. 34 and also J. Daris, p. 669.

4.

1 See P. Tannery (b), pp. 94, 95 and the rest of his paper, and J. E. Hofmann (b), I, p. 80. Detailed data also in M. Cantor, I.

2 P. Tannery (a), (b) and (d).

3 Edited by P. Tannery (d).

4 Edition and analysis of the sources by M. Folkerts.

5 P. Tannery (d), p. 513. K. Menninger, I, p. 171. Still other names were in use, so e.g. $\frac{1}{8}$, $sescuncia, \frac{1}{96}$, *drachma*, and so on. See L. Karpinski, p. 125 and G. Friedlein, pp. 33–46.

6 P. Tannery (d), p. 518.

7 P. Tannery (d), p. 527.

8 P. Tannery (d), pp. 527, 528; also p. 497. A triangle, without specification, was always an equilateral one.

9 P. Tannery (d), p. 528.

10 P. Tannery (d), p. 532. M. Cantor, I, p. 861.

11 J. E. Hofmann (a).

12 P. Tannery (d), pp. 515, 524.

13 M. Cantor, I, p. 551.

14 The number of lines in the quotations of Franco correspond to the present edition.

15 The number of lines in the passages of the *Geometry II* correspond to those stated in the edition by M. Folkerts.

16 For example, *geometricae disciplinae peritissimi*, Franco 76, 148 ≈ "Boethius" line 607; *dubitationis obscuritate … exempli luce*, Franco 87, 83 ≈ "Boethius" 324 sq; *Pythagorica subtilitas*, Franco 90, 10 ≈ "Boethius" 447 sq; *Patricius* (= *Symmachus*), Franco 90, 21 ≈ "Boethius" 2.

17 As in the correspondence between the Rhenish scholastic relations to *Geometry I* by Ps. Boethius but not to *Geometry II* can be realized, the *Geometry II* was possibly written after 1025 and before Franco's treatise appeared. Up to now the *Geometry II* was attributed without closer determination to the first half of the 11th century.

(Book I)

1 M. Clagett (b), p. 607.

2 Or do we have to suppose that this was indeed the intention? It was within his reach to verify the proposition of an exterior angle equaling the sum of two interior angles.

3 P. Tannery (d), pp. 527, 528.

4 P. Tannery (e); also (d), p. 522, where Regimboldus is writing: '*dilectus noster domnus Gvazzo*'.

5 About these figures, see P. Tannery (d), pp. 537, 538. On Radolf's demonstration, the same: pp. 498, 531, 532. Franco gives the figures with the proposition on the exterior angle to which

only figure A, of Wazo–Regimbold's relates. The other three figures, regarding also P. Tannery, certainly are again on the sum of the interior angles.

[6] M. Cantor, I, pp. 641, 642.

[7] M. Cantor, I, pp. 544, 642, 876. P. Tannery (d), p. 511. M. Cantor, I, p. 877, discussing Franco, does mention the methods a) and c) but not b).

[8] M. Cantor, I, p. 591. See A. J. E. M. Smeur (c), pp. 249–253.

[9] M. Cantor, I, p. 877: 'Ferner hält Franco selbst $\frac{9}{10}$ des Durchmessers für die Seite des dem Kreise flächengleichen Quadrates, rechnet also mit $\pi = \left(\frac{9}{5}\right)^2 = 3{,}24$.' This is not correct. Cantor refers to p. 187 of Winterberg's edition but that text is not by Franco.

[10] Or do we have to suppose that the case of a triangle with a base and height indivisible by 6, so e.g., 8 and 9, was beyond Franco's ability? We don't know if he knew how to calculate the area of a triangle; it appears nowhere in his treatise. It also has to be observed that *tetragonus* did not mean some quadrilateral, but a square; M. Cantor, I, p. 207. See also Franco, at the beginning of E, Book I.

[11] Such a division into unit-squares is mentioned in Book II, in the discussion of the equality of a rectangle and a circle; as to a circle, the division of course is impossible. See the figure on f. 90r.

[12] M. Cantor, I, p. 162. The relation has come down to us by Nicomachus and Boethius. It can easily be illustrated, by gnomons.

[13] Franco's calculation shows that he was familiar with the relation $(a + b)^2 = a^2 + 2ab + b^2$.

(Book II)

[1] So e.g. the diagonal of a unit-square is the side of a square with area 2.

[2] About 'monk B', see P. Tannery (d), p. 511 and also M. Cantor, I, p. 876 and B. Lefebvre, p. 110. He was squaring a circle with area $38\frac{1}{2}$. After Franco's treatise the Vatican manuscript continues with '*Incipit liber de eadem re*', edited also by C. Winterberg (pp. 183–187) and by P. Tannery (d), no. 9. The area of a circle with diameter 7 is found by multiplying $\frac{22}{7} \times \frac{7}{2} \times \frac{7}{2}$ = '*XXXVIII. s. ... quadratum circuli satagimus coaequari spatio, ex ipso diametro VI pedum quintaeque septimi pedis et ducentesimae partis quantitate latus quadrati facimus*', calculated as $\left(6 + \frac{1}{5} + \frac{1}{200}\right)^2 = 38 + \frac{2}{5} + \frac{1}{10}$, so an approximation. Noteworthy is the fact that the fractions are not duodecimal. $6^2 + 2 \times 6 \times \frac{1}{5} + 2 \times 6 \times \frac{1}{200} + \left(\frac{1}{5}\right)^2$ is exactly $38\frac{1}{2}$; $(6.205)^2 = 38.502025$. The value of $\frac{(6.205)^2}{(3.5)^2}$ is $3.1430224\ldots$, so less than $\frac{1}{2}°/_{00}$ too much. Remarkable also is the notation $38 + \frac{2}{5} + \frac{1}{10}$ for $38\frac{1}{2}$.

[3] E.g. a square with a side of 6 and a rectangle with sides 4 and 9, as given before.

[4] Length and width are changed in the figure of the Vatican manuscript. Franco's method is correct as a first approximation of the area of a circle. About analogue methods see M. Cantor, I, p. 656 and A. P. Juschkewitsch, p. 161.

(Book III)

[1] So 'monk B' has a better approximation.

[2] Apparently again $\left(7\frac{1}{12}\right)^2 = 7^2 + 2 \times 7 \times \frac{1}{12} + \left(\frac{1}{12}\right)^2$, '*secundum tetragonicam multiplicationem*' (f. 95r).

(Book IV)

[1] E.g. a circle with a diameter of 14 and a square with sides $11 + \sqrt{2}$.

[2] He points out that there are 21 possibilities for solid bodies of equal volumes; in addition to d and e there is also f, the height. The statement can easily be checked; to a (equal) belong the combinations def, de, df, ef, d, e, f, which also holds for b (larger) and c (smaller).

[3] In the figure both triangles are even congruent; this is not necessary.

[4] To b) also belong the last four figures (24–27) on f. 98v, which are not mentioned in the text. The 9 figures in the Edinburgh manuscript differ much from those in the Vatican manuscript and are worse.

251

(Book V)

[1] Interesting, in this place, are Franco's remarks on the remodelling of materials.

[2] C. Winterberg, p. 168, also has $MO = 11 + \sqrt{2}$ and from this $LM = \frac{14}{11}(11 + \sqrt{2})$. The sides now indeed are in proportion 11 to 14 and the area is $\frac{14}{11}(11 + \sqrt{2})^2 \approx \frac{14}{11} \times 154 = 14^2$. But the text only has MO, not the length of LM.

[3] The construction is described completely on f. 101 v. C. Winterberg, p. 169, has a faulty interpretation and does not find the value $11 + \sqrt{2}$ for the sides.

[4] Originally the 'gnomon' was a vertical bar, used as sundial. Afterwards it meant the figure formed by the difference of two squares, having in common one angular point and the adjacent sides. See, for instance, M. Cantor, I, p. 536.

[5] So in the figure, discussed before, at B.

[6] C. Winterberg, p. 171, interprets the smaller square as the inscribed one, with area 98, and then: if a is the area of the circumscribed square, i of the inscribed one and x that of the circle, $a - x = 42$ and $x - i = 56$, so $2\left(x - \frac{a-i}{2}\right) = 14$, 'welcher Werth sowohl den Überschuß des Quadrats über den ihm flächengleichen Kreis, als auch umgekehrt den des letzteren über jenes ausdrückt, so daß die Summe beider $= 28$. Ein einzelner Theil ist somit $\frac{28}{8} = 3\frac{1}{2}$.' This interpretation is illogical and also cannot be inferred from the text.

[7] So $11\left(1 + \frac{7}{12}\right)$, and the calculation must have been $11(14 - 11 - \sqrt{2}) = 11(3 - \sqrt{2}) = 11\left(1 + \frac{7}{12}\right)$, which is the area of the parts of the rectangle outside the square. Had Franco also calculated the area of the parts of the square outside the rectangle, so $(11 + \sqrt{2})\sqrt{2}$, he would have found the somewhat greater value $17\frac{7}{12}$.

[8] This figure, in the Vatican manuscript at the top of f. 102r, belongs at the bottom of f. 102v. In the Vatican manuscript, however (not in Edinburgh), it is completely out of drawing.

(Book VI)

[1] For a survey of the old names for proportions see for instance A. J. E. M. Smeur (a), pp. 130–134.

[2] Franco, of course, does not have $\sqrt{2}$, $\sqrt{3}$ and so on but first line, second line, third line and so on.

[3] C. Winterberg, p. 175, interprets: 'Wäre $3 = m \cdot 2$, $4 = m \cdot 3$, so müßte $4 = m^2 \cdot 2$ oder $m^2 = 2$ sein, was für m als ganze Zahl nicht möglich.'

[4] C. Winterberg, p. 176, has, faultily, 'Die Superparticule bedeuten offenbar das Vielfache vorher bereits gegebener Theile'.

[5] C. Winterberg, p. 176: 'Die Superpartientes sind Theile der früheren', which is wrong.

[6] Franco also dismisses, as *secunda linea* the values $\frac{5}{3}, \frac{7}{4}, \frac{9}{5}, \frac{11}{6}$ and so on. Only one of them, viz. $\frac{7}{4}$, was sometimes used for $\sqrt{3}$, as was also the mean value of $\frac{5}{3}$ and $\frac{9}{5}$, viz. $\frac{26}{15}$ See M. Cantor, I, pp. 556 and 866–867.

[7] We could not trace the relation elsewhere.

[8] One easily finds: $\ell_4 = \frac{97}{84}\ell_3 = \frac{97}{84} \times \frac{49}{40}\ell_2 = \frac{97}{84} \times \frac{49}{40} \times \frac{17}{12}\ell_1$ which can be reduced to $\frac{97}{12} \times \frac{7}{40} \times \frac{17}{12} \times \ell_1$. This Franco has not done, but, to avoid fractions, he could have taken 172800 in stead of 1209600.

[9] $\frac{1712634}{1209600} = 1{,}4158\ldots$, so a good 0.1% too much.

[10] This approximation, suggested by Franco in some way in the form of a formula, is unknown in the history of mathematics. Nor do we know how it came down to Franco.

[11] B. L. van der Waerden suggested this reduction, in his review in *Centaurus* 1970, vol. 15, nr. 1, pp. 107–108.

[12] At the upper right of fig. 38 please add the following points: $E =$ the corner, $AB =$ the smaller diagonal, $CD =$ the longer diagonal.

[13] P is the point at the lower left of fig. 39.

[14] J. Tropfke II, p. 170.

A treatise on the squaring of the circle by Franco of Liège, of about 1050   252

LIST OF LITERATURE

Allgemeine Deutsche Biographie, I, Leipzig 1875; XII, Leipzig 1880.

Biographie nationale de Belgique, I, Bruxelles 1866; VII, Bruxelles 1880–1883.

BIRKENMAJER, A., La bibliothèque de Richard de Fournival. *Académie Polonaise des Sciences et des Lettres, Faculté de Philologie, Travaux*, tome LX, no. 4, Cracow 1922 (Polish); reprinted in: *Studia Copernicana*, I, Wroclaw/Warsew/Cracow 1970.

BORMANS, S., *Ly Myreur des histors, chronique de Jean des Preis dit d'Outremeuse*, t. IV. Bruxelles 1877.

BUBNOV, N., *Gerberti postea Silvestri II papae Opera Mathematica* (972–1003). Berlin 1899, reprint Hildesheim 1963.

CANTOR, M., *Vorlesungen über Geschichte der Mathematik*, I (third edition). Leipzig 1907.

CLAGETT, M., (a) 'Gerard of Brussels, Liber de motu', *Osiris* 12, 1956, 73–175. (b) *Archimedes in the Middle Ages*, I. Madison 1964. (c) See: E. A. MOODY.

CLERVAL, A., (a) *Les écoles de Chartres au moyen-age*. Chartres 1895. (b) See: P. TANNERY (d).

DARIS, J., *Histoire du diocèse et de la principauté de Liège*. Liège 1890.

FOLKERTS, M., '*Boethius*' *Geometrie II. Ein mathematisches Lehrbuch des Mittelalters*. Wiesbaden 1970.

FORBES, E. G., 'The Crawford Collection of the Royal Observatory', *The 150th anniversary of the Royal Observatory, Edinburgh*. Edinburgh 1973.

FRIEDLEIN, G., *Die Zahlzeichen und das elementare Rechnen der Griechen und Römer und des christlichen Abendlandes vom 7. bis 13. Jahrhundert*. 1869, reprint Wiesbaden 1968.

Germanicarum rerum quatuor celebriores vetustioresque chronographi. Francoforti ad Moenum 1566.

HEIBERG, J. L., see P. TANNERY (a).

Histoire littéraire de la France, par des Religieux Bénédictins de la Congrégation de S. Maur, VII (nouv. éd.). Paris 1867.

HOFMANN, J. E., (a) Zum Winkelstreit der rheinischen Scholastiker in der ersten Hälfte des 11. Jahrhunderts. *Abhandlungen der Preußischen Akademie der Wissenschaften, Jahrgang 1942. Math.-naturw. Klasse. Nr.* 8. Berlin 1942. (b) *Geschichte der Mathematik*, I (second edition). Berlin 1963.

HUYSMANS, R., *Wazo van Luik in de ideeënstrijd zijner dagen*. Nijmegen/Utrecht 1932.

IVES, S. A., 'Corrigenda and addenda to the descriptions of the Plimpton manuscripts as recorded in the De Ricci Census', *Speculum* 17, 1942, 44–45.

JUSCHKEWITSCH, A. P., *Geschichte der Mathematik im Mittelalter*. Leipzig 1964.

KARPINSKI, L. CH., *The history of arithmetic*. Chicago/New York 1925.

KURTH, G., *Notger de Liège et la civilisation au Xe siècle*, I. Paris 1905.

LEFEBVRE, B., *Notes d'histoire des mathématiques (antiquité et moyen-âge)*. Louvain 1920.

MAI, A., *Classici auctores e Vaticanis codicibus editi*, III. Romae 1831.

MENNINGER, K., *Zahlwort und Ziffer*. Göttingen 1958.

MIGNE, J. P., *Patrologiae cursus completus . . . series latina*, CXLIII. Paris 1853.

MOODY, E. A. and M. CLAGETT, *The medieval science of weights*. Madison 1952.

NARDUCCI, E., 'Intorno a due trattati inediti d'abaco contenuti in due codici Vaticani del secolo XII', *Bullettino di bibliografia e di storia delle scienze matematiche e fisiche*, 15, 1882, 111–162.

Nieuw Nederlandsch biografisch woordenboek, IV. Leiden 1914.

PAIGE, C. LE, *Notes pour servir à l'histoire des mathématiques dans l'ancien pays de Liège*. Liège 1890.

PERTZ, G. H., *Monumenta Germaniae historica*, VI (from the first page this part VI is indicated as part VIII). Hannoverae 1844.

PEZIUS, B., *Thesaurus anecdotorum novissimus, seu veterorum monumentorum, praecipue Ecclesiasticorum, ex Germanicis potissimum bibliothecis adornata collectio recentissima*, IV. Augsburg 1723.

QUETELET, A., *Histoire des sciences mathématiques et physiques chez les Belges*. Bruxelles 1864.

SASSEN, F., *De wijsbegeerte der Middeleeuwen in de Nederlanden*. Lochem 1944.

SMEUR, A. J. E. M., (a) *De zestiende-eeuwse Nederlandse rekenboeken*. 's-Gravenhage 1960. (b) De verhandeling over de cirkelkwadratuur van Franco van Luik van omstreeks 1050. *Mededelingen van de Koninklijke Vlaamse Academie voor Wetenschappen, Letteren en Schone Kunsten van België*, XXX, 11. Brussel 1968. (c) 'On the Value Equivalent to π in Ancient Mathematical Texts. A New Interpretation', *Archive for the History of Exact Sciences*, 6, pp. 249–270. Berlin/Göttingen/Heidelberg 1970.

SMITH, D. E., *History of mathematics*, I, II. Boston/London 1951, 1953.

X

STRUIK, D. J., *A Concise History of Mathematics*. New York 1948.

TANNERY, P., (a) 'Une correspondance d'écolâtres du onzième siècle', *Extrait des Comptes rendus de l'Académie des Inscriptions et Belles-Lettres*, XXV, pp. 214–221. Paris 1897. Also in: P. TANNERY, *Mémoires scientifiques, publiés par J. L. Heiberg*, V: Sciences exactes au moyen âge, 1887–1921. Toulouse–Paris 1922 (nr. 6, pp. 103–111). (b) 'La géometrie au XIe siècle', *Revue générale internationale scientifique, littéraire et artistique*, pp. 343–357. Paris 1897. Also: *Mémoires*, nr. 5, pp. 79–102. (c) 'Notes sur la pseudo-géometrie de Boèce', *Bibliotheca mathematica* III₁, pp. 39–50. 1900. Also *Mémoires*, nr. 9, pp. 211–228. (d) 'Une correspondance d'écolâtres du XIe siècle, publiée par M. Paul TANNERY et M. L'Abbé CLERVAL', *Notices et extraits des manuscrits de la bibliothèque nationale et autres bibliothèques*, XXXVI, 2me partie, pp. 487–543. Paris 1901. Also: *Mémoires*, nr. 10, pp. 229–303. (e) 'Ragimbold (Regimbold)', *Grand Encyclopédie*, XXVIII, p. 90. Also: *Mémoires*, nr. 5, p. 350.

THORNDIKE, L. and P. KIBRE, *A Catalogue of Incipits of Mediaeval Scientific Writings in Latin*. Revised and augmented edition. London 1963.

TRITTENHEIM, J. DE, (Tritheim, Trithemius), *De Scriptoribus Ecclesiasticis*. Paris s.a. (Preface: Basel 1494).

TROPFKE, J., *Geschichte der Elementar-Mathematik*, II. Berlin/Leipzig 1933.

WINTERBERG, C., 'Der Traktat Franco's von Luettich: "De quadratura circuli"', *Abhandlungen zur Geschichte der Mathematik*, 4, pp. 135–190. Leipzig 1882.

"Rithmomachia", a Mathematical Game from the Middle Ages[*]

Rithmomachia is a mathematical game invented in the 11th century. It was popular during the whole of the Middle Ages: only in the 17th century did it become forgotten. As a board game it competed for a long time with chess and was at some times more popular.

Research into the history of *Rithmomachia* began with the articles of Peiper, 'Fortolfi Rythmimachia' (1880) and Wappler, 'Bemerkungen' (1892). A short and easy introduction to the game was written by Smith and Eaton, 'Rithmomachia' (1911); they relied mainly on early printed sources. From the 1940s on there followed further articles on particular texts on the game (Richards, 'Manuscript' and 'Pythagorean Game'; Chicco, 'Rithmomachia' and 'Giuoco') and on the purpose of the game (Evans, 'Rithmomachia'). Only since 1986, after the fundamental work of A. Borst ('Zahlenkampfspiel'), has there been any real clarity about the origin and history of the game. This article gives a survey of the elements of play, the variants of the name, the origin and purpose of the game. It presents for the first time a list of all known texts on *Rithmomachia* from the 11th to the early 17th century.

a. The name

The name of the game appears in various forms: e. g. "R(h)ythmomachia", "R(h)ythmimachia", "Rithmimachia", "Ritmomachia", "Rithmachia"; in the oldest texts no name is given for it. In some texts the word was derived from *rhythmos* (= *numerus*, number) and *machia* (= *pugna*, fight). *Rhythmos* is here apparently not a corrupt form of *arithmos*, but means, like *numerus* (e. g. in Augustine's *De musica*), numerical relationship (of two temporally connected events). *Rithmomachia* is, accordingly, a game not with numbers, but with ratios of numbers; and ratios of numbers were in medieval theory of music known by the term *rhythmus* or *numerus*.

---

[*]This English version is based upon my German articles on *rithmomachia* published in 1990 and 2001 (see the Bibliography).

b. Essential elements

Rithmomachia was played on a board longer than it was broad, the size of which was as first not specified and so apparently not standard; later the board always had 8×16 squares. There are two armies, that of the even numbers and that of the odd, each with 3×8 pieces. On the top of the pieces are the following numbers:

	even				odd			
(1)	2	4	6	8	3	5	7	9
(2)	4	16	36	64	9	25	49	81
(3)	6	20	42	72	12	30	56	90
(4)	9	25	49	81	16	36	64	100
(5)	15	45	91	153	28	66	120	190
(6)	25	81	169	289	49	121	225	361

The "even" army thus also contains odd numbers and *vice versa*, but the numbers which generate the "even" army are even and those which generate the "odd" army are odd. These generating numbers are the *digiti* (units); "one" is not present, because, according to the Pythagoreans, it is not a number, but the unit from which the numbers are generated. All elements of the game are derived from Pythagorean number theory as described in Boethius' *Arithmetic*.

Beginning with the simplest even or odd numbers (row 1), the following rows represent the Pythagorean numerical ratios: the *proportio multiplex* $mn : n$ (row 2), the *proportio superparticularis* $(n+1) : n$ (rows 3 and 4) and the *proportio superpartiens* $(n+m) : n$ with $n > m > 1$ (rows 5 and 6). In the row of the *multiplices* each number is generated from the number above by multiplying it by n (thus $m = n$), in the row of the *superparticulares* by multiplying it by $(n+1) : n = 1 + \frac{1}{n}$ and in the row of the *superpartientes* by multiplying it by $(2n+1) : (n+1) = 1 + \frac{n}{n+1}$. In this way it can be proved that in rows 2, 4 and 6 there are squares and that the numbers in rows 3 and 5 may be found by adding the two numbers directly above them.

In this way 48 numbers are produced, which are inscribed on the upper surfaces of the pieces. We notice that some numbers are present in the pieces of both players (i. e. 9, 25, 49, 81; 16, 36, 64) and that the numbers 25 and 81 occur as many as three times, in each of the three classes, i. e. the *multiplices*, the *superparticulares* and the *superpartientes*.

Two numbers enjoy a special status: the 91 in the "even" army and the 190 in the "odd" army. Since

$$91 = 1^2 + 2^2 + 3^2 + 4^2 + 5^2 + 6^2$$

and

$$190 = 4^2 + 5^2 + 6^2 + 7^2 + 8^2 \,,$$

the 91 may be considered as made as a pyramid of six squares of sides 1, 2, ..., 6 and the 190 as a truncated pyramid of five squares with sides 4, 5, ..., 8. The latter is called *pyramis tercurta*, because the first three squares (of sides 1, 2, 3) are missing. These pyramidal numbers, which represent the three-dimensional extension of "figurate numbers", are described as pyramids built up from squares of corresponding size.

The pieces on which the numbers are written have various shapes according to the ratios that they represent: in most of the older descriptions the *multiplices* are symbolized by smaller squares, the *superparticulares* by larger squares and the *superpartientes* by circles. Later the *multiplices* are symbolized by circles, the *superparticulares* by triangles and the *superpartientes* by squares. The colours of the pieces were in the early descriptions not consistent and only later was one colour assigned for one team and another for the other – in general, black and white were chosen.

The pieces of the two armies were placed on the board in a way not specified in the earliest sources and appears variously in the later texts. They could move according to rules that differed with the classes to which the numbers belong. In most texts the round pieces could move to adjacent squares, the triangular pieces to a square two squares away and the quadrangular to a square three squares away. Sometimes diagonal moves are allowed or even moves involving a broken line. A piece may not jump over other pieces. The pyramids move according to special and various rules.

For the capturing of enemy pieces there are typically four possibilities: *congressus* (a piece is placed on a square so that with its next move it could reach the square occupied by the enemy piece holding the same number); *insidiae* (two pieces are placed on squares so that with their next moves they could reach the square occupied by the enemy piece holding the number which is the sum of those of the two pieces); *eruptio* (a piece is placed on a square so that the product of the number on the piece and the number of squares between it and the enemy piece is equal to the number on the enemy piece); *obsidio* (the enemy piece is surrounded on all sides, so that it cannot move; in this case the numbers

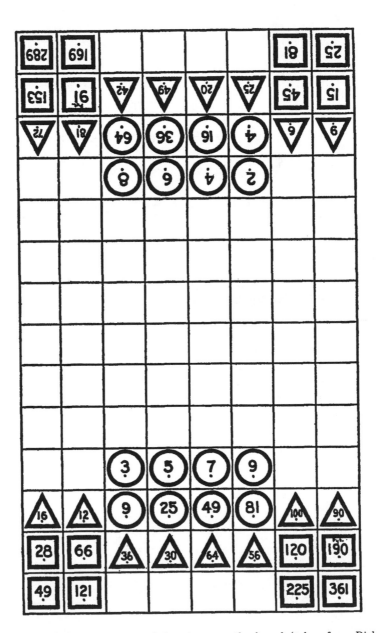

Figure 1: One possible position of the pieces on the board (taken from Richards, 'Rithmomachia', p. 98)

are not involved). To capture a pyramid, it usually suffices to capture the square forming its lowest part.

In order to win a game, it is necessary first to capture the pyramid. After this, the player must contrive to put three or four pieces into a winning position in the half of the board originally occupied by the enemy: the pieces must be placed in a straight line and must present at least one of the three Pythagorean means (arithmetic, geometric, harmonic). If two or more means are presented, the victory is correspondingly better. In *victoria magna* the three pieces display one mean; in *victoria maior* there are four pieces displaying two means; and in *victoria praestantissima* four pieces display three means. Since it was hardly possible to know all the winning positions by heart, tables were made, even in the earliest times – some of these incorporated mistakes or were incomplete.

c. History of the game

All mathematical elements of *rithmomachia* go back to the Pythagorean theory of numbers as is to be found in Nicomachus. The latter forms the basis of Boethius' *De institutione arithmetica*, which was available for scholars to the whole of the Latin Middle Ages (even before the translations from Arabic). *Rithmomachia* made it possible to learn, by playing, the arithmetic that normally had to be learnt the hard way in the cathedral and monastic schools as part of the *quadrivium*. The principal aim was the comprehension of arithmetic (i. e. the Boethian number theory), but it was also of service in learning the other subjects of the quadrivium (geometry, astronomy and music).

Medieval authors said that the inventor of the game was Pythagoras, Boethius or Gerbert. There are texts describing the game from the 11th century. The earliest descriptions are so short that only those acquainted with the game could understand them. Details of the game and how to play it were first given by Fortolf about 1130. In the later Middle Ages *rithmomachia* became detached from learning arithmetic and became a pastime for educated people. Its origin was in southern Germany – see below – and it was a little later also known in western Germany and Lorraine. From the 13th century on there were many reworkings originating in France and England. In the 14th and 15th centuries England became the centre for *rithmomachia*. In early modern times there were printed descriptions from Italian, French and German authors. In the 17th century the game fell into oblivion, perhaps because the mathematics involved had no relation to the mathematics of the time. It remained forgotten until about 1880, when Peiper, a German secondary school teacher, wrote the first modern paper on this subject. But only recently have individual texts been investigated and identified. Now we are in a position to list the texts that were written on the subject and to trace the origin and development of the game in the 700 years when it was known and played.

The known texts, with some indication of their contents and a list of the known manuscripts and printed editions, are as follows.

1. Asilo of Würzburg (about 1030)

A monk from Würzburg named Asilo invented, some time between 1022 and 1042, the game of *rithmomachia* to praise the wonder of the works of God which are manifested in number and measure. In all probability the invention was a consequence of a competition between the cathedral schools of Worms and Würzburg. It not clear who Asilo was: perhaps he was Adalbero, a pupil of the cathedral school who became bishop of Würzburg in 1045; or perhaps he was Hezilo of Hildesheim or perhaps Atto of Amorbach. In writing his text Asilo used a treatise written in Würzburg, *De aggregatione naturalium numerorum*, in which the classes of proportions found in Boethius are transformed into a series of natural numbers; these numbers were applied to the numbers on the pieces in *rithmomachia*. Asilo's text was a draft: his specifications about the board, the squares, the positions and moves of the pieces are not very clear. Even the name *rithmomachia* does not appear: the game was called *conflictus* or *altercatio*.

INCIPIT:
Quinque genera inaequalitatis ex aequalitate procedere manifestum est ...
MANUSCRIPTS:
Copenhagen, KB, Gamle Kgl. Saml., S. 277 fol., s. 13, f. 88rv (chs. 1–13)
Munich, BSB, Clm 14836, s. 11, f. 4v–6v (chs. 1–13)
Paris, BnF, lat. 7377C, s. 11, f. 17v
Vatican, Ottob. lat. 1631, s. 12, f. 25r–27r (chs. 1–13)
EDITIONS:
Wappler, 'Bemerkungen', pp. 14–17; Borst, 'Zahlenkampfspiel', pp. 330–334.
LITERATURE:
Borst, 'Zahlenkampfspiel', pp. 50–80.

2. Hermannus Contractus (about 1040)

Only one answer to Asilo's challenge is extant. Mostly for stylistic reasons Hermannus Contractus, a monk of Reichenau, has been considered the author. He did not give a special description of the game, but explained and improved Asilo's sketch. This text shows that the author had considerable didactic ability. Hermann suggests as the best winning position the "perfect harmony" formed from the numbers 12, 9, 8, 6. In appendixes he further explains unclear formulations of Asilo's text.

INCIPIT:
Qui peritus arithmeticae huius inventi noticiam curet habere ...

MANUSCRIPTS:
London, Wellcome Historical Medical Library, Ms. 364, s. 15, f. 1rv (chs. 1–10)
Montpellier, BU, Ms. 366, a. 1429, f. 42r–43r (chs. 1–10)
Munich, BSB, Clm 14836, s. 11, f. 3v–4v (chs. 1–10)
Paris, Arsenal, Ms. 830, s. 15, f. 103rv (chs. 1–10)
Paris, BnF, lat. 7185, s. 12, f. 107rv (chs. 1–10)
Paris, BnF, nouv. acq. lat. 1630, s. 11, f. 19rv (chs. 1–10)
Vatican, lat. 3101, a. 1077, f. 71v–72v (chs. 1–10)
Vatican, Ottob. lat. 1631, s. 12, f. 24r–25r (chs. 1–10)
Vatican, Reg. lat. 598, s. 11, f. 122rv

EDITIONS:
Wappler, 'Bemerkungen', pp. 12–24; Borst, 'Zahlenkampfspiel', pp. 335–339.

LITERATURE:
Borst, 'Zahlenkampfspiel', pp. 81–97.

3. Anonymus author of Liège (about 1070)

Between 1050 and 1090 an unknown monk, almost certainly in Liège, wrote a description of *rithmomachia*, apparently as an aid to elementary teaching. The author, who valued clear guidelines, gave the first time the size of the board (12 × 8 squares), determined the sizes of the various pieces, had clear ideas how to place them on the board for the initial position and gave examples for the applications of the rules for capturing. He brought *rithmomachia* close to Gerbert's abacus.

INCIPIT:
Omnis inaequalitas ex aequalitate procedit ...

MANUSCRIPTS:
Munich, BSB, Clm 28118, s. 11, f. 18v (chs. 8–10)
Paris, BnF, lat. 7377C, s. 11, f. 16r–17r

EDITION:
Borst, 'Zahlenkampfspiel', pp. 340–343.

LITERATURE:
Borst, 'Zahlenkampfspiel', pp. 98–114.

4. Odo of Tournai (about 1090)

Before the invention of printing the most copied treatise on *rithmomachia* was an anonymous text which must have been written shortly before 1100. The provenance of its sources, the early copies and the content of the work point to Odo of Tournai as the author. The author knew the works of Asilo, Hermann and the anonymous writer of Liège. Odo's composition is characterized by homogenity and clarity and combines number theory and practical calculation. The text, which seems to be unfinished, was supplied with various appendixes by Odo's pupils and later users. – On Odo's work depends a numerical table, which was drawn up in the Benedictian abbey of Thorney in eastern England about 1110. This table presents the numbers written on the pieces in a systematic arrangement and expressed by Greek capital letters.

INCIPIT:
Quinquegenam inequalitatis regulam ex equalitate procedere manifestum est ...

MANUSCRIPTS:

Avranches, BM, Ms. 235, s. 12, f. 76v–77r (main text); 77rv (appendixes a, d, g, b)

Hereford, Cathedral Library, O.I.VI, s. 12, f. 76r–77v (main text); 77v–78r (appendixes a, d)

Copenhagen, KB, Gamle Kgl. Saml., S. 277 fol., s. 13, f. 88v (main text, chs. 21–22); 88v (appendix k)

London, BL, Harley 3353, s. 14, f. 157v–159v (main text, chs. 2–12, 15–22); 159v (appendix f)

London, Wellcome Historical Medical Library, Ms. 364, s. 15, f. 1v–3r (main text, chs. 2–14); 2v (appendix b)

Montpellier, BU, Ms. 366, a. 1429, f. 43r–44v (main text, chs. 2–14); 44r (appendix b)

Oxford, BL, Ashmole 1471, s. 14, f. 48v–49v (main text)

Oxford, BL, Auct. F.1.9, s. 12, f. 1rv (main text); 1v (appendix a)

Oxford, BL, Rawl. C 270, s. 12, f. 20r–22v (main text)

Paris, Arsenal, Ms. 830, s. 15, f. 98v (main text, only chs. 12 end – 14)

Paris, Arsenal, Ms. 830, s. 15, f. 103v–104v (main text, chs. 1–15)

Paris, BnF, lat. 7185, s. 12, f. 107v–108v (main text, chs. 2–14)

Paris, BnF, lat. 14065, s. 13, f. 3v–4r (main text, chs. 15–20); 4r (appendix i); 5v–6v (chs. 1–14)

Paris, BnF, lat. 15119, s. 12, f. 9r–11v (main text); 11v–12r (appendixes a, e)

Trier, Bistumsarchiv, Abt. 95, Nr. 6, ca. 1100, f. 77v–78v (main text); 79r–82v, 91rv (appendixes b, a, c)

Vienna, ÖNB, Ms. 2503, s. 13, f. 49r (appendix h); 49r–51r (main text, chs. 2–20)

EDITIONS:
Gerbert, 'Scriptores', pp. 285–287; Borst, 'Zahlenkampfspiel', pp. 344–355 (main text), pp. 356–381 (appendixes a–k).

LITERATURE:
Borst, 'Zahlenkampfspiel', pp. 115–134.

Numerical table from Thorney:

MANUSCRIPT:
Oxford, St. John's College, Ms. 17, ca. 1100, f. 56v.

EDITIONS:
Evans, 'Difficillima', pp. 26f.; Borst, 'Zahlenkampfspiel', pp. 382f.

5. Anonymous writer of Regensburg (about 1090)

In St. Emmeram, near Regensburg, a further *Rithmomachia* was written, between 1080 and 1100, with the purpose of unravelling the intricacies of the game. Its anonymous author builds a theoretical foundation based on the texts of Asilo (no. 1) and Hermann (no. 2) and also a practical table, which follows the anonymous author from Liège (no. 3). – On the Regensburg text depends the design of a *rithmomachia* board drawn by Thiemo of Michelsberg (near Bamberg) shortly before 1100 – probably to demonstrate the different sorts of proportion.

INCIPIT:
Quinque genera inaequalitatis ex aequalitate procedere manifestum est ...

MANUSCRIPTS:
Dresden, LB, C 80, s. 15, f. 258r–259v (main text); 259v–260r (appendix c)

Düsseldorf, UB, F 13, s. 15/16, f. 22r–24r (main text, chs. 1–15); 25r (board with numbered pieces by Thiemo)

Karlsruhe, LB, Karlsruhe 504, ca. 1100, f. 87r (only board with numbered pieces by Thiemo)

Munich, BSB, Clm 28118, s. 11, f. 19r (appendix a)

Trier, StB, Ms. 1896/1438, s. 12, f. 1r (appendix a)

Vatican, lat. 3101, shortly after 1100, f. 2rv; 2v (appendix b), 41v (appendix a [before 1100])

Vienna, ÖNB, Ms. 5216, s. 15, f. 59r–61v (main text); 61v–62r (appendixes c, a)

EDITIONS:
Borst, 'Zahlenkampfspiel', pp. 384–394 (main text), pp. 395–402 (appendixes a–c). – Board with numbered pieces by Thiemo: Borst, 'Zahlenkampfspiel', pp. 403f.

LITERATURE:
Borst, 'Zahlenkampfspiel', pp. 135–151.

10   "Rithmomachia", a Mathematical Game from the Middle Ages

6. The "Franconian compilation" (about 1100)

About 1100 two hastily written compilations were made: the "Franconian compilation" and the "Bavarian compilation". The first was written between 1087 and 1110, as it seems in Moselle-Franconia, possibly in Trier. It was compiled from the texts of Odo (no. 4) and the anonymous writer of Regensburg (no. 5).

INCIPIT:
Quinquigenam inequalitatis regulam ex equalitate procedere ex libris arithmeticae manifestum est ...

MANUSCRIPT:
Rochester, Sibley Music Library, Ms. 1, s. 12, p. 31–40, 47

EDITIONS:
Richards, 'Manuscript', pp. 170–176; Borst, 'Zahlenkampfspiel', pp. 405–414.

LITERATURE:
Borst, 'Zahlenkampfspiel', pp. 152–164.

7. The "Bavarian compilation" (about 1105)

The "Bavarian compilation" was written shortly after 1100, apparently near Regensburg, Ilmmünster or Tegernsee. It is based upon the "Franconian compilation" (no. 6). It recommends a board 16 squares long. There is an appendix in which the capture of the pyramids is discussed.

INCIPIT:
Sit tabula ad latitudinem octo, ad longitudinem sedecim campis distincta ...

MANUSCRIPTS:
Karlsruhe, LB, Karlsruhe 504, ca. 1100, f. 152v (main text, chs. 12–14)
Vatican, lat. 3101, shortly after 1100, f. 3r (main text)
Vienna, ÖNB, Ms. 2503, s. 13, f. 51v–52r (main text, chs. 10–11); 52r–54r (appendix)

EDITIONS:
Gerbert, 'Scriptores', pp. 288f. (main text), 289–291 (appendix); Borst, 'Zahlenkampfspiel', pp. 415–421 (main text), 422–426 (appendix).

LITERATURE:
Borst, 'Zahlenkampfspiel', pp. 165–185.

8. Fortolf (about 1130)

About 1130 a certain "Fortolfus", who lived in the Main-Franconian area, perhaps at the bishop's court in Würzburg, wrote a textbook, with claims to literary value, to introduce laymen to *rithmomachia*. This is the first detailed description of the game. Fortolf, who used the "Franconian compilation" (no. 6), gave definitive instructions on the number of squares, on the board and the arrangement and forms (circle, small square, large square) of the pieces. He also prescribed the colours of the pieces: the three groups of the "even" team are white, red and black and those of the "odd" team are black, white and red. The pyramids are helically formed, sharper at the top. Fortolf gave detailed descriptions of the different ways of achieving a winning position. He was the first to describe a *victoria musica*, in which he proved that all numerical ratios may be represented by musical notes.

INCIPIT:
Prologus in rithmimachiam. Quoniam quidem huius artis scientia ab ignorantibus contempnitur ...

MANUSCRIPTS:
Berlin, SB, lat. fol. 246, s. 15, f. 206r–215v
Brussels, BR, 926–940, s. 15, f. 103r–110r
Jena, UB, El. fol. 71, s. 16, f. 132r–142v
Wrocław, BU, Rehd. 54, ca. 1140, f. 86r–94v (partly destroyed in or after 1945)

EDITIONS:
Peiper, 'Fortolfi Rythmimachia', pp. 169–197; Borst, 'Zahlenkampfspiel', pp. 427–470.

LITERATURE:
Borst, 'Zahlenkampfspiel', pp. 186–211.

9. Pseudo-Abaelard (2nd half of the 12th century)

A detailed scholastic commentary of Fortolf's text (no. 8) was written in the second half of the 12th century in France, perhaps in Paris. In a manuscript now lost, Richard of Fournival ascribed this text to Peter Abaelard, but there is no reason to believe it. In two appendixes the musical intervals were derived from arithmetic ratios.

INCIPIT:
Rithmimachia grece, numerorum pugna exponitur latine ...

MANUSCRIPTS:
Paris, BnF, lat. 14065, s. 13, f. 1r–3v (main text, chs. 1, 3–25); 4v–5v (appendix b)
Vienna, ÖNB, Ms. 2503, s. 13, f. 51rv (appendix a); 54r–57v (main text, chs. 1–14, 16–17, 19, 23–25)

EDITIONS:
Gerbert, 'Scriptores', pp. 291–295; 287f. (appendix a); Borst, 'Zahlenkampfspiel', pp. 471–481; 482–487 (appendixes a, b).

12　　　"Rithmomachia", a Mathematical Game from the Middle Ages

LITERATURE:
Borst, 'Zahlenkampfspiel', pp. 212–214.

10. Werinher of Tegernsee (12th century)

The author of an anonymously transmitted "Rithmachia" was very probably Werinher of Tegernsee. Before 1180 the provost Otto of Rottenbuch had asked the author to explain the rules of *rithmomachia* to him, and the "Rithmachia" was the result. The text is based on the "Bavarian compilation" (no. 7), but is filled out by material from the "Regensburg collection" (no. 5), the "Franconian compilation" (no. 6) and Fortolf's text (no. 8). Werinher prescribed a round shape for the *multiplices* pieces, a triangular for the *superparticulares* and a quadrangular for the *superpartientes*.

INCIPIT:
Nomen, materia, intentio, finis, cui parti philosophie supponitur ...

MANUSCRIPT:
London, BL, Addit. 22790, ca. 1200, f. 4r–6r, 11v

EDITION:
Folkerts, 'Werinher'.

LITERATURE:
Borst, 'Zahlenkampfspiel', pp. 216–219; Folkerts, 'Werinher'.

11. English compilation (end of 12th century)

In a manuscript written in England (Coventry?) at the end of the 12th century, a manuscript containing many texts on arithmetic, there is also a text on *rithmomachia*. It is based on the description by Odo (no. 4) and enlarged by additions that mostly refer to other, arithmetical, parts of the same codex. According to the compiler, *rithmomachia* helps the acquisition of practical knowledge of arithmetic (i. e. number theory in the Boethian sense). The author also mentions chess.

INCIPIT:
Fit tabula ad longitudinem et latitudinem distincta campis ...

MANUSCRIPT:
Cambridge, TC, R.15.16, s. 12, f. 61v–62r

EDITION:
Burnett, 'Instruments', pp. 189–201 (with English translation).

LITERATURE:
Burnett, 'Instruments'.

12. English compilation (beginning of the 13th century)

At the beginning of the 13th century an anonymous English compiler (perhaps Alexander Neckam) combined Werinher's text (no. 10) with Odo's (no. 4). From Werinher's work he took over some elements of the rules of play and limited the colours of the pieces to black and white.

INCIPIT:
Ex numeris sese respicientibus proportione multiplici ...

MANUSCRIPT:
Oxford, BL, Auct. F.5.28, s. 13, f. 15v–16r

EDITION:
No edition of the text; board reproduced in Murdoch, 'Album', p. 105.

LITERATURE:
Borst, 'Zahlenkampfspiel', p. 220.

13. English sketch (beginning of the 13th century)

The anonymous author, who calls himself *iuvenis*, points to the fun of playing the *rithmomachia* and informs other people of his age about the fundamentals of the game. He calls *rithmomachia* "pyoxaxym".

INCIPIT:
Animi adolescentum ad exercenda ludicra proni et effervescentes excogitare aggressi sunt ...

MANUSCRIPT:
Oxford, BL, Digby 67, f. 79rv

EDITION: –

14. Extracts from Pseudo-Ovid, *De vetula* (about 1240)

This work was written between 1222 and 1266 and ascribed (wrongly) to Ovid. The author may well be Richard of Fournival (1201–1260). In *De vetula* some board games are mentioned, above all *rithmomachia*, which is called the most beautiful game. The author used primarily the text of Pseudo-Abaelard (no. 9) and added details from Werinher (no. 10). There is no information about how the pieces move, methods of capture and victories.

INCIPIT:
O utinam ludus sciretur rithmimachie ...

MANUSCRIPTS:
See the editions by Klopsch and Robathan.
The verses which concern *rithmomachia* are separatedly transmitted in the following manuscripts:

14   "Rithmomachia", a Mathematical Game from the Middle Ages

London, BL, Harley 3353, s. 14, f. 147rv
London, Wellcome Historical Medical Library, Ms. 364, s. 15, f. 3rv
Montpellier, BU, Ms. 366, a. 1429, f. 44v–45v
Oxford, BL, Ashmole 1471, s. 14, f. 49v
Paris, Arsenal, Ms. 830, s. 15, f. 98v
Paris, BnF, lat. 7368, s. 14, f. 67v–68v
Regensburg, Bischöfliche Bibl., Proske Th 98, s. 15, p. 135–136
Wolfenbüttel, HAB, Cod. Guelf. 238 Extrav., f. 66r–67r, s. 17

EDITIONS:
Robathan, 'De Vetula', pp. 73–75 (I 672–721); Klopsch, 'De Vetula', pp. 217–
219 (I 649–698, 811–818, with table, partly incorrect).

LITERATURE:
Peiper, 'Fortolfi Rythmimachia', pp. 222f.; Chicco, 'Rithmomachia', pp. 84f.;
Borst, 'Zahlenkampfspiel', pp. 224f.

15. French compilation (end of the 13th century)

At the end of the 13th century an unknown compiler in France wrote a
"Edicio nova ritmomachie". It is based on Odo's rules (no. 4) and also
took material from the *Arithmetic* of Jordanus Nemorarius und Pseudo-
Ovid *De vetula* (no. 14). The author took some pains to be mathematically and linguistically clear.

INCIPIT:
Septem sunt capitula edicionem novam ritmomachie continencia ...
MANUSCRIPT:
London, BL, Harley 3353, s. 14, f. 148r–157r
EDITION: –

LITERATURE:
Borst, 'Zahlenkampfspiel', pp. 228–230.

16. Pseudo-Bradwardine (about 1330)

From Thomas Bradwardine's circle, but apparently not from his own pen,
came a compilation which was much copied. It is based upon Odo's text
(no. 4), the English compilation (no. 12) and *De vetula* (no. 14).

INCIPIT:
Quinque genera inequalitatis ex equalitate procedere ...
MANUSCRIPTS:
Dresden, LB, C 80, s. 15, f. 267v–268r
Erfurt, Ea Q 2, s. 14, f. 37rv, 1rv
Erfurt, Ea Q 325, s. 14, f. 45r, 46v
London, Wellcome Historical Medical Library, Ms. 364, s. 15, f. 3v–8r

Montpellier, BU, Ms. 366, a. 1429, f. 46r–52r
Paris, Arsenal, Ms. 830, s. 15, f. 99rv, 100r–101v
Paris, BnF, lat. 7368, s. 14, f. 65r–67r

EDITION:
No edition of the text; board reproduced in Beaujouan, 'Enseignement', p. 645.

LITERATURE:
Wappler, 'Bemerkungen', pp. 9f.; Chicco, 'Rithmomachia', p. 85; Borst, 'Zahlenkampfspiel', pp. 230–232.

17. *Exposicio artis armachie* (about 1370)

This is an relatively long, anonymous commentary on the *De vetula* (no. 14), written in England. It is divided into seven chapters. The compiler used the "French compilation" (no. 15) and also the first two chapters of Thomas Bradwardine's *Tractatus de proportionibus*.

INCIPIT:
Incipit exposicio artis armachie, id est ludi rithmimachie sive pugne numerorum, de quo tractat Ovidius libro suo secundo de vetula dicens ...

MANUSCRIPT:
Dublin, TC, Ms. 375, s. 15, f. 19r–28r

EDITION: –

LITERATURE:
Borst, 'Zahlenkampfspiel', pp. 232–236.

18. John Lavenham (beginning of the 15th century)

The monk John Lavenham from Colchester wrote a "Tractatus de ludo philosophorum", that is a *rithmomachia* "quem iuvenis quondam ex ovidio de vetula et ceteris antiquis collegeram". He wrote the treatise, he says from memory, and dedicated it to the bishop Henricus of Norwich. It consists essentially of comments on the *De vetula* (no. 14).

INCIPIT:
Venerabili in christo patri ac domino domino henrico Norwicensi episcopo ...

MANUSCRIPT:
Princeton, UL, Garrett 95, s. 15, p. 99–113

EDITION: –

19. Version Oxford, Canon. misc. 334 (beginning of the 15th century)

The anonymous author of this incompletely transmitted tractate based it, as he says, on the *De vetula* (no. 14). He tries to explain the mathematical principles of the game and to clarify the essentials by means of tables.

16   "Rithmomachia", a Mathematical Game from the Middle Ages

INCIPIT:
Pro aliquali introductione in ludum philosophorum ...

MANUSCRIPT:
Oxford, BL, Canon. misc. 334, s. 15, f. 95r–99r

EDITION: –

20. Critical commentary (about 1450)

The most extensive medieval treatise on *rithmomachia* was written in the mid-15th century. The anonymous author tries to take account of all variants and to use the material critically. His sources were Odo's text (no. 4; "vetus edicio"), the French compilation (no. 15; "nova edicio"), the commentary to *De vetula* (no. 17) and writings of Jordanus Nemorarius. Unfortunately, the text ist not clearly structured and partly confused.

INCIPIT:
Iam superest aperire numeros et proporciones numerorum ...

MANUSCRIPT:
Oxford, BL, Ashmole 344, ca. 1470, f. 40r–71r

EDITION: –

LITERATURE:
Borst, 'Zahlenkampfspiel', pp. 241–243.

21. John Shirwood (after 1474)

The *protonotarius* of the Holy Chair and later bishop of Durham, John Shirwood, wrote, some time after 1474, a *rithmomachia* from memory according to rules that he had learned as a schoolboy from a faulty manuscript (probably Oxford, Ashmole 344; no. 20). In 1480 he dedicated the work to cardinal Marco Bembo, who had it printed in 1482.

INCIPIT:
Cum, reverendissime pater ac amplissime domine, is preceptor ...

MANUSCRIPTS:
Dresden, LB, C 80, about 1485, f. 261r–265v (extracts), 266rv (additions by Johannes Widmann)
Jena, UB, El. fol. 74, about 1520, f. 54r–58v
Oxford, BL, Ashmole 344, before 1524, f. 24r–39v
Rome, Bibl. Casanatense, Ms. 791, a. 1511, f. 1r–9r
Wolfenbüttel, HAB, Cod. Guelf. 238 Extrav., s. 17, f. 26r–37r

EDITION:
Johannes Shirwod, *De ludo arithmomachiae ... epitome*. Rome 1482 (14 folios).

LITERATURE:
Murray, 'History', pp. 84–87; Chicco, 'Rithmomachia', pp. 85f.; Borst, 'Zahlen-kampfspiel', pp. 27f.

22. Italian redaction (15th century)

This Latin treatise was written in Italy. It is a reworking of the text by Werinher of Tegernsee (no. 10). The manuscript which transmits this text belonged to Johannes Widmann, who added marginal notes and diagrams.

INCIPIT:
Prelocucio De Rithmachia. Quandoquidem fortiter in theatro philosophiae sudantes ardua ...

MANUSCRIPT:
Dresden, LB, C 19, Nr. 3, s. 15, f. 1r–6r

EDITION:
Only short excerpts were edited by Wappler, 'Bemerkungen', pp. 6–8.

LITERATURE:
Borst, 'Zahlenkampfspiel', pp. 239–241.

23. Faber Stapulensis (1496)

In 1496 the well-known humanist Jacques Lefèvre d'Etaples (Johannes Faber Stapulensis) had a *rithmomachia* printed as an appendix to his reworking of Jordanus Nemorarius' *Arithmetic*. It is based on Shirwood's publication (no. 21). The work is in dialogue form: the Pythagorean Alcmeon of Croton informs his younger friends Brontinus and Bathyllus about the common rules of the game, though without going into details.

INCIPIT:
Considerasti, mi Bernarde, omnes disciplinas ad quas generoso spiritu sit annitendum, difficiles esse ...

EDITIONS:
Jacobus Faber Stapulensis, *Arithmetica decem libris demonstrata, Musica libris demonstrata quatuor, Epitome in libros Arithmeticos divi Severini Boetii, Rithmimachie ludus qui et pugna numerorum appellatur.* Paris 1496 (the *rithmomachia* on fol. i 6v – i 8v). – Practically identical is the edition Paris 1514. – Reprinted, without dedication and ending, in Boissière (1556, no. 25), f. 49v–52r.

LITERATURE:
Peiper, 'Fortolfi Rythmimachia', pp. 224–226; Smith / Eaton, 'Rithmomachia', p. 73; Beaujouan, 'Enseignement', pp. 646f.; Chicco, 'Rithmomachia', pp. 86f., 91–93; Borst, 'Zahlenkampfspiel', pp. 26f.

24. Florentine dialogue (1539)

In 1539 the Florentine humanists Carlo Strozzi, Benedetto Varchi and Luca Martini wrote, in Italian, a treatise on proportions in the Pythagorean tradition and in this connection a description of *rithmomachia* with the title "Giuoco di Pittagora". It is based on the rules given by Faber Stapulensis (no. 23). Taking part in the dialogue are: Carlo Strozzi, Cosimo Rucellai, Jacopo Vettori, Niccolo Alamanni.

INCIPIT:
Quanto io desidero piu di giovare ad altrui et compiacere agli amici ...

MANUSCRIPTS:
Fano, Bibl. Civica, Ms. Polidori 50/7, modern copy
Florence, BL, Redi 21 (135), s. 17
Florence, BN, II.II.278, s. 16, f. 1r–49v
Florence, BN, Landau Finaly 205, s. 16
Florence, BN, Magl. XI 125, s. 16
Florence, BN, Magl. XI 135, s. 17
Florence, BN, Magl. VIII 1492, fasc. 18 (*Trattato delle proportioni et proportionalita*)
Florence, Ricc., Ms. 890, s. 16, f. 84r–130r
Florence, Seminario Arcivescovile, A III 10, s. 16/17 (*Compendio del trattato delle proporzioni e proportionalità*)
Longboat Key (Florida), Schoenberg collection, no. 232, s. 16
Modena, Bibl. Estense, Ms. Campori App. 501 (Gamma T. 6. 2)

EDITIONS:
Only the Latin rules are edited in Chicco, 'Giuoco', pp. 29–31.

LITERATURE:
Chicco, 'Rithmomachia', pp. 94–96; Chicco, 'Giuoco'; Borst, 'Zahlenkampfspiel', p. 25.

25. Boissière (Buxerius) (1554/1556)

The philosopher, mathematician, astronomer and musician, Claude de Boissière, published a French description of *rithmomachia* in 1554 and another in Latin in 1556. It contains a history of the game (largely false) and an exposition of the rules, which are mainly based upon Faber Stapulensis' text (no. 23). For Boissière *rithmomachia* was an aid to teaching arithmetic. His Latin book was well known.

INCIPIT:
Non ignoras, clarissime vir, quantum utilitatis et relaxationis cognitio numerorum hominum mentibus adferat ...

EDITIONS:

C. de Boissière, *Le très excellent et ancien Jeu Pythagorique, dit Rhythmomachie*. Paris 1554 (not seen). – Claudius Buxerius: *Nobilissimus et antiquissimus ludus Pythagoreus (qui Rythmomachia nominatur)*. Paris 1556 (52 folios).

There is an English translation of the sections on how the game is played in Richards, 'Pythagorean Game'.

LITERATURE:

Peiper, 'Fortolfi Rythmimachia', pp. 224–226; Smith, 'Rara Arithmetica', pp. 271–273; Beaujouan, 'Enseignement', pp. 646–649; Chicco, 'Rithmomachia', p. 87; Borst, 'Zahlenkampfspiel', pp. 24f.

26. Abraham Ries (1562)

In 1562 Abraham Ries, a son of the famous *Rechenmeister* Adam Ries, wrote a German "Arithmomachia" in Annaberg, perhaps at the behest of the Elector August of Saxony. It is mostly based on the text by Pseudo-Bradwardine (no. 16).

INCIPIT:

Das die fünff genera Proportionis inaequalitatis aus der Proportion aequalitatis ihren vrsprung haben ...

MANUSCRIPTS:

Dresden, LB, C 433, s. 16, f. 2r–44v

Florence, BL, Ashb. 1322, s. 16, f. 57r–73v (other text than Dresden, C 433)

EDITION: –

LITERATURE:

Wappler, 'Bemerkungen', p. 11; Borst, 'Zahlenkampfspiel', p. 23.

27. Lever – Fulke (1563)

In 1563 Ralph Lever published a description of the *rithmomachia* – which he called "the philosophers game" –, which gave the reader the choice of three ways of playing the game. According to the title page, this game was "invented for the honest recreation of students, and other sober persons, in passing the tediousnes of tyme, to the release of their labours, and the exercise of their wittes".

Lever's book was extended by "W. F." (= William Fulke), who used it to contrive still more complicated games: an *OYPANOMAXIA* (London 1571) and a *METPOMAXIA* (London 1578). Just as *rithmomachia* served to demonstrate arithmetic relations, so the *OYPANOMAXIA* was to serve astronomy and *METPOMAXIA* geometry. The aim of *METPOMAXIA* was to capture a camp which was on the other side of a river.

Two hostile armies face each other in front of the river. For bombarding and storming the camp there are various war machines and special aids. As in *rithmomachia*, they are placed on a rectangular board, may be moved and may capture enemy pieces according to prescribed rules. The game included the height as a third dimension, so that the players sometimes had to have recourse to Pythagoras' theorem.

EDITION:
*The most noble auncient, and learned playe, called the philosophers game ...
by Rafe Lever and augmented by W. F.* London 1563 (<49> folios).

LITERATURE:
Chicco, 'Rithmomachia', pp. 87f.

28. Barozzi (Barocius) (1572)

In 1572 the Venetian Francesco Barozzi published a *rithmomachia* in Italian. It was essentially Boissières description (no. 25), though with elements from Fulke's book (no. 27). Barozzi introduced elements from chess. He also treated the theory of proportion.

INCIPIT (DEDICATION):
Quel bellissimo, & diletteuolissimo Giuoco ritrouato dal sapientissimo Pythagora ...

EDITION:
Francesco Barozzi, *Il nobilissimo et antiquissimo giuoco Pythagoreo nominato Rythmomachia cioè battaglia de consonantie de numeri ...* Venice 1572 (24 folios).

LITERATURE:
Smith, 'Rara Arithmetica', pp. 295, 340f.; Chicco, 'Rithmomachia', pp. 88f.; Borst, 'Zahlenkampfspiel', p. 23.

29. Eberbach (1577)

In 1577 Gottschalk Eberbach published in Latin a book on *rithmomachia* in Georg Baumann's publishing house in Erfurt. He dedicated it to Christian, a son of the Elector August of Saxony. This work is in two parts, one following almost word for word the text of Faber Stapulensis (no. 23) and the other that of Shirwood (no. 21). There follows a summary of the rules in German.

INCIPIT (DEDICATION):
Cum Arithmomachiae geminas has descriptiones in chartis multam antiquitatem prae se ferentibus reperissem ...

MANUSCRIPT:
Wolfenbüttel, HAB, Cod. Guelf. 238 Extrav., s. 17, f. 38r–54v (copy of the printed edition)

EDITION:
Rhythmomachiae sive Arithmomachiae Ludi Mathematici ingeniosissimi descriptiones duae. Ex antiquis exemplaribus nunc primum editae. Beschreibunge eines Kunstreichen alten Spiels mit Zahlen. Aus dem Latein verdeudscht, verstentlich und kurz zusammen gezogen, und jetzt newlich an tag gebracht. Erfurt 1577 (11 folios).

LITERATURE:
Chicco, 'Rithmomachia', p. 89; Borst, 'Zahlenkampfspiel', p. 23, note 33.

30. Selenus (1616)

In 1616 Duke August II of Brunswick-Luneburg edited a *rithmomachia* as an appendix to his book on chess, using the pseudonym "Gustavus Selenus". This text is almost a literal translation of the Italian treatise of Barozzi (no. 28) with some reworked additions taken from Boissière (no. 25). To compile his book, he used the printed works by Faber Stapulensis, Boissière and Barozzi and handwritten copies of the books by Shirwood and Eberbach. All this material is still to be found in the library at Wolfenbüttel as well as the autograph of Duke August's compilation, with many insertions and changes which show the different stages of his composition.

INCIPIT (PREFACE):
Es haben die Alten Weltweisen / umb die gemühter der Menschen / wan dieselben / im studiren / oder der Kopf-Arbeit / in etwas ermüdet wahren / zu belüstigen / und zu ergätzen ...

MANUSCRIPT:
Wolfenbüttel, HAB, Cod. Guelf. 221.8 Extrav., f. 265v–310 (autograph of Duke August)

EDITION:
Rythmomachia. Ein vortrefflich und uhraltes Spiel deß Pythagorae: Welches Gustavus Selenus, auß des Francisci Barozzi, Eines Venedischen Edelmans welschem Tractätlein ins Deutsche ubergesetzet ... und mit nützlichen glossen auß dem Claudio Buxero Delphinate, verbessert. 1616. In: Das Schach- oder König-Spiel. Leipzig 1617, 443–495.

LITERATURE:
Peiper, 'Fortolfi Rythmimachia', p. 227; Faber, 'Schachspiel'; Borst, 'Zahlenkampfspiel', pp. 21, 320.

22   "Rithmomachia", a Mathematical Game from the Middle Ages

31. "The phylosophers game" (17th century)

This anonymous English text with the title "The phylosophers game" names no sources. Its characteristic is a section on the algebraic symbols for irrational quantities. These symbols are assigned to some of the pieces and play a role in capturing.

INCIPIT:
Of the name of this game and what it is. The phylosophers game is called in greke rithmomachia ...

MANUSCRIPT:
London, BL, Sloane 451, s. 17, f. 1r–10r

EDITION: –

32. Funcke (1705)

Under the pseudonym "H.A.V.W." (probably: "Herzog August von Wolfenbüttel", i. e. "Duke August of Wolfenbüttel") Christian Gabriel Funcke, a teacher at the *Gymnasium* in Görlitz, published a book on *rithmomachia*. It contains shortened texts of Shirwood (no. 21) and Faber Stapulensis (no. 23) – of whom only Faber Stapulensis is mentioned by name in the introduction. In fact, this book is effectively a re-edition of the compilation of Eberbach (no. 29).

INCIPIT (DEDICATION):
Quamvis nihil fermè in praesentes pagellas mihi competat juris ...

EDITION:
H.A.V.W., *Rhythmomachiae sive Arithmomachiae Ludi Mathematici ingeniosissimi Descriptiones Duae Ex antiquis exemplaribus nunc denuo editae. Beschreibung eines vor 127. Jahren gebräuchlichen kunstreichen Spieles mit Zahlen, aus dem Lateinischen ins Teutsche übersetzet ...* Görlitz 1705 (47 pages).

LITERATURE:
Borst, 'Zahlenkampfspiel', pp. 18f., 24, 320.

Bibliography

G. Beaujouan, "L'enseignement du 'Quadrivium'", in *Settimane di studio del Centro Italiano di Studi sull'Alto Medioevo, XIX: La Scuola nell'Occidente Latino dell'Alto Medioevo, Spoleto, 15-21 aprile 1971*, Spoleto 1972, pp. 639–667.

A. Borst, *Das mittelalterliche Zahlenkampfspiel*, Heidelberg 1986.

Ch. Burnett, "The Instruments which are the proper delights of the Quadrivium: Rhythmomachy and Chess in the teaching of Arithmetic in twelfth-century England", *Viator* 28 (1997), 175–201.

A. Chicco, "La Rithmomachia", in *Bonus Socius*, 's-Gravenhage 1977, pp. 81–101.

A. Chicco, *Il giuoco di Pitagora*, Genova 1979.

G. R. Evans, "The Rithmomachia: A Mediaeval Mathematical Teaching Aid?", *Janus* 63 (1976), 257–273.

G. R. Evans, "Difficillima et Ardua: theory and practice in treatises on the abacus, 950–1150", *Journal of Medieval History* 3 (1977), 21–38.

M. Faber, "Schachspiel", in *Sammler, Fürst, Gelehrter. Herzog August zu Braunschweig und Lüneburg 1579–1666*, Wolfenbüttel 1979, pp. 172–180.

M. Folkerts, "Rithmimachia", in: *Die deutsche Literatur des Mittelalters. Verfasserlexikon*, 2nd ed., vol. 8, Berlin / New York 1990, cols. 86–94.

M. Folkerts, "Die *Rithmachia* des Werinher von Tegernsee", in M. Folkerts, J. P. Hogendijk (ed.), *Vestigia Mathematica. Festschrift für H. L. L. Busard*, Amsterdam 1993, pp. 107–142.

M. Folkerts, "Rithmimachie", in: M. Folkerts, E. Knobloch, K. Reich, *Maß, Zahl und Gewicht. Mathematik als Schüssel zu Weltverständnis und Weltbeherrschung*, 2nd ed., Wiesbaden 2001, pp. 333–340.

M. Gerbert, *Scriptores ecclesiastici de musica*, vol. 1, St. Blasien 1784.

P. Klopsch, *Pseudo-Ovidius De Vetula. Untersuchungen und Text*, Leiden / Cologne 1967.

J. E. Murdoch, *Album of Science. Antiquity and the Middle Ages*, New York 1984.

H. J. R. Murray, *A History of Board-Games other than Chess*, 2nd ed., New York 1978.

R. Peiper, "Fortolfi Rythmimachia", *Abhandlungen zur Geschichte der Mathematik* 3 (1880), 169–227.

J. F. C. Richards, "A New Manuscript of a Rithmomachia", *Scripta Mathematica* 9 (1943), 87–99, 169–183, 256–264.

J. F. C. Richards, "Boissière's Pythagorean Game", *Scripta Mathematica* 12 (1946), 177–217.

D. M. Robathan, *The Pseudo-Ovidian De Vetula*, Amsterdam 1968.

D. E. Smith, *Rara Arithmetica*, Boston / London 1908.

D. E. Smith, C. C. Eaton, "Rithmomachia, the Great Medieval Number Game", *The American Mathematical Monthly* 18 (1911), 73–80.

H. E. Wappler, "Bemerkungen zur Rythmomachie", *Zeitschrift für Mathematik und Physik, hist.-litt. Abt.* 37 (1892), 1–17.

Index of Names

This index contains all personal names (and the titles of some anonymous writings) found in the text and footnotes of this volume, including the authors of all referenced literature. Excluded are names mentioned in the titles of bibliographical items and names of publishers. Generally, Greek names have been printed in their Latin form (e. g., Menelaus for Menelaos). Names of medieval persons in the Latin West (up to ca. 1475) are usually listed under their first names, all other persons under their surnames or family names. Cross references are supplied in all cases where multiple forms of a name are in use.

Index of Manuscripts

This index contains page references to all manuscripts mentioned in this book. Pages on which a manuscript is cited by sigla and not by shelfmark are also indicated. These sigla are given in brackets after the shelfmark of the manuscript. If there are different sigla for the same manuscript in different articles, the article to which the corresponding siglum refers is added in square brackets.

Printed and bound by CPI Group (UK) Ltd, Croydon, CR0 4YY

21/10/2024

01777085-0017